U.S. CONGRESS

OFFICE OF
TECHNOLOGY
ASSESSMENT

Harmful Non-Indigenous Species in the United States

Recommended Citation:
U.S. Congress, Office of Technology Assessment, *Harmful Non-Indigenous Species in the United States,* OTA-F-565 (Washington, DC: U.S. Government Printing Office, September 1993).

For sale by the U.S. Government Printing Office
Superintendent of Documents, Mail Stop: SSOP, Washington, DC 20402-9328
ISBN 0-16-042075-X

Foreword

Non-indigenous species (NIS)—those species found beyond their natural ranges—are part and parcel of the U.S. landscape. Many are highly beneficial. Almost all U.S. crops and domesticated animals, many sport fish and aquaculture species, numerous horticultural plants, and most biological control organisms have origins outside the country. A large number of NIS, however, cause significant economic, environmental, and health damage. These harmful species are the focus of this study.

The total number of harmful NIS and their cumulative impacts are creating a growing burden for the country. We cannot completely stop the tide of new harmful introductions. Perfect screening, detection, and control are technically impossible and will remain so for the foreseeable future. Nevertheless, the Federal and State policies designed to protect us from the worst species are not safeguarding our national interests in important areas.

These conclusions have a number of policy implications. First, the Nation has no real national policy on harmful introductions; the current system is piecemeal, lacking adequate rigor and comprehensiveness. Second, many Federal and State statutes, regulations, and programs are not keeping pace with new and spreading non-indigenous pests. Third, better environmental education and greater accountability for actions that cause harm could prevent some problems. Finally, faster response and more adequate funding could limit the impact of those that slip through.

This study was requested by the House Merchant Marine and Fisheries Committee; its Subcommittee on Fisheries and Wildlife Conservation and the Environment and the Subcommittee on Oceanography and Great Lakes; the Subcommittee on Water Resources of the House Committee on Public Works and Transportation, and by Representative John Dingell. In addition, Representatives Amo Houghton and H. James Saxton endorsed the study.

We greatly appreciate the contributions of the Advisory Panel, authors of commissioned papers, workshop participants, survey respondents, and the many additional people who reviewed material. Their timely and indepth assistance enabled us to do the extensive study our requesters envisioned. As with all OTA studies, the content of the report is the sole responsibility of OTA.

Roger C. Herdman, Director

iii

Advisory Panel

Marion Cox
Chair
Resource Associates
Bethesda, MD

J. Baird Callicott
University of Wisconsin-
    Stevens Point
Stevens Point, WI

Faith Thompson Campbell
Natural Resources Defense
    Council
Washington, DC

James Carlton
Williams College-Mystic Seaport
Mystic, CT

Alfred Crosby
University of Texas
Austin, TX

Lester E. Ehler
University of California
Davis, CA

William Flemer, III
Wm. Flemer's Sons, Inc.
t/a Princeton Nurseries
Princeton, NJ

John Grandy
Humane Society of the U.S.
Gaithersburg, MD

Lynn Greenwalt
National Wildlife Federation
Washington, DC

Robert P. Kahn
Consultant
Rockville, MD

William B. Kovalak
Detroit Edison Co.
Detroit, MI

John D. Lattin
Oregon State University
Corvallis, OR

Joseph P. McCraren
National Aquaculture Association
Shepherdstown, WV

Marshall Meyers
Pet Industry Joint Advisory
    Council
Washington, DC

Robert E. Morris
Northcoast Mortgage
Eureka, CA

Philip J. Regal
University of Minnesota
Minneapolis, MN

Rudolph A. Rosen[1]
Texas Parks and Wildlife
    Department
Austin, TX

Don C. Schmitz
Florida Department of Natural
    Resources
Tallahassee, FL

Jerry D. Scribner
Attorney-at-Law
Sacramento, CA

Howard M. Singletary, Jr.
North Carolina Department of
    Agriculture
Raleigh, NC

Clifford W. Smith
University of Hawaii at Manoa
Honolulu, HI

Reggie Wyckoff
National Association of Wheat
    Growers' Associations
Genoa, CO

---

[1]Affiliation provided for identification only.

EXECUTIVE BRANCH LIAISONS

Gary H. Johnston
U.S. Department of the Interior
Washington, DC

Robert Peoples
U.S. Fish and Wildlife Services
Arlington, VA

William S. Wallace
U.S. Department of Agriculture
Washington, DC

Kenneth Knauer[2]
U.S. Department of Agriculture
Washington, DC

Katherine H. Reichelderfer[3]
U.S. Department of Agriculture
Washington, DC

Melvyn J. Weiss[4]
U.S. Department of Agriculture
Washington, DC

NOTE: OTA appreciates and is grateful for the valuable assistance and thoughtful critiques provided by the advisory panel members. The panel does not, however, necessarily approve, disapprove, or endorse this report. OTA assumes full responsibility for the report and the accuracy of its contents.

[2]Until January 1992.
[3]Panel member until August, 1991; liaison thereafter.
[4]After January 1992.

Project Staff

Walter E. Parham
Program Manager
Food and Renewable Resources
    Program

Phyllis N. Windle
Project Director

ANALYTICAL STAFF

Elizabeth Chornesky
Analyst

Peter T. Jenkins
Analyst

Steven Fondriest
Research Assistant[1]

Kathleen E. Bannon
Research Assistant[2]

Christine Mlot
Editor

ADMINISTRATIVE STAFF

Nathaniel Lewis
Office Administrator

Nellie Hammond
Administrative Secretary

Carolyn Swann
Personal Computer Specialist

[1]Until January 8, 1993.
[2]After April 12, 1993.

Contents

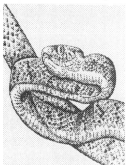

Summary, Issues, and Options | 1

T he movement of plants, animals, and microbes beyond their natural range is much like a game of biological roulette. Once in a new environment, an organism may simply die. Or it may take hold and reproduce, but with little noticeable effect on its surroundings. But sometimes a new species spreads unimpeded, with devastating ecological or economic results. This latter category—including species like the zebra mussel (*Dreissena polymorpha*) and the gypsy moth (*Lymantria dispar*)—is largely the focus of, and the reason for, this assessment. This opening chapter both summarizes the assessment and spells out the policy issues and options for Congress that emerged from the analysis.

SUMMARY OF FINDINGS

The summary portion of this chapter compiles the more detailed findings from the individual chapters that follow (box 1-A). It is organized to reflect the three focal points of the report:

- an overview of the status of harmful non-indigenous species (NIS) in the United States (chs. 2, 3);
- an analysis of the technological issues involved in dealing with harmful NIS (chs. 4, 5, 9); and
- an examination of the institutional organization in place (chs. 6, 7).

Two chapters cut across these areas. Chapter 8 presents detailed case studies for two States with particularly severe NIS-related problems—Hawaii and Florida. Chapter 10 discusses the future and the international context in which NIS issues will evolve. In each case, the pertinent chapter provides additional documentation.

1

Box 1-A—A Road Map to the Full Assessment

This assessment has three focal points: the status of harmful non-indigenous species (NIS) in the United States; technological issues regarding decisionmaking and species management; and institutional and policy frameworks. Each chapter elaborates on the findings summarized here and contains additional examples of problem species and their locations.

Chapter

1   **Summary, Issues, and Options**
     chapter findings; 8 major issues; policy options; New Zealand's approach

2   **The Consequences of Harmful Non-Indigenous Species**
     definitions and scope; benefits; economic, health, and environmental costs;extinctions and biological diversity

3   **The Changing Numbers, Causes, and Rates of Introductions**
     pathways into and within the country; numbers per taxonomic group, state, decade; new detections since 1980

4   **The Application of Decisionmaking Methods**
     uncertainty; 'clean' and 'dirty' lists; risk analysis; environmental impact assessment; benefit/cost analysis; protocols; values; new approaches; Siberian timber

5   **Technologies for Preventing and Managing Problems**
     inspection and detection; databases; quarantine and containment; control methods; eradication; environmental education; ecological restoration; FIFRA reregistration

6   **A Primer on Federal Policy**
     summary lessons; President Carter's Executive Order; Aquatic Nuisance Species Task Force; activities of 21 agencies by type of activity and organisms affected

7   **State and Local Approaches from a National Perspective**
     Federal/State relations; States' legal approaches, standards, gaps, and statutes on fish and wildlife; survey results; State laws on plants, insects, and other invertebrates; model State laws; enforcement; exemplary approaches

8   **Two Case Studies: Non-Indigenous Species in Hawaii and Florida**
     the States' uniqueness; introduction rates; critical species; affected sectors; new programs; fruit flies and brown tree snakes in Hawaii; melaleuca and Hurricane Andrew in Florida

9   **Genetically Engineered Organisms As a Special Case**
     technical and policy controversies; Federal regulation since 1984; ecological risk assessment; scale-up of releases; transgenic fish and squash; NIS vs. GEOs;

10   **The Context of the Future: International Law and Global Change**
     treaties and trade agreements; CITES as a model; technological change; impacts of current trends; future pests; climate change; worst and best case scenarios

Appendixes
     list of boxes, figures, and tables; authors, workshop participants, reviewers, and survey respondents; references

Indexes
     common and scientific names of species; general index

Table 1-1—Estimated Numbers of Non-Indigenous Species in the United States[a]

Species with origins outside of the United States		
Category	Number	Percentage of total species in the United States in category
Plants	>2,000	—[b]
Terrestrial vertebrates	142	≈6%
Insects and arachnids	>2,000	≈2%
Fish	70	≈8%[c]
Mollusks (non-marine)	91	≈4%
Plant pathogens	239	—[b]
Total	4,542	

Species of U.S. origin introduced beyond their natural ranges		
Category	Number	Percentage of total species in the United States in category
Plants	—[b]	—[b]
Terrestrial vertebrates	51	≈2%
Insects and arachnids	—[b]	—[b]
Fish	57	≈17%[c]
Mollusks (non-marine)	—[b]	—[b]
Plant pathogens	—[b]	—[b]

[a] Numbers should be considered minimum estimates. Experts believe many more NIS are established in the country, but have not yet been detected.

[b] Number or proportion unknown.

[c] Percentage for fish is the calculated average percentage for several regions. Percentages for all other categories are calculated as the percent of the total U.S. flora or fauna in that category.

SOURCES: Summarized by the Office of Technology Assessment from: J.C. Britton, "Pathways and Consequences of the Introduction of Non-Indigenous Freshwater, Terrestrial, and Estuarine Mollusks in the United States," contractor report prepared for the Office of Technology Assessment, October 1991; W.R. Courtenay, Jr., "Pathways and Consequences of the Introduction of Non-Indigenous Fishes in the United States," contractor report prepared for the Office of Technology Assessment, September 1991; K.C. Kim and A.G. Wheeler, "Pathways and Consequences of the Introduction of Non-Indigenous Insects and Arachnids in the United States," contractor report prepared for the Office of Technology Assessment, December 1991; R.N. Mack, "Pathways and Consequences of the Introduction of Non-Indigenous Plants in the United States," contractor report prepared for the Office of Technology Assessment, September 1991; C.L. Schoulties, "Pathways and Consequences of the Introduction of Non-Indigenous Plant Pathogens in the United States," contractor report prepared for the Office of Technology Assessment, December 1991; S.A. Temple and D.M. Carroll, "Pathways and Consequences of the Introduction of Non-Indigenous Vertebrates in the United States," contractor report prepared for the Office of Technology Assessment, October 1991.

■ Non-Indigenous Species Today: Numbers, Pathways, Rates, and Consequences

Many more NIS—those plants, animals, and microbes found beyond their natural geographical ranges—are in the United States today than there were 100 years ago. At least 4,500 species of foreign origin have established free-living populations in this country. These include several thousand plant and insect species and several hundred non-indigenous vertebrate, mollusk, fish, and plant pathogen species (table 1-1). Approxi- mately 2 to 8 percent of each group of organisms is non-indigenous to the United States.

Some NIS are clearly beneficial. Non-indigenous crops and livestock—like soybeans (*Glycine max*), wheat (*Triticum* spp.), and cattle (*Bos taurus*)—form the foundation of U.S. agriculture, and other NIS play key roles in the pet and nursery industries, fish and wildlife management, and biological control efforts. These and other positive contributions of NIS are largely beyond the scope of this study, however. OTA's work takes a comprehensive look at the damaging

Figure 1-1—State by State Distribution of Some High Impact Non-Indigenous Species

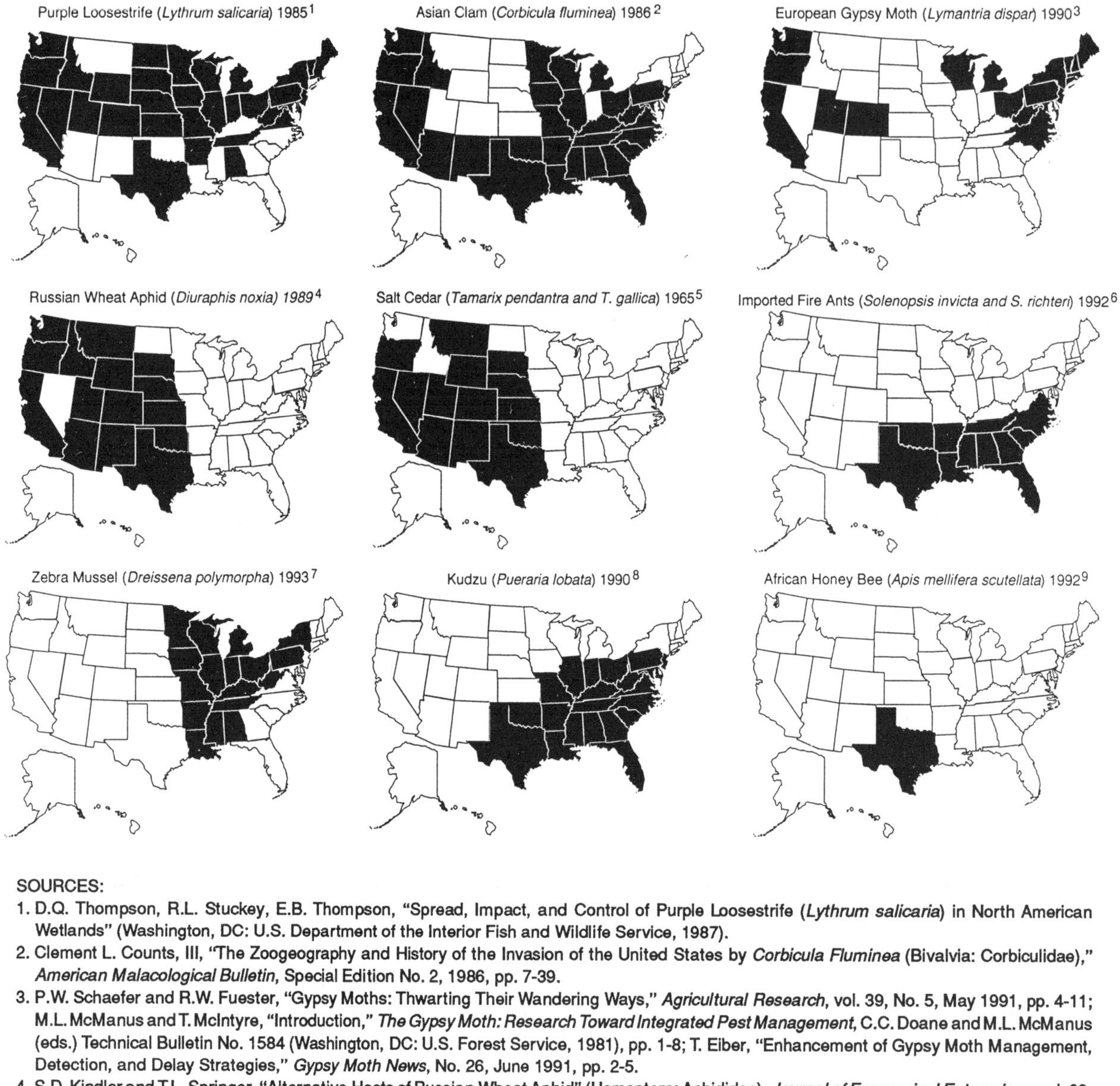

Purple Loosestrife (*Lythrum salicaria*) 1985[1]

Asian Clam (*Corbicula fluminea*) 1986[2]

European Gypsy Moth (*Lymantria dispar*) 1990[3]

Russian Wheat Aphid (*Diuraphis noxia*) 1989[4]

Salt Cedar (*Tamarix pendantra and T. gallica*) 1965[5]

Imported Fire Ants (*Solenopsis invicta and S. richteri*) 1992[6]

Zebra Mussel (*Dreissena polymorpha*) 1993[7]

Kudzu (*Pueraria lobata*) 1990[8]

African Honey Bee (*Apis mellifera scutellata*) 1992[9]

SOURCES:

1. D.Q. Thompson, R.L. Stuckey, E.B. Thompson, "Spread, Impact, and Control of Purple Loosestrife (*Lythrum salicaria*) in North American Wetlands" (Washington, DC: U.S. Department of the Interior Fish and Wildlife Service, 1987).
2. Clement L. Counts, III, "The Zoogeography and History of the Invasion of the United States by *Corbicula Fluminea* (Bivalvia: Corbiculidae)," *American Malacological Bulletin*, Special Edition No. 2, 1986, pp. 7-39.
3. P.W. Schaefer and R.W. Fuester, "Gypsy Moths: Thwarting Their Wandering Ways," *Agricultural Research*, vol. 39, No. 5, May 1991, pp. 4-11; M.L. McManus and T. McIntyre, "Introduction," *The Gypsy Moth: Research Toward Integrated Pest Management*, C.C. Doane and M.L. McManus (eds.) Technical Bulletin No. 1584 (Washington, DC: U.S. Forest Service, 1981), pp. 1-8; T. Eiber, "Enhancement of Gypsy Moth Management, Detection, and Delay Strategies," *Gypsy Moth News*, No. 26, June 1991, pp. 2-5.
4. S.D. Kindler and T.L. Springer, "Alternative Hosts of Russian Wheat Aphid" (Homoptera: Aphididae), *Journal of Economical Entomology*, vol. 82, No. 5, 1989, pp. 1358-1362.
5. T.W. Robinson, "Introduction, Spread and Areal Extent of Saltcedar (Tamarix) in the Western States," Studies of Evapotranspiration, Geological Survey Professional Paper 491-A (Washington, DC: U.S. Government Printing Office, 1965).
6. V.R. Lewis et al., "Imported Fire Ants: Potential Risk to California," *California Agriculture*, vol. 46, No. 1, January-February 1992, pp. 29-31; D'Vera Cohn, "Insect Aside: Beware of the Fire Down Below, Stinging Ants From Farther South Have Begun to Make Inroads in Virginia, Maryland," *Washington Post*, June 2, 1992, p. B3.
7. U.S. Department of the Interior, Fish and Wildlife Service, briefing delivered to the Senate Great Lakes Task Force, May 21, 1993.
8. Anonymous, *National Geographic Magazine*, 'Scourge of the South May be Heading North," vol. 178, No. 1, July 1990.
9. M.L. Winston, "Honey, They're Here! Leaning to Cope with Africanized Bees," *The Sciences*, vol. 32, No. 2, March/April 1992, pp. 22-28.

Table 1-2—Estimated Cumulative Losses to the United States From Selected Harmful
Non-Indigenous Species, 1906-1991

Category	Species analyzed (number)	Cumulative loss estimates (millions of dollars, 1991)	Species not analyzed[a] (number)
Plants[b]	15	603	—
Terrestrial vertebrates	6	225	>39
Insects	43	92,658	>330
Fish	3	467	>30
Aquatic invertebrates	3	1,207	>35
Plant pathogens	5	867	>44
Other	4	917	—
Total	79	96,944	>478

a Based on estimated numbers of known harmful species per category (figure 2-4).

b Excludes most agricultural weeds; these are covered in box 2-D.

NOTES: The estimates omit many harmful NIS for which data were unavailable. Figures for the species represented here generally cover only one year or a few years. Numerous accounting judgments were necessary to allow consistent comparison of the 96 different reports relied on; information was incomplete, inconsistent, or had other shortcomings for most of the 79 species.

SOURCE: M. Cochran, "Non-Indigenous Species in the United States: Economic Consequences," contractor report prepared for the Office of Technology Assessment, March 1992.

species: how they get here, their impacts, and what can be done about them.

Distinguishing between "good" and "bad" NIS is not easy. Some species produce both positive and negative consequences, depending on the location and the perceptions of the observers. Purple loosestrife (*Lythrum salicaria*), for example, is an attractive nursery plant but a major wetland weed. Approximately 15 percent of the NIS in the United States cause severe harm. High-impact species—such as the zebra mussel, gypsy moth, or leafy spurge (*Euphorbia esula*) (a weed)—occur throughout the country (figure 1-1). Almost every part of the United States confronts at least one highly damaging NIS today. They affect many national interests: agriculture, industry, human health, and the protection of natural areas.

The number and impact of harmful NIS are chronically underestimated, especially for species that do not damage agriculture, industry, or human health. Harmful NIS cost millions to perhaps billions of dollars annually. From 1906 to 1991, just 79 NIS caused documented losses of $97 billion in harmful effects, for example (table 1-2). A worst-case scenario for 15 potential high-impact NIS puts forth another $134 billion

in future economic losses (table 1-3). The figures represent only a part of the total documented and possible costs—that is, they do not include a large number of species known to be costly but for which little or no economic data were available, e.g., non-indigenous agricultural weeds. Nor do they account for intangible, nonmarket impacts.

Harmful NIS also have had profound environmental consequences, exacting a significant toll on U.S. ecosystems. These range from wholesale ecosystem changes and extinction of indigenous species (especially on islands) to more subtle ecological changes and increased biological sameness. The melaleuca tree (*Melaleuca quinquenervia*) is rapidly degrading the Florida Everglades wetlands system by outcompeting indigenous plants and altering topography and soils. In Hawaii, some NIS have led to the extinction of indigenous species, and the brown tree snake (*Boiga irregularis*) may further this process.

Naturally occurring movements of species into the United States are uncommon. Most new NIS arrive in association with human activity, transport, or habitat modification that provides new opportunities for species' establishment. Numerous harmful species arrived as unintended byproducts of cultivation, commerce, tourism, or travel.

Table 1-3—Worst Case Scenarios: Potential Economic Losses From 15 Selected Non-Indigenous Species[a]

Group	Species studied	Cumulative loss estimates (in millions, $1991)[b]
Plants	melaleuca, purple loosestrife, witchweed	4,588
Insects	African honey bee, Asian gypsy moth, boll weevil, Mediterranean fruit fly, nun moth, spruce bark beetles	73,739
Aquatic invertebrates	zebra mussel	3,372
Plant pathogens	annosus root disease, larch canker, soybean rust fungus	26,924
Other	foot and mouth disease, pine wood nematodes	25,617
Total	15 species	134,240

[a] See index for scientific names.

[b] Estimates are net present values of economic loss projections obtained from various studies and reports on selected potentially harmful NIS. Many of the economic projections are not weighted by the probability that the invasions would actually occur. Thus, the figures represent worst case scenarios. The periods of the projections range from 1 to 50 years.

SOURCE: M. Cochran, "Non-Indigenous Species in the United States: Economic Consequences," contractor report prepared for the Office of Technology Assessment, March 1992.

For example, they arrived as contaminants of bulk commodities, packing materials, shipping containers, or ships' ballast. Weeds continue to enter the country as contaminants in seed shipments; both plant and fish pathogens have arrived with diseased stocks. Some NIS stow away on cars and other conveyances, including military equipment.

Other harmful NIS were intentionally imported as crops, ornamental plants, livestock, pets, or aquaculture species—and later escaped. Of the 300 weed species of the western United States, at least 36 escaped from horticulture or agriculture. A number of NIS were imported and released for soil conservation, fishing and hunting, or biological control and later turned out to be harmful. A few illegal introductions also occur.

Different groups of organisms arrive by different pathways. Some fish are imported intentionally to enhance sport fisheries; others are illegally released by aquarium dealers or owners or escape from aquaculture facilities. Most foreign terrestrial vertebrates are intentional introductions. Insects (except for biological control organisms) and aquatic and terrestrial mollusks usually hitchhike with plants, commercial shipments, baggage, household goods, ships' ballast water, or aquarium and aquaculture shipments.

Far more unintentional introductions of insects and plant pathogens have had harmful effects than have intentional introductions. For terrestrial vertebrates, fish, and mollusks, however, intentional introductions have caused harm approximately as often as have unintentional ones, suggesting a history of poor species choices and complacency regarding their potential harm.

Far more is known about pathways of foreign NIS into the United States than the routes by which NIS have spread beyond their natural ranges within the country. Once here, NIS spread both with and without human assistance. A few of these pathways have no international counterpart, e.g., the release of bait animals like the sheepshead minnow (*Cyprinodon variegatus*). Known or potentially harmful NIS that are commercially distributed or officially recommended for various applications can spread especially quickly.

OTA found no clear evidence that the rate of harmful NIS imports has climbed consistently over the past 50 years. The ways and rates at which species are added from abroad fluctuate widely because of social, political, and technological factors, e.g., new trade patterns and innovations in transportation. Such factors have had major significance in the past and will continue to operate. For example, State and Federal plant quarantine laws slowed rates of introduction of insect pests and plant pathogens after 1912. However, rates rarely reach zero and they have been higher throughout the 20th century than in the preceding one.

More than 205 NIS from foreign countries were first introduced or detected in the Unites States since 1980, and 59 of these are expected to cause economic or environmental harm. There may be limits to the acceptable total burden of harmful NIS in the country. This consideration has yet to be incorporated into policy decisions such as setting tolerable annual levels of species entry.

OTA has carefully examined the best available evidence on the numbers, rates, pathways, and impacts of NIS. Six scientists prepared background papers on the pathways and consequences of NIS within their area of expertise. Another 36 experts from industry, academia, and government reviewed their work. OTA supplemented this work with its own analysis of the science and policy literature.

Based on this extensive review of the status of NIS, OTA concludes that the total number of harmful NIS and their cumulative impacts are creating a growing economic and environmental burden for the country. This conclusion leads to certain policy issues discussed later in this chapter. These address:

- the merits of prompt congressional action to create a more stringent national policy (pp. 15-19), and
- ways to provide funding for new or expanded efforts and to increase accountability for actions that lead to damage (pp. 40-45).

▊ Technological Issues: Decisionmaking About NIS, Pest Management, and the Special Case of Genetically Engineered Organisms

Some of the most harmful NIS—like kudzu (*Pueraria lobata*), water hyacinth (*Eichhornia crassipes*), and feral goats (*Capra hircus*)—were imported and released intentionally, with their negative effects unanticipated or underestimated. The central issues for NIS and genetically engineered organisms (a special subset) are the same: deciding which to keep out, which to release, and how to control those that have unexpected harmful effects. Consequently, part of OTA's study

Federal laws helped decrease the number of harmful non-indigenous insect pests, plant pathogens, and weeds imported with crop seeds and plants.

focused on the kinds of decisionmaking tools available.

Uncertainty in predicting risks and impacts of NIS remains a problem. Generally, the impact of new species cannot be predicted confidently or quantitatively. Risk can be reduced, or at least made explicit, using methods such as risk analysis, benefit/cost analysis, environmental impact assessment, and decisionmaking protocols. Expert judgment, however, is most broadly feasible. By and large, three interrelated problems remain largely unsolved:

1. determining levels of acceptable risk;
2. setting thresholds of risk or other variables above which more formal and costly decisionmaking approaches are invoked; and

Table 1-4—Lag Times Between Identification of Species' Pathway and Implementation of Prevention Program.

Species	Pathway	Date pathway identified	Date prevention program implemented	Remaining gaps
Mediterranean fruit fly (*Ceratitis capitata*)	Fruit shipped through first-class domestic mail from Hawaii	mid 1930s	1990, mail traveling from Hawaii to California inspected	First-class mail from elsewhere or other potential pathways (e.g., Puerto Rico to California)
Aquatic vertebrates, invertebrates, and algae	Ship ballast water	1981	1992, Coast Guard proposes guidelines for treating ballast water into the Great Lakes	International shipping into other U.S. ports; ship ballast water from domestic ports
Asian tiger mosquito (*Aedes albopictus*)	Imported used tires	1986	1988, protocols established for imported used tires	Interstate used tire transport
Forest pests	Unprocessed wood (including dunnage, logs, wood chips, etc.)	1985	1991, first restrictions imposed on log imports from Siberia	Wood imports other than from Siberia

SOURCES: Bio-environmental Services Ltd., *The Presence and Implication of Foreign Organisms in Ship Ballast Waters Discharged into the Great Lakes*, vol I, March 1981; C.G. Moore, D.B. Francy, D.A. Eliason, and T.P. Monath, "*Aedes albopictus* in the United States: Rapid Spread of a Potential Disease Vector," *Journal of the American Mosquito Control Association*, vol. 4, No. 3, September 1988, pp. 356-361; I.A. Siddiqui, Assistant Director, California Department of Food and Agriculture, Sacramento, CA, testimony at hearings before the Senate Committee on Governmental Affairs, Subcommittee on Federal Services, Post Offices, and Civil Services, *Postal Implementation of the Agricultural Quarantine Enforcement Act*, June 5, 1991; United States Department of Agriculture, Animal and Plant Health Inspection Service, "Wood and Wood Product Risk Assessment," draft, 1985.

3. identifying tradeoffs when deciding in the face of uncertainty.

Federal methods and programs to identify risks of potentially harmful NIS have many shortcomings—including long response times (table 1-4). Procedures vary in stringency throughout the Animal and Plant Health Inspection Service (APHIS) in the Department of Agriculture (USDA), risks to nonagricultural areas are often ignored, and generally, new imports are presumed safe unless proven otherwise. Even with these flaws, APHIS's risk assessments are more rigorous than those conducted by the Fish and Wildlife Service (FWS) in the Department of the Interior. Most regulatory approaches to NIS importation and release use variations of "clean" (allowed) or "dirty" (prohibited) lists of species or groups. Combining both kinds of lists, with a "gray" list of prohibited-until-analyzed species would reduce risks.

Nevertheless, preventing new introductions of harmful species is the first line of defense. Various methods can help decisionmakers avert unintentional and poorly planned intentional introductions that are likely to cause harm. Port inspection and quarantine are imperfect tools, though, so prevention is only part of the solution. Some organisms are more easily controlled than intercepted. Aiming for a standard of "zero entry" has limited returns, especially when prevention efforts come at the expense of rapid response or essential long-term control.

When prevention fails—for technical or political reasons—rapid response is essential. Then managers can choose among a variety of methods for eradication, containment, or suppression (table 1-5); these choices are not necessarily easy or obvious. For example, the choice may be not to control already widespread organisms, or those for which control is likely to be too expensive and/or ineffective. For any management program,

Table 1-5—Examples of Control Technologies for Non-Indigenous Species

	Physical control	Chemical control	Biological control
Aquatic plants	Cutting or harvesting for temporary control of Eurasian watermilfoil (*Myriophyllum spicatum*) in waters	Various glyphosate herbicides (Rodeo is one brand registered for use in aquatic sites) for controlling purple loosestrife (*Lythrum salicaria*)	Imported Klamathweed beetle (*Agasicles hygrophila*) and a moth (*Vogtia malloi*) to control alligator weed (*Alternanthera philoxeroides*) in southeastern United States
Terrestrial plants	Fire and cutting to manage populations of garlic mustard (*Alliaria petiolata*) in natural areas	Paraquat for the control of witchweed (*Striga asiatica*) in corn fields	Introduction of a seed head weevil (*Rhinocyllus conicus*) to control musk thistle (*Carduus nutans*)
Fish	Fencing used as a barrier along with electroshock to control non-indigenous fish in streams	Application of the natural chemical rotenone to control various non-indigenous fish	Stocking predatory fish such as northern pike (*Esox lucius*) and walleye (*Stizostedion vitreum*) to control populations of the ruffe (*Gymnocephalus cernuus*)
Terrestrial vertebrates	Fencing and hunting to control feral pigs (*Sus scrofa*) in natural areas	Baiting with diphacinone to control the Indian mongoose (*Herpestes auropunctatus*)	Vaccinating female feral horses (*Equus caballus*) with the contraceptive PZP (porcine zona pellucida) to limit population growth
Aquatic invertebrates	Washing boats with hot water or soap to control the spread of zebra mussels (*Dreissena polymorpha*) from infested waters	In industrial settings, chlorinated water treatments to kill attached zebra mussels	No known examples of successful biological control of non-indigenous aquatic invertebrates (Target specificity is a major concern)
Insects/mites	Various agricultural practices, including crop rotation, alternation of planting dates, and field sanitation practices	Mathathion bait-sprays for control of the Mediterranean fruit fly (*Ceratitis capitatis*)	A parasitic wasp (*Encarsia partenopea*) and a beetle (*Clitostethus arcuatus*) to control ash whitefly (*Siphoninus phillyreae*)

SOURCE: Office of Technology Assessment, 1993.

accurate and timely species identification is essential but sometimes not available.

Eradication of harmful NIS is often technically feasible but complicated, costly, and subject to public opposition (box 1-B). Chemical pesticides play the largest role now in management. They will remain important for fast, effective, and inexpensive control. In the future, an increased number of biologically based technologies will probably be available. Genetic engineering will increase the efficacy of some. Development of biological and chemical pesticides entail the same difficulties, however—ensuring species specificity, slowing the buildup of pest resistance to the pesticide, and preventing harm to nontarget organisms. So there are no "silver bullets" for NIS control and some troublesome gaps may appear in the next 10 years. Pests have already developed resistance to some microbial pesticides, one alternative to chemical methods. A number of chemical pesticides are being phased out for regulatory or environmental reasons. And new alternatives are slow to come online. Ecological restoration, by changing the conditions

Box 1-B—Failure and Success: Lessons From the Fire Ant and Boll Weevil Eradication Programs

Imported Fire Ant Eradication:

Two species of imported fire ants are assumed to have entered at Mobile, Alabama, in dry ship ballast: *Solenopsis richteri* in 1918, and, around 1940, *Solenopsis invicta.* The ants became a public health problem and had significant negative effects on commerce, recreation, and agriculture in the States where they were found. In late 1957, a cooperative Federal-State eradication program began. It exemplifies what can go wrong with an eradication program.

Funding was provided to study the fire ants, but information on the biology of the species was lacking, and the ant populations increased and spread. Various chemicals (heptachlor and mirex) were used to control and eradicate the ants over a 30-year period. Although they did kill the ants, the chemicals caused more ecological harm than good. Their widespread application, often by airplane, destroyed many non-target organisms, including fire ants' predators and competitors, leaving habitats suitable for recolonization by the ants.

The chemicals eventually lost registration by the Environmental Protection Agency, leaving few alternatives available. In the 5 years after 1957, fire ant infestations increased from 90 million to 120 million acres.

Boll Weevil Eradication:

The boll weevil, *Anthonomus grandis*, a pest of cotton, naturally spread into Texas, near Brownsville, from Mexico, in the early 1890s and crossed the Mississippi River in 1907. By 1922, it infested the remainder of the southeastern cotton area. Unlike the imported fire ant eradication program, boll weevil eradication does not rely solely on chemicals.

The eradication program centers around the weevil's life cycle and uses many different techniques. Part of the boll weevil population spends the winter in cotton fields. Insecticides are used to suppress this late season population. In spring and early summer, pheromone bait traps and chemical pesticides reduce populations before they have a chance to reproduce. Still other control technologies (e.g., sterile male release or insect growth regulators) limit the development of a new generation of boll weevils.

Boll weevil eradication trials were conducted from 1971-1973 (in southern Mississippi, Alabama, and Louisiana) and from 1978-1980 (in North Carolina and Virginia). Although results of the trials were mixed, cotton producers in the Carolinas voted in 1983 to support the boll weevil eradication program in their area and to provide 70 percent of the funding. The USDA Animal and Plant Health Inspection Service was charged with overall management of the program.

By the mid-1980s, the boll weevil was eradicated from North Carolina and Virginia. This 1978-1987 eradication program achieved a very high rate of return, mainly from increased cotton yields and lower chemical pesticide spending and use. In 1986, pesticide cost savings, additions to land value, and yield increases amounted to a benefit of $76.65 per acre. The benefit was $78.32 per acre for the expansion area in southern North Carolina and South Carolina.

SOURCES: G.A. Carlson, G. Sappie, and M. Hamming, "Economic Returns to Boll Weevil Eradication," U.S. Department of Agriculture, Economic Research Service, September 1989, p. 31; W. Klassen, "Eradication of Introduced Arthropod Pests: Theory and Historical Practice," Entomological Society of America, Miscellaneous Publications, No. 73, November 1989; E.P. Lloyd, "The Boll Weevil: Recent Research Developments and Progress Towards Eradication in the USA," *Management and Control of Invertebrate Crop Pests*, G.E. Russell (ed.) (Andover, England: Intercept, 1989), pp. 1-19; and C.S. Lofgran, W.A. Banks, and B.M. Glancey, "Biology and Control of Imported Fire Ants," *Annual Review of Entomology* vol. 30, pp. 1-30, 1975.

that may make a habitat suitable for NIS, shows promise for preventing or limiting the establishment or spread of some harmful NIS. Continued research and development on new ways to manage harmful NIS remain essential.

OTA commissioned 3 papers on decisionmaking methods for this study, submitted those papers to peer review by 20 experts, held a workshop for the papers' authors and several additional specialists, and added a staff review of control methods and biotechnology policy, along with another expert paper on genetic engineering—each with extensive informal input from technical and policy specialists.

Based on this work regarding technical aspects, OTA concludes that some continued unintentional introductions are inevitable, as are illegal ones, and ones with unexpected effects. Perfect screening, detection, and control are technically impossible and will remain so for the foreseeable future. These results lead to certain of the congressional policy issues discussed later in this chapter. These include the need for:

- more effective screening for fish, wildlife, and their diseases (pp. 22-24);
- more stringent evaluations of new plant introductions for their potential as weeds (pp. 28-30); and
- more rapid response to emergencies and better means for setting priorities (pp. 36-40).

Continued intentional introductions of certain species are, of course, desirable. None of the policy options are intended to stop them.

▌ Institutional Issues: the Federal and State Policy Patchwork

The current Federal effort is largely a patchwork of laws, regulations, policies, and programs. Many only peripherally address NIS, while others address the more narrowly drawn problems of the past, not the broader emerging issues.

The need for a more restrictive national policy on introductions and use is widely acknowledged. Development of such a policy is impeded by historical divisions among agencies, user groups, and constituencies. Technical barriers also obstruct accurate and consistent Federal policy. For example, terms and definitions differ greatly among NIS-related statutes, regulations, policies, and publications.

At least 20 Federal agencies work at researching, using, preventing, or controlling desirable and harmful NIS (table 1-6), with APHIS playing the largest role. Federal agencies manage about 30 percent of the Nation's lands, some of which have severe problems with NIS. Yet management policies regarding harmful NIS range from being nearly nonexistent to stringent. The National Park Service has fairly strict policies. However, removal or control of unwanted NIS is not keeping pace with invasions, and concerns are growing that NIS threaten the very characteristics for which the Parks were established.

Federal agencies do not uniformly evaluate the effects of NIS before using them for federally funded activities. However, a Federal interagency group is planning to coordinate work on noxious weeds. Another interagency task force is developing a major program on aquatic nuisance species.

Federal laws leave both obvious and subtle gaps in the regulation of harmful NIS. Most State laws have similar shortcomings. Significant gaps in Federal and State regulation exist for nonindigenous fish, wildlife, animal diseases, weeds, species that affect nonagricultural areas, biological control agents, and vectors of human diseases. Many of these gaps also apply to genetically engineered organisms (GEOs), which are commonly regulated under the same laws. Commercial development is imminent for several such categories of GEOs.

Pre-release evaluations for certain GEOs have been more stringent than for NIS—reflecting past underestimates of NIS risks. Some of these stricter GEO-related methods might be used for NIS. So far, APHIS has only evaluated proposals

Table 1-6—Areas of Federal Agency Activity Related to NIS

Agency[a]	Movement into U.S. Restrict	Movement into U.S. Enhance	Interstate movement within U.S. Restrict	Interstate movement within U.S. Enhance	Regulate product content or labeling	Control or eradication programs	Fund or do introductions	Federal land management Prevent eradication or control	Federal land management Introduce or maintain	Fund or do research Prevention control eradication	Fund or do research Uses of species	Aquaculture development	Biocontrol development
APHIS	✓		✓		✓	✓	✓			✓	✓		✓
AMS	✓		✓		✓								
FAS	b												
USFS	✓	✓		✓		✓	✓	✓	✓	✓	✓		✓
ARS	✓	✓	✓	✓		✓	✓			✓	✓	✓	✓
SCS		✓	✓			✓		✓		✓	✓	✓	✓
ASCS							✓						
CSRS													
FWS	✓		✓		✓	✓	✓	✓	✓	✓	✓	✓	✓
NPS						✓		✓	✓	✓	✓		✓
BLM						✓		✓	✓	✓			✓
BIA													
BOR	✓		✓			✓	✓	✓		✓	✓	✓	
NOAA	✓					✓		✓		✓	✓	✓	✓
DOD	✓	✓	✓	✓		✓		✓				✓	✓
EPA	✓		c		✓			e		✓	d		
PHS	✓							e		✓			
Customs	✓												
USCG	✓									✓			
DOE								e	e				
DEA	✓												

a Acronyms of Federal Agencies: **Department of Agriculture**—Animal and Plant Health Inspection Service (APHIS); Agricultural Marketing Service (AMS); Foreign Agricultural Service (FAS); Forest Service (USFS); Agricultural Research Service (ARS); Soil Conservation Service (SCS); Agricultural Stabilization and Conservation Service (ASCS); Cooperative State Research Service (CSRS). **Department of the Interior**—Fish and Wildlife Service (FWS); National Park Service (NPS); Bureau of Land Management (BLM); Bureau of Indian Affairs (BIA); Bureau of Reclamation (BOR). **Department of Commerce**—National Oceanic and Atmospheric Administration (NOAA). **Department of Defense (DOD)**. Environmental Protection Agency (EPA). **Department of Health and Human Services**—Public Health Service (PHS). **Department of Energy (DOE)**. Department of the Treasury—Customs Service (Customs). Department of Transportation—Coast Guard (USCG). **Department of Justice**—Drug Enforcement Agency (DEA).

b Monitors animal diseases abroad.

c Monitors spread of human disease vectors within the United States.

d Regulates experimental releases of microbial pesticides.

e DOE lacks policies on NIS.

SOURCE: Office of Technology Assessment, 1993.

for releasing low risk GEOs. Setting acceptable risk levels for higher risk GEOs will be more difficult, a problem the agency has not solved for NIS. Experience with NIS shows overwhelmingly that organisms' effects and ecological roles can change in new environments. Thus, caution is warranted when extrapolating from small to large-scale GEO releases and when exporting GEO's to other countries.

State laws on NIS vary from lax to exacting and use a variety of basic legal approaches (table 1-7). They are relatively comprehensive for agricultural pests but only spotty for invertebrate and plant pests of nonagricultural areas.

States play a larger role than the Federal Government in the importation and release of fish and wildlife. Several States present exemplary approaches. Yet many State laws are weak and their implementation inadequate. For example, most State fish and wildlife agencies rate their own resources for implementing and enforcing their own NIS laws as "less" or "much less" than adequate; they would need, on average, a 50-percent increase in resources to match their responsibilities. States' evaluations of new releases are not stringent: no States require the use of scientific protocols for evaluating proposed introductions, and about one-third do not even require a general determination of potential negative impacts. States prohibit a median of only eight potentially harmful fish and wildlife species or groups; about one-third of the agency officials OTA surveyed believe their own lists of prohibited species are too short. About one-fourth of the States lack legal authority over the importation or release of at least one major vertebrate group. About 40 percent of the agency officials would like additional regulatory authority from their State legislatures.

Federal and State agencies cooperate on many programs related to agricultural pests, but their policies can also conflict, e.g., when agencies manage adjacent lands for different purposes. Sometimes Federal law preempts State law, more often regarding agriculture than fish and wildlife.

Conflicts between States also occur, often without forums for resolving the disputes. Regional approaches—used mostly to evaluate aquatic releases—provide means for States to affect their neighboring States' actions. Such approaches are promising but limited by the fact that participation is not mandatory.

For the section on institutional issues, OTA commissioned 3 background papers, on the Department of Interior, USDA generally, and APHIS in particular; 20 people took part in the papers' external peer review. Also, OTA did extensive internal research on the missions and activities of Federal agencies. In addition, OTA compiled State laws and regulations relating to NIS, with assistance from an expert group, and surveyed the heads of State fish and wildlife agencies.

Based on this institutional analysis, OTA concludes that Federal and State efforts are not protecting national interests in certain important areas. Thus, OTA highlights congressional policy issues on:

- needed changes to the Lacey Act for fish and wildlife (pp. 19-24);
- new roles for the States in fish and wildlife management (pp. 24-25);
- needed changes to the Federal Noxious Weed Act (pp. 25-28); and
- improved weed management on Federal lands (pp. 30-31);
- other gaps in legislation and regulation (pp. 45-50).

■ The Special Cases of Hawaii and Florida

Virtually all parts of the country face problems related to harmful NIS, but Hawaii and Florida have been particularly hard hit because of their distinctive geography, climate, history, and economy. In both States, natural areas and agriculture bear the brunt of harm and certain NIS threaten the State's uniqueness. As a set of islands, Hawaii is particularly vulnerable to sometimes devastating ecological impacts. More than one-half of Hawaii's free-living species are non-indigenous.

Table 1-7—Basic Legal Approaches Used by States for Fish and Wildlife Importation and Release

Basic approach	Importation[a][b]		Release	
	Number	States	Number	States
All species are prohibited unless on allowed ("clean") list(s).	2 + 1pt[c]	HI, IDpt, VT[d]	1 + 5pt	AKpt, FLpt, GApt, HI, IDpt, KYpt
All species may be allowed except those on prohibited ("dirty") list(s).				
Prohibited list(s) have 5 or more identified species or groups.	20 + 3pt	AL, AR, CO, CT, FL, IL, KS, KY, MI, MN, MTpt, NC, NE, NY, OH, PA, SCpt, SD, TN, TXpt, UT, WA, WY	14 + 6pt	AL, AR, CO, CT, FLpt, GApt, IL, KS, KYpt, MN, NE, NY, OHpt, PA, SCpt, TN, TXpt, UT, WA, WY
Prohibited list(s) have fewer than 5 identified species or groups.	11 + 3pt	AK, DE, IN, LApt, MD, ME, MS, NH, NV, NJ, ORpt, RI, VA, WVpt	11 + 6pt	AKpt, IN, LApt, NC, NDpt, NJ, MD, MN, MS, NH, NV, OR, RIpt, SD, VA, VTpt, WVpt
All species may be allowed; there is no prohibited list.	11 + 7pt	AZ, CA, GA, IDpt, IA, LApt, MA, MO, MTpt, ND, NH, NM, OK, ORpt, SCpt, TXpt, WI, WVpt	12 + 9pt	AZ, CA, DE, IDpt, IA, LApt, MA, ME, MI, MO, MT, NDpt, NM, OHpt, OK, RIpt, SCpt, TXpt, VTpt, WI, WVpt

[a] State regulation of "possession" of a group or groups is considered here as regulation of both "importation" and "release," since neither act can be done without having possession. For the few States that specifically regulate "importation with intention to release (or introduce)," it is not treated here as comprehensive regulation of "release" because it covers only acts of importation done with a specific intent.

[b] Many States that regulate importation of particular groups exempt mere transportation through the State. These are not distinguished here.

[c] Some States treat different groups of vertebrates differently. This is designated, where applicable, by using the abbreviation "pt" after the State initial to indicate the entry covers only "part" of the vertebrates regulated. They are totaled separately.

[d] The summary classifications are general; in many States there are limited exemptions, such as for scientific research, and other minor provisions which are not covered here. The extensive State regulation of falconry is excluded.

SOURCES: Office of Technology Assessment, 1993 and Center for Wildlife Law, University of New Mexico Law School, "Selected Research and Analysis of State Laws on Vertebrate Animal Importation and Introduction," contractor report prepared for the Office of Technology Assessment, Washington, DC, April 1992.

New species played a significant role in past extinctions of indigenous species and continue to do so. In Florida, several non-indigenous aquatic weeds and invasive trees seriously threaten the Everglades wetlands system.

Hawaii's isolation makes it most in need of a comprehensive policy to address NIS. Differing Federal and State priorities have made this difficult to achieve, however. Cooperative efforts have sprung up in both States among State and Federal agencies, nongovernmental organizations, agricultural interests, and universities. Increasingly, these groups see harmful NIS as a unifying threat and public education as an important tool to address it. The situation in Hawaii and Florida, while unusual in some ways, nevertheless heralds what other States face as additional harmful NIS

enter and spread throughout the United States and people become more aware of their damage.

For this chapter, OTA commissioned a background paper on each State and 12 experts reviewed this work. Two contractors conducted extensive interviews and site visits in Hawaii and OTA staff did the same in Florida. Also, OTA commissioned a survey and assessment of U.S. environmental education programs.

Based on this work, OTA concludes that the situation in Hawaii and Florida, while unusual in some ways, nevertheless heralds what other States face as additional harmful NIS enter and spread throughout the United States and people become more aware of their damage. These results lead to the policy options discussed later in this chapter on:

- better protection for National Parks and other natural areas throughout the country (pp. 31-34), and
- the role of information and environmental education in preventing future problems in these States and elsewhere (pp. 34-36).

■ The Look of the Future

Increasing international trade, including commerce in biological commodities, will open new pathways for NIS. International regulation of NIS has a poor track record and is not likely to stem this flow. Technology is likely to open additional pathways as well as provide better ways to detect, eradicate, and manage harmful NIS. Many observers expect increasingly negative impacts from NIS introductions—a world of increasing biological sameness. Climate change is the wild card: it would require re-thinking definitions of indigeneity and could drastically change patterns of species movement. These are forecasts, based on analyzable and nearly irreversible trends already underway. Visions, however, are about the desirable and imagined. OTA's Advisory Panelists envisioned a future in which beneficial NIS contributed a great deal to human well-being and indigenous species were preserved (box 1-C). Deciding this vision's worthiness is not a question for science. Which species to import and release and which to exclude are ultimately cultural and political choices—choices about the kind of world in which we want to live.

POLICY ISSUES AND OPTIONS

In this section, OTA sets out the major policy issues that emerged from its analysis. Related congressional options seem straightforward in some cases, e.g., changes to the Lacey Act[1] or the Federal Noxious Weed Act (FNWA).[2] In other cases, policy actions are not so apparent. Therefore, the policy options that follow vary in their specificity and the degree to which OTA has evaluated their implications and alternatives. Few prior reports on NIS have addressed policy changes. OTA's work is, in effect, exploratory—a first step in highlighting policy needs and a few of the means to fill them. The discussion is organized around these eight policy issues:

Issue 1: Congress and a More Stringent National Policy

Issue 2: Managing Non-Indigenous Fish, Wildlife, and Their Diseases

Issue 3: The Growing Problem of Non-Indigenous Weeds

Issue 4: Damage to Natural Areas

Issue 5: Environmental Education as Prevention

Issue 6: Emergencies and Other Priorities

Issue 7: Funding and Accountability

Issue 8: Other Gaps In Legislation and Regulation

■ Issue 1: Congress and a More Stringent National Policy

The most fundamental issue is whether the United States needs a more stringent and comprehensive national policy on the introduction and management of harmful NIS. General agreement exists that the United States has no such policy now. The United States has, through various Federal and State laws and President Carter's Executive Order 11987, attempted to prevent and manage the impacts of harmful NIS. However, applicable legislation has significant gaps and the Executive Order has not been implemented fully (55,70) (ch. 6). Invasive NIS continue to enter, spread, and cause economic and environmental harm, despite governments' collective efforts (chs. 2, 3). In one of the most extensive State studies to date, the Minnesota Interagency Exotic Species Task Force noted:

---

[1] Lacey Act (1900), as amended (16 U.S.C.A. 667 *et seq.*, 18 U.S.C.A. 42 et seq.)

[2] Federal Noxious Weed Act of 1974, as amended, (7 U.S.C.A. 2801 *et seq.*)

Box 1-C—OTA's Advisory Panel Envisions the Future

OTA's Advisory Panelists (p. iv) have been dealing with NIS for much of their professional lives and are more expert than most in assessing what the future might hold. Following are some of the fears and hopes they identified when asked to ponder the best and worst that might be ahead.

Life Out of Bounds . . .

"The future will bring more reaction to zebra mussels (*Dreissena polymorpha*) and inaction to the massive alteration of natural habitats and natural flora and fauna . . . By the mid-21st Century, biological invasions become one of the most prominent ecological issues on Earth . . . A few small isolated ecosystems have escaped the hand of [humans] and in turn NIS . . . One place looks like the next and no one cares . . . The homogeneity may not be aesthetically or practically displeasing, but inherently it diminishes the capacity of the biotic world to respond to changing environments such as those imposed by global warming . . . The Australian melaleuca tree (*Melaleuca quinquenervia*) continues its invasive spread and increases from occupying half a million acres in the late 1980s to more than 90 percent of the Everglades conservation areas."

. . . Or Life In Balance

"An appropriate respect for preserving indigenous species becomes a national goal by consensus . . . All unwanted invasions are treated with species-specific chemicals or by vast releases of 100 percent sterile triploids (created quickly) that depress the exotic populations. Invasions slow to a trickle and fade away like smallpox . . . Jobs for invasion biologists fade away . . . [There is] an effective communication network, an accessible knowledge base, a planned system of review of introductions, and an interactive, informed public . . . Native [species] are still there in protected reserves . . . The contribution of well-mannered NIS—for abuse-tolerant urban landscaping, for ornamentals in gardens, for biological control of pests, for added interest, for increased biodiversity, for new food and medicine—is appreciated. The overarching criterion for judging the value of a species is its contribution to the health of its host ecosystem."

SOURCE: Advisory Panel Meeting, Office of Technology Assessment, July 29-30, 1992, Washington, DC.

Needed is a plan to address all [non-indigenous species], changes in the laws that provide closer monitoring of new introductions, and coordination among all State and Federal agencies that control [non-indigenous] species. (70)

Gaps in the Federal, regional, and State system arise from several sources. First, Federal and many State agencies lack broad authority over NIS as a whole, e.g., to protect against NIS' negative effects on biological diversity, or to ensure that environmental impact assessments take potentially harmful NIS into account (box 1-D). In turn, the agencies have been reluctant to exert authority where statutes are not clear. Consequently, NIS issues often receive governmental attention on a piecemeal basis *after* major infestations, such as that of the zebra mussel. Attention wanes between harmful episodes.

Second, the lack of information on the origins, numbers, distribution, and potential impacts of many NIS hampers the design of appropriate responses (chs. 2, 4). Distinguishing indigenous species from NIS and beneficial NIS from harmful ones is difficult in some cases yet these are crucial distinctions for regulatory and control efforts. Some NIS escape detection at ports-of-entry and ordinary quarantines cannot contain them because of inadequate scientific knowledge and detection technologies.

Third, the U.S. system for dealing with harmful NIS involves a complex interplay of Federal and State authorities, with numerous Federal, State, and regional coordinating bodies attempting to enhance consistency and resolve conflicts. Sometimes the respective Federal and State roles are not adequately defined (1), especially for problems that cross State boundaries.

Certain trends specific to NIS are likely to continue—trends that shape public policy. These point to increased public and scientific awareness of the damage some NIS cause and a concomitant caution toward importing new ones (46). The U.S. press is giving more attention to NIS-related problems caused by single species, e.g., zebra mussels, African honey bees (*Apis mellifera scutellata*), or cheatgrass (*Bromus tectorum*).

At the same time, many forces are elevating the visibility of harmful NIS on a broader, ecosystem basis. Some Federal and State agencies—e.g., the National Park Service, the Bureau of Land Management, the Minnesota Department of Natural Resources, and the Illinois Department of Conservation—are considering and in some cases adopting, more stringent policies (chs. 6, 7). In addition, the use of indigenous (native) plants and animals is increasingly popular in public and private landscaping, reforestation, fisheries management, wildlife enhancement, and other projects (96,130). These trends suggest that management of at least some harmful NIS is likely to improve even without congressional action.

On the other hand, the current situation provides considerable cause for concern (ch. 2). A *status quo* approach comes with certain, sizable risks—for example, that important resources such as the Everglades and Haleakala National Parks will lose their uniqueness (ch. 8); that western U.S. forests will be threatened by a more virulent gypsy moth (ch. 4); and that, in the absence of unifying Federal action, private firms importing or shipping live organisms will face increasingly inconsistent State and local regulations (ch. 7).

Environmental groups, professional organizations of scientists, and individual biologists are among those urging far stronger efforts to restrict the entry and spread of NIS. Participants in a conference sponsored by the U.S. Environmental Protection Agency (EPA) recommended that the United States aim for no new introductions of non-indigenous aquatic nuisances (132). One of the Nonindigenous Aquatic Nuisance Prevention and Control Act's several goals is similar: "to prevent unintentional introduction and dispersal of nonindigenous species into waters of the United States through ballast water management and other requirements."[3] The North American Native Fishes Association recommends banning all introductions of non-native fish (79). Some credible scientific sources—especially those with first-hand knowledge of the worst U.S. problems—have recommended bans on biological control introductions in natural areas or against indigenous pests; on the release of non-indigenous big game animals into public natural areas; on particularly risky types of imports such as unprocessed wood; or on all further intentional introductions for whatever purposes (25,61,69,100).

Usually, though, suggestions fall short of a ban on all new NIS introductions because broad-brush bans risk handicapping entry of desirable NIS that cause no harm. The International Union for the Conservation of Nature and Natural Resources (44) formulated a model national law on NIS and suggested that:

- release of NIS be considered only if clear and well-defined benefits to humans or natural communities can be foreseen;
- release be considered only if no indigenous species is suitable;
- no NIS be deliberately released into any natural area and releases into seminatural areas not occur without exceptional reasons; and
- planned releases, including those for biological control, include rigorous assessment of desirability, controlled experimental releases, then careful post-release monitoring and pre-arrangement for control or eradication, if necessary.

---

[3] Nonindigenous Aquatic Nuisance Prevention and Control Act of 1990, as amended (16 U.S.C.A. 4701 *et seq.*, 18 U.S.C.A. 42)

Box 1-D—The National Environmental Policy Act and Non-Indigenous Species

The National Environmental Policy Act (NEPA), which mandates environmental impact assessment, has rarely been applied to decisions about introductions of non-indigenous species (NIS) (ch. 7). NEPA makes no explicit mention of NIS. Many potentially significant actions, such as allowing wood imports from risky new sources, have not been considered sufficient to trigger NEPA review. A recent exception, however, is the environmental impact statement prepared regarding the New Jersey Division of Fish, Game, and Wildlife's proposal to introduce chinook salmon (*Oncorhynchus tshawytasha*) from the Pacific coast into the Delaware Bay. A number of NIS-related Federal activities are categorically excluded from NEPA review, including:

- low-impact range management activities, such as . . . seeding (U.S. Forest Service).
- all activities of the Plant Materials Centers, such as comparative field plantings, release of cooperatively improved conservation plants, production of limited amounts of foundation seed and plants, and assisting nurseries in plant production (Soil Conservation Service).
- the reintroduction (stocking) of native or established species into suitable habitat within their historic or established range (U.S. Fish and Wildlife Service).
- highway landscaping (Federal Highway Administration)

Full NEPA application to problems of NIS is unlikely without explicit direction from Congress. Various measures are available. In the most rigorous application, Congress could declare that new, unanalyzed releases of NIS are, per se, potentially significant environmental impacts that require analysis. Or Congress could require that NIS concerns be specified in the checklists used for preliminary environmental assessments and for making decisions regarding the need for further evaluation. Or Congress could limit related exclusions (see also ch. 7.)

Recently, a Federal court ruled that NEPA applied to the North American Free Trade Agreement—for which no environmental impact statement had been prepared. That decision has been appealed so NEPA's application remains legally unclear (ch. 10). Any eventual application of NEPA is likely to highlight concerns regarding NIS. International trade is a major pathway for the movement of potentially harmful NIS yet related issues have received little consideration in free trade discussions so far.

A comprehensive environmental impact assessment would address, among other possible impacts, the extent to which risks from harmful NIS would increase with any introduction and the capability of U.S. agencies to respond to any such increase. In the past, these agencies often have lacked the institutional and financial flexibility to anticipate and respond quickly to new risks (chs. 4, 6).

SOURCES: J. Kurdila, "The Introduction of Exotic Species Into the United States: There Goes the Neighborhood!" *Environmental Affairs*, vol. 16, 1988, pp. 95-118; U.S. Department of the Interior, Fish and Wildlife Service, *Administrative Manual: Environment, NEPA Handbook*, Part 516, April 30, 1984; Versar, Inc., "Introduction of Pacific Salmonids into the Delaware River Watershed," draft environmental impact statement prepared for the U.S. Fish and Wildlife Service and the New Jersey Division of Fish, Game and Wildlife, July 25, 1991; 23 CFR 771.117(7), as amended (Aug. 28, 1987) (Federal Highway Administration); 56 *Federal Register* 19718 (U.S. Forest Service); 7 CFR 613, 650.6 (Soil Conservation Service).

The nursery, pet, aquaculture, and agriculture industries have traditionally been strong advocates for further introductions of desirable NIS and have noted the burdens of more time-consuming and complex evaluations of their potential risks. These groups can be expected to be cautious about any congressional action that would make U.S. policy more stringent. For example, those in the nursery industry fear that banning NIS and requiring the use of indigenous plants would create complex definitional problems regarding which species are indigenous; outlaw the hardy non-indigenous plants most suitable for urban landscapes; require using indigenous plants that are less resistant to diseases and pests than their close foreign relatives; and eliminate highly ornamental plants that many people prefer to less showy indigenous ones (52).

However, pressures on Congress and Federal and State agencies to enact some partial measures are likely to increase as NIS-related issues receive more attention. Florida has prohibited any releases of non-indigenous marine plants or animals into State waters.[4] The New Mexico State Legislature recently considered a bill that would have led to the eradication of several ''exotic'' non-indigenous game animals and required the Department of Game and Fish to ban further game introductions (101). (State game officials considered the legislation extreme and opposed it, whereas hunting and environmental groups were divided.) Several local ordinances require landscape architects, designers, and contractors to use a percentage of indigenous plants in their projects (52).

Bans are intended to slow the intentional introduction of organisms into and within the United States. Even the strictest ban could not stop unintentional introductions. Nor could it limit damage caused by the continuing spread of harmful NIS already in the country. Therefore, even the most restrictive policies regarding new introductions would not solve all problems associated with harmful NIS.

New Zealand, a small island nation with NIS problems as severe as Hawaii's, is often cited as the country that addresses NIS most effectively (77). Its approach merits consideration here (box 1-E). New Zealand's recent policy changes illustrate an attempt to be comprehensive, forward looking, fair to importers, and responsible. However, New Zealand is much smaller and less diverse than the United States. In this country, States play an important role in setting and implementing U.S. national priorities. Therefore only some of New Zealand's approaches would be feasible here.

Attempts to formulate a similarly comprehensive and more stringent national policy on harmful NIS would need to account for the following seven issues. In most of these areas, OTA suggests possible statutory changes. These should be approached with one caution. The release of NIS and GEOs is regulated by many of the same statutes. Legislative changes intended to affect harmful NIS could inadvertently apply to GEOs if definitions are not crafted with care.

■ Issue 2: Managing Non-Indigenous Fish, Wildlife, and Their Diseases

Federal and State governments presently divide responsibilities for introductions of fish, wildlife, and their diseases. The Lacey Act is the primary Federal vehicle for excluding harmful imports. Under the Lacey Act, the U.S. Fish and Wildlife Service (FWS) restricts importation into the country of fish or wildlife that pose a threat ''to humans, agriculture, horticulture, forestry, or to wildlife or the wildlife resources of the United States.''[5] Current regulations restrict only 2 taxonomic families of fish (1 to prevent entry of 2 fish pathogens), 13 genera of mammals and shellfish, and 6 species of mammals, birds, and reptiles.[6] The USDA's APHIS and the Public Health Service prohibit entry of a several additional wildlife species (reptiles, birds, and mammals) to prevent entry of pathogens affecting poultry or livestock or because they pose human health threats.[7]

The Nonindigenous Aquatic Nuisance Prevention and Control Act of 1990 authorized FWS and the National Oceanic and Atmospheric Administration (NOAA) to issue regulations related to the prevention of unintentional introductions of aquatic nuisance species, like the zebra mussel.[8] Al-

---

[4] 28 Fla. Stat. Annot. sec. 370.081(4)

[5] 18 U.S.C.A. 42(a)(1)

[6] 50 CFR 16 (Jan. 4, 1974)

[7] 9 CFR 92, as amended (Aug. 2, 1990)

[8] 6 U.S.C.A. 4722

Box 1-E—How New Zealand Addresses Non-Indigenous Species

New Zealand's legal and institutional framework and the nature of its programs are key to its current successes managing harmful non-indigenous species (NIS). As in the United States, however, protecting agriculture has received higher priority than safeguarding the indigenous flora and fauna. Some aspects of New Zealand's approach that are absent or rare in the United States are given here:

Legal and Institutional Aspects:

- Agency performance standards implemented through agency "contracts" to provide specified governmental services and through detailed annual reports.
- Detailed national standards for animal imports and strong authority to require bonds for potential costs of escape and to impose other conditions.
- A "user pays" approach to cover most costs of inspection, surveillance, scientific analysis, and enforcement against violators.

Programmatic Aspects:

- Intensive inspection of arriving passengers, baggage, and goods with random checks to evaluate interception rates.
- 100 percent treatment of arriving aircraft with insecticide.
- Computerized tracking of imports, from arrival to unloading.
- Detailed surveillance of and contingency planning for forest pests.
- Extensive enlistment of public support for pest surveys and monitoring.

Recently, New Zealand determined that its more than a dozen major acts and several hundred subsidiary regulations pertaining to agriculture needed consolidation and revamping. The new approach will regulate all potentially harmful imports through an appointed Hazards Control Commission.

An independent professional staff will advise the Commission, with input from expert advisory committees. Proposals for imported and genetically engineered organisms will be advocated by private or governmental proponents. Countervailing arguments will be presented by the Department of Conservation. The law provides for full economic and ecological consideration, public hearings, and opportunities for appeal. Known low-risk organisms will receive less scrutiny. Decisions must balance "the benefits which may be obtained from . . . new organisms against the risks and damage to the environment and to the health, safety and economic, social and cultural wellbeing of people and communities." If this new approach succeeds, it could provide a broad model for the United States.

SOURCES: Anonymous, "Biosecurity Bill: Update," *Sentinel*, New Zealand Ministry of Agriculture and Fisheries, Wellington, No. 19, Feb. 1, 1992, p. 3; Director of the Law Commission, "VIII. Public Welfare Emergencies," *Final Report on Emergencies*, Law Commission Report No. 22, Wellington, New Zealand, December 1991, pp. 230-246; Office of the Minister of Agriculture, Ministry of Agriculture and Fisheries, Wellington, New Zealand, memorandum regarding Agricultural Regulation Reform, to Chairman, Cabinet Strategy Committee, undated; A. Moeed, Chairperson, Interim Assessment Group, Ministry for the Environment, Wellington, New Zealand, letter to P.T. Jenkins, Office of Technology Assessment, Feb. 10, 1992; D. Towns, IUCN Regional Member, Department of Conservation, Aukland Conservancy Office, Aukland, New Zealand, letter to P.T. Jenkins, Office of Technology Assessment, Oct. 29, 1991.

though none have been issued to date, eventual regulations under the Act could impose additional restrictions on the importation of harmful aquatic NIS (30).

In practice, then, the Federal Government places only a few piecemeal constraints on the *importation* of fish, wildlife, and their diseases.

Tens of thousands of different species (most of the world's fauna, excluding insects) potentially could be legally imported into the United States (81). Well over 300 non-indigenous fish and wildlife species of foreign origin have established here already, approximately 122 of which are known to cause harm (ch. 2) (8,23,104).

The Federal Government currently plays a small role in restricting *interstate* transfers of non-indigenous fish and wildlife (ch. 6). FWS does not impose regulations or quarantines to prevent interstate transfers of harmful fish, wildlife, or fish diseases, since neither are authorized under the Lacey Act. APHIS sometimes quarantines wildlife to prevent the spread of pathogens, but only for those causing significant diseases of poultry or livestock. Amendments to the Lacey Act in 1981 authorized the FWS to enforce State laws prohibiting transport of species into a State,[9] but FWS enforcement is understaffed, underfunded, and has numerous other pressing responsibilities (74,121). Future implementation of the Nonindigenous Aquatic Nuisance Prevention and Control Act could impose domestic regulations or quarantines for aquatic species (30).

States play the prominent role in many areas related to fish and wildlife. They vary in how rigorously they guard their own borders or prevent releases of harmful species. States prohibit relatively few injurious species; their standards of review for predicting harm are low; and enforcement is weak (55) (ch. 7). The same conditions apply to the States' roles in releasing fish and wildlife within their borders.

Taken together, these Federal and State gaps constitute a serious threat to the Nation's ability to exclude, limit, and rapidly control harmful fish and wildlife. For example, importation and transfer of zebra mussels within much of the United States remained legal for approximately 2 years after they had inadvertently entered the United States and demonstrated their devastating potential. An opportunity to slow their spread was lost. The potential for spread of pathogens of fish and aquatic invertebrates is another example. Federal regulations under the Lacey Act require accurate labeling of shipping containers for species identity and numbers. Screening for contamination by pathogens is not required. There is no Federal quarantine of diseased fish stocks and in many

RON PEPLOWSKI

*The Nonindigenous Aquatic Nuisance Prevention and Control Act of 1990 authorized new regulations and programs for aquatic species like the costly zebra mussel (*Dreissenna polymorpha).

States diseased fish and invertebrates can be legally imported and released.

Some observers have called for an increased Federal presence to fill gaps like those above. Julianne Kurdila (55), for example, suggested either implementing President Carter's 1977 Executive Order 11987 (box 6-B) or the passage of new legislation to correct the Lacey Act's deficiencies, recommendations passed along by the Minnesota Interagency Exotic Species Task Force (70). USDA officials see the need to screen fish for diseases, like they do for livestock (56).

Proposals to expand the Federal role have engendered considerable controversy in the past. However, OTA's survey of State fish and wildlife agencies asked whether they would like to see the Federal role "increase," "decrease," or "stay about the same" in the regulation of non-indigenous fish and wildlife (ch. 7). A clear majority—63 percent—favored an increased Federal role; 23 percent favored keeping the role about the same; only one State (Wisconsin) preferred to see the Federal role decreased (3 percent were not sure and 8 percent did not answer). Peter Schuyler conducted a separate survey of 271 resource managers and others

[9] 16 U.S.C.A. 3372

involved with issues related to non-indigenous animals. Of the 265 U.S. respondents, 65 percent perceived the problem's biological aspects to have international significance (92, 93)—clearly beyond local or State scope.

Two areas in which the Federal Government might strengthen its role are in:

1. increasing the rigor of screening before importation and release of fish and wildlife; and
2. defining new State roles.

The first area arises from widespread criticism that the Lacey Act is failing to protect the United States from entry of harmful new NIS; also, many decisions to introduce NIS are made without thorough risk assessment (ch. 4). The second area regarding State roles emerges from OTA's analysis of State laws and regulations regarding fish and wildlife (ch. 7).

TIGHTENING FISH AND WILDLIFE SCREENING

Option: Congress could amend the Lacey Act to lengthen its list of excluded injurious wildlife and to speed the process by which new listings are added.

Option: Congress could require that Federal agencies and others using Federal funds to introduce non-indigenous fish and wildlife develop and adopt specific, rigorous decisionmaking methods for screening species prior to release.

A number of problems have been documented with the Lacey Act and its implementation by FWS (55,83). The most commonly acknowledged problem is that regulation and enforcement hinge on a short and noncomprehensive list of "injurious" wildlife and adding new species to the list is time-consuming (116). The Lacey Act

is also criticized for not providing comprehensive regulation of interstate transport of federally listed species and for not being clear regarding its application to hybrid and feral animals. FWS enforcement of the Act's sparse interstate transport provisions is limited and programs to control or eradicate non-indigenous fish and wildlife are piecemeal, lack emergency measures, and have no proactive components to catch problems early.

Only five new species or taxonomic groups were added over the 7-year period from 1966 to 1973, with one more addition over the next 15 years. Several potentially injurious species are under consideration in 1993 for listing, on a species by species basis. Efforts to list the mitten crab (*Eriocheir* spp.) took at least 2 years, with some evidence that they were successfully introduced during this time (83). This means that organisms are unregulated when they are most amenable to control and eradication, i.e., shortly after entry when their populations are small.

The greatest potential for the Lacey Act is to reduce problems related to NIS used in the pet and aquarium trades, "exotic" non-indigenous game ranching, and aquaculture.[10] The potential risks of species in these groups are relatively well known and most of these NIS can be readily identified and detected at ports of entry. However, greater use of the Lacey Act would require aggressive efforts to expand the Act's list of injurious species (6). This has not been tried since 1977. The current FWS approach remains largely reactive, with little outside pressure to change or increase the list of species (83).

Congressional action to amend the Lacey Act (box 1-F) could address some concerns without changing the basic, Federal "dirty list" regulatory approach. The dirty list approach prohibits certain unacceptable species and allows unlisted species to be imported. This puts the burden on

---

[10] The Federal interagency Aquatic Nuisance Task Force has concluded that the escape, accidental release, or improper disposal of intentionally introduced organisms is "virtually inevitable" and that these should not be considered unintentional (122). By this interpretation, non-indigenous aquaculture species could be listed under the Lacey Act. The newer Nonindigenous Aquatic Nuisance Prevention and Control Act of 1990 would not apply, because it covers only unintentional introductions.

Box 1-F—How To Improve the Lacey Act

The following changes to the Lacey Act would provide more comprehensive protection and management of the Nation's resources. The U.S. Fish and Wildlife Service (FWS) would need additional staff and other resources to make these changes. The FWS currently spends approximately $3 million annually for port inspections for fish and wildlife. In contrast, the U.S. Department of Agriculture's Animal and Plant Health Inspection Service (APHIS) spends approximately $80 million for agricultural port inspections. The two agencies do not need comparable budgets but clearly an amended Lacey Act would require budgetary changes for the FWS.

Lengthen the list of injurious wildlife. Congress could provide the FWS with increased guidance on the purpose of this list and the specific criteria for adding species to it. Proposed amended criteria would be discussed with outside experts and be as comprehensive as possible. One possibility would be to include harmful species indigenous to the United States, but established outside their range, as injurious. A quite different alternative would be to supplement this current approach with a "clean list" approach (ch. 4).

Speed the listing process. Congress could add provisions to: 1) eliminate, reduce, or expedite the most time-consuming parts of the listing process (public notice and comment, etc.), 2) use emergency listing procedures more often, or 3) give FWS authority to impose emergency control, with monitoring, while the usual listing process takes place. Eliminating requirements for public notice and comment could have unintended negative effects: decreasing officials' accountability, limiting access by stakeholders, and excluding broad expert participation from an already-limited group of decisionmakers. If Congress gave FWS emergency authority, reasonable time limits could be set for study and reaching decisions on final listings. FWS and APHIS might together streamline their listing processes to ensure procedural consistency between the Lacey Act and the Federal Noxious Weed Act.

Consider whether FWS should assist with enforcement of State injurious wildlife lists and provide FWS with authority for emergency quarantine and emergency actions. First, the respective Federal and State responsibilities would need to be clarified. Then, Congress could take any of several steps: direct FWS to strengthen its role; provide additional resources to States for enforcement; and/or amend the Lacey Act to provide for Federal quarantines on interstate movement of injurious wildlife.

SOURCES: M.J. Bean, "The Role of the U.S. Department of the Interior in Nonindigenous Species Issues," contractor report prepared for the Office of Technology Assessment, November 1991; J. Kurdila, "The Introduction of Exotic Species into the United States: There Goes the Neighborhood!" *Environmental Affairs*, vol. 16, 1988, pp. 95-118; H.A. Peoples, Jr., J.A. McCann, and L.B. Starnes, "Introduced Organisms: Policies and Activities of the U.S. Fish and Wildlife Service," *Dispersal of Living Organisms Into Aquatic Ecosystems*, A. Rosenfield and R. Mann (eds.) (College Park, MD: Maryland Sea Grant, 1992), pp. 325-352; U.S. Department of the Interior, Fish and Wildlife Service, internal memorandum, 1987.

---

regulators to determine whether a species is harmful. Commonly cited alternatives to dirty lists are ''clean lists'' or combinations of clean and dirty list approaches (ch. 4). The clean list approach prohibits all species unless they are determined to be acceptable, that is, unless they merit being on the clean list. This puts the burden on the importer to prove a species is not harmful. States, such as Hawaii, that are most concerned about NIS are moving from simple dirty list regulatory approaches toward using both clean and dirty lists.

Clean lists can only be used for certain kinds of organisms. Many pathogens and invertebrates are too little known to classify their impacts as acceptable or not. Generally, though, clean lists represent a more stringent, proactive policy, especially when dirty lists are short and noncomprehensive. What is ''clean'' in one part of the United States is not necessarily so elsewhere, however. Therefore, any new policy using clean lists would need regional flexibility.

Some contend that any Federal clean list is infeasible because of lingering opposition from FWS's earlier attempts to adopt this approach

(83) (box 4-A). The pet industry, along with portions of the zoological and scientific communities, spearheaded opposition in the 1970s (55). Marshall Meyers, general counsel for the Pet Industry Joint Advisory Council, articulates the industry's continuing opposition to regulations viewed as overly restrictive, vague, or poorly justified (14), as they found previous clean list proposals. On the other hand, the pet industry recently joined environmental groups in supporting tighter regulation of importation of wild-caught birds.[11]

Both clean and dirty lists require determining whether species pose acceptable risks. Formal decisionmaking protocols, risk analysis, cost-benefit analysis, and other techniques attempt to accomplish this goal (ch. 4). Each has advantages and disadvantages. For example, protocols like the American Fisheries Society's for the release of fish (51) represent a high level of decisionmaking rigor and best suit the most potentially risky types of introductions. Typically, these methods require large amounts of highly technical information and are therefore demanding in financial and scientific terms. Also, these methods are controversial because their usefulness has not been established clearly.

No single method is ideal for assessing all Federal and federally funded introductions of non-indigenous fish and wildlife. However, formal decisionmaking methods designed to more carefully assess and decrease risks are considered to be prudent alternatives to banning all potentially risky introductions (83). Congress could require that agencies develop and adopt either a recognized decisionmaking protocol or another formal and rigorous method suited to their situations. This was the approach taken in the proposed Species Introduction and Control Act of 1991 regarding non-indigenous fish and wildlife.[12]

DEFINING NEW STATE ROLES IN FISH AND WILDLIFE INTRODUCTION

Option: Congress could address weaknesses in some States' fish and wildlife laws by implementing national minimum standards. These standards would provide legal authority to regulate harmful NIS and be linked to funding for States to implement them.

Option: Alternately, Congress could encourage wider adoption of a federally developed model State law to make legal authority among States more comprehensive.

The strength of the U.S. Federal system is that the 50 States provide a testing ground for new ideas. Such new ideas turn up in the exemplary approaches discussed in chapter 7. On the other hand, federalism leads to duplication of efforts and highly variable, and sometimes conflicting, regulations (72). This has been the case for non-indigenous fish and wildlife.

States' standards vary considerably regarding which species and groups are regulated and how carefully they are regulated; many State efforts to regulate importation, possession, introduction, and release are inadequate (ch. 7) (55). In some cases, the weaknesses of State programs stem from incomplete legal authority.

The Lacey Act leaves decisions on almost all intentional introductions of fish and wildlife to the States; only the relatively few organisms on the list of injurious wildlife are prohibited. Thus, correcting problems would entail full exercise of State prerogatives (83). However, Federal programs support many State-sponsored introductions, so the Federal Government has a strong interest in this area.

A variety of approaches could be used to encourage improved State performance. Federal pre-emption of State NIS laws is unlikely to be justifiable or politically feasible. Two more

---

[11] The Wild Bird Conservation Act of 1992, Public Law 102-440, Title I, Section 102, Oct. 23, 1992; 106 stat. 2224.

[12] H.R. 5852, introduced by Rep. H. James Saxton.

tenable and often-suggested methods are national minimum standards and wider use of model State laws. Either method could ensure that State fish and wildlife laws provide adequate authority for more comprehensive regulation.

Box 1-G illustrates a national minimum standards approach. Three elements would be needed:

1. a process to determine whether State laws are consistent with the new national minimum standards,
2. a program of incentives for States to adopt or retain laws meeting the national minimum standards and to provide sanctions against States that do not, and
3. a means to provide reliable sources of revenue to fund these efforts.

Also, careful individual State review is needed in several other areas: quarantine requirements; containment specifications; responsibility for control of escapees; and regulation of live bait fish and invertebrates affecting nonagricultural areas.

Incentives could include Federal grants or matching funds to States for initial reviews of their fish and wildlife laws. Also, Federal funds could be made available for NIS control or eradication for States whose NIS laws meet the national minimum standard. Sanctions would most reasonably include denial of Federal funds for fish and wildlife restoration and/or other Federal aid-to-States programs. Sanctions could be phased in over a suitable period, such as 5 years.

A national minimum standards program could be administered by FWS, another existing agency, or a new Federal office or commission. Its duties would include: monitoring and reporting on State compliance; processing requests for State funding; and maintaining up-to-date, publicly available compilations of States' fish and wildlife statutes, regulations, quarantines, and other important information.

An alternate approach would be to provide incentives for States to adopt a federally developed, comprehensive model State law. Voluntary examples already have been used to some extent for fish and wildlife.

The Southeast Cooperative Wildlife Center's model law combined laws on endangered species, injurious wildlife, disease control, public health, wildlife management, humane care, and interstate control. The model was reviewed by all States and parts of it used by a few. Missouri used part of the model, while Utah considered it but adopted their own approach (ch. 7). This specific model State law, however, received substantial criticism for being overly broad and creating excessive administrative rules and paperwork (67).

Generally, voluntary approaches for environmental compliance are receiving increased attention for a number of problems. Industry groups often support such initiatives, claiming that voluntary programs are more effective and cut costs (99). Few environmental groups have endorsed voluntary programs, however (88).

∎ Issue 3: The Growing Problem of Non-Indigenous Weeds

The continuing entry and spread of non-indigenous weeds in the United States raises serious concerns in many quarters. State agriculture and natural resource officials, Federal land managers, members of conservation organizations, and scientists have expressed their concern that existing Federal weed laws are flawed, their implementation incomplete, and too few resources have been directed toward weed problems (chs. 2, 3, 6). In some cases, listing prohibited weeds under State noxious weed and seed acts may reduce the interstate spread of non-indigenous weeds otherwise allowed by Federal laws and regulations. However, the States can only partially compensate for insufficient Federal presence.

Three areas seem to call for a strengthened Federal role:

1. improving the Federal Noxious Weed Act (FNWA), by broadening its coverage and simplifying its procedures;

Box 1-G—National Minimum Standards for State Fish and Wildlife Laws

OTA finds in chapter 7 that States need the following types of legal authority and decisionmaking procedures to ensure comprehensive treatment of non-indigenous fish and wildlife:

1. Each State needs statutory or regulatory provisions that allow the State to regulate the importation, possession, and release of all classes of non-indigenous animals (including ferals and non-indigenous hybrids). This authority could allow for appropriate exemptions. The authority over importation would apply to NIS originating in foreign countries and to that from other parts of the United States. The authority over introduction would apply to both public and private property.
2. State laws need to provide authority to regulate intrastate stocking of species where hybridization with indigenous species or other harmful impacts may occur.
3. All States need legal authority to list potentially harmful NIS in all taxonomic groups as prohibited from importation, possession, and/or release. Their lists would supplement the Lacey Act list. In this and other listing processes, States would actively solicit expert technical advice and public comment. However, under extraordinary circumstances States would also have emergency authority to prohibit species without administrative delays.
4. States' decisions regarding importation, possession, and release of NIS would be based on defined and rigorous standards of review that comprehensively consider the new releases' environmental impacts. Detailed studies, equivalent to an environmental impact statement, would be required in cases of potentially significant impacts.
5. All decisions to approve new releases would be conditioned on the following: a) notification and comment given to other potentially affected States, the Federal Government, and Canada and Mexico if they are potentially affected; b) stipulations for follow-up monitoring and review; and c) provisions governing public and/or private responsibility for the costs of control or eradication and for damages if unanticipated negative impacts occur.

SOURCE: Office of Technology Assessment, 1993.

---

2. increasing weed management on public lands; and
3. tightening screening before the release of new, potentially weedy non-indigenous plants.

The first area arises from concerns that FNWA is an inadequate tool for preventing the problems now facing resource managers. The second area arises from existing massive and spreading weed problems, especially on western public lands, and the view that the Federal Government has not fully met its responsibility here. Finally, those responsible for introducing new plants for horticulture and soil conservation have been reluctant to recognize the importance of rigorous screening for weediness before a plant's release.

THE FEDERAL NOXIOUS WEED ACT AND FEDERAL SEED ACT

Option: Congress could amend and expand the Federal Noxious Weed Act to rectify several widely acknowledged problems regarding definitions, interpretation, and its relationship to the Federal Seed Act.

The Federal Noxious Weed Act and the Federal Seed Act[13] provide the main authority for APHIS to restrict entry and spread of noxious weeds. The FNWA prohibits importation of listed noxious weeds and provides authority to quarantine species already in the country. The Act has been criticized by the Weed Science Society of America, environmental groups, State and some industry representatives, and scientific experts (60,

---

[13] Federal Seed Act (1939), as amended (7 U.S.C.A. 1551 *et seq.*)

112,113). Commonly cited shortcomings include: problems with the definition of a "noxious weed;" confusion between this Act and the Federal Seed Act; the inadequacy of the list of prohibited species and the cumbersome nature of the listing process; and APHIS' interpretation limiting the restriction of interstate weed transfer to only those species under quarantine (36,60,70,98).

A major shortcoming is that the Act is applied to too few species. APHIS took 8 years to place 93 species on the current list of Federal noxious weeds, yet at least 750 weeds meeting the Act's definition remain unlisted (98). Unlisted weeds can continue to be legally imported, although their potential for causing damage is known. APHIS' narrow interpretation of the definition of a Federal noxious weed has kept it from regulating clearly harmful NIS with wider distributions, including those meriting restriction to prevent further spread (86). Purple loosestrife, Brazilian pepper (*Schinus terebinthifolius*), and Eurasian watermilfoil (*Myriophyllum spicatum*) are prominent unlisted weeds. Moreover, the requirement that a noxious weed be of foreign origin means FNWA does not cover plants like the western wetland invader smooth cordgrass (*Spartina alterniflora*), which originated in the eastern United States. Difficulties make the listing process slow (36,98), yet FNWA has no emergency mechanism to allow rapid action on unlisted species causing incipient problems.

APHIS has barely implemented FNWA's Section 4, which requires a permit for moving listed species between States. Under APHIS's interpretation of the Act's legislative history, this restriction only applies when the agency has imposed a specific quarantine under Section 5. Yet in 18 years, APHIS has imposed only one quarantine for a noxious weed. As a result, at least nine Federal noxious weeds were sold in interstate commerce as of 1990 (98). APHIS has maintained this interpretation in the face of steady pressure from some State officials to change it (49).

APHIS has traditionally emphasized insect and disease problems and lacked professional weed

LEWIS WATERS

*Purple loosestrife (*Lythrum salicaria*) is among the prominant weeds not listed by the Federal Noxious Weed Act.*

scientists in key positions (128), contributing to the low priority of weed management among its various responsibilities (ch. 7). Then Administrator Glosser contended, however, that lack of funding—not priority setting—limits APHIS' weed control programs (36).

Some gaps in FNWA might eventually be filled under the recently enacted Nonindigenous Aquatic Nuisance Prevention and Control Act. NOAA and the FWS could eventually move to regulate importations or impose quarantines of aquatic or wetland weeds, although no such regulations are either in place or planned.

The Federal Seed Act provides for accurate labeling and purity standards for seeds in commerce. Only 12 species have been listed under the Federal Seed Act, with "tolerances" set for contamination by small amounts of their seed.

Just one of these species is listed among the 93 prohibited entry under FNWA (62). It has not been clear whether species prohibited under FNWA could be legally imported and transported within the country as part of seed shipments. In 1988, APHIS initially allowed importation of grass seed contaminated by serrated tussock (*Nassella trichotoma*)—a weed listed under the Federal Noxious Weed but not the Federal Seed Act. In 1992, a Federal district court judge ruled that the Federal Noxious Weed Act applied to seed shipments; however, the case is on appeal at this writing.[14]

A second limitation of the Federal Seed Act is it only applies to agricultural and vegetable seed. The Act's requirements for truth in advertising do not cover horticultural seeds, including "wildflower" and "native grass" mixtures. Such commercial mixtures are increasingly popular, especially for use in suburban and seminatural areas. The use of "wildflower" and "native" may be misleading, because the mixtures frequently contain plants that do not grow naturally in the wild, either in the United States or in the region for which they are promoted (62). Some even contain Federal or State listed noxious weeds. State laws on consumer protection and accurate weights and measures could provide States with general authority to address horticultural seed mixtures, but little indication exists that they have done so (50).

Commonly suggested changes to improve FNWA include those in box 1-H. Some of these are included in amendments that Senator Byron Dorgan anticipates introducing in fall, 1993.

In 1990, APHIS attempted to consolidate its plant protection statutes into one piece of legislation. While that attempt failed, the Agency expects to try again. Any such consolidation could address the concerns raised here, without amending FNWA and the Federal Seed Act. It could also address the need for emergency and proactive measures discussed in a later section. Congress would need to ensure that no important functions were dropped in the consolidation process, however. Consolidated legislation would include many additional complex and potentially controversial issues. Its passage is not likely to be straightforward or rapid.

TIGHTENING PLANT SCREENING

Option: Congress could require that all entities introducing non-indigenous plant material conduct pre-release evaluations of its potential for invasiveness.

Option: Congress could require that APHIS conduct periodic evaluations of its port and seed inspection systems to test their adequacy and provide feedback for improvements.

At a minimum, Congress could ensure that current laws and regulations are adequately enforced. This requires that APHIS report on the effectiveness of its inspection system and regularly seek improvements. Also, a minimal approach would ensure that all new, potentially damaging introductions be screened for invasiveness. Past experiences show that releasing unscreened introductions is asking for trouble. Specifying methods to use for such screening, including review under NEPA (box 1-D), would require congressional intervention.

Intentional introductions of plants are almost entirely unregulated, unlike certain other categories of potentially harmful NIS that require permits or receive some Federal scrutiny. Yet some of the worst U.S. weeds were intentionally introduced by people who thought that they would be beneficial: kudzu, water hyacinth, and multiflora rose (*Rosa multiflora*) (60), and experts express concern about the possible invasiveness of some contemporary releases (ch. 6).

---

[14] Memorandum Opinion in *Pennington Enterprises, Inc.* v. *United States*, Civil Action No. 90-1067 (U.S. District Court, District of Columbia), on appeal to the D.C. Circuit Court of Appeals, Case No. 92-5179.

Box 1-H—How to Improve the Federal Noxious Weed Act

Change the definition of a "noxious weed." Redefine so that plant pests of nonagricultural areas and weeds of U.S. origin—but outside their natural ranges—are clearly included. (These definitional weaknesses commonly apply to State noxious weed laws, too.) The 1990 FNWA amendments directed Federal agencies to undertake several actions against "undesirable plant species" on Federal lands. These were defined to include noxious, harmful, exotic, injurious, or poisonous plants pursuant to Federal or State law but not including plants "indigenous to an area where control measures are to be taken." Thus, a precedent exists for basing definitions on U.S. ranges of plants.

Address weeds widespread within the United States. The lack of an approach to deal with widespread weeds is serious enough that APHIS should be asked to prepare a strategic plan for dealing with pests of this type. Then, other policy questions could be addressed, including whether to change the number of States that determine when APHIS ends its involvement. (APHIS presently interprets the Act to mean found in no more than two States).

Address the inconsistency between the Federal Noxious Weed Act and the Federal Seed Act. This could be done by deleting the provision in Section 12 that prohibits the application of FNWA to seed shipments regulated under the Seed Act; or by amending the Seed Act to make its list of excluded species identical to that of FNWA, whichever is more extensive.

Provide for emergency listing of weeds. Streamline the listing process or grant APHIS emergency authority to exclude those plants that meet the definition of a Federal noxious weed but have not yet been listed as such. As in the Lacey Act, current requirements for public notice and comment are important. However, they can create inordinate delay when time is essential. Therefore, strengthening the agency's authority to take emergency action before listing might be more desirable. APHIS and the Fish and Wildlife Service might develop emergency listing processes together to ensure their procedural consistency.

Clarify APHIS' role in regulating the interstate transport of weeds. This may require an amendment; Congress has conducted oversight in this area in the past and problems remain. One possibility would be to: **Make planting, distributing, and possessing noxious weeds with intent to distribute them illegal under almost all circumstances.** This would make interstate distribution of Federally listed weeds clearly illegal regardless of the existence of an APHIS quarantine. Minnesota recently took a stricter approach by prohibiting most instances of transport, possession, sale, purchase, import, propagation, or release of approximately 30 species of plants and animals.

Increase resources for control programs, including those on Federal lands. APHIS allocates few resources to the control and eradication of noxious weeds and other Federal agencies face similar shortfalls. (See Issue 7 for means to increase resources.)

SOURCES: D.H. Kludy, "Federal Policy on Non-Indigenous Species: The Role of the United States Department of Agriculture's Animal and Plant Health Inspection Service," contractor report prepared for the Office of Technology Assessment, Washington, DC, December 1991; R.N. Mack, Professor and Chair, Department of Botany, Washington State University, Pullman, WA, letter to P. Windle, OTA, Aug. 4, 1992; Minnesota Rules Chapter 6216, "Ecologically Harmful Exotic Species," St. Paul, MN, effective Aug. 12, 1993; D.C. Schmitz, Florida Department of Natural Resources, Tallahassee, FL, statement submitted at hearings before the Senate Subcommittee on Agricultural Research and General Legislation, Committee on Agriculture, Nutrition, and Forestry, "Preparation for the 1990 Farm Bill: Noxious Weeds," Mar. 28, 1990, pp. 357-360; H.M. Singletary, Director, Plant Industry Division, North Carolina Department of Agriculture, Statement submitted before the Senate Subcommittee on Agricultural Research and General Legislation, Committee on Agriculture, Nutrition, and Forestry, Mar. 28, 1990, pp. 354-356; Weed Science Society of America, "WSSA Position Statement on Changes in the Federal Noxious Weed Act," Davis, CA, May 8, 1990.

Current Federal restrictions on importation and interstate transport of plants (other than noxious weeds listed under FNWA) relate to preventing transfers of plant pests and pathogens—not evaluating the plant itself for harmful qualities. The USDA's Agricultural Research Service (ARS) annually imports large quantities of foreign plant material to develop new species or varieties for horticulture, soil conservation, or agriculture. Neither the Soil Conservation Service (SCS) nor

ARS specifically evaluates plants for invasiveness before their release for soil conservation or horticulture. These plants undergo little or no systematic evaluation for weediness and risk to nonagricultural systems (ch. 3). Evaluation of horticultural varieties developed abroad and imported for commercial sale is similarly lax.

More careful and consistent pre-release screening is needed. Some screening methods are already in place. Usually these methods are applied only to agricultural threats, however. APHIS initially used an expert panel, the Technical Committee to Evaluate Noxious Weeds (TCENW), to designate species for the Federal list of noxious weeds.[15] These or similar screening methods could serve as models for the ARS Germplasm Resources Laboratory to evaluate plant material. Possibilities include the use of risk analysis, benefit/cost analysis, safe minimum standards, and review under NEPA (ch. 4).

Harmful NIS commonly present insidious, long-term, low-probability, but high-risk problems. Under these circumstances, many standard decisionmaking methods fit only partially. For example, eventual costs may be impossible to predict, making economic projections of little use. Any new screening methods should be adopted on a test basis and evaluated before broader implementation. Certain additional decisionmaking steps are fairly clear now, however:

- increasing the role of technical advisory groups (98);
- expanding the scope of scientific and other expertise available to these advisory groups to include evolutionary and conservation biologists and ecologists (46);
- ensuring that decisionmaking processes are documented, clear, open to public scrutiny, and periodically evaluated;

- guaranteeing input from industries, States, other Federal agencies, and special interest groups that may be affected by the decision (49); and
- ensuring that the final decision is implemented effectively (61).

WEED MANAGEMENT ON PUBLIC LANDS

Option: Congress could monitor and evaluate closely the weed control efforts undertaken by Federal agencies as a result of FNWA amendments to the 1990 Farm Bill.

Management of non-indigenous weeds is a growing problem involving local, State, and Federal agencies (113). Most land management agencies now acknowledge the problems of noxious weeds and are beginning to attempt control. However, these programs generally are small, underfunded, and need additional support (chs. 6, 7). The Bureau of Land Management (BLM), for example, identified seven major deficiencies in its programs: funds and staff; policy guidance and awareness of the problem; basic information on expansion of weed populations; attention to nonrangelands; active and preventive programs; training beyond pesticide application; and coordination with other Federal, State, and county agencies (115). Many areas with severe non-indigenous weed problems are among the most protected categories of federally managed lands. Their problems are distinct enough to be discussed separately in the next section.

Congress gave weed control on Federal lands an important stimulus in 1990. Amendments to the Federal Noxious Weed Act[16] included in the 1990 Farm Bill[17] require that each Federal land management agency establish and fund an undesirable plant management program for lands under its jurisdiction (6). Sustained congressional

---

[15] The Committee was disbanded in 1983 after suggesting an additional 750 Federal noxious weeds and developing 261 statements of harm for the *Federal Register*. Its recommendations were not followed.

[16] 7 U.S.C.A. 2814

[17] The Food, Agriculture, Conservation, and Trade Act of 1990, Public Law 101-624

interest is needed now, along with preparations for a thorough evaluation of these amendments' effectiveness within the next few years. Such an evaluation might assess the degree to which each program met its goals; the speed with which agencies responded to new weed problems; the extent and adequacy of interagency Federal-State cooperation, and so on.

Many Federal lands with serious non-indigenous weed problems are vast, remote, and have low economic value. These features make chemical control costly and difficult and biological control an attractive alternative. Biological control organisms are non-indigenous and also capable of harm if not properly screened. Of the Federal land management agencies, only BLM has clearly defined policies for evaluating the safety of non-indigenous biological control agents before their release onto public lands. Comparable policies are needed by other agencies (see biological control section below).

Managers complain that suitable biological control agents are difficult to obtain. Similarly, indigenous germplasm and products are in short supply. The agencies or Congress could ease such technical bottlenecks.

The use of non-indigenous plants for applications such as landscaping and erosion control sometimes comes about because of the high cost or unavailability of indigenous species. For example, farmers cut planting costs per acre by 17 percent when they chose non-indigenous rather than indigenous grasses for acreage enrolled in the Federal Conservation Reserve Program (20). However, a cooperative State-Federal program in Illinois demonstrated that propagation of indigenous plants for large-scale uses is economically and technically feasible (39) (box 7-E).

An indigenous perennial clover (*Trifolium carolinianum*) has been found to be a better and less expensive ground cover than many newly developed non-indigenous varieties (2). However, lack of commercial sources is a barrier to its use in the Federal Conservation Reserve Program. Managers of national parks similarly find that

indigenous plants are not readily available from nurseries (33). Such problems stimulated a successful collaboration in which SCS propagates indigenous plants for park restoration (118).

Wider availability of indigenous plants at comparable costs, along with public education, could go far towards increasing their use—especially if combined with new requirements for truthful reporting of plant origins for commercially sold seeds and plants. The Federal Government could play a significant role in encouraging the use of indigenous plants. Current USDA programs of ARS (the National Plant Germplasm System) and SCS (Plant Materials for Conservation Program) collect plant germplasm and make it widely available for use by plant breeders and producers (ch. 7). Congress could require an increased emphasis on the collection, development, and distribution of indigenous germplasm by these programs.

■ Issue 4: Damage to Natural Areas

Option: Congress could assign broad and explicit responsibility for the control of non-indigenous species that damage natural areas to APHIS, the Forest Service, or another agency and provide resources for its implementation.

Option: Congress could require that the National Park Service commit, in measurable ways, to elevating the priority of natural resource management.

Option: Congress could appropriate additional funds for the Park Service to implement large-scale control and eradication programs for those natural areas most damaged by NIS. Alternately, Congress could provide more funds for these purposes by changing the amount or structure of park entrance or user fees.

A variety of Federal (and State and local) agencies manage protected areas. Among the

STEVE KEMP

Great Smoky Mountains National Park is among the U.S. parks and protected areas hard hit by harmful non-indigenous species.

most "natural" of federally owned lands are the National Parks and other areas managed by the National Park Service (NPS). These represent a small fraction (approximately 3 percent) of U.S. land, but their significance in preserving and protecting natural and cultural resources goes far beyond their relatively small acreage. The U.S. Forest Service, BLM, and FWS manage more modified, yet largely undeveloped, lands—as much as 23 percent of U.S. land.

These areas are significant for maintaining indigenous animals and plants—the biological diversity of the United States. Also, these lands can harbor troublesome NIS that degrade resources and move to private land.

No Federal agency clearly sees its mission as protecting natural areas from harmful NIS. Although some protection incidentally arises from Federal coverage of other areas, it is noncomprehensive and misses many harmful species. State coverage varies and is similarly incomplete. The harmful effects of NIS in natural areas tends to be poorly documented—a cause and a consequence of the lack of focused Federal and State attention. For example, the significance of harmful non-indigenous insects in natural areas can only be guessed, since the U.S. fauna is so poorly known. The effects of at least one-third of the non-

indigenous insects in the country are undocumented (ch. 3) (48). Nevertheless, harmful NIS clearly threaten nonagricultural areas like the National Parks (chs. 2, 8).

State efforts do not compensate for the lack of Federal attention (ch. 7). State regulation of fish and wildlife is patchy. State coverage of invertebrates outside of agriculture varies from spotty to nonexistent.

The Federal Government historically has had a small and erratic role in assisting the States with control programs. The recent Nonindigenous Aquatic Nuisance Prevention and Control Act sought to remedy this with a program for Federal funding of State programs to eradicate or control harmful aquatic species that were unintentionally introduced. In the 3 years since its authorization, no funds have yet been appropriated. Moreover, the rocky start of its Federal interagency Aquatic Nuisance Species Task Force makes its future potential uncertain.

Responsibility for studying, regulating, and controlling harmful NIS in nonagricultural areas such as parks and protected areas is a large enough problem that it needs to be assigned explicitly to some agency or institution. This could be APHIS, although it lacks expertise in this area. Such responsibility would entail a substantial expansion of duties, which could conflict with APHIS' traditional mission to protect agriculture. APHIS, at least, should consider the impact of NIS on natural areas when listing weeds under FNWA (49), when restricting other NIS, and if the agency begins to screen fish for pathogens.

Alternately, the Forest Service might be able to assume responsibility for non-indigenous weed control in nonagricultural areas, with its approach to forest pests serving as a model for nonforest organisms. This would require developing authority for interagency cooperative programs to act outside National Forest System lands.

Others have suggested that control of NIS on nonagricultural lands be assigned to an agency outside USDA, perhaps to BLM, EPA, or a new institution that would take over a majority of NIS-related functions. The efficiency, cost-savings, effectiveness of government re-organizations is far from clear (105). Undoubtedly, NIS control on nonagricultural lands should be the responsibility of an organization with an interest in protecting biological diversity and ecological expertise.

Of all Federal land management agencies, the National Park Service (NPS) has the most restrictive and elaborate policies regarding NIS (ch. 6). Despite these policies, harmful NIS are causing fundamental changes inside and nearby some National Parks. As early as 1980, a NPS report to Congress cited encroachment of NIS as one of the threats to the Parks (117). The changes prompted by NIS are large enough now to jeopardize some Parks' abilities to meet the goals for which the Parks were established (41,60). In a survey done in 1986 and 1987, respondents rated non-indigenous plants as the most common threat to park natural resources while non-indigenous animals ranked fourth (41).

Threats to Hawaii's National Parks are probably worst, although many other Parks are damaged by NIS, such as wild hogs (*Sus scrofa*) in Great Smoky Mountains National Park, a non-indigenous thistle (*Cirsium vulgare*) in Yosemite National Park, and gypsy moths in Shenandoah National Park (6); feral rabbits (*Oryctolagus cuniculus*) in Channel Islands National Park, salt cedar (*Tamarix* spp.) in Canyonlands and Big Bend National Parks, and non-indigenous vines on Theodore Roosevelt Island (59) (table 2-4). Although the Parks face many threats, harmful NIS are considered more pervasive, subtle, and harder to rectify than other disturbances that threaten biological diversity (27).

A growing recognition exists that NPS' funding priorities will have to shift if it is to address degradation of the Parks' natural resources, including funding related to NIS (76, 102). Natural resource management generally has low priority.

The Park Service allocates no more than 2 percent of its annual budget to research, management, and control of NIS and the backlog of unmet needs is growing (6,45).

Ambiguity in the NPS Organic Act[18] is partly responsible for the lack of focus in NPS management; neither the 1970 nor 1978 amendments defined or set priorities for use, versus preservation, of the Parks (94). Further amendments could clarify these sometimes conflicting goals, but disagreement exists as to their necessity. A major recent report—prepared by an independent steering committee for the NPS Director drawing on a 700-participant symposium—recommended that protection of Park resources from internal and external impairment be NPS' primary responsibility. The authors saw this choice as within the current authority of NPS leaders (102).

Park Service officials seem less willing to make such a choice without legislative change. An internal NPS workshop on protecting biological diversity in the Parks, for example, recommended new legislation to make such protection an explicit statutory responsibility and to secure a mandate for restoration of extirpated or degraded ecosystems (27). Specifically, this group called for reducing the densities of harmful NIS within and around Parks to levels where their influence is minimized or eliminated.

New NIS control and eradication efforts, along with other priority resource management tasks, would require additional funds. The steering committee, in their 1992 report, suggested a variety of funding mechanisms in addition to regular congressional appropriations: funding the Land and Water Conservation Fund Act to the full extent authorized; a ''modest'' gasoline tax; returns from concessions and extractive operations; small levies on activities and equipment; voluntary income tax check-offs; sale of tokens and passes for admission; and returning 50 percent of visitor fees to Park units (102).

---

[18] National Park Service Organic Act of 1916, as amended (16 U.S.C.A. 1 *et seq.*)

The Park Service alone cannot solve its pressing resource management problems. Up to 70 percent of the external threats to Parks result from actions by other Federal agencies or by State or local governments (75). This suggests NPS must work closely with adjacent land managers. Specifically, Congress could require that NPS initiate agreements for managing those NIS that threaten park lands from outside their boundaries. Those projects that serve multiple goals, e.g., NIS removal and recovery of endangered species, are the best candidates for top priority (6).

A Keystone Center Policy Dialogue on biological diversity (47) suggested an agency-by-agency approach to NIS on public lands. Participants recommended that each agency: prohibit potentially harmful new releases of NIS, including any intended to control indigenous species; identify, control, or replace already established NIS; eliminate any newly discovered NIS; and maintain those beneficial NIS that do not interfere with biological diversity.

Congress' 1990 amendments to the FNWA took a similar approach, requiring each agency to develop plans for weed control on lands under its jurisdiction. The FNWA could further protect natural areas if this function were more explicit (98). The definition of a Federal noxious weed includes species affecting ''fish and wildlife resources.'' Nevertheless, critics complain that APHIS has been slow or failed to act on weeds of natural areas such as melaleuca and Australian pine (*Casuarina equisetifolia*) (ch. 8). At least one State—Washington—has recently provided more complete protection for natural areas from weeds (box 7-D) (124).

Improved implementation of the Lacey Act and future implementation of the Nonindigenous Aquatic Nuisance Prevention and Control Act might go far towards protecting natural areas from harmful, non-indigenous fish and wildlife (including aquatic invertebrates). Today, however, protection of natural areas from these NIS is almost nonexistent. For example, mollusks that harm natural areas continue to arrive in the country (ch. 3) (8). APHIS may screen out some mollusks during inspection of plant imports, but only if they are potential agricultural pests. Just one species would be stopped due to a prohibition under the Lacey Act—the well-known zebra mussel, which was listed far too late to stop its spread across the country.

Congress might delay further legislation on harmful aquatic NIS until the 1990 Nonindigenous Aquatic Nuisance Prevention and Control Act is fully implemented, although the Federal interagency Aquatic Nuisance Species Task Force has been slow to fulfill its required assignments (table 6-1). Instead, Congress might evaluate the Task Force program to date, urge faster implementation, and ensure that funds are provided for State control in a timely manner.

■ Issue 5: Environmental Education as Prevention

Option: Congress could require that the 20-some Federal agencies involved with NIS develop broadly based environmental education programs to increase public awareness of problems caused by damaging or unpredictable NIS.

Option: Alternately, Congress could develop a smaller scale initiative to take greater advantage of current programs and information.

Option: Congress could require that airlines, port authorities, and importers intensify their public educational efforts regarding harmful NIS.

Although public appreciation of U.S. biological diversity is increasing (ch. 4), the difference between indigenous and NIS in natural surroundings is not commonly perceived—thus the neglect of a coherent public policy regarding harmful NIS.

Lack of awareness on the part of the public and policymakers is mutually reinforcing. Many, including OTA's expert contractors and its Advisory Panelists, believe this cycle of ignorance must be broken (22,46,49,60,104). Also, this theme surfaces frequently in recommendations by nongovernmental groups (46) and scientists and managers (83,93).

Education on NIS ranks low in priority in most State and Federal agencies and private organizations that are involved with natural resources, receiving an estimated less than 1 percent of most organizations' budgets (96). Numerous activities are under way, but efforts are fragmented, uncoordinated, with little formal institutional backup.

In 1989, a coalition of at least 100 environmental groups recommended a sweeping approach to environmental education, including

1. re-establishing an Office of Environmental Education in the U.S. Department of Education,
2. appointing a National Advisory Council on Environmental Education within that Department, and
3. requiring that USDA, the Department of the Interior, and EPA develop and distribute environmental programs and materials (15).

The first two activities were estimated at an additional $20 million annually. In part, they were seen as fulfilling unmet goals of the 1970 Environmental Education Act, which expired in 1982.

The North American Association for Environmental Education (NAAEE) suggested a less sweeping strategy, based on its survey for OTA of current NIS-related programs. Previous education campaigns have not been systematically evaluated, which made recommending definitive changes difficult (96). NAAEE's suggestions included: cooperative government-private programs for groups working on similar NIS; improved exchange of already-developed educational materials; designation of specialized "centers of excellence" for particular species or approaches; teacher training; and improved links between

Agricultural items that can harbor foreign pests are prohibited from entry but these banned items arrived with international travellers on just one flight.

scientists (who often are charged with designing education campaigns) and educators (who have more expertise in programs' effectiveness) (96).

Regardless of approach, program evaluations should be incorporated from their beginning.

The public has the greatest need for education related to non-indigenous animals, according to survey responses of 271 U.S. resource managers and others involved with these issues (93). However, few environmental education campaigns are initiated for the general public for logistical reasons; efforts are more realistically focused on particular groups of people (96). Education regarding harmful NIS will be more effective if focused on people whose incentives for harmful introductions or other actions are weak and for whom the information is likely to tip the balance of their behavior. Little research has been done on why people bring plants and animals into the United States illegally or why they dump NIS outside their property. Also, careful quantitative analysis of the pathways by which NIS reach the United States and the rate at which these pathways lead to serious problems has not been linked to educational efforts for the people using these pathways. Such an analysis could be a highly effective way to set priorities for educational programs.

Few NIS are introduced intentionally and illegally (smuggled), with the exception of sport fish (ch. 3). For smugglers, steep fines may be more appropriate than education. On the other hand, Ralph Elston, from the Battelle Marine Sciences Laboratory in Sequim, WA, suggests that commercial groups transporting aquatic NIS can be expected to respond to education and self-enforcement (31). For other vertebrates, people may intentionally release animals believing they are doing the right thing, or at least not understanding the possible harmful effects of their actions. Educational efforts aimed at buyers at the point of import or sale might effectively change this behavior. Warnings on packages or special forms describing dangers might alert importers. Horticulturist Gary Koller (52) of the Arnold Arboretum, for example, suggests that plants like running bamboo species,[19] which are known to be highly invasive, be sold with individual warning labels so that gardeners recognize their danger and prevent their spread.

International travelers' baggage is often cited as an important source of unintentional (but illegal) introductions (11). This suggests that airline crews, immigrants, and departing or returning residents should receive intensified education. Also, foreign travel might automatically trigger certain steps: handouts from travel agents, enclosures with airline tickets, visas or passports (77), or videos on aircraft that graphically portray

the potential damage from NIS. Similar attempts sometimes failed in the past because too little care was taken in developing a clear message; the support of the Advertising Council was not secured for media saturation; travel agents and air carriers were reluctant to distribute information; and APHIS usually did not include other inspection agencies (64). These lessons need to be heeded in the future.

■ Issue 6: Emergencies and Other Priorities

Option: Congress could ensure that all Federal agencies conducting NIS control on public lands have adequate authority—via existing or new legislation—and funding to handle emergency infestations of damaging NIS.

Option: Congress could set deadlines for APHIS' completion and implementation of comprehensive regulations for the importation of unprocessed wood.

Option: Congress could specify that APHIS and FWS conduct high-level, strategic reviews of how the agencies balance resources directed to excluding, detecting, and managing harmful NIS.

For agricultural pests, Federal and State statutes are relatively comprehensive. Many problems in this area are due to slow or incomplete implementation, difficulties coordinating Federal and State roles, or a tendency to inadequately address larger strategic questions.

In 1991 and 1992, APHIS allowed entry of several shipments of timber or wood chips from Chile, New Zealand, and Honduras without careful analysis (57). Critics complained that APHIS was ill-prepared and slow to recognize the risks that such shipments could carry significant new pests to U.S. forests (see ch. 4, box 4-B). Moreover, when APHIS moved to regulate ''logs,

---

[19] Koller's reference to running bamboo species includes plants in 15 different genera. The most invasive in northern North America are *Arundinaria* spp., *Phyllostachys* spp., *Pleioblastus* spp., and *Sasa* spp. (53).

lumber, and certain other wood products'' in 1992,[20] these proposed regulations were incomplete, failing to address not only crates, pallets, or packing material made from unprocessed wood but also the control of ships and containers coming to the United States from high-risk areas.

Also, an unwillingness by APHIS to see localized problems as potential national concerns has been a source of continuing tension between the agency and State departments of agriculture (chs. 7, 8). APHIS has several times failed to act on significant pests because they were considered local problems. For example, the agency ignored Florida's 1987 problems with infestations of varroa mites (*Varroa jacobsoni*) in honey bee (*Apis mellifera*) colonies (1)—only to see the pest spread to at least 30 States by 1991 (73). Similar situations have arisen regarding plant pests and providing APHIS with emergency powers under the Federal Noxious Weed Act could clarify APHIS' role and speed responses (86).

EMERGENCY RESPONSES

Rapid response requires: careful monitoring for invasive or potentially invasive species to ascertain incipient problems; quickly deciding whether to attempt eradication, and, if so, being willing to eliminate more species than might eventually prove hazardous; and having the resources to implement that or other control decisions quickly.

The current situation contrasts sharply with the ideal (ch. 6). APHIS systematically monitors for a number of agricultural pests in various parts of the country, e.g., African honey bees, Mediterranean fruit flies (*Ceratitis capitata*), cotton boll weevils (*Anthonomus grandis*), and gypsy moths (49). However, improvements to the U.S. detection system are recommended by many scientists for plant pathogens (89), additional insects (48), weeds (60), and mollusks and other aquatic invertebrates (8). No centralized list of recently detected or potential new pests exists (ch. 3, 10).

And databases that might provide such information have received sporadic support (ch. 5).

In contrast, New Zealand's forest industries conducted a detailed benefit/cost analysis of different levels of pest detection surveys. Maximum benefits were achieved by aiming to detect 95 percent, not 100 percent, of new introductions (13) (figure 4-3). Relatively few detailed economic studies of this kind are available to guide U.S. NIS programs (ch. 4).

Federal and State agencies are capable of rapid response after eradication decisions are made. A cooperative Federal-State program to eradicate chrysanthemum rust (*Puccinia chrysanthemi*) in the early 1990s was rapid and successful (90). Joint action in 1992 by APHIS and the Forest Service with the Oregon and Washington Departments of Agriculture eradicated infestations of the *Asian* gypsy moth. Forest Service expenditures for *European* gypsy moth suppression and eradication on Federal, State, and private lands in the eastern United States averaged $10,322,000 annually from 1987 to 1991 (126). Entomologists are concerned that the Asian gypsy moth, if established, could require a similar scale of effort.

On the other hand, Donald Kludy, a former official of the Virginia Department of Agriculture, cites three cases where regulatory changes to quarantines were delayed, sometimes repeatedly: Mexican citrus (*Citrus* spp.), fruit from Bermuda, and the Federal gypsy moth quarantine (49). S.A. Alfieri (1), a Florida agricultural official, also was less sanguine about the Federal-State partnership and its effectiveness in responding quickly to small infestations. He recommended that funds be set aside for emergency pest problems and that action plans be developed and continuously updated for each serious potential pest and disease, accompanied by cost-benefit analyses.

For fast response and eradication, safe and effective chemical pesticides are needed. Classical biological control cannot take their place, although it can be feasible for long-term control

---

[20] 57 *Federal Register* 43628-43631 (Sept. 22, 1992)

of widespread infestations, e.g., noxious weeds on western rangelands. By design, however, classical biological control allows pest populations to persist at tolerable levels. This is counterproductive in a rapid response program aimed at completely eradicating incipient pest populations.

Major concerns exist whether chemicals that are considered safe and effective now are likely to remain available because of regulatory changes (ch. 5). Many registered chemical pesticides are due for renewal under the Federal Insecticide, Fungicide, and Rodenticide Act (FIFRA).[21] Most herbicides for agricultural use are expected to be re-registered. Manufacturers are not expected to seek reregistration for many of the minor use insecticides, rodenticides, avicides, and fungicides. Reregistration is time-consuming and expensive, especially for chemicals with small markets. Chemicals used to control nonagricultural pests, including aquatic plants and large vertebrates, fall into this group. Manufacturers' decisions, as well as government policy, will have important implications. For example, costs of aquatic weed control could jump from $10 to at least $100 per hectare if 2,4-D amine is not reregistered; because many weed control budgets are capped, higher herbicide costs will translate into fewer areas controlled (34).

Section 18 of FIFRA does, however, provide for emergency use of unregistered pesticides. According to the General Accounting Office, Section 18 exemptions were intended for several situations, including the quarantine of pests not previously known in the United States.

Two Federal programs might prove instructive regarding policies on NIS-related minor use pesticides. The Interregional Research Program Number 4 (IR- 4), in USDA's Cooperative State Research Service, develops and synthesizes data to clear existing pesticides for minor uses on food and feed crops. However, it is heavily burdened

APHIS

*Although officials anticipated that the Asian gypsy moth (*Lymantria dispar*) could accompany timber imports, grain ships brought an early infestation and State and Federal agencies cooperated to quickly eradicate it.*

and unlikely to meet reregistration deadlines (ch. 5) (110). Nor does it address problems of new pesticide development. Congress used the Orphan Drug Act[22] to address similar problems with developing limited-use pharmaceutical products. This Act provides pharmaceutical companies with 7 years' exclusive marketing rights and tax credits for developing drugs for rare diseases. The Act has successfully prompted new drug development (3), although controversy regarding several drugs' high profitability has prompted Congress to consider modifications.

SETTING PRIORITIES

Decisions about which organisms to prevent, eradicate, or control are not always made systematically or strategically, despite the large amounts of money involved. This risks wasting money, given the biology of invasions. The APHIS line-item budget directs most NIS-related funds to particular species and different programs compete against each other for priority. Highly visible programs with strong support of industry, States, or the public receive highest priority. As a

---

[21] Federal Insecticide, Fungicide, and Rodenticide Act (1947) as amended, (7 U.S.C.A. 136, *et seq.*)

[22] Orphan Drug Act of 1983, as amended Public Law 97-414, Public Law 100-290.

result, potential new diseases and pests often lack attention, although money could be well invested at an early stage (49). State officials express confusion as to how APHIS decides whether and when to begin and end its programs.

James Glosser, former APHIS Administrator, stated that: ''Probably the greatest problem confronting us in noxious weed control is identifying what constitutes a noxious weed and how to establish priorities for control efforts'' (36). Managers tend to set priorities based on either species' impact or the likelihood of successful control. USDA's Noxious Weed Technical Advisory Group suggested criteria based on potential economic damage, size of infestation, and support for a control or eradication program (80).

Ranking current and potential plant pests was a major task of the Minnesota Interagency Exotic Species Task Force (70). Florida's Exotic Plant Pest Council is also developing an extensive, prioritized list of harmful non-indigenous plants (26). The McGregor Report (64) was among the Federal Government's first attempts to rank agricultural pests and diseases, although it had limited impact. The seven western States participating in BLM's research plan for restoring diversity on degraded rangelands listed four high priority non-indigenous weeds[23] (114).

Others would give highest priority to harmful NIS in their earliest stages of invasion. Plant invasions are typical of many NIS in that their populations do not spread at steady rates. Weeds are easiest to control or eradicate immediately after detection, before their population growth accelerates (71). Richard Mack, Professor of Botany at Washington State University, suggests that eradication aimed at already well-established, widespread weeds is likely to produce only temporary gains unless control is permanently maintained. This is costly and difficult. The most aggressive plant pest control program ever conducted in the United States succeeded in restricting, but not eradicating, barberry (*Berberis vulgaris*) (62). Nor, according to Mack, could all possible weeds be prevented from entering the United States at a tolerable cost: society would not accept the expense and delays involved in inspecting all arriving cargo, luggage, and passengers. For these reasons, he would increase resources for detecting newly established weeds, add species to the Federal Noxious Weed Act, but keep quarantine, port inspection, and control of widespread weeds near current levels (62).

Richard Mack's recommendations are a clear strategic statement that could guide policy. However, those advocating higher priority for control of widespread weeds would sharply disagree with his approach and they can also make a strong case (see preceding section on non-indigenous weeds). A large proportion (39 percent) of those involved in issues related to non-indigenous animals feel that the length of time a population has existed should bear little influence on the decision to remove or control it (93). However, significantly more administrators than other types of workers supported using length of time in making decisions about non-indigenous animals (93). Such fundamental disagreements on priorities highlight the lack of information, dialogue, and consensus on managing harmful NIS.

Approaches to setting priorities may vary, depending on the type of organisms involved and the state of scientific knowledge. Containment of non-indigenous fish and other aquatic species is difficult. Once released, large aquatic invertebrates and fish spread easily within river systems, and their larval, sub-adult and adult forms may each be disruptive (44). Attempts to eradicate fish after they have developed a substantial range are often a waste of time and resources (22). Thus, groups like the American Fisheries Society have often focused on the need for stricter pre-introduction screening.

---

[23] Medusa head (*Taenniatherum asperum*), cheatgrass (*Bromus tectorum*), diffuse knapweed (*Centaurea diffusa*), and spotted knapweed (*Centaurea maculosa*) (114).

For plant pathogens, overseas screening by commodity, along with inspection at ports of exit, might be most effective (91). USDA has focused on identifying foreign pathogens likely to be damaging in the United States (89). With a list of potential pathogens running to 1,000 pages and limited detection methods for micro-analysis, complete exclusion at ports of entry is impossible. Pathogens tend to be insidious—they may become apparent only *after* populations are beyond what would amount to ''early detection'' for larger and less mobile NIS. Pathogen hosts must be eradicated to eliminate diseases, but many hosts are valuable commodities, and their destruction can be costly and controversial.

Others have recommended alternative criteria for setting priorities. For example, Walter Westman, of Lawrence Berkeley Laboratory in Berkeley, CA, suggested that priorities might be based on severity of impact on indigenous biota, with wilderness areas receiving higher priority than urban recreation areas. Also, control might be emphasized for more easily contained NIS (e.g., those with slow rates of spread, localized occurrence, and susceptibility to available methods) and/or those that threaten endangered species. Those NIS that play a role in ecosystem function (e.g., controlling soil erosion control or supporting wildlife) and cannot be readily replaced could be given lower priority (129). Stanley Temple, a zoologist at the University of Wisconsin, likewise suggests NIS that threaten endemic species on remote islands deserve special, high-priority treatment (103). The International Union for the Conservation of Nature and Natural Resources (IUCN) took a similar approach. Its Species Survival Commission counseled that special efforts should be made to eradicate harmful NIS in: islands with a high percentage of endemic plants and animals, centers of biological uniqueness, areas with high species or ecological diversity, and in places where a NIS jeopardizes a unique and threatened plant (44).

In the long-term, strategic decisionmaking, like better detection and more rapid response, requires solid databases (with information from foreign sources) and substantial taxonomic expertise. The inadequacy of the former and the dwindling of the latter are common concerns in the scientific community (ch. 5) (24,60,63).

■ Issue 7: Funding and Accountability

Option: Congress could increase user fees that relate directly to the evaluation, use, and management of potentially or actually harmful NIS. Also, Congress could require that recreational fees collected by Federal land management agencies be made available for management of harmful NIS on public lands.

Option: Congress could examine the adequacy of Federal and State fines related to illegal and poorly planned introductions. If necessary, Congress could develop additional mechanisms to recoup an increased proportion of the costs for preventing and minimizing damage from NIS that become public nuisances.

Option: Congress could change the Aid-to-States program to encourage projects that limit damage from non-indigenous fish and wildlife.

Many small-scale efforts related to NIS could be improved without large funding increases. Some of the options suggested for issues above fall into that category. However, some initiatives are large enough to require additional money. These needs are likely to grow as the number and impact of harmful NIS also grows.

Options that give additional responsibilities to Federal or State agencies—e.g., for more complex risk assessment or earlier pest detection—need to be matched with increased funding if they are to be effective. The problems faced by the Federal interagency Aquatic Nuisance Species Task Force—delays in reporting to Congress, lack of funding and staff—illustrate what happens when new obligations are assigned without the resources to implement them. Some Federal officials find that funding is the primary factor in agencies' ability to proactively deal with harmful NIS (17). In a survey of those working with issues related to non-indigenous animals, for example, respondents listed funding problems as the single largest contributing factor to the lack of success in control programs (93).

This problem is not confined to NIS. Both Federal and State environmental legislation has multiplied during the 1980s and early 1990s (32,84). At the same time, the funding available to States and localities has been decreasing (32,95). Clearly, questions of funding will be crucial for new or improved efforts to succeed.

To date, the total costs of harmful NIS to the national interest have not been tabulated. Quarantine containment can fail; a newly imported species can become unexpectedly invasive; a previously innocuous pathway can become a conduit for a major new pest. However, little explicit accountability exists for the damage caused in such cases, especially as compared with other areas of potential environmental harm. Federal, State, and local governments have borne significant costs that could be more appropriately assigned to individuals and industries, e.g., for the Asian gypsy moth and the zebra mussel.

Expensive and time-consuming lawsuits provide virtually the only avenue for assigning liability and recovering control or eradication costs. In part, this may be because many damaging NIS have been associated with agriculture and agriculture has engendered less Federal intervention with respect to its environmental consequences than other industries (84).

Long lag times between the action of the responsible party (if that party can be determined) and the impacts of NIS are typical. For example, witchweed (*Striga asiatica*) probably arrived in North Carolina with military equipment from Africa after World War II; it was detected some 20 years later. The APHIS eradication program in North and South Carolina cost $5.2 million in fiscal year 1991 (90). Often the effects, as well as origin, of a given NIS will be uncertain and undocumentable. And one area or economic sector could be severely harmed by a NIS while another might benefit. Relying solely on U.S. courts to assign damages and to recoup costs is an ineffective policy under these circumstances.

FEES AND OTHER FUNDING

Fees are a prevalent means of raising funds for matters directly and indirectly related to NIS and Federal and State governments are expanding user fees. Typically, fees are structured to raise revenue, not to recoup damages or to change people's behavior (85). As of the late-1980s, Evelyn Shields, in a report for the National Governors' Association, (95) found that 43 States used fees to fund local, State, and Federal environmental programs, generating roughly $240 million. In fiscal year 1991, State parks and similar areas alone produced approximately $433 million from entrance and user fees (119).

However, the more public organizations rely on funding that is independent of the appropriations process, the more independent they are of congressional control (105). This has been a common issue in the continuing debate in Congress regarding fees.

Relating user or other fees[24] directly to harmful NIS or services associated with them has an advantage since management of harmful NIS otherwise suffers when funding drops and populations outstrip control. For example, 1993 funding cuts to the South Florida Water Management District mean reduced melaleuca control in the Everglades conservation areas; Donald Schmitz (87), an aquatic weed specialist with the Florida Department of Natural Resources, anticipates some past gains in melaleuca control will be lost

---

[24] The definition of a "user fee" varies, depending on the author. Doyle (28) describes 4 general types of fees: impact fees, user fees, and fees for services and discharges. The agencies discussed here distinguish user and entrance fees for reporting to Congress. GAO (109) appears to have grouped all FWS fees as "user fees."

DON SCHMITZ

*Some funding for melaleuca (*Melaleuca quinquenervia*) control is dropping while associated problems are increasing—the type of situation that user fees are intended to prevent.*

and future efforts made more difficult as a result. Ideally, NIS funding would be predictable and increase if NIS-related problems do. User fees can be tailored so that this occurs.

In 1991 and 1992, APHIS published regulations implementing the user fees for international inspection services authorized in the 1990 Farm Bill;[25] these range from $2.00 for air passengers and commercial trucks, $7 for loaded commercial railroad cars, to $544 for commercial vessels of at least 100 tons (49). User fees for agricultural inspection, issuance of plant health certificates, animal quarantines and disease tests, and export health certificates were also authorized and are expected to be in place by the beginning of fiscal year 1994.[26] In contrast, Congress struck down APHIS' attempt to institute a domestic quarantine user fee between Hawaii and the mainland (ch. 8). In fiscal year 1992, user fees provided 80.7 percent of program funding for APHIS' Agricul-

tural Quarantine Inspection program; this was estimated at 78.6 percent for fiscal year 1993 (78).

Additional opportunities exist to more closely match fees to NIS use and the prevention and minimization of NIS damage. For example, private parties in New Zealand pay all costs associated with risk analysis and port inspection for imported NIS. In contrast, those commercial interests advocating Siberian timber imports to the United States spent about $200,000 to develop Russian contacts and promote imports. The U.S. Government spent approximately $500,000 more to analyze associated risks. These were not additional appropriations but came from U.S. Forest Service contingency funds.

Seven Federal land management agencies[27] are authorized by Congress to charge entrance or user fees under the Land and Water Conservation Fund Act of 1965 (LWCFA), as amended.[28] Fees generated by the LWCFA account for amounts ranging from 1 percent (BLM) to 85 percent (NPS) of the agencies' total receipts from sale and use of land and resources (4).

Congress has considered numerous amendments to the LWCFA since 1965 to prohibit, authorize, or re-establish various agencies' ability to charge fees, to change the amount of different fees, and to change the purposes to which fees can be put (9,108). Legislative changes generally have expanded and increased fees to meet the agencies' growing needs for operating and maintenance funds. Making entrance or user fees available for NIS-related programs would likely require further changes in this legislation.

Changes to the LWCFA have been controversial, in part because of the tradition of free public access to Federal recreational lands (9). Other specific user fees, e.g., grazing permits on Federal

---

[25] 56 *Federal Register* 14844 (Apr. 12, 1991); 57 *Federal Register* 769, 770 (Jan. 9, 1992); 57 *Federal Register* 62472, 62473 (Dec. 31, 1992)

[26] Proposed regulations are in 56 *Federal Register* 37481-37493 (Aug. 7, 1991)

[27] Bureau of Land Management, Bureau of Reclamation, Army Corps of Engineers, Forest Service, Fish and Wildlife Service, National Park Service, and Tennesee Valley Authority.

[28] 16 U.S.C.A. 4601-6.

lands, also have been highly controversial, as is the general issue of charging full market value for Federal services. However, sizable amounts of potential revenue are involved. For five Federal land management agencies, 80 to 99 percent of recreational visits are to sites for which no fees are charged; the National Park Service, on the other hand, charges fees for about 65 percent of visits (119). In some cases, agencies consider sites too dispersed for ready fee collection; in other cases, Congress or the agency has designated particular units as nonfee areas. Internal audits estimated that approximately $24 million could be collected annually with new or increased fees by NPS, BLM, FWS, and the Minerals and Management Service (120). The Forest Service estimates that charging full value for its recreational services would generate $5 billion annually (85).

A variety of additional means—besides increases in fees—could fund various NIS-related activities. For up-front funding, Congress could levy taxes on those who use the pathways by which harmful NIS enter the United States and move within the country. Such users include importers, retailers, and consumers of foreign seeds, nursery stock, and timber, exotic pets and wildlife, and non-indigenous aquaculture and aquarium stock. Similarly, a tax could appropriately be applied to international airline and train tickets, docking fees, and gasoline. The Minnesota Exotic Species Task Force (70), focusing on NIS pathways, suggested these sources of new revenue:

- establish a surcharge on boat trailer licenses;
- establish a tax on the sale of non-indigenous nursery products such as trees, shrubs, and flowers;
- establish a ballast tax on foreign ships;
- require licenses and license fees for importers; and

- continue and expand the surcharge on boat licenses.

State and Federal Governments use tax policy— excise taxes,[29] exclusions and other modifications to income taxes, and tax credits—to meet a variety of environmental goals and provide funding for targeted programs (111). Most tax policies have little relationship to NIS. However, sales taxes are collected on pets and nursery plants and excise taxes are imposed on airline tickets for the Airport and Airway Trust Fund (67).

Also, the Federal Government collects a 10 to 11 percent manufacturers' excise tax on firearms and hunting and fishing supplies (111). These funds are returned, in the next fiscal year, to States for fish and wildlife management projects (ch. 6; fig. 6-1). In fiscal year 1991, payments to States totaled more than $320 million (107).

These funds are intended for projects that benefit wildlife. They have been used to introduce NIS and for projects that indirectly affect wildlife, e.g., restoration of wetlands. States could be encouraged to fund projects that repair damage from past introductions of harmful nonindigenous fish and wildlife. Alternately, Congress could amend the program to set aside funds for eradication and control of harmful NIS or restoration of indigenous species' habitats. Such projects are already eligible for funding. A set-aside, however, could further encourage States to undertake such efforts without removing State control of the program's money. Attempts to do so could provoke considerable State resistance. Currently, only State agencies qualify for these funds. Some observers have suggested that the program be changed so Federal projects might be eligible for a portion of these funds.

INCREASING ACCOUNTABILITY

Responsibility for the costs of harmful introductions could be shifted to those who benefit

---

[29] Excise taxes are collected on commodities—their manufacture, sale, or consumption—or a privilege. The latter are often assessed as licenses.

from the relatively open U.S. system of importation. At the same time, the benefits of introductions could be preserved without unduly burdening private individuals or groups. Those engaged in intentional introductions are most easily assigned certain costs—for example, fees for pre-release risk assessments and fines for illegal releases. For unintentional introductions, all users of high-risk pathways (e.g., shippers using ballast water) could be charged for their pooled risk with funds paid into a trust fund.

The Species Survival Commission of IUCN recommended that each nation have legislation to ensure that persons or organizations introducing harmful NIS, not the public, bear costs for their control. Further, the Commission stated that parties responsible for illegal or negligent introductions should be legally liable for damages, including costs of eradication and habitat restoration, if needed. F.C. Craighead, Jr. and R.F. Dasmann, two wildlife biologists, made a similar recommendation regarding non-indigenous big game animals that spread onto public lands (25). A number of States have programs to hold game breeders, private owners, or importers liable for controlling escapees and for damages (ch. 7).

A host of mechanisms is available to increase accountability. Bonding and insurance, for example, could be required of importers, but have been little used. Permits and fines are most commonly used now.

The Federal Government imposes fines for bringing foreign material into the United States illegally, e.g., international, interstate, and intrastate violations of the Plant Pest Act,[30] the Plant Quarantine Act,[31] and the Lacey Act.[32] Both civil and criminal sanctions are involved. The 1981 Lacey Act amendments increased maximum penalties and jail sentences for violations ($20,000, imprisonment for up to 5 years) and provided for forfeiture of wildlife;[33] fines were further increased by the 1987 Omnibus Crime Control Act[34] (55). Hawaii's recently amended laws provide some of the largest fines for violating its importation permit laws—up to $10,000 for a first offense and up to $25,000 for subsequent offenses within 5 years of a prior offense (ch. 7).

Agricultural inspectors (APHIS) can fine violators up to $10,000 but most civil penalties are under $1,000. Officials estimate about 30,000 actions per year, with almost all settled for less than $100 immediately (40). In fiscal year 1990, APHIS found 1,303,000 baggage violations and assessed $723,345 in penalties for 23,676 of these (37), for an average of approximately $30.

Release of organisms into National Parks is a citable offense.[35] The BLM has a policy to hold people responsible for damages and control costs for unauthorized introductions of ''exotic wildlife;'' however, no law or regulation specifies such liability beyond the common law, so the policy's implications are not clear (6).

For fines to be effective deterrents, enforcement must be a priority. A recent advisory commission found that FWS' law enforcement division was seriously understaffed and underfunded, lacked clear priorities, provided inadequate staff supervision, and had insufficient technical expertise to identify species (121). The U.S. General Accounting Office (109) concurred,

---

[30] Federal Plant Pest Act (1957), as amended (7 U.S.C.A. 147a *et seq.*),

[31] Nursery Stock Quarantine Act (1912), as amended (7 U.S.C.A. 151 *et seq.*.; 46 U.S.C.A. 103 *et seq.*

[32] The Lacey Act's 1981 amendments allow FWS agents to use the Act when enforcing any Federal law, treaty, regulation, or tribal law. It provides for warrantless search and seizure and allows prosecution regardless of whether offenders crossed State lines. These provisions compensate for weaknesses in the authority of other Federal wildlife laws. FWS agents prefer the Lacey Act for these reasons and because its allows larger fines (109).

[33] 16 U.S.C.A. 3373, 3374.

[34] Omnibus Crime Control Act (1987), as amended (18 U.S.C.A. 3571).

[35] 36 CFR Part 2.1(a)(2) (June 30, 1983).

FISH AND WILDLIFE SERVICE

The Fish and Wildlife Service confiscated these cockatoos under a treaty banning their import. The agency's efforts to enforce both international and domestic laws may be inadequate to deter violators.

finding that the number of investigations is too low to minimally deter crime, that FWS is increasingly unable to assist States with investigations, and that FWS has no reliable direct measures of their law enforcement's effectiveness. Many States also lack adequate law enforcement resources (ch. 7). Thus, fines could only be a larger source of revenue and a greater disincentive for illegal behavior if enforcement is improved. However, fines are just one means of creating disincentives for wrong doing—and they carry with them the potential for ''fund raising through harassment'' (67). Generally, prosecutions for environmental crimes are climbing (54) but critics charge that their deterrent potential is far from clear (12).

Taxes, fees, fines, and other tools are designed to achieve one of several aims, i.e., to increase the benefits or decrease the costs of doing right, to increase the costs or decrease the benefits of doing wrong, or to increase the probability that such benefits and costs will occur (72). The overall trend in U.S. public policy is toward greater use of incentives for doing right, according to Stuart Nagel, a political scientist at the University of Illinois.

However, little attention has been directed toward creating positive incentives regarding harmful NIS, e.g., for encouraging adequate containment of aquaculture species. In some cases, bounties are paid for removing harmful NIS, rewards are provided for tips leading to successful prosecutions, and the Lacey Act's 1981 amendments included provisions[36] for rewarding those who provide information leading to enforcement against or conviction of violators (55). Increasing other types of incentives may require new statutes and/or regulations.

■ Issue 8: Other Gaps in Legislation and Regulation

As a result of the Federal and State patchwork of laws, regulations, and programs, important types of non-indigenous organisms remain potential sources of damaging introductions. The most serious gaps are discussed above. Additional organisms are not adequately covered by Federal and/or State laws, however, and are the basis for a second tier of possible options. In priority order, these gaps pertain to:

1. vectors of human diseases;
2. sale and release of biological control organisms;
3. live organisms moved by first-class mail, shipping services, and catalog sales;
4. hybrid and feral animals;
5. NIS used in research; and
6. new strains of already established harmful NIS.

Some of these gaps require legislative change to fill; others need more adequate implementation by Federal agencies.

---

[36] 16 U.S.C.A. 1531-1543.

VECTORS OF HUMAN DISEASES

Option: Congress could lay groundwork by investigating the adequacy of the Nation's response to NIS that pose significant threats to human health. This might begin with a General Accounting Office investigation of APHIS and the Public Health Service's respective roles.

Non-indigenous human health threats are largely beyond the scope of this study. Two cases, however, illustrate continuing, significant problems with Federal management.

The Centers for Disease Control and Prevention of the Public Health Service (PHS) responded slowly to the threat posed by the Asian tiger mosquito (*Aedes albopictus*), a potential vector for several serious viral diseases. These non-indigenous mosquitoes apparently entered the United States in 1985 in used automobile tires and have now spread to 22 States (ch. 3; box 3-A). The Centers' lack of action to stop the insects' spread raises questions regarding its effectiveness in dealing with NIS new to the United States.

The African honey bee poses a public health threat and a threat to U.S. agriculture. Because of the latter, APHIS is responsible for developing responses to control the bee's spread from Mexico. However, APHIS cannot fully address the human health issues.

Researching and preventing acute infectious diseases, many of which have non-indigenous mammal or insect vectors, have received a reduced national commitment since the 1950s, according to a recent report by the Institute of Medicine (58). This report, on emerging microbial threats, recommends increased surveillance for infectious diseases and their vectors. It also calls for enhancing information data bases and improving the structure of PHS and inter-agency cooperation.

These seem to be matters of improving Federal implementation. The first step might be congressional oversight designed to provide increased public scrutiny.

THE SALE AND RELEASE OF BIOLOGICAL CONTROL ORGANISMS

Option: Congress could either create new legislation or amend existing law to more comprehensively regulate biological control agents.

Option: Congress could increase the level of environmental review required for importations of biological control agents by making them subject to NEPA.

Biological control agents used in the United States include non-indigenous microbes, insects, and other animals that damage, or eat, undesirable plants or insects. Congress has never directly addressed biological control. No single Federal statute requires that biological control agents be reviewed before introduction (69) or regulates importation, movement, and release of biological control agents (19). Instead, potential risks are dealt with by existing regulations, supplemented with a complex system of voluntary protocols or guidelines (19).

Federal regulation of biological control agents—like genetically engineered organisms—uses several laws designed for other purposes, e.g., laws on quarantine, product registration, and environmental protection. EPA regulates the commercial sale and release of pesticidal microbes under FIFRA. Biological control agents that are not microbes are exempt from FIFRA and fall under APHIS's jurisdiction, although the agency has not yet promulgated regulations specifically for such biological control agents. Instead, APHIS requires researchers and producers to follow procedures and permitting requirements developed for plant pests under authority of the Federal Plant Pest Act and the Plant Quarantine Act (10). NEPA, along with the Endangered Species Act, also affects importation and research on biological control organisms (19), although NEPA's application has been uneven and poorly defined.

Several aspects of commercial distribution and sale of biological control agents are among the topics not addressed by current statutes or regulations. No requirements exist for clear and accurate labeling of insects or other animals (e.g., nematodes) used for biological control. No law specifically gives APHIS authority to regulate the labeling, purity, or disease status of these insects and animals. Nor are those who release improperly screened or tested agents accountable for any resultant damage. It is unclear whether current statutory authority covers all the categories of biological control agents APHIS is seeking to regulate. Specifically, it is questionable whether beneficial insects that prey on insect pests fit under the Federal Plant Pest Act's definition of "plant pest."

Opinion is divided regarding the suitability of the current system and how its weaknesses should be corrected. Peter Kareiva, an ecologist at the University of Washington, expressed a particular concern about APHIS' lack of formal criteria for approving releases of biological control agents (46). Francis Howarth and Arthur Medeiros, from the Bishop Museum in Honolulu and Haleakala National Park, in Makawao, HI, respectively, suggested requiring formal environmental impact statements or environmental assessments to ensure the widest possible public review (42). Ecologist Gregory Aplet and attorney Marc Miller (69) contend that current laws do not—and cannot be amended to—fill critical gaps. They propose a Federal Biological Control Act that would ensure public participation in decision-making and correct what they see as serious shortcomings in the current review process:

• harm to noneconomic species and ecosystems is ignored;
• repeated introductions are allowed when a given organism is approved, even into new ecological settings with different, potentially damaging consequences;

• transfers of biological controls within the United State or within States are disregarded; and
• no formal, enforceable requirements are required for research and follow-up to determine whether detrimental impacts have occurred (69).

The Species Survival Commission of IUCN (44) recommended that biological control organisms should be subject to the same care and procedures as other NIS.

On the other hand, USDA biological control experts such as J.R. Coulson and Richard Soper prefer the current voluntary system for assessing risks of new introductions, updated by biological control and quarantine specialists (19). U.S. biological control programs have excellent safety and environmental records, they maintain, and have accommodated needs to consider impacts on nontarget species. Therefore, environmental impact statements are not only unnecessary but also would demand superfluous or frivolous studies, slowing or halting the use of many biological control agents. Coulson and Soper hope that further development of informal guidelines can limit adverse effects on existing biological control programs and preempt stricter legislation or regulations developed by nonspecialists. Miller, Aplet, Coulson, Soper, and Howarth all agree that more post-release evaluations are needed.

Federal and State protocols for introductions protect only a limited part of the United States but eventually need to address all of North America (19). Miller and Aplet describe laws in seven States that encourage the development and application of biological control.[37] They consider Wisconsin's provisions the most protective. An earlier survey found just three States with particular laws addressing biological control species and only one—North Carolina—addressed issues related to commercial sales (66).

---

[37] Arizona, California, Connecticut, Florida, Minnesota, New York, and Washington encourage biocontrol generally, for specific pests, or as part of integrated pest management (69).

Eventually, specific biological control legislation may be the vehicle to extend needed protection throughout the country. States could potentially deal with problems related to product labeling and performance through their weights and measures or consumer protection statutes, although a complaint would be necessary to trigger action (50). For example, the Pennsylvania State Bureau of Consumer Protection recently brought a lawsuit against the manufacturer of a biological control product when it was discovered that the product contained no trace of the active pesticidal microbe (16).

Regardless of the approach Congress takes, issues associated with biological control are likely to be increasingly visible and controversial as public interest grows. Biological control's popularity increases the risk of unwise introductions by amateurs (19). The potential danger of biological control releases has been scrutinized more closely in conjunction with proposals for releases of genetically engineered organisms.

LIVE ORGANISMS MOVED BY FIRST CLASS MAIL, SHIPPING SERVICES, AND CATALOGUE SALES

Since the time when Benjamin Franklin lived in Europe, Americans have sent attractive or promising NIS home (125). In the early part of this century, the Commissioner of Patents used congressional franking privileges to distribute foreign seeds to farmers (125). Domestic and international mail is also a known pathway for the spread of harmful non-indigenous plants and prohibited agricultural pests however (49,61) (ch. 3). Some introductions of Mediterranean fruit flies in California are thought to have originated in tropical produce mailed first-class from Hawaii (97).

The Constitution and Federal laws protect domestic first class private and commercial mail against unreasonable searches. On the other hand, most international mail is subject to unrestricted searches, but funding and personnel to do this are scarce.

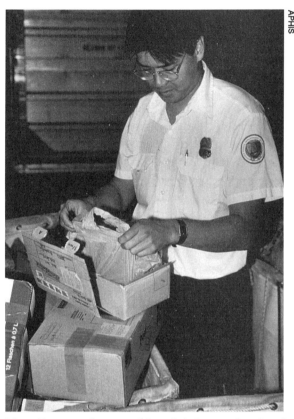

APHIS

Many live organisms are shipped via international and domestic mail; only limited searches are allowed for domestic first-class mail.

In 1990, APHIS and the U.S. Postal Service began a trial program in Hawaii using trained dogs to identify outgoing packages containing agricultural products. This evidence is then used to obtain warrants to open the package to determine whether the products are illegal. The program reportedly has been quite successful (106). It is cumbersome, however, which may justify easing the warrant requirements.

Congress recently passed a law specific to Hawaii, the Alien Species Prevention and Enforcement Act,[38] which is to allow the same sort of inspection for mail coming into Hawaii as for outgoing mail (ch. 8). The Federal and State agencies involved have fallen behind schedule in

---

[38] Alien Species Prevention and Enforcement Act (1992), Public Law 102-383, section 631.

setting up a cooperative agreement for the inspection, however, because of the agencies' differing regulatory authorities regarding inspections and types of organisms.

Similar programs do not exist for other areas where first-class mail poses pest risks, e.g., from Puerto Rico into California (97). Donald Kludy (49), a former official with the Virginia Department of Agriculture, suggests that mail shipments are a serious enough problem to extend the Hawaii U.S. Postal Service pilot program to items mailed from Puerto Rico and other U.S. territories or to pass new legislation for all mail originating outside the contiguous 48 States. Congress might evaluate the Hawaii inspection program and, based on this information, consider whether its application to other areas is warranted and feasible.

Many live organisms now are available through catalogue sales, including insects and other animals for biological control, as well as a wide variety of plants and seeds. Adherence to Federal or State laws that limit areas to which species may be shipped is largely voluntary. Catalogue sales do not present the same inspection and regulatory opportunities that are available in the case of ordinary retail outlets. Nurseries and aquatic plant dealers sell several federally listed noxious weeds through the mail, such as the rooted water hyacinth (*Eichhornia asurea*), which can clog waterways and cause a navigation hazard (127). Packages sent via private delivery services are not protected from inspection as is first-class mail. However, they are unlikely to be inspected unless the package is broken or leaking.

This opens the possibility that commercial distribution may provide a pathway for spread of potentially harmful NIS, including pathogens and parasites. The wasp parasite (*Perilitus coccinellae*) of the indigenous convergent lady beetle (*Hippodamia convergens*), for example, already has been spread in this manner (43). The 16-member expert Working Group on Non-*Apis* Bees expressed similar concerns regarding the movement of bumble bee (*Bombus* spp.) colonies between eastern and western North America.

Rental and sale of bee colonies has increased in the past 5 years, along with the potential spread of accompanying non-indigenous nematodes, mites, diseases, and parasites (131).

HYBRID AND FERAL ANIMALS

Option: Congress could amend the Lacey Act so that it clearly applies to harmful hybrid and feral animals and they could be included in any new Federal initiatives for States' roles.

Non-naturally occurring hybridization with NIS can present a serious threat to indigenous species by diluting gene pools (59) and causing other genetic harm (38). Most Federal and State laws that protect indigenous species, or prohibit harmful NIS, lack clarity in their application to hybrids. This can lead to controversy, such as the dispute over a policy adopted by FWS, that narrowly interpreted the protection of hybrids offered by the Endangered Species Act (82). Unclear or disputed taxonomy, particularly in the delineation of subspecies, can contribute to the ambiguity (35).

Non-indigenous hybrids require flexible policies, adaptable to each case. Hybrids can represent important genetic diversity to be preserved—this applies to economically and ecologically important species such as the endangered Florida panther (*Felis concolor coryi*) (ch. 2). In contrast, hybrids between dogs (*Canis familiaris*) (non-indigenous) and wolves (*Canis lupus*) (indigenous), which are popular as pets, are not only dangerous to humans, they also obstruct recovery of endangered wolves in the wild (5,7). They often escape or are released by owners unable to manage them. An international group of wolf experts has called for governments to prohibit or tightly restrict wolf-dog hybrid ownership and breeding (65).

Most Federal laws are silent in their treatment of feral animals—wild populations of formerly domestic animals. Few State laws covering the accidental or intentional introduction of such animals or responsibility for damage they may cause.

Yet feral animals continue to cause significant damage. In a recent survey, managers of national parks and other reserves named feral cats (*Felis cattus*) and feral dogs to be two of the three most common subjects of wildlife control efforts. The other was wild pigs (*Sus scrofa*), many populations of which are feral (29). Feral cats kill large numbers of small mammals and birds, dogs attack livestock and indigenous wildlife, and pigs destroy indigenous plants and do other damage (123).

Federal or State laws could be amended to more clearly apply to hybrid and feral animals.

NON-INDIGENOUS SPECIES USED IN RESEARCH

Scientific researchers initially introduced several very harmful NIS, including gypsy moths, African honey bees (in South America), and peanut stripe virus (48,89). The rapid spread of the Asian clam (*Corbicula fluminea*), a serious fouler of power plant pipes, is thought to have been assisted by inadvertent research releases (21).

Research organisms are not generally subject to the same scrutiny as those for other applications. The Lacey Act allows certain organisms to be imported or moved interstate for research and many State laws allow research imports of otherwise prohibited species. Microbes can be freely imported for research if they do not pose a risk to agriculture or human health.

Some Federal and federally funded research on NIS is evaluated for the risk of species escape or potential effects. ARS has extensive protocols governing its research on biological control agents (19). The Federal interagency Aquatic Nuisance Species Task Force recently issued protocols for research on harmful aquatic NIS. These protocols will be mandatory for any research funded under the Nonindigenous Aquatic Nuisance Species Prevention and Control Act and have been voluntarily adopted by agencies on the Task Force (18,122). However, most of the research protocols developed by Federal agencies do not apply to research funded by outside sources (ch. 6).

NEW STRAINS OF ALREADY ESTABLISHED HARMFUL NON-INDIGENOUS SPECIES

APHIS does not consistently prevent repeated importation of pest species that are already established here. New, different strains of some species potentially may be imported, worsen effects, and spread into areas where the pest is not yet well-established. Regulating strains would pose significant technical difficulties; rapid identification would be difficult, for example. Nevertheless, some pest experts express concerns that new strains of widespread pests like the Russian wheat aphid (*Diuraphis noxia*) and bromegrasses (*Bromus* spp.) are allowed continued entry (48,60,68).

CHAPTER REVIEW

This chapter summarized what we know about harmful NIS in the United States: their growing numbers and impacts, their routes of entry and movement, the methods by which they are evaluated and managed, and related State and Federal policies.

This chapter also presented policy options on 8 issues—those most in need of attention, according to OTA. Each issue allows for a range of options, demanding greater or fewer resources. If each area is not addressed in some form problems are likely to worsen, with no assurance that the biological resources of the United States will be protected. Only Congress can decide how stringent national policy should be. Everyday management of non-indigenous fish, wildlife, and weeds, though, falls to many Federal and State agencies and they need better guidance and support. Also, natural areas must be better safeguarded if they are to retain their unique character. Emergencies must be handled more quickly to keep problems from snowballing. And the public needs better education so their actions prevent, rather than cause, problems.

To reach these conclusions, OTA gathered an array of data. The next chapter lays out OTA's methods, then begins to present results.

The Consequences of Harmful Non-Indigenous Species | 2

C hapters 2 and 3 examine basic aspects of non-indigenous species (NIS)—their effects, how many there are, and how they get here. Technologies to deal with harmful NIS, including decisionmaking methods and techniques for preventing and managing problem species, are covered in chapters 4 and 5. Chapters 6, 7, and 8 assess what various institutions at the Federal, State, and local levels do, or fail to do, about NIS. Finally, chapters 9 and 10 place NIS in a broader context by examining their relationships to genetically engineered organisms, to international relations, to other prominent environmental issues, and to choices regarding the future of the nation's biological resources.

WHAT'S IN AND WHAT'S OUT: FOCUS AND DEFINITIONS

Although considerable benefits accrue from the presence of many NIS in the United States, others have caused significant harm. This report's goal is to identify where and how such problems arise, and how these problems can be avoided or minimized. This ''problem-oriented'' approach requires that beneficial introductions get limited attention throughout the assessment. They are summarized only briefly in this chapter. The emphasis is on harmful NIS, encompassing terrestrial and aquatic ecosystems and also most types of organisms (figure 2-1). An important consideration is whether a species can establish free-living populations beyond human cultivation and control. Non-indigenous species within this category—those living beyond human management—cause most harmful effects.

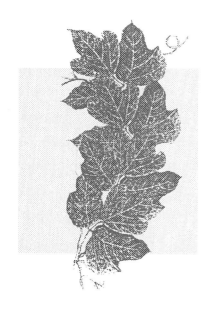

Figure 2-1—Scope of Study

Species Central to the Assessment
(to be given full consideration)

Potentially or already harmful non-indigenous species:

- not yet in the United States (e.g., certain weedy bromegrasses)
- in the United States, but in a captive or managed state (e.g., some tilapia in aquaculture)

Potentially or already harmful non-indigenous species established as free-living populations in the United States:

- of non-U.S. origin (e.g., zebra mussel)
- originating in one area of the United States, but non-indigenous in another (e.g., certain salt marsh grasses on the West Coast)
- feral species (e.g., wild hogs)

Species Not Central to the Assessment
(to be considered only when they raise important ecological or economic issues)

Beneficial non-indigenous species:

- of non-U.S. origin not yet in the United States (e.g., new crops)
- of non-U.S. origin presently in the United States in a captive (e.g., elephants), managed (e.g. alfalfa), or free-living state (e.g., several earthworms)
- originating in one area of the United States, but non-indigenous in another (e.g., Pacific salmon in the Great Lakes)
- except those also having the potential to escape and/or cause harm

Indigenous species, including those:

- naturally expanding their ranges into the United States (e.g., Old World blackheaded gull)
- previously extirpated, but presently being reintroduced (e.g., Californian condors)
- stocked or planted within their natural ranges (e.g., southern-pine plantations) naturally occurring hybirds between indigenous species (e.g., grey wolf/coyote hybrids)

Species of unknown origin (e.g., dogwood anthracnose)

Bioengineered orgnisms (e.g., transgenic fish)-- but central in chapter 9

Structural pests (e.g., cockroaches)

Human diseases (e.g., swine flu)

NOTE: When the word "species" occurs above, "subspecies" and "recognized variants" may be substituted. Our emphasis is species-level issues first, then subspecies and variants in decreasing priority. See index for species' scientific names.

SOURCE: Office of Technology Assessment, 1993.

■ Definitions

Finding:

Terms and definitions pertaining to NIS differ greatly among various laws, regulations, policies, and publications, making direct comparisons misleading. A need exists for uniform definitions to ensure accurate assessments of problems and consistent applications of policies.

Movements of people and cargo across the Earth provide routes by which species spread to new locales. "Exotic," "alien," "introduced," "immigrant," "non-native," and "non-indige-nous" have all been used to refer to these species. No universally accepted or standard terminology exists.

OTA has chosen "non-indigenous" as the most neutral, inclusive, and unambiguous term. OTA's definition of non-indigenous (box 2-A) avoids some common sources of confusion. It sets spatial limits based on a species' ecology rather than on national or State boundaries. Other definitions of non-indigenous and related terms, like exotic, vary greatly as to whether they include only species foreign to the United States, or additionally incorporate species of U.S. origin

---

Box 2-A—Terms Used by OTA

- **Non-indigenous**—The condition of a species being beyond its natural range or natural zone of potential dispersal; includes all domesticated and feral species and all hybrids except for naturally occurring crosses between indigenous species.
- **Indigenous**—The condition of a species being within its natural range or natural zone of potential dispersal; excludes species descended from domesticated ancestors.
- **Feral**—Used to describe free-living plants or animals, living under natural selection pressures, descended from domesticated ancestors.
- **Natural range**—The geographic area a species inhabits or would inhabit in the absence of significant human influence.
- **Natural zone of potential dispersal**—The area a species would disperse to in the absence of significant human influence.
- **Introduction**—All or part of the process by which a non-indigenous species is imported to a new locale and is released or escapes into a free-living state.
- **Established**—The condition of a species that has formed a self-sustaining, free-living population at a given location.

OTA's definitions of "indigenous" and "non-indigenous" are based on species' ecology rather than on national, State, or local political boundaries. Thus, if a species' natural range is only in west Texas, it would be non-indigenous when imported to east Texas. A species is indigenous to its entire natural range, even to areas it previously but no longer occupies due to human influence.

The definition of "natural range" incorporates the idea of a "significant human influence." This acknowledges that species can have natural ranges even when affected by humans so long as humans are not a *major* determinant of the range. The concept of "natural zone of potential dispersal" incorporates naturally occurring expansions and contractions of species ranges. For example, a shore bird that shifts naturally over time from being an "accidental" visitor to the United States to being a breeding resident would be indigenous.

Domesticated and feral species and their variants are all non-indigenous. They are products of human selection and lack natural ranges. For similar reasons, all hybrids except for naturally occurring crosses between indigenous species are also non-indigenous.

OTA will explicitly indicate where this report's discussion is limited to species non-indigenous to the United States rather than to all non-indigenous species. Similarly, the terms "indigenous" and "non-indigenous" also can apply to subspecies, recognized variants, and other biological subdivisions beneath the level of species. Uses in these contexts also will be clearly identified.

SOURCE: Office of Technology Assessment, 1993.

---

living beyond their natural ranges (48,92). OTA's definition also does not include arbitrary time limits. Some definitions classify as native or indigenous all species established in the United States by a certain date, commonly before European settlement (53). Under other definitions, NIS eventually become ''naturalized'' after a certain period has elapsed (97).

Several important categories of organisms are comprised wholly or in part of NIS. Experts estimate that at least half of U.S. weeds are non-indigenous to the country (19). A similarly large proportion of economically significant insect pests of agriculture and forestry is non-indigenous: 39 percent (67). Federal laws restrict or prohibit importation of plants and animals considered to be "noxious weeds"[1] and "injurious wildlife"[2]—species that are all non-indigenous.

■ Other Efforts Under Way

Several efforts related to this assessment are under way or were recently completed.[3] Passage of the Nonindigenous Aquatic Nuisance Prevention and Control Act of 1990[4] created the interagency Aquatic Nuisance Species Task Force. This task force is required to develop a program to prevent, monitor, and control unintentional introductions of non-indigenous aquatic nuisance species and to provide for related public education and research. A draft of the program was released for public comment November 12, 1992, and is expected to be presented to Congress in 1993 (14). The task force also is conducting a review of policies related to the intentional introduction of aquatic species. The task force's activities parallel, to some extent, portions of OTA's study.

DO WE KNOW ENOUGH TO ASSESS THE SITUATION?

Finding:

The information on NIS is widely scattered and often anecdotal. It emphasizes species having negative effects on agriculture, industry, or human health. The numbers and impacts of harmful NIS in the United States are chronically underestimated, especially for organisms lacking such economic or health effects.

■ Information Gaps

Although much information on NIS exists, overall it is widely scattered, sometimes obscure, and highly variable in quality and scientific rigor. No governmental or private agency keeps track of new NIS that enter or become established in the country, unless they also are considered a potential pest to agriculture or forestry or a human health threat, and even these databases are not comprehensive. Summary lists of NIS do not exist for most types of organisms (7,33,43,72,79). This gap is especially large for non-indigenous insect and plant species, which number in the thousands in the United States (ch. 3) (33,43). It also plagues attempts to quantify the numbers and effects of plant pathogens, since the origin of most is unknown (72). Even for known NIS, the effects of many have never been studied, especially those without clear economic or human health impacts. Information on effects is similarly lacking for the numerous as-yet-undetected NIS that many of OTA's contractors and advisory panelists believe are already established in the country.

Because of the poor documentation, presently available information provides an incomplete picture of NIS in the United States. Consequently, whatever we do know about harmful NIS surely

---

[1] "Noxious weeds" are defined under the Federal Noxious Weed Act of 1974, as amended (7 U.S.C.A. 2801-2814) as "any living stage (including but not limited to, seeds and reproductive parts) of any parasitic or other plant of a kind, or subdivision of a kind, which is of foreign origin, is new to or not widely prevalent in the United States, and can directly or indirectly injure crops, other useful plants, livestock, or poultry or other interests of agriculture, including irrigation, or navigation or the fish or wildlife resources of the United States or the public health."

[2] "Injurious wildlife" is defined under the Lacey Act (1900), as amended (16 U.S.C.A. 667 *et seq.*) as several named species "and such other species of wild mammals, wild birds, fish (including mollusks and crustacea), amphibians, reptiles, or the offspring or eggs of any of the foregoing which the Secretary of the Interior may prescribe by regulation to be injurious to human beings, to the interests of agriculture, horticulture, forestry, or to wildlife or the wildlife resources of the United States."

[3] The Hawaii office of the Nature Conservancy in collaboration with the Natural Resources Defense Council released *The Alien Pest Species Invasion in Hawaii: Background Study and Recommendations for Interagency Planning* in July 1992 (60). This report examines the causes, consequences, and solutions to harmful NIS in Hawaii. A report on NIS in Minnesota was issued by the Minnesota Interagency Exotic Species Task Force in April 1991 (53). In addition, the National Research Council (NRC) approved the concept for a broad study of science and policy issues related to marine NIS in 1991. The study was not undertaken, however, because of inadequate funding.

[4] Nonindigenous Aquatic Nuisance Prevention and Control Act of 1990, as amended (16 U.S.C.A. 4701-4751).

Table 2-1—Groups of Organisms Covered by OTA's Contractors[a]

Category examined by contractor	Number of species analyzed for summary of NIS consequences[b]	Percent of total known U.S. NIS analyzed per category by OTA's contractors
Plants—free-living plants and algae dwelling on land and in fresh water; excludes those under human cultivation	—[c]	—[c]
Terrestrial vertebrates—free-living vertebrate animals dwelling on land (birds, reptiles, amphibians, mammals); excludes strictly domesticated species	125 NIS of foreign or U.S. origin	65%
Insects—insects and arachnids (ticks, mites, spiders)	1,059 NIS of foreign origin from 149 taxonomic families	53%
Fish—free-living finfish that dwell for all or part of their lives in fresh water	111 NIS of foreign or U.S. origin	88%
Mollusks—snails, bivalves, and slugs living on land, in fresh water, and in estuaries	88 NIS of foreign origin	97%
Plant pathogens—viruses, bacteria, fungi, nematodes, and parasitic plants that cause diseases of plants	54 NIS of foreign origin from selected host plants (potato, rhododendron, citrus, wheat, Douglas fir, kudzu, five-needled pines, chestnut)	23%

[a] Major categories not covered include: exclusively marine plants and animals; organisms causing animal diseases (viruses, bacteria, etc.); worms; crustaceans (crayfish, water fleas); free-living bacteria and fungi.

[b] See figures 2-2, 2-4, and 2-5.

[c] Contractor could not quantitatively analyze effects of non-indigenous plants because of the large numbers of species (>2,000) and lack of previous summary material.

SOURCES: Summarized by OTA from: J.C. Britton, "Pathways and Consequences of the Introduction of Non-Indigenous Fresh Water, Terrestrial, and Estuarine Mollusks in the United States," contractor report prepared for the Office of Technology Assessment, October 1991; W.R. Courtenay, Jr., "Pathways and Consequences of the Introduction of Non-Indigenous Fishes in the United States," contractor report prepared for the Office of Technology Assessment, September 1991; K.C. Kim and A.G. Wheeler, "Pathways and Consequences of the Introduction of Non-Indigenous Insects and Arachnids in the United States," contractor report prepared for the Office of Technology Assessment, December 1991; C.L. Schoulties, "Pathways and Consequences of the Introduction of Non-Indigenous Plant Pathogens in the United States," contractor report prepared for the Office of Technology Assessment, December 1991; S.A. Temple and D.M. Carroll, "Pathways and Consequences of the Introduction of Non-Indigenous Vertebrates in the United States," contractor report prepared for the Office of Technology Assessment, October 1991.

underestimates their numbers and the magnitude of their effects. Even from this baseline estimate, however, a picture emerges of current and impending problems that require action. OTA's approach is to provide such a baseline estimate.

■ OTA's Approach for Chapters 2 and 3

To attempt a quantitative analysis, OTA asked experts to assess the numbers of known NIS in the country, what their effects have been, and how they entered or spread within the nation. The OTA contractors categorized impacts of established NIS by type (harmful, beneficial, neutral, or unknown); nature of effect (economic, ecological,

and other); and magnitude (high, medium, low). Six reports were prepared, one each for plants, terrestrial vertebrates, insects, fish, mollusks, and plant pathogens (table 2-1). This selection, while covering most important terrestrial and freshwater organisms, is not all-inclusive. It reflects a balance between comprehensiveness and feasibility. For example, no identifiable expert could summarize information on all aquatic invertebrate animals (e.g., mollusks, worms, crustaceans, etc.), in part because many groups are only poorly known.

In preparing background reports, the contractors reviewed available publications, surveyed or interviewed numerous other experts, and incorporated their own judgments. Their resulting summaries are the most complete and up-to-date available. Chapters 2 and 3 draw on these background summaries, additional published information, and additional expert opinions to develop a broad overview of harmful NIS in the United States. The effects of NIS—both beneficial and harmful are covered in this chapter. Chapter 3 examines the pathways by which NIS enter and spread in the United States, their rates of arrival, and current numbers in the country.

BENEFITS OF INTRODUCTIONS

Finding:

Cultivation of non-indigenous crops and livestock is the foundation of U.S. agriculture. NIS also play a key role in other industries and enterprises, many of which are based on the U.S. market for biological novelty, e.g., ornamental plants and pets.

NIS are essential to many U.S. industries and enterprises. Their benefits are great, and include economic, recreational, and social effects.

Almost all economically important crops[5] and livestock in the United States are of foreign origin (43). Non-indigenous plants have a similarly important role in horticulture and include such familiar horticultural mainstays as iris (*Iris* spp.), forsythia (*Forsythia* spp.), and weeping willow (*Salix* spp.) (26). Many plants used to prevent erosion are also non-indigenous, such as Bermuda grass (*Cynodon dactylon*) and lespedeza (*Lespedeza* spp.) (93). Importation of new species and strains continues for the development of new varieties for agriculture, horticulture, and soil conservation (65).

Non-indigenous insects also have important functions in agriculture. The European honey bee (*Apis mellifera*) forms the basis for the U.S. apiculture industry, providing bees to pollinate orchards and many other agricultural crops.

Non-indigenous organisms of many types have beneficial uses as biological control agents, frequently for control of non-indigenous pests. Insects and pathogens of plants and animals are most commonly used for control of weeds and insect pests. For example, a rust fungus (*Puccinia chondrillina*) was successfully introduced into California to control skeletonweed (*Chondrilla juncea*) in 1975 (72). Fish have been introduced in some places to control aquatic weeds, mosquitoes, gnats, and midges (23). Some consider the introduction of barn owls (*Tyto alba*) to Hawaii to control mice and rats a success, although the use of land-dwelling vertebrates for biological control has generally caused great environmental damage (79).

A number of fish and shellfish cultured in the growing aquaculture industry are non-indigenous. Virtually the entire West Coast oyster industry is based on the Pacific oyster (*Crassostrea gigas*), originally from Japan. Fish species of *Tilapia*, from Africa and the Middle East, are now commonly grown throughout the United States (10), and shrimp farmers in southeastern and other regions of the country commonly raise Pacific white shrimp (*Penaeus vannamei*), a shrimp originally from Asia.

Sport fishing often means fishing for non-indigenous fish. The rainbow trout (*Oncorhynchus mykiss*), striped bass (*Morone saxatilis*), and varieties of largemouth bass (*Micropterus salmoides*), although indigenous to the United States, have been widely introduced beyond their natural ranges for fisheries enhancement (10). A frequently stocked sport fish, the brown trout (*Salmo trutta*), originated in Europe. The Great Lakes salmon fishery is based on species indigenous to the Pacific coast of North America. Additional fish have been introduced to provide forage for game fish. Sport fishing not only provides recreational opportunities, but also stimulates the development of related businesses,

---

[5] Crops originating in the United States include cranberry (*Vaccinium macrocarpon*), pecan (*Carya illinoensis*), tobacco (*Nicotiana tabacum*), and sunflower (*Helianthus annuus*).

such as boat rentals, charter fishing, and sales of fishing equipment and supplies (10).

Some of the most widely hunted game species, such as the chukar partridge (*Alectoris chuckar*) and ring-necked pheasant (*Phasianus colchieus*), originated outside of the United States (95). Sizable businesses exist to provide supplies and services for recreational hunting (79). Some non-indigenous big-game animals, like Sika deer (*Cervus nippon*) from Asia and South African oryx (*Oryx gazella gazella*), are grown on private ranches for hunting, and also to satisfy the growing market for ''exotic'' game meats (81). Non-indigenous fur-bearing animals support both the trapping industry and fur-bearer farms (79).

Most pet and aquarium industries are based on domesticated and other NIS, including cats, dogs, hamsters, goldfish, snakes, turtles, and chameleons. These animals are valued by owners for companionship, protection, and recreation. A number of non-indigenous animals, such as the African clawed frog (*Xenopus laevis*), are used in biomedical fields for experimental work or testing (79).

Restoration of habitats degraded by pollution, mining, and other human disruptions sometimes includes planting stress-tolerant NIS. Several trees, like the ginkgo from China (*Ginkgo biloba*), are common in urban landscaping, where few indigenous species can grow. Some non-indigenous sport fish serve a similar role in reservoirs and other artificial habitats less hospitable to indigenous species. Efforts to remedy environmental contamination from oil or other substances sometimes involve the release of non-indigenous microbes that accelerate contaminant degradation (88). Certain microbes help make nutrients available to plants through nitrogen fixation. These microbes also have been widely transferred and released around the world.

Paradoxically, introductions of NIS are increasingly seen by some conservationists as a means to preserve certain endangered and threatened species that cannot be saved in their native habitats (79). Some conservationists have even

suggested that introduction of large ungulates from Africa onto the American plains may be some species' best chance at survival (74).

WHEN NON-INDIGENOUS SPECIES CAUSE PROBLEMS

Despite the clear benefits of many NIS, numerous others continue to cause great harm in the United States. Many are familiar. They range from nuisances like crabgrass (*Digitaria* spp.), dandelions (*Taraxacum officinale*), and German cockroaches (*Blattella germanica*), to species annually costing millions of dollars to agriculture and forestry, such as the Mediterranean fruitfly, or medfly (*Ceratitis capitata*), and the European gypsy moth (*Lymantria dispar*). Some pose human health risks, such as the African honey bee (*Apis mellifera scutellata*) and the imported fire ant (*Solenopsis invicta, S. richteri*). Still others, like the paper bark tree (*Melaleuca quinquenervia*) and zebra mussel (*Dreissena polymorpha*), threaten widespread disruption of U.S. ecosystems and the displacement or loss of indigenous plants and animals.

■ A Major Consideration: High Negative Impacts Are Infrequent

Finding:

A minority of the total NIS cause severe harm. However, such high-impact NIS occur in almost all regions of the country. Individually and cumulatively, they have had extensive negative impacts in the United States.

Relatively few NIS cause great harm. Estimates range from 4 to 19 percent of the NIS analyzed by OTA's contractors, depending on the type of organism (figure 2-2). Included here are NIS that are significant and difficult-to-control pests of agriculture, rangelands, or forests; seriously foul waterways, irrigation systems, and power plants; cause wide-scale disruption of indigenous ecosystems; or threaten indigenous species with extinction. At least 200 well-known, high-impact NIS presently occur in the United

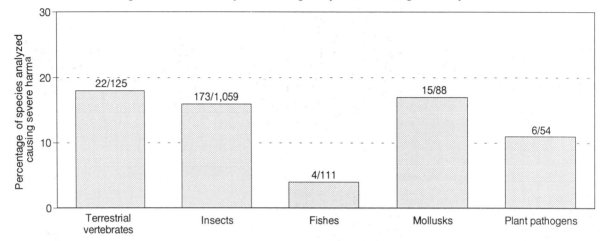

Figure 2-2—How Frequent Are High-Impact Non-Indigenous Species?[a]

a Species judged by OTA's contractors as causing severe economic or environmental harm. Numbers of severely harmful and total species are listed above each bar.

SOURCES: Summarized by OTA from: J.C. Britton, "Pathways and Consequences of the Introduction of Freshwater, Terrestrial, and Estuarine Mollusks in the United States," contractor report prepared for the Office of Technology Assessment, October 1991; W.R. Courtenay, Jr., "Pathways and Consequences of the Introduction of Non-Indigenous Fishes in the United States," contractor report prepared for the Office of Technology Assessment, September 1991; K.C. Kim and A.G. Wheeler, "Pathways and Consequences of the Introduction of Non-Indigenous Insects and Arachnids in the United States," contractor report prepared for the Office of Technology Assessment, December 1991; C.L. Schoulties, "Pathways and Consequences of the Introduction of Non-Indigenous Plant Pathogens in the United States," contractor report prepared for the Office of Technology Assessment, December 1991; S.A. Temple and D.M. Carroll, "Pathways and Consequences of the Introduction of Non-Indigenous Vertebrates in the United States," contractor report prepared for the Office of Technology Assessment, October 1991.

States (7,10,33,72). Even though relatively few NIS are highly damaging, they occur in almost all regions of the country (figure 2-3). Moreover, the summed impacts of even one disastrous species can be substantial. Estimated U.S. losses from 1987 to 1989 attributable to the Russian wheat aphid (*Diuraphis noxia*) alone exceeded $600 million (1991 dollars) (8).

■ Time Lags and Unknown Effects Are Common

Effects of many NIS remain undetected for extended periods following their establishment. Such time lags can reflect an initial period during which a species' population is too small to cause noticeable impacts. Over time, changing environmental conditions cause some previously rare NIS to become abundant and cause harmful effects. Other previously benign NIS become problems after additional NIS enter the country. For example, an Asian fig plant (*Ficus micro-*

carpa) widely planted as an ornamental in Florida only became a pest about 45 years after introduction, when its natural pollinator—a fig wasp (*Parapristina verticillata*)—was introduced (50). Similarly, at least a decade elapsed between establishment of the Asian clam (*Corbicula fluminea*) and appearance of its harmful effects; 12 years for chestnut blight (*Cryphonectria parasitica*) (see "Forestry" below); and 4 years for the cereal leaf beetle (*Oulema melanopus*) (7,33,72).

Some harmful species are mistakenly thought to have neutral consequences until other effects are detected. Thus, in many cases, "neutral" NIS are better characterized as having unknown effects. Unknown effects and time lags are common for NIS affecting non-agricultural areas, since these tend to be poorly studied. OTA's contractors found between 6 and 53 percent of the NIS examined had neutral or unknown effects (figure 2-4). Given that time delays are common, some of these eventually will cause harmful impacts.

Figure 2-3—State by State Distribution of Some High-Impact Non-Indigenous Species

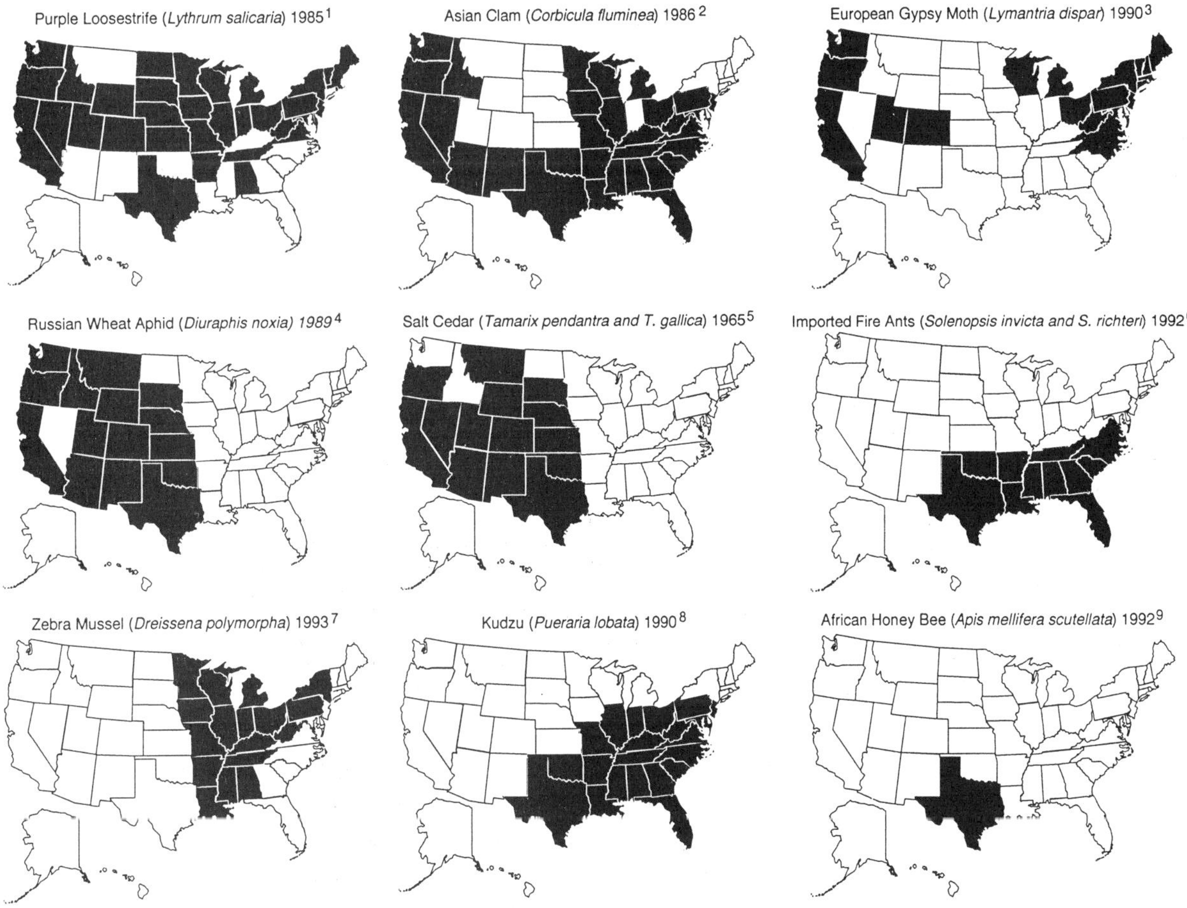

Purple Loosestrife (*Lythrum salicaria*) 1985[1]

Asian Clam (*Corbicula fluminea*) 1986 [2]

European Gypsy Moth (*Lymantria dispar*) 1990[3]

Russian Wheat Aphid (*Diuraphis noxia*) 1989[4]

Salt Cedar (*Tamarix pendantra and T. gallica*) 1965[5]

Imported Fire Ants (*Solenopsis invicta and S. richteri*) 1992[6]

Zebra Mussel (*Dreissena polymorpha*) 1993[7]

Kudzu (*Pueraria lobata*) 1990[8]

African Honey Bee (*Apis mellifera scutellata*) 1992[9]

SOURCES:
1. D.Q. Thompson, R.L. Stuckey, E.B. Thompson, "Spread, Impact, and Control of Purple Loosestrife (*Lythrum salicaria*) in North American Wetlands" (Washington, DC: U.S. Department of the Interior Fish and Wildlife Service, 1987).
2. Clement L. Counts, III, "The Zoogeography and History of the Invasion of the United States by *Corbicula Fluminea* (Bivalvia: Corbiculidae)," *American Malacological Bulletin*, Special Edition No. 2, 1986, pp. 7-39.
3. P.W. Schaefer and R.W. Fuester, "Gypsy Moths: Thwarting Their Wandering Ways," *Agricultural Research*, vol. 39, No. 5, May 1991, pp. 4-11; M.L. McManus and T. McIntyre, "Introduction," *The Gypsy Moth: Research Toward Integrated Pest Management*, C.C. Doane and M.L. McManus (eds.) Technical Bulletin No. 1584 (Washington, DC: U.S. Forest Service, 1981), pp. 1-8; T. Eiber, "Enhancement of Gypsy Moth Management, Detection, and Delay Strategies," *Gypsy Moth News*, No. 26, June 1991, pp. 2-5.
4. S.D. Kindler and T.L. Springer, "Alternative Hosts of Russian Wheat Aphid" (Homoptera: Aphididae), *Journal of Economical Entomology*, vol. 82, No. 5, 1989, pp. 1358-1362.
5. T.W. Robinson, "Introduction, Spread and Areal Extent of Saltcedar (Tamarix) in the Western States," Studies of Evapotranspiration, Geological Survey Professional Paper 491-A (Washington, DC: U.S. Government Printing Office, 1965).
6. V.R. Lewis et al., "Imported Fire Ants: Potential Risk to California," *California Agriculture*, vol. 46, No. 1, January-February 1992, pp. 29-31; D'Vera Cohn, "Insect Aside: Beware of the Fire Down Below, Stinging Ants From Farther South Have Begun to Make Inroads in Virginia, Maryland," *Washington Post*, June 2, 1992, p. B3.
7. U.S. Department of the Interior, Fish and Wildlife Service, briefing delivered to the Senate Great Lakes Task Force, May 21, 1993.
8. Anonymous, *National Geographic Magazine*, 'Scourge of the South May be Heading North," vol. 178, No. 1, July 1990.
9. M.L. Winston, "Honey, They're Here! Leaning to Cope with Africanized Bees," *The Sciences*, vol. 32, No. 2, March/April 1992, pp. 22-28.

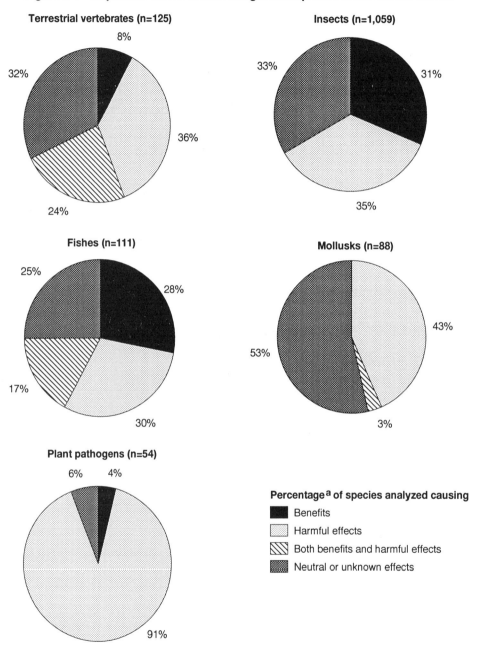

Figure 2-4—Reported Effects of Non-Indigenous Species in the United States

Terrestrial vertebrates (n=125)

8%
32%
36%
24%

Insects (n=1,059)

33%
31%
35%

Fishes (n=111)

25%
28%
17%
30%

Mollusks (n=88)

53%
43%
3%

Plant pathogens (n=54)

6% 4%
91%

Percentage[a] of species analyzed causing
- ■ Benefits
- ▢ Harmful effects
- ▨ Both benefits and harmful effects
- ▦ Neutral or unknown effects

[a] Percentages do not always total 100% due to rounding.

SOURCES: Summarized by OTA from: J.C. Britton, "Pathways and Consequences of the Introduction of Freshwater, Terrestrial, and Estuarine Mollusks in the United States," contractor report prepared for the Office of Technology Assessment, October 1991; W.R. Courtenay, Jr., "Pathways and Consequences of the Introduction of Non-Indigenous Fishes in the United States," contractor report prepared for the Office of Technology Assessment, September 1991; K.C. Kim and A.G. Wheeler, "Pathways and Consequences of the Introduction of Non-Indigenous Insects and Arachnids in the United States," contractor report prepared for the Office of Technology Assessment, December 1991; C.L. Schoulties, "Pathways and Consequences of the Introduction of Non-Indigenous Plant Pathogens in the United States," contractor report prepared for the Office of Technology Assessment, December 1991; S.A. Temple and D.M. Carroll, "Pathways and Consequences of the Introduction of Non-Indigenous Vertebrates in the United States," contractor report prepared for the Office of Technology Assessment, October 1991.

■ How Problems Arise

NIS problems have several origins. Some NIS introduced for beneficial purposes unexpectedly produce harmful consequences. Many other harmful species arrived or spread within the country unintentionally. A complicating factor is that numerous NIS cause both beneficial and harmful effects.

POOR CHOICES: INTENTIONAL INTRODUCTIONS THAT GO AWRY

Many harmful introductions probably would not have occurred had the damage they caused been anticipated in advance. But little advance evaluation of potential harmful effects was performed for many NIS intentionally released in the past. Even when advance evaluations have been performed, however, they often have done a poor job of anticipating effects. Scientists generally agree that predicting the role and effects of a species in a new environment is extremely difficult (56). Each introduction creates a novel combination of organism and environment. Detailed information about both is necessary to anticipate the result, and such information usually is lacking.

Nevertheless, some continue to use a simplistic approach to evaluating introductions. An erroneous concept still widely applied by fisheries managers is the "vacant niche." This concept holds that some ecological roles may not be filled in a community, and species can be selectively introduced to fill these voids. Application of this approach to natural communities is inappropriate both because few species can fit the narrow ecological vacancies identified by managers, and because it is virtually impossible to predetermine the role a species will assume after it has been released (28). Numerous examples exist where a species' ecological role was mistakenly understood before its release. For example, many insect parasites and predators introduced to Hawaii for biological control of pests unexpectedly expanded their diets to include indigenous species (29).

*Kudzu (*Pueraria lobata*) was initially promoted by the U.S. Department of Agriculture for erosion control and forage, but it has overgrown other vegetation throughout the southeastern United States.*

Problems also arise when a species moves into new habitats beyond the intended area of introduction. A recent example is the cactus moth (*Cactoblastis cactorum*). Introduced to the West Indies to control prickly pear cactus (*Opuntia* spp.), the moth has since spread northward into Florida. Conservationists fear it may eventually threaten indigenous prickly pear cacti throughout the United States, 16 species of which are rare and under review for listing under the Endangered Species Act (31). The seven-spotted ladybeetle (*Coccinella septempunctata*), an aphid predator, has dispersed throughout much of the United States. It appears to be outcompeting the native nine-spotted ladybeetle (*C. novemnotata*) and has displaced that species in alfalfa fields (33).

Species that escape from human cultivation, in a sense, also move beyond their anticipated distributions. Feral populations of domesticated mammals, such as goats (*Capra hircus*) and pigs (*Sus scrofa*), cause great ecological damage and erosion in natural areas by trampling, uprooting, and consuming plants. Many weeds, such as crabgrass, originally were cultivated for agriculture (26). Some ornamental plants also cause harm when they escape and form free-living populations. English ivy (*Hedera helix*) and

Figure 2-5—How Often Do Intentional Versus Unintentional Introductions Have Harmful Effects?

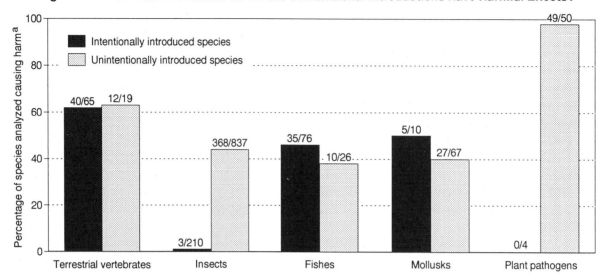

a Includes species reported by OTA's contractors as causing any economic or environmental harm. Some species may also have beneficial effects. Analysis excludes species that have been introduced both intentionally and unintentionally or for which the mode of introduction is unknown. Numbers of harmful and total species are listed above each bar.

SOURCES: Summarized by OTA from: J.C.Britton, "Pathways and Consequences of the Introduction of Freshwater, Terrestrial, and Estuarine Mollusks in the United States," contractor report prepared for the Office of Technology Assessment, October 1991; W.R. Courtenay, Jr., "Pathways and Consequences of the Introduction of Non-Indigenous Fishes in the United States," contractor report prepared for the Office of Technology Assessment, September 1991; K.C. Kim and A.G. Wheeler, "Pathways and Consequences of the Introduction of Non-Indigenous Insects and Arachnids in the United States," contractor report prepared for the Office of Technology Assessment, December 1991; C.L. Schoulties, "Pathways and Consequences of the Introduction of Non-Indigenous Plant Pathogens in the United States," contractor report prepared for the Office of Technology Assessment, December 1991; S.A. Temple and D.M. Carroll, "Pathways and Consequences of the Introduction of Non-Indigenous Vertebrates in the United States," contractor report prepared for the Office of Technology Assessment, October 1991.

Japanese honeysuckle (*Lonicera japonica*) overgrow and eventually kill trees and understory plants and have fundamentally altered the character and structure of some eastern forests (82). Among the 300 non-indigenous weeds of the western United States, at least 8 were formerly cultivated as crops and 28 escaped from horticulture (100).

THE SURPRISE OF UNINTENTIONAL INTRODUCTIONS

Many NIS currently in the United States arrived and spread as unintended stowaways on human transport. For example, in the past, many weeds moved as contaminants of agricultural seed, and many plant pathogens arrived in the soil of potted plants (43,72) (see also ch. 3).

In contrast to most intentional introductions, unintentionally introduced species have not been

chosen for any beneficial characteristics. Thus, a logical expectation might be that unintentionally introduced species are more likely to cause harmful effects than intentionally introduced species. Evaluation of the 1,483 NIS examined by OTA's contractors would seem to support this, since only 12 percent of the intentionally introduced species had harmful effects compared to 44 percent of the unintentionally introduced species (10,33,72,79). However, when specific groups of organisms are examined separately, clear differences appear (figure 2-5). Far more unintentional introductions of insects and plant pathogens have had harmful effects than have intentional introductions of these organisms. For terrestrial vertebrates, fish, and mollusks, however, intentional introductions have caused harm approximately as often as have unintentional introductions, suggesting a history of poor choices of species for introduction and complacency regarding their potential harm.

MANY SPECIES HAVE BOTH BENEFITS AND HARMFUL EFFECTS

Finding:

Certain NIS have both positive and negative consequences, especially species occurring across several regions or States. In addition, perceived effects of NIS can vary in relation to the observer's perspective. Different constituencies can hold widely divergent and deep-seated views of the potential effects and desirability of even a single species.

Many NIS simultaneously have benefits as well as harmful effects (figure 2-4). Even some NIS known for their harmful effects can also have some benefits. For example, imported fire ants, which sting people and damage crops, also suppress populations of agricultural pests and enhance available soil nutrients (73). Some non-indigenous ("exotic") game animals grown on ranches have potential economic benefits. Ranching may also help preserve animals endangered in their native ranges. Ranched non-indigenous game, however, sometimes hybridize with and dilute the gene pools of related indigenous species, or carry and spread new animal diseases (77).

The effects of some species also change as they enter new environments—a factor making prediction of harm difficult. Predators, competitors, parasites, and pathogens that keep a species' population small in one locale may be absent in another. Also, new environments may affect rates of reproduction, susceptibility to disease, and other features that affect a species' success. Consequently, a NIS that causes little damage to agriculture or natural ecosystems in one area may cause significant problems in another. Melaleuca, the paper bark tree, is a harmless ornamental in California, but causes great ecological harm in the Florida Everglades. Non-indigenous cheatgrass (*Bromus tectorum*) occurs in all 50 States, but is only a serious weed in the Midwest and West (44).

Even garden flowers like baby's breath (*Gypsophila paniculata*) can be difficult-to-control weeds in some areas (100).

The *perceived* effects of a species can also vary with the eye of the beholder (85). While many State fish and wildlife managers firmly support continued stocking with certain non-indigenous fish, some experts consider the practice to be detrimental (box 2-B). Similarly, managers of natural areas view purple loosestrife (*Lythrum salicaria*), originally from Eurasia, as a highly damaging plant because it grows prolifically in wetlands, displacing indigenous plants and providing lower quality habitat and food for wild animals. In contrast, some horticulturists in the nursery trade see purple loosestrife as a desirable plant. It also is a source of nectar for honey production.

The perceived desirability of certain NIS has changed over time, as human values and popular views have changed. The intentional introduction of songbirds, like the English sparrow (*Passer domesticus*) in the mid-1800s probably would not be allowed today, because a higher value is placed on indigenous birds. Kudzu (*Pueraria lobata*) was widely promoted for erosion control in the 1940s (89); yet the very characteristics considered beneficial then—rapid growth, ease of propagation, and wide adaptability—cause it to be considered a pernicious weed today.

ECONOMIC COSTS

Finding:

Harmful NIS annually cost the Nation hundreds of millions to perhaps billions of dollars. Economically significant species occur in all groups of organisms examined by OTA and affect numerous economic sectors. Available accountings tend to underestimate losses attributable to NIS, since they omit many harmful species and inadequately account for intangible, nonmarket impacts.

A conservative estimate is harmful NIS cause annual losses of hundreds of millions of

Box 2-B—The Case of the Brown Trout: Opposing Views of Fish Introductions

In Favor . . .

by Bruce Schmidt, Chief of Fisheries
Utah Department of Natural Resources
Salt Lake City, Utah

The introduction of non-indigenous fishes is neither all good nor all bad; judgments must be made individually. Introductions can affect pristine ecosystems, but sport fish management frequently must deal with far-from-pristine environments. Given the human species' penchant for modifying the environment, it is unrealistic to set a standard that demands no alteration of indigenous fauna. In Utah, most fish habitats are artificial reservoirs or tail waters, or are altered by water diversion, siltation, agriculture run-off, unstable banks or pollution, conditions outside the control of fisheries managers. Only four sport fish are indigenous to Utah, and none are adapted to most of these altered systems, so providing sport fishing requires introductions.

The benefits are widespread. Many species have produced excellent sport fishing when introduced into new waters in nearly all States. Sport fishing is a multibillion dollar industry, directly through input to local economies ($2.8 billion expended nationwide in 1985; $154 million by resident anglers in Utah alone in 1991) and indirectly through mental and physical benefits to people. Introductions play a significant role in this success.

Brown trout (*Salmo trutta*) are one example. They grow large, are aggressive, and are among the most prized sport fish in North America, supporting a massive recreational fishery. Brown trout have significant advantages over indigenous trout species in some situations. They can tolerate somewhat degraded environments with warmer temperatures and decreased water quality and are more resistant to intense angling pressure. Thus, they are better suited to many of the actual conditions existing today. Although brown trout would be inappropriate where they affect rare indigenous fishes, they play a major role in satisfying public demand for quality fishing opportunities.

and Against . . .

by Walter Courtenay, Professor of Biology
Florida Atlantic University
Boca Raton, Florida

The brown trout is widely regarded as a successful introduction of a non-indigenous fish, first made in 1883. Since then, the introduction of numerous other fishes, both of foreign and U.S. origin, has become a standard management tool. Negative impacts have rarely been considered before the introductions. Overall, very few introductions can be considered successful from both human and biological standpoints. As a management tool, introductions have shown minor to major negative biological impacts, including extinctions of indigenous species.

Management agencies are mostly constituent-oriented and thus are political pawns. Although agency names often contained the words "conservation" and, more recently, "natural resources," agencies are largely blind to conserving natural resources. Agency biologists often are not practicing biology, but are forced into managing, and the two are not synonymous.

Fortunately, the brown trout mostly occupies waters not preferred by indigenous trouts. In many waters, however, it is rarely as popular as transplanted rainbow trout (*Oncorhynchus mykiss*) or indigenous trouts. The positives can be counterbalanced, in part, by negatives. California, in concert with the U.S. Forest and National Park Services, has spent almost $1 million since 1965 to eradicate brown trout from the Little Kern River to save the golden trout (*Oncorhynchus aquabonita*), California's "state fish," from almost certain extermination there. Despite at least a century of fishery experience with introductions, managers seem intent on improving on nature without understanding it.

dollars to U.S. agriculture, forests, rangelands, and fisheries. Losses could reach as high as several billion dollars, especially in high-impact years. Massive expenditures on pesticides and other control and prevention technologies prevent potential additional losses of millions to billions more. Rough estimates are that the United States annually expends about $7.4 billion for pesticide applications (box 2-C), a significant proportion of which goes to control non-indigenous pests. Weeds and insects are the most costly groups, corresponding to their high numbers when compared with other NIS groups (see ch. 3).

■ Types of Economic Impacts

Harmful NIS affect numerous economic sectors. These include agriculture, forestry, fisheries and water use, utilities, buildings, and natural areas.

AGRICULTURE

Non-indigenous weeds, insects, mollusks, birds, and pathogens reduce crop and livestock production, increase production costs, and cause post-harvest crop losses. Managing the array of agricultural pests requires costly research, development, and application of control technologies.

Weeds can outcompete or contaminate crops. They have other effects as well. Johnson grass (*Sorghum halepense*) hybridizes with cultivated sorghum (*Sorghum bicolor*), producing worthless "shattercane" (43). Some weeds are either poisonous or rejected as forage by livestock (100). They reduce the value of rangelands (100); much public land has been lost for grazing because of weed infestations (43). For example, unpalatable leafy spurge (*Euphorbia esula*) has spread to 1.5 million acres of rangeland in the northern Great Plains. Direct livestock production losses together with indirect economic effects due to this species alone approached $110 million in 1990 (2). Annual U.S. losses because of weeds amount to billions of dollars (box 2-C).

The cotton boll weevil (Anthonomis grandis) *caused estimated cumulative losses of at least $50 billion for 1909-1949.*

Some weeds do not directly harm agriculture, but instead are hosts for agricultural pests. Barberry (*Berberis vulgaris*) harbors the wheat rust fungus (*Puccinia recondita*), and large losses of wheat production can occur where the plant is present (43). Wheat rust has caused approximately $100 million worth of crop losses annually over the last 20 years (37), and it caused even more significant losses before barberry was largely eradicated earlier in this century. Tumbleweed (*Salsola* spp.) similarly is a host for the curly top virus, a pathogen of crops such as sugar beets and tomatoes (102). Crested wheatgrass (*Agropyron desertorum*), widely planted for soil conservation, harbors the Russian wheat aphid (*Diuraphis noxia*), itself a significant wheat pest.

Scores of non-indigenous insect species pose serious threats to agriculture. The boll weevil (*Anthonomus grandis*), a pest of cotton, historically has the highest documented impacts—at least $50 billion (in 1991 dollars) of cumulative losses estimated for the years 1909-1949 (8). Repeated outbreaks of the medfly in California necessitate costly control programs to avert projected annual losses of up to $897 million in damaged produce, control, and reduced export revenues (34). Some other estimates of annual

---

Box 2-C—Economic Losses Caused by Non-Indigenous Weeds

The Weed Science Society of America (WSSA) recently published the report *Crop Losses Due to Weeds—1992*, covering all States but Alaska. The report relies on crop loss estimates for 46 major crops (including field crops, fruits, nuts, and vegetables) obtained through survey responses by cooperating weed scientists. The scientists estimated the cumulative value of average losses to be $4.1 billion annually, under current appropriate herbicide control strategies. They also estimated that if no herbicides were available the crop losses would total $19.6 billion.

The WSSA figures have several limitations for OTA's purposes: they only characterize a 3-year period (1989-1991); they do not cover weeds of forestry, grazing lands, horticulture, and other agricultural sectors; and they include indigenous weeds. However, indigenous weeds are less important economically than NIS, which are known to comprise the majority of weeds for most crops. For example, 23 of 37 major soybean weeds, or 62 percent, are NIS. Experts estimate that 50 percent to 75 percent of major crop weeds overall are NIS. Based on these percentages, the portion of the $4.1 billion of annual crop losses attributable to non-indigenous weeds would be approximately $2 billion to $3 billion. According to the Environmental Protection Agency, U.S. farm expenditures on pesticides amount to about $5.1 billion annually, 60 percent of which is for herbicides. Thus, roughly $1.5 billion to $2.3 billion spent annually for herbicides would be attributable to NIS.

A ballpark range for total direct non-indigenous weed costs is $3.6 billion to $5.4 billion annually. The environmental, human health, regulatory, and other indirect costs of using herbicides on non-indigenous weeds have not been adequately calculated, but rough estimates exceed an additional $1 billion annually.

SOURCES: D.C. Bridges (ed.), *Crop Losses Due to Weeds in the United States — 1992* (Champaign, IL: Weed Science Society of America, 1992); D.T. Patterson, "Research on Exotic Weeds," in *Exotic Plant Pests and North American Agriculture*, C.L. Wilson and C.L. Graham (eds.) (New York, NY: Academic Press, 1983), pp. 381-93; D. Pimentel et al., "Environmental and Economic Effects of Reducing Pesticide Use," *Bioscience*, vol. 41, No. 6, June 1991, pp. 402-9; U.S. Environmental Protection Agency, "EPA's Pesticide Programs," Publication No. 21T-1005, Washington, DC, May 1991; T.D. Whitson et al., *Weeds of the West* (Jackson, WY: Pioneer of Jackson Hole, 1991).

---

losses from insect pests compiled for OTA by the Animal and Plant Health Inspection Service of USDA include: $500 million (in 1990) for the alfalfa weevil (*Hypera postica*); $172.8 million (in 1988) for the Russian wheat aphid; and $16.6 million (annual average for 1960-1988) for the pink bollworm (*Pectinophora gossypiella*) in California (17).

The honey bee industry currently faces two new pests, the tracheal mite (*Acarapis woodi*) and the varroa mite (*Varroa jacobsoni*), which parasitize and kill honey bees. The National Association of State Departments of Agriculture estimates potential annual losses of $160 million—due to lost honey production, lost pollination fees, and costs of replacing bees—should each pest have nationwide effects similar to those reported in Michigan (1990) and Washington (1989) (59).

FORESTRY

In the early 1900s, the chestnut blight, brought in on diseased horticultural stock from China, all but eliminated the American chestnut (*Castanea dentata*), killing as many as a billion trees. American chestnut had been the most economically important hardwood species in eastern forests (91). It was widely used in urban plantings and had been a significant food source for wild animals (72). Dutch elm disease (*Ceratocystis ulmi*) also devastated vast numbers of shade trees following its U.S. discovery in 1930—an aesthetic loss for many U.S. cities as well as an expense to replace the 40 million elms estimated to have died (91).

Several other NIS currently threaten U.S. forests, including insects like the balsam wooly adelgid (*Adelges piceae*) and pathogens such as white pine blister rust (*Cronartium ribicola*). Pear thrips (*Taeniothrips inconsequens*) damaged 189,000 hectares of Vermont sugar maple in 1988 and is expected to spread throughout the Appalachians (35). The European gypsy moth exacts the greatest measurable losses and expenditures for research, control, and eradication. The USDA estimated losses of $764 million from the European gypsy moth in 1981 alone, although that figure so far has been the all-time high (17). The Asian strain of the moth recently necessitated a $14 to $20 million eradication program in the Pacific Northwest (see ch. 4, box 4-B).

FISHERIES AND WATERWAY USE

Both wild fisheries and aquaculture have been damaged by harmful NIS. Some fisheries have been decimated. In the mid-1900s, the eel-like, parasitic sea lamprey (*Petromyzon marinus*) migrated via the newly constructed Welland Canal from Lake Ontario to other Great Lakes. It caused tremendous economic losses to commercial and recreational Great Lakes fisheries. Today, about $10 million is spent annually on control and research to reduce its predation, plus roughly an equal amount annually on fish stocking (86). If control were terminated and populations of the lamprey expanded again, the total value of the lost fishing opportunities plus indirect economic impacts could exceed $500 million annually (75).

The European ruffe (*Gymnocephalus cernuus*), a fish that entered the Great Lakes via expelled ballast water in the early 1980s, poses a new threat. Based on experience in Scotland and Russia, and preliminary assessment of North American impacts, experts predict the ruffe will cause populations of commercially valuable fish to decline. The Great Lakes Fishery Commission estimates that annual losses of more than $90 million could occur if it is not controlled (24).

Several non-indigenous aquatic weed species clog waterways. An estimated $100 million is spent nationally each year to control aquatic weeds, a majority of which are non-indigenous (20). Hydrilla (*Hydrilla verticillata*) in the Southeast blocks irrigation and drainage canals, en-hances sedimentation in flood control reservoirs, interferes with public water supplies, impedes navigation, and generally restricts public water uses (32). At high densities, it also reduces productivity of recreational fisheries (32).

UTILITIES

Fouling of water pipes by zebra mussels has imposed large expenses on the electric power industry and its customers. Costs have been incurred for the development and implementation of antifouling technologies, application of control techniques to remove zebra mussels already present, and plant shut-downs. Another mollusk, the Asian clam, has had similar effects (box 2-D). Zebra mussels and the Asian clam also clog water pipes for municipal and irrigation water supplies.

BUILDING STRUCTURES

Non-indigenous pests damage commercial and residential structures, threaten the health of occupants, and reduce property values. The full effects of structural pests—cockroaches, rats, and others that are non-indigenous—are beyond the scope of this report. However, they contribute significantly to the national market for pest control inside buildings, which totals roughly $6 billion dollars in annual sales of extermination services, retail products, and associated items (63).

NATURAL AREAS

Millions of dollars are spent annually to address the harmful effects of NIS on natural ecosystems, mostly by public agencies (see ch. 6). Expenditures are required for the development and application of control and eradication measures, as well as for ecological restoration. Indirect economic effects result from reduced recreational opportunities in areas invaded by harmful NIS, and the loss of indigenous species. Because of the absence of clear financial incentives, such as exist in agriculture, many NIS problems in natural areas remain unaddressed. The cost of back-logged control or eradication projects is difficult to estimate, but is very likely higher than for any

---

Box 2-D—Case Study of an Affected Industry: The "One-Two Punch" of Asian Clams and Zebra Mussels on the Power Industry

Two harmful non-indigenous species—the Asian clam, *Corbicula fluminea*, and the zebra mussel, *Dreissena polymorpha*—have and will continue to have significant and lasting effects on the U.S. power industry and electricity consumers.

The Asian clam entered North America some time before 1924. This small clam grows and reproduces rapidly, producing massive numbers of shells shortly after entering new waterways. Its harmful effects received little attention until the 1950s, when it was found clogging California irrigation systems as well as condensers of the Shawnee Steam Electric Power Station at Paducah, Kentucky. Populations of *Corbicula* grew explosively during the 1960s and 1970s. During that period it disrupted the operations of numerous steam and at least three nuclear electric generating stations, with down-time, corrective actions, and maintenance costing millions of dollars. In 1980, the Arkansas Nuclear One power plant was forced to shut down because of waterline clogging by Asian clams, prompting the Nuclear Regulatory Commission to issue a directive requiring the nuclear electric industry to determine whether *Corbicula* fouling was a hazard at each nuclear facility in the nation. The estimated cost of compliance with this directive was $4.5 million. One estimate put total losses at $1 billion annually in the early 1980s. More recently, populations of the Asian clam have begun to decline for unknown reasons. Nevertheless, it remains a serious fouling pest.

The industry was dealt a second blow by entry of another mollusk. The zebra mussel entered the Great Lakes by way of discharged ballast water during the mid-1980s and has since spread as far as the Hudson, Susquehanna, Mississippi, and Illinois river basins. Like Asian clams, zebra mussels are highly fertile, enabling populations to quickly reach large sizes. Zebra mussels adhere to water pipes by tough threads, clogging water flow and increasing sedimentation and corrosion. One expert from the New York Sea Grant Extension Service estimated costs for the power industry of up to $800 million for plant redesign and $60 million for annual maintenance. Fouling by zebra mussels of cooling or other critical water systems in power plants can require shut-down, costing as much as $5,000 per hour for a 200-megawatt system. Some experts expect total costs to the power industry from zebra mussels to match those for the Asian clam, perhaps reaching $3.1 billion (1991 dollars) over a 10-year period.

SOURCES: J.C. Britton, "Pathways and Consequences of the Introduction of Freshwater, Terrestrial, and Estuarine Mollusks in the United States," contractor report prepared for the Office of Technology Assessment, October 1991; M. Cochran, "Non-Indigenous Species in the Unites States—Economic Consequences," contractor report prepared for the Office of Technology Assessment, March 1992; B.G. Isom, "Historical Review of Asiatic Clam (*Corbicula*) Invasions and Biofouling of Waters and Industries in the Americas," *American Malacological Bulletin*, special edition No. 2, pp. 1-5, 1986.

---

other sector. For example, removal of all of the damaging salt cedar (*Tamarix* spp.) infestations bordering the lower Colorado River, and restoration of the indigenous vegetation, would cost an estimated $45 million to $450 million (94).

■ Cumulative Losses

OTA summarized some of the estimated economic losses to the United States from introductions of 79 harmful NIS between 1906 and 1991 (table 2-2). The species range from the brown tree snake (*Boiga irregularis*) (the costs of keeping it out) to hog cholera virus. The estimated total of $97 billion (1991 dollars) provides a minimum benchmark for true losses during the 85 years. This total is likely a fraction of the total costs during the period. Only about 14 percent of NIS known to be harmful are included, because comparable estimates of economic effects for the remaining 86 percent were unavailable; one of the most costly groups—non-indigenous agricultural weeds (see box 2-C)—is omitted.

■ Under-Counted Effects

The economic data on NIS are heavily weighted toward direct market effects and government

Table 2-2—Estimated Cumulative Losses to the United States From Selected Harmful
Non-Indigenous Species, 1906-1991

Category	Species analyzed (number)	Cumulative loss estimates (millions of dollars, 1991)	Species not analyzed[a] (number)
Plants[b]	15	603	—
Terrestrial vertebrates	6	225	>39
Insects	43	92,658	>330
Fish	3	467	>30
Aquatic invertebrates	3	1,207	>35
Plant pathogens	5	867	>44
Other	4	917	—
Total	79	96,944	>478

a Based on estimated numbers of known harmful species per category (figure 2-4).

b Excludes most agricultural weeds; these are covered in box 2-D.

NOTES: The estimates omit many harmful NIS for which data were unavailable. Figures for the species represented here generally cover only one year or a few years. Numerous accounting judgments were necessary to allow consistent comparison of the 96 different reports relied on; information was incomplete, inconsistent, or had other shortcomings for most of the 79 species.

SOURCE: M. Cochran, "Non-Indigenous Species in the United States: Economic Consequences," contractor report prepared for the Office of Technology Assessment, March 1992.

control costs. Past accountings generally incorporated little information on several other important effects, such as research and private control costs (8). The latter are especially significant in agriculture, where farmers bear much of the cost of control. Even outside of farming, control costs can be substantial; North Carolina homeowners spent an estimated $11 million annually to protect residential trees from the European gypsy moth (12). Accounting for nonmarket effects may be the only way to capture the full economic impacts of NIS affecting natural areas. Chapter 4 discusses such accounting difficulties and the disputed role of economics in NIS decisionmaking. Harmful NIS have numerous other health and environmental costs that are difficult to count in dollars.

HEALTH COSTS

Non-indigenous diseases of humans are beyond the scope of this assessment (figure 2-1). A number of other NIS directly affect human health, however. African honey bees and imported fire ants sting, and can also cause severe allergic reactions in sensitive people (78,90). African honey bees have in addition a propensity to sting with little provocation and repeatedly. The Bra-

zilian pepper tree (*Schinus terebinthifolius*), currently spreading throughout Florida, produces allergens that cause respiratory difficulty in many people and contact dermatitis in many more (43). Approximately half of the poisonous plants found in non-agricultural areas of eastern North America are non-indigenous (98), including foxglove (*Digitalis purpurea*) and tansy (*Tanacetum vulgare*) (101). Hybrids (*Canis lupus* x *C. familiaris*) between dogs (*Canis familiaris*) (non-indigenous) and wolves (*Canis lupus*) (indigenous), although popular as pets, are dangerous to humans (5).

Human health may also be indirectly influenced by some NIS. For example, non-indigenous aquatic weeds growing en masse provide a sheltered habitat for mosquito larvae, which spread human diseases when they mature (21). Several NIS currently in the United States are vectors for human diseases, although some of the diseases are not yet present in this country. For example, the snail *Biomphalaria*, presently in Florida and Texas, can carry the blood fluke (*Schistosoma* spp.) that causes schistosomiasis, although the populations in the United States do not yet harbor the flukes (7). The Asian tiger mosquito (*Aedes albopictus*) entered the United

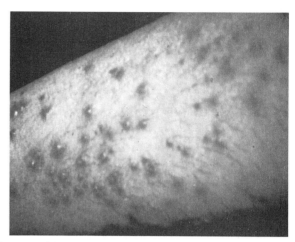

Imported fire ants (Solenopsis *spp.*) *probably reached the United States in dry ship ballast; they have negative health as well as economic effects.*

States in 1985 and is now established in 21 States (see ch. 3; box 3-A) (55). This insect can transmit several human diseases not yet present in the United States, including dengue and yellow fever, as well as a virulent form of encephalitis already present (55).

ENVIRONMENTAL COSTS

Finding:

Harmful NIS threaten indigenous species and exact a significant toll on U.S. ecosystems. Numerous declines in populations of indigenous species have been attributed to NIS, a signal of their diverse and growing impacts across the country. The worst NIS have caused species extinctions and wholesale transformations of ecosystems.

Populations of many NIS expand rapidly upon reaching new habitats where the competitors, predators, pathogens, and parasites that formerly kept them in check are no longer present. Some of these NIS become harmful by competing with, preying upon, parasitizing, killing, or transmitting diseases to indigenous species. They may also alter the physical environment, modifying or destroying habitats of indigenous species. In places, NIS that outcompete indigenous species

have, to some extent, replaced them. Abundant evidence shows declines in indigenous species resulting from NIS introductions, in some cases causing or contributing to a species' endangerment or extinction. At the worst, such processes have caused fundamental—and perhaps irreversible—changes in the functioning of U.S. ecosystems (11).

The popular press and environmentalists frequently stress the role of NIS in species extinctions (1,16,40,46). However, much of the supporting evidence is anecdotal or equivocal, in part because demonstrating the cause of an extinction after the fact is difficult. Also, NIS introductions in many cases may be just one of several factors contributing to a species' demise, and the exact role of NIS is therefore hard to evaluate (42).

Overemphasizing the significance of extinction as a consequence of NIS tends to divert attention from their other very significant and unambiguous environmental effects. Species extinctions do not have to occur for biological communities to be radically and permanently altered. Nor are extinctions necessary for the United States to experience a significant decline in the abundance, diversity, and aesthetic value of its biological resources as populations of indigenous species shrink and numbers of NIS increase.

▮ Decline of Indigenous Species

Many examples exist of declines in populations of indigenous species resulting from NIS introductions. Such declines occur across a broad array of ecosystems and as a result of diverse NIS.

Some NIS displace indigenous species by out-competing them. Throughout the American West, several non-indigenous grasses, including the widely planted crested wheatgrass, have been shown to suppress the of seedlings of oaks, pines, and other indigenous plants by reducing light, water, and nutrients (11). At least 10 indigenous plant species are less common in parts of Arizona where African lovegrass (*Eragrostis lehmanniana*) occurs (11).

Competition from the introduced house sparrow and European starling (*Sturnus vulgaris*) caused dramatic declines in the numbers of eastern bluebirds (*Sialia sialis*) and other indigenous birds (79). Presence of the mosquitofish (*Gambusia affinis*) has been associated with localized declines in populations of at least 15 indigenous fishes found in desert rivers and springs (71). The non-indigenous crayfish *Orconectes rusticus* competes with the indigenous *O. virilis* and caused its local disappearance from several Wisconsin lakes during the 1980s (38). Introduction of a periwinkle (the snail *Littorina littorea*) to U.S. shores in the late 1800s pushed the mud snail (*Ilyanassa obsoleta*) out of many near shore habitats (6).

Non-indigenous diseases, parasites, and predators have driven down populations of some indigenous species. The brown-headed cowbird (*Molothrus ater*), a bird indigenous to the eastern United States, parasitizes other birds by placing its eggs in their nests, where young cowbirds compete aggressively for food. Its range expansion following the growth of U.S. agriculture contributed to a drop in populations of migratory songbirds such as Kirtland's warbler (*Dendroica kirtlandia*) (80). Predation by non-indigenous fishes on young razorback suckers (*Xyrauchen texanus*) has contributed to its decline in the Colorado River basin (45). Introduced predatory rosy snails (*Euglandina rosea*) have been observed decimating populations of indigenous tree snails in Hawaii (25). The balsam woolly adelgid has killed almost all of the adult fir trees in Great Smoky Mountains National Park, formerly the repository of about 74 percent of all spruce-fir forest in the southern United States (35).

Some introduced NIS are not harmful themselves, but carry diseases or other organisms that harm indigenous species. Widespread concerns exist among State wildlife biologists that non-indigenous game raised on ranches can be a source of diseases affecting indigenous wild animals (36). Sika deer, for example, can harbor meningeal worms (*Parelaphostrongylos tenuis*) and numerous other parasites and pathogens that can infect wild animals and livestock. The Asian tapeworm (*Bothriocephalus opsarichthydis*) was inadvertently released in the United States via infected grass carp (*Ctenopharyngodon idella*) from China and now infects indigenous fishes in North America (22).

Some NIS are closely enough related to indigenous species to hybridize with them. Hybridization results in a loss of successful reproduction when the offspring are less viable. It can also genetically "swamp" and eliminate an indigenous species when successive generations of offspring become increasingly genetically similar to the NIS, as has occurred with certain indigenous trout in western locales (13). Hybridization with NIS can impair recovery of endangered species. An international group of experts has called for governments to prohibit or tightly restrict ownership and breeding of wolf/dog hybrids because they can interbreed with endangered wolves (52).

▮ Species Extinction

The introduction of NIS has been closely correlated with the disappearance of indigenous species in Hawaii and other islands (29,79). Some observers consider competition by non-indige-

Table 2-3—Contribution of Non-Indigenous Species to Threatened and Endangered Species Listings by the U.S. Fish and Wildlife Service[a]

	Total threatened and endangered species (number)	Category of impact on threatened and endangered species		
		Species where NIS contributed to listing (number, percent)	Species where NIS are *a* major cause of listing (number, percent)	Species where NIS are *the* major cause of listing (number, percent)
Plants	250	39(16%)	—	14 (6%)
Terrestrial vertebrates	182	47(26%)	3(2%)	19(10%)
Insects[b]	25	7(28%)	—	2 (8%)
Fish	86	44(51%)	8(9%)	5 (6%)
Invertebrates[c]	70	23(33%)	1(1%)	1 (1%)
Total	613	160	12	41

c Includes species listed through June 1991.

b Includes arachnids.

c Includes mollusks and crustaceans.

SOURCE: M. Bean, "The Role of the U.S. Department of the Interior in Non-Indigenous Species Issues," contractor report prepared for the Office of Technology Assessment, November 1991.

nous weeds and predation by non-indigenous animal pests to be the single greatest threat to Hawaii's indigenous species (60). There, introduced biological control agents have been implicated in the extinction of 15 indigenous moth species (29). Similarly, scientists believe predation by the introduced brown tree snake in Guam has caused the extinction of 5 species or subspecies of birds and the decline of numerous others (15,68).

The U.S. Fish and Wildlife Service considers NIS to have been a contributing factor in the listing of 160 species as threatened or endangered under the Endangered Species Act[6] (3). Of these, approximately one-third are from island ecosystems in Hawaii or Puerto Rico. Non-indigenous species are considered to have been the *major* cause of listing for 41 species, of which 23 are from Hawaii or Puerto Rico (table 2-3).

Direct evidence that a NIS has caused the extinction of an indigenous species in the continental United States is lacking. However, even in the continental United States, patchy environments like forest remnants, lakes, hot springs, and

artesian springs form habitat "islands." Species whose distributions are limited to such islands tend to have small localized populations and narrow ecological requirements. Consequently, they are more vulnerable to extinction than are widespread species. Effects of introductions under such conditions can mirror those on true islands. For example, the snail *Elimia comalensis* lives only in several springs and spring-fed rivers in Texas. Introduction of two non-indigenous snail species in the late 1960s has caused populations of *E. comalensis* to reach precariously low levels several times (7).

NIS clearly have caused population declines of indigenous species in mainland habitats. When other stresses such as pollution and habitat destruction adversely affect a population in concert with NIS, populations may be pushed to dangerously low levels (57). The combination of water projects and introductions of species better adapted to altered habitats is considered to be the major cause of declines in California's indigenous fishes, 76 percent of which are now declining, threatened, endangered, or extinct (58).

---

6 Endangered Species Act of 1973, as amended (7 U.S.C.A. 136, 16 U.S.C.A. 4601-9 *et. seq.*).

RON PEPLOWSKI

*Zebra mussels (*Dreissena polymorpha*), one of the most costly recent accidental imports, clog intake pipes, coat equipment, and are expected to significantly alter aquatic ecosystems.*

■ Transformation of Ecological Communities and Ecosystems

Some NIS transform ecosystems by modifying basic physical and chemical features of the environment. These NIS ''don't merely compete with or consume native species, they change the rules of the game by altering environmental conditions or resource availability'' (11). Zebra mussels, for example, rapidly filter water, decreasing the food available for other aquatic animals and increasing light penetration. This, coupled with the zebra mussel's dense, bottom-dwelling populations, is expected to cause major changes in the biological communities found within U.S. lakes, rivers, and streams—including the possible extinction of part of the rich indigenous mussel fauna in the United States (7).

The Australian melaleuca tree is rapidly modifying large areas of the Florida Everglades by changing soil characteristics and topography. Dense, pure stands of melaleuca displace indigenous vegetation and provide poorer habitat and forage for wildlife (70). Salt cedar, now abundant along the lower Colorado River, was originally introduced as an ornamental and for erosion control (61). It forms thickets along waterways, crowding out indigenous plants, banking up

sediments, and altering water flow (39). Certain non-indigenous plants, like cheatgrass in northwestern States and bunchgrass (*Schizachyrium condensatum*) and molasses grass (*Melinis minutiflora*) in Hawaii, burn easily and recover rapidly from fires, unlike indigenous plants of these areas. Where abundant, they increase the frequency of brush fires, seriously offsetting the normal ecological processes by which indigenous plant communities become established. Bunchgrass and molasses grass now comprise 80 percent of the plant cover in parts of Hawaii Volcanoes National Park (11).

Wild hogs, descended from animals that escaped from hunting enclosures in 1912, in Great Smoky Mountains National Park now eat, uproot, or trample at least 50 species of herbaceous plants and can reduce the cover of understory plants in forests by 95 percent (64). Their rooting displaces animals like voles and shrews, which depend on undisturbed leaf litter for habitat. It also increases soil erosion and the resulting turbidity of small streams. Hogs consume small animals, including potentially threatened salamanders and snails, and compete with several indigenous species for food. Aquatic equivalents of hogs are the grass (*Ctenopharyngodon idella*) and common carps (*Cyprinus carpio*), widely introduced to control aquatic weeds. These fishes indiscriminately consume aquatic vegetation, destroying habitats for young fish and increasing water turbidity (57).

Some NIS have major effects on ecosystems because they affect indigenous species that play a pivotal ecological role. Initial effects of the NIS on one species then cascade throughout the system, like a line of falling dominoes. Recent introduction of the opossum shrimp (*Mysis relicta*) into the Flathead River-Lake ecosystem of Glacier National Park caused populations of many other animals to drop. Because of feeding by the shrimp, zooplankton became less numerous. This decline, in turn, contributed to a drop in forage fish, ultimately driving away the area's fish predators—including eagles, otters, coyotes and bears (76).

Declines in indigenous plants can have important repercussions because they change the physical structure of the environment and reduce available habitat for the insects, birds, or other organisms that normally dwell in the vegetation. Chestnut blight virtually eliminated stands of the American chestnut in about 91 million hectares of eastern U.S. forests, where, in places, it previously constituted up to 25 percent of the trees (96). Loss of the American chestnut is thought to have caused at least five indigenous insect species to disappear and also to have contributed to an increase in oak wilt disease (*Ceratocystis fagacearum*) because of subsequent changes in the density and distribution of red oak (*Quercus rubra*) (41). Several vines, including kudzu and Oriental bittersweet (*Celastrus orbiculatus*), overgrow and eventually pull down trees, and have changed parts of some eastern forests from open canopies to dense thickets (51, 82). The spread of purple loosestrife to wetlands in 41 States has been called an ''ecological disaster'' (83). In some areas, it has displaced half of the previous biomass of indigenous plants—many of which are important sources of food for other species—and has further contributed to the decline of bird and turtle species by destroying their habitats (83). European leafy spurge, now widespread on U.S. rangelands, attracts few insect grazers, diminishing food supplies for insect-eating birds (4).

■ Special Consideration of NIS in the National Parks

Finding:

Increasing numbers of NIS are causing ecological disruption in the U.S. National Parks. Removal or control of NIS is not keeping pace with species' invasions and spread. Concerns are increasing that the ecological changes overtaking the parks may be so severe that they will eliminate the very characteristics for which the parks were originally established.

The conservation mandate of the U.S. National Park Service has resulted in the development of restrictive policies related to introductions of NIS. Consequently, NIS seen as beneficial in some locales are considered harmful in the National Parks. For example, rainbow trout (*Oncorhynchus mykiss*) and brown trout widely stocked for sport fisheries are being eradicated in the Great Smoky Mountains National Park because of their harmful effects on indigenous brook trout (*Salvelinus fontinalis* (10).

National Parks in all areas of the United States are experiencing problems with NIS in spite of the restrictive policies and eradication efforts (table 2-4) (27,41). A backlog of unfunded NIS control programs continues to expand (30). Increasing concern exists among scientists, environmentalists, and others that the threats from NIS in some National Parks are so severe that park ecosystems will be permanently altered if large-scale control and eradication efforts are not undertaken (43). In the Everglades Conservation Areas near Everglades National Park, the spread of melaleuca is rapidly changing the wetlands—known as a ''river of grass''—into a stand of non-indigenous trees. Unchecked, such changes eventually will eliminate the National Parks' role as a caretaker of U.S. ecosystems and indigenous species.

These concerns are not confined to National Parks. NIS threaten many State parks as well. In Missouri's Cuivre River State Park, one of the State's largest and most rustic parks, European buckthorn (*Rhamnus cathartica*) has spread widely, forming impenetrable thickets throughout the forest understory (54). A 1991 Missouri study concluded NIS are among the State's parks' 10 most serious and widespread threats (54).

RELATIONSHIP TO BIOLOGICAL DIVERSITY

The preservation of biological diversity is of growing concern among the public, Con-

Table 2-4—Examples of Non-Indigeneous Species Problems in the National Parks

Park	Impacts
Channel Islands National Park, California	Feral mammals, like the European rabbit (*Oryctolagus cuniculus*), are thought to have caused irreversible loss of topsoil by destroying vegetation and causing erosion. Introduced ice plant (*Mesembryanthemum crystallinum*) accumulates salt, changes soil salt content, and excludes indigenous vegetation.
Everglades National Park, Florida	Australian pine (*Casuarina equisetifolia*) causes development of steeper shorelines thereby impairing nesting by loggerhead sea turtles (*Caretta caretta*).
Canyonlands National Park, Utah	Salt cedar (*Tamarix* spp.) replaces indigenous vegetation, banks up sediments, reduces channel width, and increases overbank flooding. Non-indigenous grasses largely replace indigenous grasses and are thought to have increased the frequency of fire on grasslands.
Big Bend National Park, Texas	Salt cedar lowers the water table and dries up springs, contributing to the decline of desert bighorn sheep (*Ovis canadensis*).
Theodore Roosevelt Island, Washington, DC	Japanese honeysuckle (*Lonicera japonica*) and English ivy (*Hedera helix*) inhibit growth of new trees and understory plants. They also overgrow and kill adult trees.
Hawaii Volcanoes National Park, Hawaii	Non-indigenous plants (fire tree *Myrica faya* and leucaena *Leucaena leucocephala*) elevate nutrient levels on young lava flows, potentially enhancing invasion by other NIS. Non-indigenous grasses, like crested wheatgrass (*Agropyron desertorum*), increase the frequency and intensity of wildfires.

SOURCE: I.A.W. MacDonald et al., "Wildlife Conservation and the Invasion of Nature Reserves by Introduced Species: Global Perspective," *Biological Invasions: A Global Perspective*, J.A. Drake et al. (eds.) (New York, NY: John Wiley and Sons, Ltd., 1989), pp. 215-255.

gress,[7] scientists, and conservationists. Biological diversity[8] encompasses the biological variation occurring within and among species as well as among ecological communities and ecosystems. Processes that reduce this variation at any level negatively affect biological diversity. Many harmful NIS clearly impair biological diversity by causing population declines, species extinctions, or simplification of ecosystems. Moreover, the very establishment of a NIS diminishes global biological diversity: as NIS like starlings, grass carp, and crabgrass spread to more places, these places become more alike biologically.

The relationship between NIS and biological diversity is not always straightforward, however. Under certain circumstances, such as those listed below, NIS *may* actually enhance biological diversity although negative counter-examples exist for each category. The same NIS, under other circumstances, may diminish biological diversity. Thus, each situation requires careful case-by-case analysis (see ch. 4).

- **Where Indigenous Species Utilize or Depend on NIS**—Certain indigenous birds appear to reside almost exclusively in eucalyptus (*Eucalyptus* spp.)—introduced to California over 135 years ago (99). Monarch butterflies (*Danaus plexippus*) also prefer eucalyptus to the native woodlands. In Florida, heavy human use of beaches disturbs nesting by the American oystercatcher (*Haematopus palliatus*). Some achieve greater nesting success within stands of introduced Australian pine (84).

- **Where Altered Environments Are Inhospitable to Indigenous Species**—Non-indigenous fishes may be the only ones able

---

[7] For example, U.S. Congress, 102nd Congress, 1st Session, H.R. 585, proposed the National Biological Diversity Conservation and Environmental Research Act (1991).

[8] A previous OTA study defined biological diversity as "the variety and variability among living organisms and the ecological complexes in which they occur" (87).

to live in the new reservoir habitats created when rivers are dammed (69). Some introduced plants, like red bromegrass (*Bromus rubens*) in southern California, may prove to be more suited to heavily polluted areas than indigenous ones (99). In such cases, "artificial diversity" may be the only feasible option unless the underlying human disturbance is eliminated or modified.

- **Where NIS Hybridize with Certain Endangered Indigenous Species**—Only 30 to 50 individuals remain of the Florida panther (*Felis concolor coryi*), a critically endangered subspecies. Some carry genes from a Central or South American subspecies, probably from captive animals released into the Everglades decades ago (18). Commentators have argued that this should not detract from the panther's protected status under the Endangered Species Act (62). Similarly, some endangered indigenous trout species in the Southwest have heavily hybridized with introduced cutthroat (*Oncorhynchus clarki*) and rainbow trouts (13). Eradicating these hybrids could destroy the only remaining vestiges of the indigenous fish.

- **Where the NIS Itself Represents Valuable Genetic Diversity**—Feral hogs on Ossabaw Island, Georgia (*Sus scrofa domesticus*) are descendants of animals introduced by Spanish explorers in the 16th and 17th centuries. They appear to have evolved certain unique biochemical features (47). Eradication of the hogs would mean a loss of this genetic diversity.

- **Where a Species Must be Introduced at New Locales to Ensure Its Survival**—The brown tree snake, now well established in Guam, has driven the Guam rail (*Rallus owstoni*) near extinction. Introduction of the bird outside its natural range (e.g., in Hawaii) may be better than allowing it to become extinct or to survive only in captivity (9).

- **Where a NIS Removes Harvesting Pressure From Indigenous Species**—The Washington State Department of Fisheries actively promotes the shad (*Alosa sapidissima*), which was introduced decades ago, to reduce fishing pressure on the low numbers of indigenous salmon (49).

Management decisions, under circumstances like those listed above, may be controversial, even among experts seeking to maximize biological diversity. They raise legitimate concerns about whether short-term solutions (e.g., introducing pollution-tolerant plants) are acceptable or counterproductive over the long term. Although contentious cases are relatively uncommon, they sometimes command the lion's share of resources and attention. For example, "hundreds of other exotics and naturalized aliens go unattended in California parks" while much of the budget for NIS control is devoted to the controversial fight against eucalyptus (99).

CHAPTER REVIEW

This chapter is the first of two that, taken together, paint a picture of harmful NIS in the United States today. This chapter defined NIS and described the impacts that distinguish beneficial from harmful species, e.g., those that cut agricultural or other productivity, those with high control and eradication costs, and those associated with the decline of indigenous species or ecosystems. Not all NIS cause damage; nor does each have the same positive or negative impacts every place it occurs. Yet harmful NIS generate substantial economic, health, and environmental costs for the Nation—costs often uncounted in the past. With highly damaging species in virtually every State, the sketch that emerges from this chapter is worrisome.

Chapter 3 completes the picture. It traces the various pathways by which NIS enter the United States and spread from state to state and estimates the numbers of species involved.

The Changing Numbers, Causes, and Rates of Introductions | 3

Non-indigenous species (NIS) arrive by way of two general types of pathways (figure 3-1). First, species having origins outside of the United States enter the country and become established either as free-living populations or under human cultivation—for example, as pets or in agriculture, horticulture, or aquaculture. Some cultivated species subsequently escape or are released and also become established as free-living populations. Second, species already within the United States, of U.S. or foreign origin, can spread to new locales. Pathways of both types include intentional as well as unintentional species transfers.

This chapter first identifies current pathways that either are known or can be reasonably inferred to have been routes of introduction for NIS since 1980. Included are routes for both harmful and beneficial introductions; effects of NIS can change over time or as they enter new environments, and some introductions that appear benign today may eventually cause harm (ch. 2). The chapter goes on to assess the growing numbers of NIS in the United States, their geographical distribution, and the various factors affecting rates and pathways of species transfers.

PATHWAYS: HUMANS INCREASE THE MOVEMENT OF SPECIES

Finding:

Naturally occurring movements of species into the United States are uncommon. Most arrive in association with human activities or transport. Species can be brought into the country and released intentionally, or their movement and

Figure 3-1—Generic Pathways of Species Entry and Spread

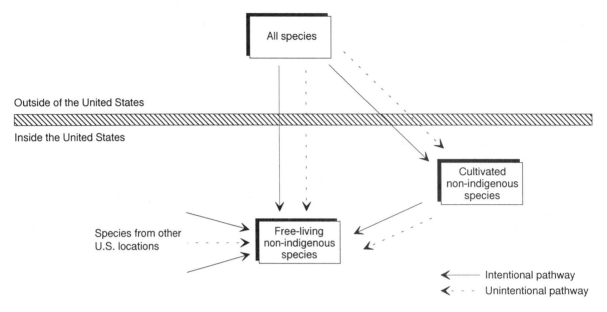

SOURCE: Office of Technology Assessment, 1993.

release can be an unintentional byproduct of cultivation, commerce, tourism, or travel. In addition, human modification of natural habitats continues to provide new opportunities for species establishment.

Geographic distributions of species naturally expand or contract. However, over historical time intervals (tens to hundreds of years), species' ranges rarely expand thousands of miles or across physical barriers like mountains or oceans (12,26,53,63,82). Such large-scale movements have become commonplace today, driven by human transformations of natural environments as well as the continual transport of people and cargo around the globe. Resulting rates of species movement dwarf natural rates in comparison.

■ The Role of Habitat Change

Habitat modification can create conditions favorable to the establishment of NIS. Soil disturbed in construction and agriculture is open for colonization by non-indigenous weeds. Non-indigenous plants, in turn, may provide habitats for the

non-indigenous insects that evolved with them. For example, European viper's bugloss (*Echium vulgare*), a weed common along roads and railroad tracks, provides a habitat for the Eurasian lace bug (*Dictyla echii*). Non-indigenous plants that would not tolerate dry conditions flourish in newly irrigated parts of arid regions, such as the American Southwest (63). Other human-generated changes in fire frequency, grazing intensity, soil stability, and nutrient levels similarly facilitate the spread and establishment of non-indigenous plants (47).

Thermal effluents from power generating stations and industrial installations create suitable environments for tropical non-indigenous fish and snails (12). Gardens as well are common habitats for non-indigenous snails and slugs (12). Pollution and habitat degradation have made some environments inhospitable to certain indigenous species. Such changes encourage fisheries managers and others to introduce NIS more tolerant of the degraded conditions (26).

When human changes to natural environments span large geographical areas, they effectively create conduits for species movement between previously isolated locales. Such modifications have an important role in facilitating the spread of NIS within the country. The rapid spread of the Russian wheat aphid (*Diuraphis noxia*) to 15 States in just 2 years following its 1986 arrival has been attributed, in part, to the prevalence of alternative host plants that are available when wheat (*Triticum* spp.) is not. Many of these are non-indigenous grasses recommended for planting on the 40 million or more acres enrolled in the U.S. Department of Agriculture's Conservation Reserve Program (54) (see also ch. 6). Many newly introduced weeds followed railway construction across the continent to the American West because they can grow in disturbed land beside the tracks (63). Roads and backcountry trails have helped to spread non-indigenous grasses within Glacier National Park, Montana (98). The 1829 construction of the Welland Canal in the Great Lakes provided a route for the sea lamprey (*Petromyson marinus*), alewife (*Alosa pseudoharengus*), and rainbow smelt (*Osmerus mordax*) to migrate upstream from Lake Ontario (26). The Asian clam (*Corbicula fluminea*) expanded its range following irrigation and drinking water canals in California and Arizona (12). The growth of agriculture, urbanization, pollution, and a host of other human habitat modifications have enhanced the movement of many species across the country.

■ Present Pathways Into the United States

More than 205 NIS were introduced or first detected in the United States since 1980. (See table 3-1 at the end of this chapter.) Fifty-nine of these are expected to cause economic or environmental harm. These NIS followed many different pathways into the country.

A number of factors confound quantitative evaluation of the relative importance of various entry pathways. Time lags often occur between

NIS establishment and detection, and tracing the pathway for a long-established species is difficult (65). One expert estimates that non-indigenous weeds usually have been in the country for 30 years or have spread to 10,000 acres before they are detected (65). In addition, Federal port inspection is a major source of information on NIS pathways, especially for agricultural pests. However, it provides data only on whether NIS enter via scrutinized routes, not on whether and how many NIS enter via as-yet-undetected pathways. Finally, some comparisons between pathways defy quantitative analysis—for example, which is more "important": the entry pathway of one very harmful NIS or one by which many less harmful NIS enter the country? For these reasons, OTA has chosen not to rank the pathways according to relative significance.

UNINTENTIONAL PATHWAYS

Many species enter the United States each year as unintentional contaminants of commodities. Agricultural produce, nursery stock, cut flowers, and timber sometimes harbor insects, plant pathogens, slugs, and snails (12,53). Of 23 non-indigenous insect species that became established in California since 1980, 20 arrived on imported plants, 2 on fruit, and 1 on infested wood (35). At least 45 percent of the snails and slugs intercepted by agricultural inspectors between 1984 and 1991 were found on plants or plant products (12). Bulk commodities like gravel, iron ore, sand, wool, and cotton (*Gossypium hirsutum*) can contain hidden weed seeds (63,106). Commodities were the single greatest source (81 percent) of noxious weed Federal interceptions from October 1987 through mid-July 1990 (106).

Weeds continue to enter the United States as seed contaminants even though the content of imported seed is regulated under the Federal Seed Act (63,106).[1] These weed seeds ultimately can be widely distributed and then planted in favorable environments along with the desired agricul-

---

[1] Federal Seed Act (1939), as amended (7 U.S.C.A. 1551 *et seq.*).

tural or other seed. For example, serrated tussock (*Nassella trichotoma*)—a noxious weed that degrades rangelands and pastures—was repeatedly found in 1988 in seed from Argentina of tall fescue (*Festuca arundinacea*) a lawn and pasture grass. Contaminated seed ultimately was distributed to at least five States and sold through such popular retailers as K-Mart, Walmart, and Ace Hardware. Over 58,000 pounds were sold before the seed was recalled in 1989 (103).

Despite Federal requirements for inspection and quarantine, plant pathogens sometimes arrive as unintended contaminants of plant materials. Importation of seeds and other plant germ plasm for propagation and breeding was a pathway for at least three plant pathogens entering the country between 1982 and 1991 (82) (table 3-1).

A number of fish and shrimp pathogens and parasites have similarly entered the country in infected stock for aquaculture or fishery enhancement (42,60). The introduction of the Pacific oyster (*Crassostrea gigas*) to the West Coast in the 1920s brought with it a Japanese snail (*Ocenebra japonica*) that preys on oysters, a flatworm (*Pseudostylochus ostreophagus*), and possibly also a copepod parasite (*Mytilicola orientalis*) (104). The Asian tapeworm (*Bothriocephalus opsarichthydis*) was found infecting several indigenous fishes in North America during the 1970s; it entered the country earlier, probably in infected grass carp (*Ctenopharyngodon idella*) (42,48).

Today, most imported freight is packed into standardized, boxcar-sized containers for ease of shipping and handling (70). Containerized freight is difficult to inspect, requiring costly unloading and reloading of the contents (61). Consequently, inspections tend to occur only when there is good cause to suspect illegal imports or contamination by pests. Decreased inspection increases the possibility that NIS will go undetected (82).

Freight containers can sit idle at ports for weeks or longer before loading, during which time organisms can board and become hidden (12,63). Also, containers generally are not cleaned be-

tween shipments (70). Containerized freight is thus thought to be a significant pathway for the entry of insects, weed seed, slugs, and snails into the country (12,53,63). Containerized shipments of used tires were the source for introductions of the Asian tiger mosquito (*Aedes albopictus*) from 1985 to 1988, until new U.S. Public Health Service regulations required tires to be mosquito free (30) (box 3-A). At least 15 percent of the snails and slugs intercepted by Federal agriculture inspectors between 1984 and 1991 were in freight containers (12). Since containers frequently are not unloaded until they reach their inland destinations, any species they contain are released within the country rather than at a port of entry. This reverses the historical pattern wherein species generally first appeared at ports of entry (53).

Crates were the source of at least 11 percent of the mollusks intercepted by Federal inspectors from 1984 to mid-1991 (12). The crating and packing material itself poses additional risks. A threatening new bark beetle (*Tomicus piniperda*), discovered near Cleveland, Ohio in 1992, is believed to have entered the country in ship's dunnage (wood packing material) (78). Packing material used to ship dishes from Greece is suspected to have been the pathway for the new weed early millet (*Milium vernale*) (65). Unprocessed wood and wood products have been a source of forest pests and pathogens in the past (11); current concerns center on their potential to convey pests from Asia to forests in the Pacific Northwest (101) (see also box 4-B). Wooden crates carrying oysters have been suggested, although not proven, as a possible source of wood-boring aquatic animals as well (19).

Some NIS stow away on cars and other conveyances. This is thought to be a pathway by which weed seeds spread, including across national borders from Mexico and Canada into the United States (63). Noxious weed seeds have been intercepted in aircraft, automobiles, railway cars, ships, tractor trailers, and other vehicles entering the country (106). The Asian gypsy moth (*Lymantria dispar*), a new strain of this destruc-

Box 3-A—The Unwelcome Arrival of the Asian Tiger Mosquito

On August 2, 1985, the Asian tiger mosquito (*Aedes albopictus*) was discovered in Houston, apparently imported in containerized shipments of used tires. An aggressive biter and prolific breeder, this species is a vector of several serious viral diseases such as dengue fever, LaCrosse encephalitis, and eastern equine encephalitis. The last has a 30 percent mortality rate in humans. As of 1991 the mosquito had been found in 22 States. Experts predict that rapid evolution of cold-tolerant and heat-tolerant strains may eventually allow the mosquito to occupy an even broader range. The mosquito thrives in used tires—it breeds in the small, protected pools of water often found inside. Unfortunately, more than 2 *billion* scrap tires are now piled up in the country, usually close to large population centers, with 250 million more tires added each year.

Official response was slow and inadequate to stop the mosquito. Not until 1988 did the Centers for Disease Control and Prevention (CDC) of the Public Health Service impose regulations requiring that used tire imports be dry and free of mosquito eggs or larvae. According to one expert, inspection to ensure compliance with the regulations is minimal. Further, in early 1987, CDC rejected the recommendation of its own expert panel to develop a $20 million research and control plan, citing fiscal constraints. The American Mosquito Control Association officially censured CDC's rejection of the control plan.

Although CDC has done significant research, formulating responses has been largely left to State and local governments. Their uncoordinated, uneven control efforts have been no match for the problem. Meanwhile, at a major Florida tire dump nine miles from Disney World, scientists recently isolated eastern equine encephalitis from the Asian tiger mosquito for the first time since the mosquito was discovered in the country.

SOURCES: G. Craig, Professor of Biology, University of Notre Dame, letter to P.T. Jenkins, Office of Technology Assessment, March 14, 1992; R.B. Craven et al., "Importation of *Aedes albopictus* and Other Exotic Mosquito Species Into the United States in Used Tires from Asia," *Journal of the American Mosquito Control Association*, vol. 4, No. 2, 1988, pp. 138-142; C.J. Mitchell et al., "Isolation of Eastern Equine Encephalitis Virus From *Aedes albopictus* in Florida," *Science*, vol. 257, July 24, 1992, pp. 526-527.

tive forest pest, is thought to have recently found its way to the Pacific northwest on grain ships (31). Cargo in planes and trucks are important pathways for insects entering the country (53).

Military freight enters the United States continuously, periodically in high volume. The geographic origin depends on the location of recent military action. Equipment and supplies sometimes are covered with dirt or mud from the field (5). These can be an unintended source of insects and plant pathogens if not properly washed. Military cargo and equipment historically has resulted in several introductions of harmful species, like the golden nematode (*Globodera rostochiensis*). This process was vividly demonstrated in the spread of the brown tree snake (*Boiga irregularis*) across islands of the Pacific by military cargo planes after World War II (41) (see also box 8-B). In 1992, concerns again surfaced that military transport of equipment might pro-

vide a pathway for non-indigenous pests—this time from Operation Desert Storm in the Middle East (5).

Establishment of the harmful zebra mussel (*Dreissena polymorpha*) in the Great Lakes during the 1980s focused attention on ballast water as an unintentional pathway by which aquatic species enter the country. Ballast water is taken on by large cargo ships when they are empty to provide stability at sea. It is then dumped when the ship is loaded at a different port. If environments at the two ports are similar, species taken up in the water at one may become established at the other. Since 1980, at least eight new NIS entered U.S. waters by way of dumped ballast water (71) (table 3-1). These include the potentially damaging European ruffe (*Gymnocephalus cernuus*) and two non-indigenous clams newly established in California bays (*Theora lubrica* and *Potamocorbula amurensis*) (12,21). The potential for species transfers by ballast water is great; at least 367 distinctly identifiable taxonomic groups of plants and animals have been found in the ballast water of ships arriving in Oregon from Japan (22).

INTENTIONAL PATHWAYS

Large amounts of plant germ plasm arrive annually for use in the breeding and development of plants for agriculture, horticulture, and soil conservation. Plants for pasture and range improvement and wildlife forage may be directly planted in uncultivated areas. Some notable pests have been introduced in the past for soil conservation including kudzu (*Pueraria lobata*) and multiflora rose (*Rosa multiflora*). Scotch broom (*Cytisus scoparius*) was introduced to California as an ornamental plant, and also used by the U.S. Soil Conservation Service for preventing erosion. It now has spread to at least 500,000 acres in the State, where it displaces indigenous flora and fauna and is a serious weed of tree plantations (10). Concerns continue today regarding the pest potential of new species deliberately released for preventing erosion (84).

Although most plant introductions are legal, some do occur illegally. Often these involve species for ornamental horticulture smuggled into Hawaii (63). Some seeds are sent to plant breeders in the United States through international first-class mail to avoid inspection or quarantine at the port of entry (8). Baggage accompanying individuals visiting or returning to the United States is a common pathway for the illegal transport of NIS. At least 82 percent of the plants or seeds of noxious weeds intercepted at U.S. ports of entry between October 1987 and mid-July 1990 occurred in baggage (106). The ultimate fate of organisms entering in baggage is unknown, but it is likely some have been grown or otherwise released by their owners. For example, Asian water spinach (*Ipomoea aquatica*) is a Federal noxious weed and a prohibited aquatic weed under Florida State regulations. Yet, from 1979 through 1990, Florida State officials recorded 20 cases of illegal possession of seeds or deliberate plantings (83).

Intentional importation and release for biological control of pests has been a source of non-indigenous insects, snails, fish, plant pathogens, and nematodes (12,26,53,82). Estimates are that a total of 722 non-indigenous insect species have been purposely introduced in the United States for biological control of pests. Of these, 237 have become established (44). Since 1980, at least 6 insect species have been newly introduced in the country for biological control (table 3-1). Insects also have been purposely released for plant pollination; researchers from the U.S. Agricultural Research Service working in California released several thousand mason bees (*Osmia cornuta*) from Spain in experimental tests from 1976 to 1984 (96).

During the late 1980s, two plant pathogens were introduced for biological control: a nematode (*Subanguina picridis*) from Russia to control Russian knapweed (*Centaurea repens*) and a rust fungus (*Puccinia carduorum*) from Turkey to control musk thistle (*Carduus nutans*) (82). Two illegal introductions of plant pathogens in 1990 were a smut fungus (*Ustilago esculenta*), which is grown on Manchurian rice (*Zizania latifolia*) to produce edible galls, and the chrysanthemum white rust (*Puccinia horiana*), which is used by hobbyists to produce unusual flowers (82). In both cases of illegal introduction the infected plants were located by authorities and subsequently destroyed (82).

Although generally less common today than in the past, State wildlife managers continue to import and release non-indigenous birds for game hunting. Between 1985 and 1988 the State of Michigan imported 3,600 eggs of the Sichuan pheasant from China—a subspecies of the already established ring-necked pheasant (*Phasianus colchicus*) (97). Like its predecessor, the bird is expected to cause few problems; nevertheless, the Sichuan's broad habitat range and "unbelievable adaptability" (97) suggest its introduction should be carefully evaluated.

USDA

The advent of containerized freight allows direct introduction of harmful non-indigenous species throughout the country—instead of just at U.S. ports of entry.

Intentional introductions of fishes from abroad also are less common today, but continue still. The State of Texas tried unsuccessfully to introduce the Nile perch (*Lates niloticus*) and bigeye lates (*Lates mariae*) in 1979 and 1983, respectively (26). North Dakota recently proposed to introduce the European zander (*Stizostedion lucioperca*), which critics feared might transmit diseases to or hybridize with indigenous fish like the walleye (*S. vitreum*) (28).

Some non-indigenous clams and oysters have been intentionally imported and released for commercial exploitation (12). Among these is the Pacific oyster, imported from Japan, which now is successfully grown and harvested in West Coast bays from Washington to California (46). Recent proposals to transfer the Pacific oyster to the East Coast have been controversial, and the introduction has not occurred thus far (see ch. 7).

ESCAPE OR RELEASE FROM CONFINEMENT

Species imported to be held in captivity sometimes subsequently escape or are released. Often, determining which of the two has occurred is difficult (i.e., whether the introduction is intentional or accidental). For example, the source of bighead carp (*Hypophthalmichthys nobilis*) re-

cently established in Mississippi is unclear. Some contend it escaped from aquaculture facilities, while others believe it was illegally released in order to establish free-living populations (27).

Many plants and seeds of foreign origin are directly marketed in the United States, especially for ornamental horticulture. Quarantine of imported species primarily guards against unintentional importation of insects, pathogens, and other pests, rather than the noxious qualities of the plant itself. Thus, specialized nurseries can offer ''ivies of the world'' (7), even though English ivy (*Hedera helix*) is known to cause ecological damage in deciduous forests of the eastern United States.

Significant numbers of non-indigenous plants have escaped from human cultivation. Among the 300 weed species of the western United States, at least 28 escaped from horticulture and 8 from agriculture (107). Baby's breath (*Gysophila elegans*), foxglove (*Digitalis purpurea*), and creeping bellflower (*Campanula rapunculoides*) all are horticultural species that become weeds outside of gardens (107). Some 300 established non-indigenous plant species in California are escapees from ornamental horticulture (68). These include a number of invasive weeds of native vegetation, such as European gorse (*Ulex europaeus*), Andean pampas grass (*Cortaderia jubata*), and Scotch broom (68). A new addition is oleander (*Nerium oleander*), now well established along the Sacramento River and in the northern Central Valley (14). The edible fig (*Ficus carica*), has recently escaped from agriculture and become established in some riparian woodlands (14).

Several NIS imported for medical diagnostic or research purposes have escaped in the past. The recent spread of African honey bees (*Apis mellifera scutellata*) to the United States was set in motion by escape of bees from a research facility in Brazil in 1957 (52). The giant tiger shrimp (*Penaeus monodon*), originally from the Indo-Pacific, escaped into South Carolina's coastal waters from the Waddell Research Facility in

1988 (19). The African clawed frog (*Xenopus laevis*) was originally imported in the 1930s for use in diagnostic pregnancy tests, but had established free-living populations in California by 1969 (69). The Asian Amur maple (*Acer ginnala*)—a potential weed of midwestern natural areas—has now become common in woods and fields surrounding the Lincoln, Missouri, plant testing center of the U.S. Soil Conservation Service, from where it apparently escaped (36).

A different kind of research introduction involved peanut (*Arachis hypogaea*) germ plasm imported from China in 1978 that was unknowingly contaminated with the peanut stripe virus (82). In 1983, the virus was found in peanut breeding lines at university experimental farms from Texas to Virginia to Florida—it had inadvertently been introduced by distribution of the diseased germ plasm to numerous researchers.

Throughout a number of States, ranchers have introduced non-indigenous, big-game animals onto private lands for ranching, to enhance hunting opportunities, or for other purposes. The more than 450 members of the Exotic Wildlife Association combined own an estimated 200,000 head of some 125 NIS (92). Many of the game animals are held in fenced enclosures, but some eventually escape. Indeed, a committee from the State of Wyoming considers such escapes "inevitable" (57). Texas has the highest numbers of non-indigenous big-game animals; in 1989 the State was home to 164,257 free-ranging animals of 123 species (94). The State government, however, treats these animals as livestock and not as wildlife (94).

About 23 percent of the vertebrate species of foreign origin that currently live in the wild were originally imported as cage birds or other wildlife pets (95). Given the high U.S. rates of pet imports—estimated to be hundreds of thousands to millions of wild birds, aquarium fish, and reptiles annually (33,59)—the potential for pet escapes and releases is great. Illegal imports further expand the total numbers and types of organisms brought into the country. In one recent

*Snails commonly enter the United States unintentionally on plants or agricultural produce but the African giant snail (*Achatina fulica*) was smuggled into the country and sold in Florida and Virginia pet stores.*

example, perhaps as many as hundreds of fist-sized African giant snails (*Achatina fulica*) were smuggled into the country from Nigeria and sold in Florida and Virginia pet stores (3,4).

The Massachusetts Division of Fisheries and Wildlife recently summarized the frequent reports of pet escapes in that State (16). Escaped or recovered pets in that State from 1988 through 1992 included: a 20-pound crocodile (*Caiman crocodilus*); three Boa constrictors (*Boa constrictor*); a Nile monitor lizard (*Varanus niloticus*); several hundred birds (various species of cockatoos, cockatiels, parrots, parakeets, and macaws); three wallabies; a bobcat from Texas (*Felis rufus*); and nine European fallow deer (*Dama dama*). Escaped monk and black-hooded parakeets (*Myiopsitta monachus* and *Nandayus nenday*) are known to have established free-living populations in the northeast (16). More anecdotal accounts of escaped pets generally are common in the popular press (2).

Fish and aquatic invertebrates such as shrimp frequently escape from confinement. The peacock cichlid (*Cichla ocellaris*) was intentionally stocked in Florida's warm water canals during the mid-1980s. It subsequently escaped (110), despite detailed analysis by the State before stocking that concluded the fish would remain limited to the canals (81).

The aquarium trade remains a significant pathway by which snails enter the United States. During the past few decades at least three snail species entered U.S. waters when they were discarded by aquarium dealers or their customers (12). Some plants also are distributed for use in aquaria. Hydrilla (*Hydrilla verticillata*), an aquatic weed that causes a significant navigation hazard and ecological harm, first entered U.S. waters sometime after 1956, it is thought, when it was released by aquarium dealers to create a domestic source (111). Release from aquaria was the source of at least 7 non-indigenous fish species that have become established since 1980 (27). Some were found in remote natural areas, like the green swordtail (*Xiphophorus helleri*) and zebra danio (*Brachydanio rerio*), which were discovered in the 1980s living in warm springs of Grand Teton National Forest (26). The aquarium fish trade is thought to be the source of at least 27 non-indigenous fish species now established in the continental United States (29).

Pessimism about the ability to keep aquaculture species confined is so great that, according to some, including the Federal interagency Aquatic Nuisance Species Task Force, species maintained for this purpose are virtually guaranteed of eventually escaping to the wild (26,89,99). Potentially free-living non-indigenous shrimp are grown in at least four coastal States (79), and the commonly cultured Pacific white shrimp (*Penaeus vannamei*) was captured in 1991 off the coast of South Carolina (1). Escape from aquaculture facilities is thought to have been a major source of the many tropical aquarium species now found in Florida's waters (29).

If an NIS imported into confinement harbors any other species, these also may eventually escape. Escape from a fish aquaculture facility is thought to have been the source of the freshwater snail (*Potamopyrgus antipodarum*) found in the Snake River in 1987 and now threatening indige-

nous mollusks in the region (12). Numerous fish pathogens and parasites have accompanied introductions for aquaculture and fishery enhancement (42). Five non-indigenous shrimp viruses entered the United States in contaminated shrimp stock and have become widely distributed in the aquaculture industry (60). Fish imported into the aquarium trade commonly harbor parasites. One 1984 study of hundreds of fish shipped from southeast Asia and South America found infestation rates of from 61 to 98 percent (90). Whether and how many pathogens and parasites have escaped from aquaculture facilities or aquaria is unknown.

■ Present Pathways of Spread Within the United States

Many NIS have continued to spread within the United States long after they entered and became established, sometimes even after the pathway by which they entered the country was closed. This is true for European gypsy moth (*Lymantria dispar*) and purple loosestrife (*Lythrum salicaria*), which continue to spread and cause harm at new locations (figure 3-2). For such species, the means of transport *within* the country is more important from a management or regulatory perspective than how they originally entered. Pathways of species movement within the country also are significant for U.S. species that have been transported beyond their natural ranges.

However, there is relatively little quantitative information about the pathways and rates of species movement within the country. Systematic reporting of regional species transfers is virtually non-existent. In part this results from a definitional inconsistency. Many resource managers do not consider U.S. species moved outside of their natural ranges to be non-indigenous. In some cases, particularly in fisheries management, a distinction is made between ''exotic'' species (i.e., non-indigenous to the United States) and ''transplanted'' ones (i.e., species indigenous to the United States but moved beyond their natural

Figure 3-2—State by State Spread of Four Harmful Non-Indigenous Species

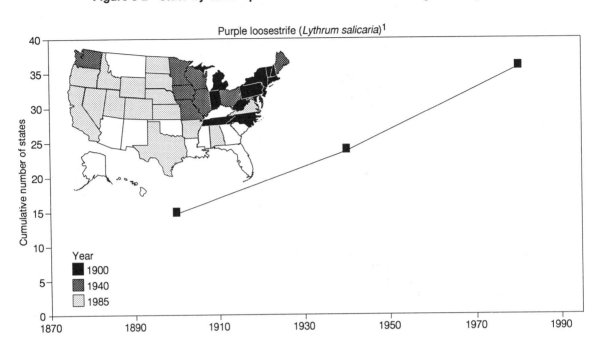

Purple loosestrife (*Lythrum salicaria*)[1]

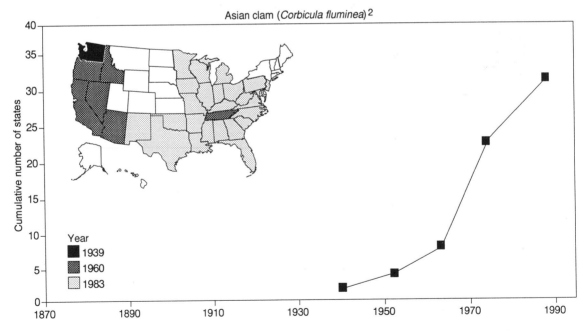

Asian clam (*Corbicula fluminea*)[2]

SOURCES:

1. D.Q. Thompson, R.L. Stuckey, and E.B. Thompson, "Spread, Impact, and Control of Purple Loosestrife (*Lythrum salicaria*) in North American Wetlands (Washington, DC: U.S. Department of the Interior, Fish and Wildlife Service, 1987).
2. C.L. Counts, III, "The Zoogeography and History of the Invasion of the United States by *Corbicula fluminea* (Bivalvia: Corbiculidae), *American Malacological Bulletin*, Special Edition No. 2, 1986, pp. 7-39.

Figure 3-2—State by State Spread of Four Harmful Non-Indigenous Species—Continued

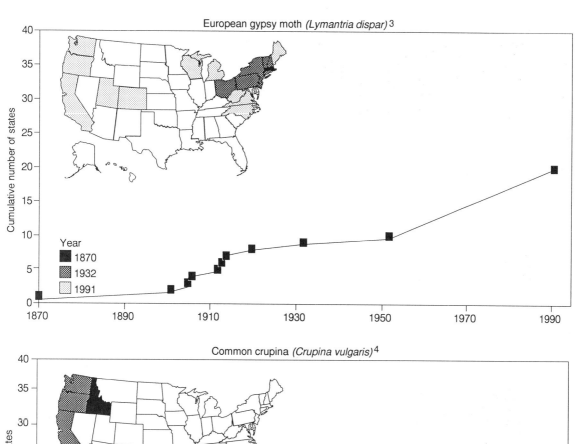

European gypsy moth *(Lymantria dispar)* [3]

Year
- 1870
- 1932
- 1991

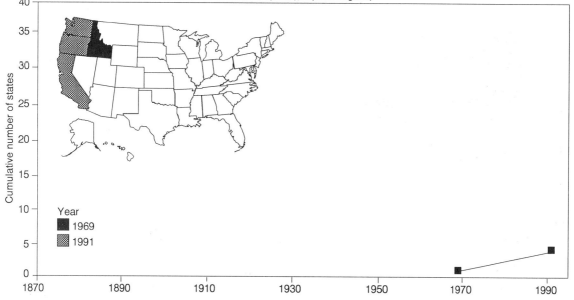

Common crupina *(Crupina vulgaris)* [4]

Year
- 1969
- 1991

SOURCES:

3. P.W. Schaefer and R.W. Fuester, "Gypsy Moths: Thwarting Their Wandering Ways," *Agricultural Research*, May 1991, pp. 4-11; M.L. McManus and T. McIntyre, "Introduction," *The Gypsy Moth: Research Toward Integrated Pest Management*, C.C. Doane and M.L. McManus (eds.) (Washington, DC: U.S. Forest Service, Technical Bulletin no. 1584, 1981), pp. 1-8; T. Eiber, "Enhancement of Gypsy Moth Management, Detection, and Delay Strategies," *Gypsy Moth News*, No. 26, June 1991, pp. 2-5.

4. T.S. Prather et al., "Common Crupina: Biology, Management, and Eradication," University of Idaho, Agricultural Experiment Station, Current Information Series No. 880, 1991.

ranges) (66). Introduction dates are largely unrecorded for most transplanted fish (26). Systematic reporting also is lacking for continued restocking of NIS already established in an area or of new introductions of NIS in common use elsewhere in the United States. Several generalizations can be made despite these limitations.

UNASSISTED SPREAD

Once established, some NIS of foreign origin disperse even in the absence of human activities. Few geographic barriers block the transcontinental expansion of some NIS, like the African honey bee and Asian tiger mosquito. The American elm bark beetle (*Hylurgopinus rufipes*) can be a vector of Dutch elm disease (*Ceratocystis ulmi*) (56). Plants like the Brazilian pepper tree (*Schinus terebinthifolius*) in Florida have been spread by wildlife that consume the tree's seeds (111). The range of certain fish parasites has expanded as infected fish have migrated within and between watersheds (42).

Natural disasters provide new opportunities for the establishment of certain NIS. The 1992 passage of Hurricane Andrew through Florida knocked down indigenous trees, spurring the growth of non-indigenous vines in some natural areas; State officials fear this "window of opportunity" may result in permanent domination of certain indigenous plant communities by NIS (45). A similar situation exists in Hawaii, where Hurricane Iniki in 1992 cleared the way for expansion of several harmful plants like banana poka (*Passiflora mollissima*) (37). A recent aquatic example is the explosive population growth by an Asian clam (*Potamocorbula amurensis*) in San Francisco Bay following a major flood that eliminated other species more vulnerable to reduced salinity (75).

UNINTENTIONAL AND INTENTIONAL PATHWAYS

In contrast to these unassisted types of spread, a significant number of NIS expand throughout the United States via pathways associated with human activities. Some of these are the same

pathways that bring new species into the country, like ballast water (71). Others are unique to the domestic movement of species, such as the releases of non-indigenous bait animals like the sheepshead minnow (*Cyprinodon variegatus*) and the Asian clam (12,26).

A number of these domestic pathways are linked to national distribution systems that enable a NIS to become widely disseminated and introduced many times throughout the country. Such multiple introductions speed NIS dispersal and have significant consequences for the choice of appropriate management strategies (see ch. 5).

Species that are sold commercially, for example, have great potential to be transported throughout a broad geographic area. Commercial distribution in the 19th century seed trade aided the spread of at least 28 non-indigenous weeds, including several of the nation's worst weeds, like Johnson grass (*Sorghum halepense*), salt cedar (*Tamarix africana* and *T. gallica*), water hyacinth (*Eichhornia* spp.), and kudzu (62,64). Sales of harmful non-indigenous plants continue today. At least six non-indigenous plant species on the Federal noxious weed list—hydrilla, for example—were sold in interstate commerce in 1990 (105). Of Illinois's 35 weeds of natural areas, 21 are legally sold in the nursery trade throughout the State (85). Seed of both federally and State-listed noxious weeds—e.g., animated oats (*Avena sterilis*) and dyer's woad (*Isatis tinctoria*)—currently can be bought at retail stores in Washington State (65).

Species recommended for specific applications can become widely distributed. Various agencies and organizations currently recommend a number of invasive plants. At least seven cultivars released by the U.S. Soil Conservation Service since 1980 are potentially invasive, according to one weed expert (65). Other examples of recommended species include: autumn olive (*Elaeagnus umbellata*), a plant that displaces indigenous vegetation in natural areas of the Midwest, by the Army Corps of Engineers; sawtooth oak (*Quercus serrata*), an Asian tree currently invading southeastern forests, by the South Carolina Department of Fish and Game; and leuceana (*Leucaena leucocephala*), a rapidly growing tree from Central America that invades disturbed lowlands in Hawaii, by the Arbor Day Foundation (77).

Current popular interest in "wildflowers" for ornamental uses and "native grasses" for livestock and wildlife forage (86) may inadvertently be fueling widespread planting of NIS in natural and semi-natural areas. In one 1992 "wildflower" seed catalog, only about 60 percent of the listed species were indigenous, and at least 80 percent of the NIS listed have escaped cultivation in the United States—plants like cornflower (*Centaurea cyanus*), crimson clover (*Trifolium incarnatum*), and dame's rocket (*Hesperis matronalis*), all originally from Europe (109). Plants marketed as "native grasses" in seed catalogs sometimes are non-indigenous and may even be known to be potentially invasive, like Bermuda grass (*Cynodon dactylon*), Russian wild rye (*Psathyrostachys junceus*), and Japanese millet (*Echinochloa crus-galli* var. *frumentacea*) (65,87,108).

Non-indigenous plants, including both those sold in the horticultural trade and known weeds, find their way into natural areas through various pathways. Rock Creek National Park in the District of Columbia now has 33 invasive NIS, some of which spread from adjacent gardens or landscape plantings; rooted from discarded yard refuse; entered as seed in topsoil, root balls, riprap, and lawn-legume mixtures; or were carried in by animals (39). Garlic mustard (*Alliaria petiolata*), a weed of natural areas, was first recorded in Illinois in 1918. It has since spread throughout 42 counties in the State, carried by flood waters; automobiles; trains; mowers; and the boots, clothes, and hair of hikers (76).

Numerous highly damaging weeds, such as cheatgrass (*Bromus tectorum*) and spotted knapweed (*Centauraea maculosa*), were spread as contaminants of agricultural seed before the enactment of seed purity laws early in this century

(9). The extent to which contamination of seed currently not covered by these laws, such as flower seed, is a pathway for harmful NIS is unknown.

Shipments of live plants can also inadvertently harbor NIS. A 1989 survey found that cabbage (*Brassica oleracea*) seedlings transported to New York from Georgia, Maryland, and Florida were infested with an average of up to eight larvae of the diamondback moth (*Plutella xylostella*) per hundred plants (88). A tree frog (*Hyla cinerea*), an anole (*Anolis* spp.), and a scarlet kingsnake (*Lampropeltis triangulum elapsoides*) were some of the finds in recent plant shipments to Massachusetts (16). The high volume of traffic in nursery stock and landscaping plants is thought to play an important role in moving non-indigenous insects throughout the United States (53). Between 1989 and 1992, three of the six non-indigenous insect species from elsewhere in the United States that became established in California arrived on plants (35).

Inadvertent transfers of animals can occur when plants are transplanted for restoration or wildlife enhancement. In 1957, shoal grass (*Diplanthera wightii*) was shipped from Texas to the California Salton Sea to provide waterfowl forage. The plants carried a number of aquatic invertebrates (like the crustaceans *Gammarus mucronatus* and *Corophium louisianum*), which subsequently became established there (19).

Agricultural produce shipped interstate sometimes harbors non-indigenous pests. This is the basis for many of the U.S. Department of Agriculture's domestic quarantines.[2] Some of the costly infestations of Mediterranean fruit flies (*Ceratitis capitata*) in California might have originated in tropical produce sent via first-class mail from Hawaii (91). A recent cooperative warrant system for inspection of first-class mail between Hawaii and the mainland has reduced such pest transfers, although not in other areas of the country.

---

[2] 7 CFR 301.

U.S. FISH AND WILDLIFE SERVICE

States frequently stock non-indigenous fish to enhance sport fisheries, and this has been an important pathway for the entry and spread of non-indigenous species historically.

Various animals are available through the mail for wildlife enhancement nationwide, including water fleas (*Daphnia* spp.), freshwater shrimp, crayfish, fresh water clams, turtles, and bull frogs (108); whether these species are non-indigenous in some regions where they are marketed is impossible to determine, since species names are not always listed. The 1989 ''Buyer's Guide'' in *Aquaculture Magazine* lists 82 species of freshwater and marine fish, invertebrates, and algae available for sale in the United States (20). Sales of the European fish the rudd (*Scardinius erythrophthalmus*) for use as bait eventually resulted in its capture in eight States (13).

Interstate shipments of fish and wildlife sometimes harbor NIS other than the intended species. Reported incidents include inadvertent introductions to California of the Texas big-scale logperch (*Percina macrolepida*) and rainwater killifish (*Lucania parva*) from New Mexico with shipments of largemouth bass (*Micropterus salmoides*) (73). The distribution of the stickleback (*Gasteosteus aculeatus*) in regions of Southern California where it is non-indigenous may be due to its unintended presence in trout stocks used to enhance sport fisheries (73). Fish shipped interstate sometimes carry larvae of freshwater mussels (*Anodonta* spp.) (93). Containers of the Pacific oyster from California to the East Coast in 1979 contained numerous stowaway mussels, worms, and crustaceans (19). A fish parasite, the Asian copepod *Argulus japonicus*, is thought to have spread throughout the country via the aquarium trade (71).

Indigenous and non-indigenous insects, snails, and fish have been transferred within the United States for biological control (12,53). Since the 1970s, the non-indigenous snail *Rumina decollata* has been raised, sold, and distributed throughout an estimated 50,000 acres of citrus groves in California as a biological control for non-indigenous snail pests (38). The grass carp, originally from Asia, has been widely propagated and sold for biological control of aquatic weeds (26).

Although largely unmonitored today, interstate shipments of biological control agents are a potential source of insect pathogens and parasites; according to an expert on the species, the wasp *Perilitus coccinellae*, a parasite of the indigenous convergent lady beetle (*Hippodamia convergens*) already is spread in this manner (51). In international transit, by contrast, such pests would probably be intercepted through inspection and quarantine.

Interstate transfers of honey bee (*Apis mellifera*) colonies inadvertently facilitated the rapid spread of honey bee parasites (varroa mites—*Varroa jacobsoni*—and tracheal mites—*Acarapis woodi*) (74). According to a 1982 survey, about a quarter of all commercially operated colonies (500,000) are moved south each winter to prevent losses from the cold, and about 2 million colonies are rented each year for pollination. The result is large-scale movements of colonies throughout the country that helped spread the damaging varroa mite to 30 States in just 4 years following its 1987 detection in Florida and Wisconsin (74). The honey bee industry has concerns that such interstate transfers may similarly enable rapid spread of the African honey bee which recently arrived in Texas (74).

Researchers working on NIS have been the source of several introductions throughout the country. The rapid spread of the Asian clam, a serious fouler of pipes in power plants, is thought to have been assisted by inadvertent research releases (25). The California sea squirt (*Botrylloides diegense*, a marine animal) was released by a scientist at Woods Hole, Massachusetts, in 1972 and is now an abundant fouler of rocks, piers, and boat hulls throughout southern New England (19). Plant breeders regularly trade germ plasm for breeding purposes—some from potentially invasive species. One reported having acquired the salt- and drought-tolerant ruby salt bush (*Enchylaeua tomentosa*), originally from Australia "from a nursery in Tucson who got it from Soil Conservation Service, who decided not to officially release it since it was such a potential pest, which it is" (8).

Even shipments of inanimate objects and vehicles can harbor NIS. The European gypsy moth can travel long distances clinging to household articles, lawn furniture, firewood, lawn mowers, and recreational vehicles such as motor homes, campers, and boats (32). Since 1984, California border inspectors have intercepted imported fire ants (*Solenopsis invicta* and *S. ritcheri*) along State lines, in decreasing order of frequency, in nonagricultural shipments (e.g., pallets, roofing materials, carpets); empty trucks; agricultural shipments; automobiles; U-Hauls; and nursery stock (58). At least 3,000 Japanese beetles (*Popillia japonica*) were found in cargo planes landing at Ontario, California, from the eastern United States in 1986 (34). The Asian cockroach (*Blattella asahinai*) has spread in Florida mainly by hitching rides on cars leaving infested areas (72). The tiny Argentine ant (*Iridomyrmex humilis*)—an inadvertent 1906 introduction to New Orleans—has dispersed widely by way of the dirt on truck mud flaps, among other means (23).

Dumped ballast water, known to be a significant pathway for harmful introductions from

USDA

*Several harmful non-indigenous species have hitchhiked into the country with returning military equipment, e.g., the brown tree snake (*Boiga irregularis*), witchweed (*Striga asiatica*), and the golden nematode (*Globodera rostochiensis*). Similarly, motor homes, automobiles, and boats help spread harmful NIS within the United States.*

abroad, has also provided a means for species spread within the country. Since 1980, at least three NIS entered the Great Lakes from other U.S. locales in ballast water: the four-spine stickleback fish (*Apeltes quadracus*), an aquatic worm (*Ripistes parasitica*), and a green alga (*Nitellopsis obtusa*) (71). In the absence of effective control or containment, the ruffe—a harmful Eurasian fish (see ch. 2)—is expected to spread via ships' ballast and other means perhaps are far as the Ohio, Mississippi, and Missouri River drainage basins (43).

HOW MANY NON-INDIGENOUS SPECIES ARE THERE?

Finding:

Estimated numbers of NIS in the United States increased over the past 100 years for all groups of organisms OTA examined. At least several thousand non-indigenous insect and plant species occur in this country, as do several hundred non-indigenous vertebrate, mollusk, fish, and plant pathogen species.

Table 3-2—Estimated Numbers of Non-Indigenous Species in the United States[a]

Species with origins outside of the United States		
Category	Number	Percentage of total species in the United States in category
Plants	>2,000	—[b]
Terrestrial vertebrates	142	≈6%
Insects and arachnids	>2,000	≈2%
Fish	70	≈8%[c]
Mollusks (non-marine)	91	≈4%
Plant pathogens	239	—[b]
Total	4,542	

Species of U.S. origin introduced beyond their natural ranges		
Category	Number	Percentage of total species in the United States in category
Plants	—[b]	—[b]
Terrestrial vertebrates	51	≈2%
Insects and arachnids	—[b]	—[b]
Fish	57	≈17%[c]
Mollusks (non-marine)	—[b]	—[b]
Plant pathogens	—[b]	—[b]

[a] Numbers should be considered minimum estimates. Experts believe many more NIS are established in the country, but have not yet been detected.

[b] Number or proportion unknown.

[c] Percentage for fish is the calculated average percentage for several regions. Percentages for all other categories are calculated as the percent of the total U.S. flora or fauna in that category.

SOURCES: Summarized by the Office of Technology Assessment from: J.C. Britton, "Pathways and Consequences of the Introduction of Non-Indigenous Freshwater, Terrestrial, and Estuarine Mollusks in the United States," contractor report prepared for the Office of Technology Assessment, October 1991; W.R. Courtenay, Jr., "Pathways and Consequences of the Introduction of Non-Indigenous Fishes in the United States," contractor report prepared for the Office of Technology Assessment, September 1991; K.C. Kim and A.G. Wheeler, "Pathways and Consequences of the Introduction of Non-Indigenous Insects and Arachnids in the United States," contractor report prepared for the Office of Technology Assessment, December 1991; R.N. Mack, "Pathways and Consequences of the Introduction of Non-Indigenous Plants in the United States," contractor report prepared for the Office of Technology Assessment, September 1991; C.L. Schoulties, "Pathways and Consequences of the Introduction of Non-Indigenous Plant Pathogens in the United States," contractor report prepared for the Office of Technology Assessment, December 1991; S.A. Temple and D.M. Carroll, "Pathways and Consequences of the Introduction of Non-Indigenous Vertebrates in the United States," contractor report prepared for the Office of Technology Assessment, October 1991.

■ Current Numbers

An estimated total of at least 4,500 NIS of foreign origin presently are established in the United States (table 3-2). This estimate is based on analysis of six categories of organisms, omitting several others such as animal pathogens and crustaceans (see ch. 2, table 2-1). It also does not capture most marine species, like the majority of the 96 species of sponges, worms, crustaceans, and other non-indigenous marine invertebrates now found in San Francisco Bay (17). Also, numbers shown in table 3-2 are minimum esti-mates for each category. For example, about half of the U.S. insect fauna is unknown, suggesting information on a similar proportion of non-indigenous insects may be lacking (53). Studies of plant pathogens focus on species of economic importance; species affecting only natural areas are chronically under-reported (82). Newly estab-lished species that have not yet been detected also do not figure in table 3-2.

Numbers of NIS vary among the categories. Plants and insects total in the thousands, while NIS in other categories range from tens to hundreds (table 3-2). This is at least in part because there simply are more plants and insects than fish or terrestrial vertebrates. Despite these differences in absolute numbers, the proportion of NIS is relatively constant among most categories, ranging from 2 to 8 percent.

Origins of most plant pathogens are unknown, making evaluation of the contribution of NIS to the current U.S. total difficult (82). A survey of six potential host plants (potato, rhododendron, citrus, wheat, Douglas fir, kudzu) found that an average of at least 13 percent of their pathogens are non-indigenous (82). Non-indigenous pathogens are least common on indigenous or newly introduced plant hosts (82).

Very little information exists on how many species of U.S. origin have been transplanted within the country beyond their natural ranges. Estimates are approximately 2 percent of the U.S. fauna for terrestrial vertebrates and 17 percent for fish (table 3-2).

∎ Past Numbers

The number of NIS of foreign origin has grown in the United States over the past 200 years. Figure 3-3 shows how the totals have expanded for the six categories of organisms. The major increase occurred during the past 100 years for all categories.

GEOGRAPHIC DISTRIBUTION

Finding:

Non-indigenous species are unevenly distributed across the country. Higher concentrations occur around international ports of entry, in areas of active commerce, and in altered habitats. Nevertheless, NIS having significant negative impacts can be found in most regions of the country.

Non-indigenous species are more common in some places than others. Differences occur both among States (table 3-3), and also among regions within individual States. Ports of entry often harbor high numbers of NIS. This is especially true for plants, insects, snails, and slugs that arrive undetected in incoming ships and planes (12,53,63). The type of material arriving at a port influences the specific NIS that become established nearby. For example, numerous European insects were first detected in Rochester, New York, when the city supported an extensive nursery industry and large numbers of plants were routinely unloaded there (53).

Existing patterns of higher densities of NIS surrounding port areas developed over the past 200 years during colonization of the United States. The emergence of containerized freight since the 1950s may change this pattern, since freight containers often are not unloaded until reaching their destination well away from a port.

Areas of frequent commerce away from ports also tend to have higher numbers of NIS. For example, extensive agriculture and related trade and shipping in the Intermountain West (northern Utah and the Columbia Plateau) over the past 100 years have provided abundant opportunities for NIS associated with agriculture to enter and spread within the region (63).

Certain NIS tend to cluster around human population centers. High concentrations of escaped non-indigenous pets occur around Los Angeles and Miami (95). Disproportionately high numbers of non-indigenous snails and slugs similarly occur in populous areas, reflecting their association with greenhouses, gardens, and agricultural commerce (12). Areas, such as Hawaii, supporting human populations with international origins tend to have larger numbers of NIS, because the species imported and released mirror the human population's diversity of tastes and experience (63).

Urban centers often are an important site for the discovery of non-indigenous insect pests. For example, in California 85 percent of non-indigenous scale insects and whiteflies were first reported in cities (40). However, in this case proximity to ports of entry and the enhanced detection potential may also have been factors. Detection of NIS sometimes may be greater in

Figure 3-3—Estimates of the Cumulative Numbers of Non-Indigenous Species of Foreign Origin in the United States[a]

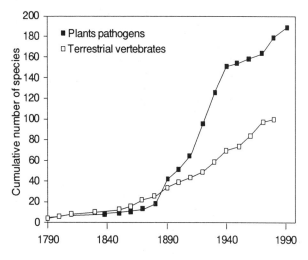

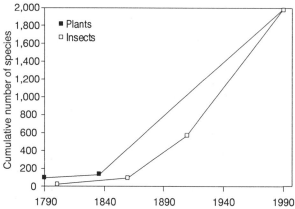

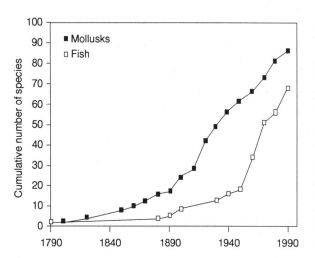

SOURCES: Summarized by the Office of Technology Assessment from: J.C. Britton, "Pathways and Consequences of the Introduction of Non-Indigenous Freshwater, Terrestrial, and Estuarine Mollusks in the United States," contractor report prepared for the Office of Technology Assessment, October 1991; W.R. Courtenay, Jr., "Pathways and Consequences of the Introduction of Non-Indigenous Fishes in the United States," contractor report prepared for the Office of Technology Assessment, September 1991; K.C. Kim and A.G. Wheeler, "Pathways and Consequences of the Introduction of Non-Indigenous Insects and Arachnids in the United States," contractor report prepared for the Office of Technology Assessment, December 1991; R.N. Mack, "Pathways and Consequences of the Introduction of Non-Indigenous Plants in the United States," contractor report prepared for the Office of Technology Assessment, September 1991; Sailer, R.I., "History of Insect Introductions," *Exotic Plant Pests and North American Agriculture*, C.L. Wilson and C.L. Graham (eds.) (New York, NY: Academic Press, 1983), pp. 15-38; C.L. Schoulties, "Pathways and Consequences of the Introduction of Non-Indigenous Plant Pathogens in the United States," contractor report prepared for the Office of Technology Assessment, December 1991; S.A. Temple and D.M. Carroll, "Pathways and Consequences of the Introduction of Non-Indigenous Vertebrates in the United States," contractor report prepared for the Office of Technology Assessment, October 1991.

[a] Figure only includes data on species with known introduction dates for plant pathogens (n = 188), terrestrial vertebrates (n = 100), mollusks (n = 85), and fish (n = 68). Graphs for plants and insects are based on rough estimates.

more densely populated areas simply because collection and observation intensity is higher (12,63).

Regions naturally depauperate in fish and game have been the sites of numerous intentional introductions. A lack of indigenous game animals in the arid State of Nevada prompted State managers to introduce numerous species including the chukar partridge (*Alectoris chukar*), ring-

necked pheasant, Himalayan snow cock (*Tetraogallus himalayensis*), and Rocky Mountain goat (*Oreamnos americanus*) (102). State agencies have released many non-indigenous fish in the American West for similar reasons, where 28 percent of the current fish species are non-indigenous to the region (26).

Intrinsic vulnerability to the establishment of NIS varies among regions in complex ways. The

Table 3-3—Estimated Numbers of Non-Indigenous Species in Selected States[a] [b]

State	Plants		Terrestrial vertebrates		Mollusks	
Alaska	170	(12%)	[c]	(1%)	0	[c]
California	975	(16%)	[c]	(2%)	31	[c]
Florida	≈925	(27%)	53	(6%)	46	(19%)
Illinois	814	(28%)	[c]	(2%)	12	[c]
Maine	[c]		[c]	(1%)	15	[c]
Massachusetts	[c]		[c]	(2%)	27	[c]
Minnesota	[c]		[c]	(2%)	2	[c]
New Mexico	231	(6%)	[c]	(2%)	5	[c]
Oregon	[c]		[c]	(2%)	7	[c]
Texas	443	(9%)	[c]		28	[c]
Utah	580	(23%)	[c]	(2%)	2	[c]
Virginia	427	(17%)	[c]		17	[c]
West Virginia	400	(19%)	[c]	(2%)	2	[c]
Great Plains	354	(13%)	[c]		[c]	
New England	821	(29%)	[c]		[c]	

[a] Numbers should be considered minimum estimates. Experts believe many more NIS are established in the country, but have not yet been detected.

[b] Data reported as the number with percent of species in the State in parentheses. Includes only species non-indigenous to the United States.

[c] Number not reported in source material.

SOURCES: Summarized by the Office of Technology Assessment from: J.C. Britton, "Pathways and Consequences of the Introduction of Non-Indigenous Freshwater, Terrestrial, and Estuarine Mollusks in the United States," contractor report prepared for OTA, October 1991; R.N. Mack, "Pathways and Consequences of the Introduction of Non-Indigenous Plants in the United States," contractor report prepared for OTA, September 1991; M. Rejmanek, C.D. Thomsen, and I.D. Peters, "Invasive Vascular Plants of California," R.H. Graves and F. DiCastri (eds.), *Biogeography of Mediterranean Invasions* (Cambridge University Press); pp. 81-101; S.A. Temple and D.M. Carroll, "Pathways and Consequences of the Introduction of Non-Indigenous Vertebrates in the United States," contractor report prepared for OTA, October 1991. See also sources for tables 8-1, 8-5.

tropical and semi-tropical environments of Hawaii and Florida are favorable to greater numbers of non-indigenous plants than climatically harsher regions experiencing winter frost and freezing (63). Escaped fish from aquaculture are more likely to establish in the benign environment of "sun-belt" States, where warm temperatures allow outdoor aquaculture year-round (26).

Disturbed areas are particularly likely to have large numbers of NIS, as are human modified habitats. For example, livestock increase disturbance by trampling and grazing. In some rangelands, livestock create conditions unfavorable to indigenous grasses, allowing colonization by non-indigenous plants (63).

Combined effects of several of the above factors especially favor NIS. In New England, proximity to ports, extensive agriculture, and removal of indigenous forests have created a region where 29 percent of the plant species are non-indigenous (63).

■ Are Rates of Movement and Establishment Increasing?

Finding:

OTA found no clear evidence that the rates at which NIS are added from abroad to the Nation's flora and fauna have consistently increased over the past 50 years. Instead, rates have fluctuated widely over time in response to an array of social, political, and technological factors.

A common assertion is that rates of species movement into the United States are increasing dramatically. OTA tested this by examining the numbers of NIS added each decade over the past 50 years for terrestrial vertebrates, fish, mollusks,

Table 3-4—Number of New Species of Foreign Origin Established Per Decade[a]

	1940-1950	1950-1960	1960-1970	1970-1980	1980-1990
Terrestrial vertebrates ..	3	11	13	3	—[b]
Fish	2	15	18	5	12
Mollusks	5	5	6	10	4
Plant pathogens	3	5	4	16	7

[a] Numbers should be considered minimum estimates. Experts believe many more NIS are established in the country, but have not yet been detected.
[b] Data unavailable.

SOURCES: J.C. Britton, "Pathways and Consequences of the Introduction of Non-Indigenous Freshwater, Terrestrial, and Estuarine Mollusks in the United States," contractor report prepared for the Office of Technology Assessment, October 1991; W.R. Courtenay, Jr., "Pathways and Consequences of the Introduction of Non-Indigenous Fishes in the United States," contractor report prepared for the Office of Technology Assessment, September 1991; C.L. Schoulties, "Pathways and Consequences of the Introduction of Non-Indigenous Plant Pathogens in the United States," contractor report prepared for the Office of Technology Assessment, December 1991; S.A. Temple and D.M. Carroll, "Pathways and Consequences of the Introduction of Non-Indigenous Vertebrates in the United States," contractor report prepared for the Office of Technology Assessment, October 1991.

and plant pathogens. No consistent increase occurred for any of the categories (table 3-4). Instead, the rate of NIS addition fluctuated. The greatest numbers of terrestrial vertebrates and fish were added during the 1950s and 1960s. The 1970s saw the most mollusks and plant pathogens arrive. A limitation of this analysis is that recently established species may not yet be detected. Thus numbers for the period 1980 to 1990 are likely underestimates.

Suitable data for comparable analyses of plants and insects are unavailable. However, a previous study of agricultural pests (insects and other invertebrates) in California showed the numbers of species established each year similarly varied greatly between 1955 and 1988 from zero to a high of 17 (figure 3-4) (34).

Even though rates of species addition tend to change over time, it is important to note that they rarely reach zero. NIS are continually being added to the nation's flora and fauna, and the cumulative numbers are climbing (figure 3-3). Also, rates throughout the 20th century have been consistently higher than those during the preceding century.

FACTORS AFFECTING PATHWAYS AND RATES

Pathways and rates of species entry to the United States vary because they are influenced by many factors (table 3-5). Many pathways that

were significant sources of NIS in the past have either declined in importance or ceased to operate. Such pathways, nevertheless, frequently are mentioned in discussions of NIS and can confuse attempts to identify present-day problems (boxes 3-B and 3-C).

Some technological innovations enhance introduction rates. For example, the advent of commercial air traffic in the 1930s greatly facilitated the transport of small birds and fish that previously had been difficult to keep alive and healthy on longer voyages (67,95). It had a similar effect on the successful number of insect introductions for biological control (44).

Other new technologies have slowed rates. Many important weeds, such as tumbleweed (*Salsola iberica*), entered and spread throughout the United States as contaminants of agricultural seed in the 1700s and 1800s (63). Improvements in threshing and harvesting machinery beginning in the 1800s decreased seed contamination (63).

Changing fashions in species preferences can drive importation, especially of organisms valued for their aesthetic qualities. Preferences for potted plants in Hawaii support an active illicit commerce in NIS from other tropical and subtropical areas (112). Rates of introduction of aquatic snails accelerated during the 1970s, apparently because of expansion of the aquarium trade and renewed interest in freshwater aquaculture (12). Some preferences relate to patterns of human

Figure 3-4—Numbers of New Insect and Other Invertebrate Species Established in California 1955-1988

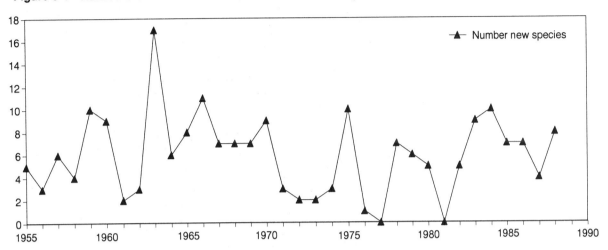

SOURCE: R.V. Dowell and R. Gill, "Exotic Invertebrates and Their Effects on California," *Pan-Pacific Entomologist*, vol. 65, No. 2, 1989, pp. 132-145.

immigration; increased immigration to California from Asia since the 1970s has led to growing importation of Asian foods and associated pests (34).

Political and economic factors are also significant. The location and size of military actions determine their potential for species transfer. Several agricultural pests returned from Europe with military cargo and supplies following World War II. Several aquatic invertebrates from southeast Asia are thought to have entered lagoons and bays of California during the Vietnam War (18).

State and Federal plant quarantine laws slowed rates of introduction of insect pests and plant pathogens after 1912 (80,82). A reversal of this trend for plant pathogens after 1970 (figure 3-3; table 3-4) may relate to globalization of agriculture and increased plant imports (82). The Federal Seed Act, diminished the flow of weed species into the United States that previously had entered as seed contaminants.

Actions of interested constituencies can have an effect insofar as they influence laws and regulations restricting species flow. Conferences, position statements, and other activities of the American Fisheries Society since 1969 helped motivate States to regulate releases of non-indigenous fish (26,55). Conversely, effective lobbying by the Pet Industry Joint Advisory Council helped halt Federal efforts to tighten regulation of fish and wildlife imports during the 1970s (26) (see also box 4-A).

Finally, the "bias of opportunity" (63)—the arbitrary aspect of where pathways happen to appear—always plays a role. For the past 30 years or more, the primary pathway for aquatic species into the Great Lakes has been through shipping—corresponding to the opening of the St. Lawrence Seaway in 1959 (71). As the shipping industry has grown in this region, so too has the number of NIS introductions; shipping was the pathway for 29 NIS introduced between 1960 and 1990 (71). Construction of roads into new areas similarly increased the opportunity for species movement. Urbanization around Tucson, Arizona, contributed to an increase in the non-indigenous plants established in the area between 1909 and 1983, from 3 to 52 species (63).

HOW MANY IS TOO MANY?

Finding:

In the United States, the total number of harmful NIS and their cumulative impacts will continue to grow. An important question is

Table 3-5—Factors Affecting Species Movements

Illustrative Technological Innovations	
Innovation	Effect
Switch from dry to wet ballast in 1800s	Changed from transport of insects, seeds, and plant pathogens to transport of fish and invertebrates
Increased rate of transit via steam ships and airplanes	Increased survival of insects, mammals, birds, and fish during transfer; increased success of introductions
Improvements in threshing and harvesting machinery	Decreased contamination of seed lots and entry and spread of weeds
Styrofoam coolers	Increased number of fish species amenable to transfer and their survival
Containerized shipping of freight	Created new mechanism for unintentional transfer of plant, insect, snail, and slug species; direct route to country interior (i.e., away from shipping port)
Importation of used tires for retreading	Created new pathway for entry of mosquitoes
Illustrative Social and Political Factors	
Social or political factor	Effect
New patterns of immigration and tourism	Change pathways for spread of species
Wars and military movements	Create new pathways for species spread
Globalization of trade	Create new pathways for species spread
Free trade agreements	Increase opportunity for species entry
Increased interest in exotic pets	Affect kind and number of species imported in the pet trade
Continued interest in new ornamental plants	Provide incentive for continued plant exploration and importation

SOURCE: Office of Technology Assessment, 1993.

whether there are limits to the acceptable total burden of harmful species in the country. Such long-term considerations need to be incorporated into shorter term regulatory decisions, for example, in determining the annual level of species entry that will be tolerated.

Even at current rates of species introduction, the total number of NIS in the United States will continue to grow. More than 205 NIS of foreign origin have been introduced or first detected in the United States since 1980, 59 of which are expected to cause economic or environmental harm (table 3-1). Past and projected losses attributable to just two of these are great. The Russian wheat aphid caused losses of over $600 million (1991 dollars) during 1987 through 1989 (24). Projected losses from the zebra mussel by the end of the century are expected to be from $1.8 billion to $3.4 billion (1991 dollars) (24). Both the zebra

mussel and the newly arrived snail *Potamopyrgus antipodarum* from Europe are expected to seriously threaten the country's unique indigenous fauna of freshwater mussels (12).

Numbers of species new to the United States give only a partial account of how many new NIS a given State or area may need to deal with. For example, between 1984 and 1986, an early detection program identified 26 plant species new to Idaho; 12 of these were new to the Pacific Northwest, but only one was new to North America (113). Of 208 invertebrate pests that became established in California between 1955 and 1988, 47 percent originated somewhere in the mainland United States (34).

Even some harmful NIS long-established in the country continue to spread (figure 3-2), taking several decades or more to reach their full geographic range and impact. Dutch elm disease only reached Sacramento County, California, in

---

Box 3-B—Importations for Fish and Wildlife Management Have Decreased

Spencer Fullerton Baird, the First Commissioner of the U.S. Fish Commission (a predecessor of the U.S. Fish and Wildlife Service and National Marine Fisheries Service) strongly supported introductions of non-indigenous species to enhance U.S. fishery resources. Numerous species were imported or transferred across the country and released under his administration. However, introductions of new non-indigenous fish from abroad have lost favor among fisheries managers over the past two decades.

Proposals today are more likely to raise controversy than in the past. A recent proposal by the State of North Dakota to introduce the European zander (*Stizostedion lucioperca*) engendered considerable controversy among other States and the U.S. Fish and Wildlife Service over the potential for disease transmission and hybridization with the indigenous walleye. As introductions of foreign origin decline, transfers of indigenous or established non-indigenous fish to new locales within the United States have increased and probably will continue to do so.

A similar pattern holds for introductions of terrestrial vertebrates. Wide support existed for introductions of species from abroad in the past. Numerous private organizations purposely imported and released wildlife species. For example, the Brooklyn Institute successfully introduced the house sparrow (*Passer domesticus*) in the 1850s, and the Cincinnati Acclimatization Society did the same for 20 additional bird species in the 1870s. The U.S. Fish and Wildlife Service's program in foreign game investigations introduced at least 32 new game species from abroad between 1948 and 1970.

The importation and release of new game species by State managers has declined over the past few decades. This has resulted from a decrease in perceived need and greater awareness of potential risks, rather than from Federal legislation or regulation and could revert should prevailing attitudes change. At the same time, rates of importation by private individuals and game ranchers have increased. Also, NIS already established in the United States continue to be propagated and introduced at new locations, and interstate transfers of indigenous species are on the rise.

SOURCES: W.R. Courtenay, Jr. "Pathways and Consequences of the Introduction of Non-Indigenous Fishes in the United States," contractor report prepared for the Office of Technology Assessment, September 1991; S.A. Temple and D.M. Carroll, "Pathways and Consequences of the Introduction of Non-Indigenous Vertebrates in the United States," contractor report prepared for the Office of Technology Assessment, October 1991.

1990, although it was first detected in the United States in 1930 (15). Imported fire ants became established in Alabama between 1918 and 1945, but only began being intercepted along California borders in 1984—39 to 66 years later (58).

Moreover, the harmful impacts of a NIS in a given State or region can also grow as its distribution and abundance increase. The paper bark tree (*Melaleuca quinquenervia*), originally introduced into Florida in 1906, has spread explosively across the State since the 1960s (49). The predicted range expansion of leafy spurge (*Euphorbia esula*) in Montana, Wyoming, and the Dakotas between 1990 and 1995 is expected to cost an additional $32 million due to diminished grazing capacity (6).

Summed effects of a single harmful species can be staggering over periods of decades. The European gypsy moth has been defoliating trees in a growing area of the eastern United States for at least 120 years (50). In 1990, despite a suppression program costing approximately $20 million, it defoliated an estimated 7.4 million acres (100).

Affected sectors face not just newly introduced species, but all those which arrived before and proved impossible to eradicate. American agriculture alone must deal with at least 235 economically significant insect pests that are non-indigenous to the United States (80). Planning for the future will require assessing not just how many new introductions will be tolerated each

---

Box 3-C—Dry Ballast Has Ceased to be a Pathway

Ships arriving in the United States used to carry dry ballast in the form of rocks, soil, and debris. The ballast was loaded abroad then off-loaded around wharves in the United States to provide cargo space. By one estimate, 1,180 tons of ballast were loaded onto ships bound for America at just one English port in 1815.

Ballast shipped between England and the United States was one of the most significant sources of unintentional insect introductions until the 1880s. It also was the pathway for many plants, including purple loosestrife (*Lythrum salicaria*) which now occurs throughout many northern and midwestern States and causes significant harm to natural areas. Increasing commerce with South America after the Civil War, and consequent ballast shipments, led to the introduction of several pests including fire ants (*Solenopsis invicta* and *S. richteri*), southern mole crickets (*Scapteriscus acletus*), and tawny mole crickets (*S. vicinus*).

Large modern ships use water for ballast instead of dry materials like soil and rock. Thus, the dry ballast pathway has closed. Fire ants discovered in Mobile, Alabama, in 1941 are thought to be the last important pest conveyed by this route. The switch from dry to wet ballast accounts, in part, for the current prominence of the latter as an unintentional pathway for aquatic species entry.

SOURCES: R.I. Sailer, "History of Insect Introductions," *Exotic Plant Pests and North American Agriculture*, C.L. Wilson and C.L. Graham (eds.) (New York, NY: Academic Press, 1983), pp. 15-38; K.C. Kim and A.G. Wheeler, "Pathways and Consequences of the Introduction of Non-Indigenous Insects and Arachnids in the United States," contractor report prepared for the Office of Technology Assessment, December 1991.

---

year, but whether there are limits to the cumulative burden of harmful NIS as well.

CHAPTER REVIEW

This chapter traced the pathways—foreign and domestic, intentional and unintentional—by which non-indigenous species arrive in U.S. locales. Some pathways remain open at all times. The nature and relative importance of other pathways change with time and technology. Combined, they allow sizable numbers of new harmful NIS to flourish here. More than 205 NIS of foreign origin were introduced or first detected in the United States since 1980, and 59 are expected to cause economic or environmental harm. These will join the more than 4,500 foreign NIS already here, a number that is certainly an underestimate. Given that the United States faces increasing numbers and costs of harmful NIS, OTA next turns to the technical questions surrounding their management and control.

Table 3-1—Some Species of Foreign Origin Introduced or First Detected in the United States From 1980 to 1993

Common name	Scientific name	Pathway[a]	Harmful[b]
Plants (9)			
Corn brome	*Bromus squarrous*	Seed contaminant	Yes
Early millet	*Milium vernale*	Stowaway in packing	—
Feather-head knapweed	*Centaurea trichocephala*	Escaped ornamental *or* stowaway in packing material	Yes
Forked fern	*Dicranopteris flexuosa*	Unassisted spread	—
Japanese dodder	*Cuscata japonica*	Seed contaminant	Yes
Lepyrodiclis	*Lepyrodiclis holosteoides*	Seed contaminant	—
Little lovegrass	*Eragrostis minor*	Seed contaminant material	—
Poverty grass	*Sporobolus vaginiglorus*	Stowaway of commerce	—
Serrated tussock	*Nassella trichotoma*	Seed contaminant	Yes
Insects and arachnids[c] (158)			
African honey bee	*Apis mellifera scutellata*[d]	Escape from research facility then spread to U.S.	Yes
Ambrosia beetle	*Xyleborus pelliculosus*	—	—
Ambrosia beetle	*Xylelborus atratus*	—	—
Ambrosia beetle	*Ambrosiodmus lewisi*	—	—
Anobiid beetle	*Lasioderma haemorrhoidale*	—	—
Anobiid beetle	*Priobium carpini*	—	—
Ant	*Pheidole teneriffana*	—	—
Ant	*Technomyrmex albipes*	—	—
Ant	*Gnamptogenys aculeaticoxqe*	—	—
Aphid	*Greenidia formosana*	—	—
Apple ermine moth	*Yponomeuta malinellus*	—	Yes
Apple pith moth	*Blastodacna atra*	Stowaway on plants	Yes
Apple sucker	*Psylla mali*	—	Yes
Ash whitefly	*Siphoninus phyllyreae*	Stowaway on plants	Yes
Asian cockroach	*Blattella asahlnal*	Stowaway on ship or plane	Yes
Asian gypsy moth	*Lymantra dispar*[d][e]	Stowaway on ship	Yes
Asian tiger mosquito (forest day mosquito)	*Aedes albopictus*	Stowaway in used tires	Yes
Avocado mite	*Oligonychus persae*	Stowaway on plants	Yes
Bahamian mosquito	*Aedes bahamensis*	—	—
Baileyana psyllid	*Acizzia acaciae-baileyanae*	Stowaway on plants	Yes
Banana moth	*Opogona sacchari*	Stowaway on plants	Yes
Bark beetle	*Pityogenes bidentatus*	Nursery stock	—
Bark beetle	*Chramesus varius*	—	—
Bark beetle	*Pseudothysanoes securigerus*	—	—
Bark beetle	*Coccotrypes robustus*	—	—
Bark beetle	*Coccotrypes vulgaris*	—	—
Bark beetle	*Theoborus solitariceps*	—	—
Bark beetle	*Araptus dentifrons*	—	—
Beach fly	*Procanace dianneae*	—	—
Black parlatoria scale	*Parlatoria ziziphi*	Stowaway on plants	Yes
Blow fly	*Chrysomya megacephala*	Introduced outside of U.S. then spread into country	Yes
Blue gum psyllid	*Ctenarytaina eucalypti*	Stowaway on plants	Yes
Bostrichid beetle	*Heterobostrychus hamatipennis*	—	—
Burrower bug	*Aethus nigritus*	—	—
Cactus moth	*Cactoblastis cactorum*	—	Yes
Cactus moth	*Ozamia lucidalis*	—	—
Carabid beetle	*Trechus discus*	—	—

(continued on next page)

Table 3-1—Continued

Common name	Scientific name	Pathway[a]	Harmful[b]
Case-bearer moth	*Coleophora deauratella*	Stowaway on plants	Yes
Case-bearer moth	*Coleophora colutella*	Stowaway on plants	—
Click beetle	*Anchastus augusti*	—	—
Cockroach	*Ischnoptera bilunata*	—	—
Cockroach	*Ischnoptera nox*	—	—
Cockroach	*Epilampra maya*	—	—
Cockroach	*Neoblattella detersa*	—	—
Cockroach	*Symploce morsei*	—	—
Collembolan	*Xenylla affiniformis*	—	—
Delphacid planthopper	*Delphacodes fulvidorsum*	Stowaway on plants	—
Delphacid planthopper	*Sogatella kolophon*	Stowaway on plants	—
Dermestid beetle	*Anthrenus pimpinellae*	—	—
Dusky cockroach	*Ectobius lapponicus*	—	—
European barberry fruit maggot	*Rhagoletis meigenii*	—	—
European violet gall midge	*Dasineura affinis*	Stowaway on plants	Yes
European yellow underwing moth	*Noctua pronuba*	Stowaway on plants into Nova Scotia then spread to U.S.	—
Eucalyptus longhorn borer	*Phoracantha semipunctata*	Stowaway in wood	Yes
Eucalyptus psyllid	*Ctenarytaina* sp.	Stowaway on plants	Yes
Eugenia psyllid	*Trioza eugensae*	Stowaway on plants	Yes
Eulophid wasp	*Tetrastichus haitiensis*	—	—
Flea beetle	*Longitarsus luridus*	Stowaway on plants	—
Flea beetle	*Chaetocnema concinna*	Stowaway on plants	—
Flower fly	*Syritta flaviventris*	—	—
Flower fly	*Eristalinus taeniops*	—	—
Forest cockroach	*Ectobius sylvestris*	—	—
Fuchsia mite	*Aculops fuchsiae*	Stowaway on plants	Yes
Green wattle psyllid	*Acizzia* nr. *jucunda*	Stowaway on plants	Yes
Ground beetle	*Harpalus rubripes*	—	—
Ground beetle	*Trechus quadristriata*	—	—
Ground beetle	*Notiophilus biguttatus*	—	—
Ground beetle	*Bembidion properans*	—	—
Ground beetle	*Bembidion bruxellense*	—	—
Guava fruit fly	*Bactrocera* (=*Dacus*) *correcta*	Stowaway in fruit	Yes
Hairy maggot blow fly	*Chrysomya rufifacies*	Introduced outside of U.S. then spread into country	Yes
Honey bee mite	*Acarapis woodi*	—	Yes
Honey bee varroa mite	*Varroa jacobsoni*	—	Yes
Lady beetle	*Decadiomus bahamicus*	—	—
Lady beetle	*Harmonia quadripunctata*	—	—
Lady beetle	*Harmonia axyridis*	—	—
Lady beetle	*Stethorus nigripes*	—	—
Lady beetle	*Scymnus suturalis*	—	—
Lauxaniid fly	*Lyciella rorida*	—	—
Leaf beetle	*Chrysolina fastuosa*	—	—
Leafhopper	*Eupteryx atropunctata*	Stowaway on plants	—
Leafhopper	*Grypotes puncticollis*	—	—
Lichen moth	*Lycomorphodes sordida*	—	—
Longhorn beetle	*Tetrops praeusta*	—	—
Mealybug	*Allococcus* sp.	Stowaway on plants	—
Mediterranean mint aphid	*Eucarazzia elegans*	Stowaway on plants	Yes

Common name	Scientific name	Pathway[a]	Harmful[b]
Megachilid bee	*Chelostoma campanularum*	Stowaway in transported twigs and wood	—
Megachilid bee	*Chelostoma fuliginosum*	Stowaway in transported twigs and wood	—
Mite	*Melittiphis alveartus*	Stowaway on plants	—
Moth	*Agonopterix alstroemeriana*	Stowaway on plants	—
Moth	*Grapholita delineana*	—	—
Moth	*Athrips mouffetella*	—	—
Moth	*Athrips rancidella*	—	—
Nesting whitefly	*Paraleurodes minei*	Stowaway on plants	—
Noctuid moth	*Noctua comes*	Stowaway on plants into Canada then spread to U.S.	—
Noctuid moth	*Rhizedra lutosa*	Stowaway on plants	—
Paper wasp	*Polistes dominulus*	—	—
Peach fruit fly	*Bactrocera (=Dacus) zonata*	Stowaway in fruit	Yes
Pepper tree psyllid	*Calophya schini*	Stowaway on plants	Yes
Pine shoot beetle	*Tomicus piniperda*	Stowaway on dunnage	Yes
Plant bug	*Ceratocapsus nigropiceus*	—	—
Plant bug	*Prepops cruciferus*	—	—
Plant bug	*Jobertus chrysolectrus*	—	—
Plant bug	*Psallus lepidus*	Nursery stock	—
Plant bug	*Orthocephalus saltator*	—	—
Plant bug	*Hyalopsallus diaphanus*	Stowaway in tropical fruit	—
Plant bug	*Stheneridea vulgaris*	Stowaway in tropical fruit	—
Plant bug	*Psallus variabilis*	Stowaway on plants	—
Plant bug	*Psallus albipennis*	Stowaway on plants	—
Plant bug	*Paracarnus cubanus*	Stowaway in tropical fruit	—
Plant bug	*Proba hyalina*	Stowaway in tropical fruit	—
Plant bug	*Rhinocloa pallidipes*	—	—
Pirate bug	*Brachysteles parvicornis*	—	—
Poinsettia whitefly (sweetpotato whitefly)	*Bemisia tabaci* [d f]	—	Yes
Potter wasp	*Delta campaniforme rendalli*	—	—
Potter wasp	*Zeta argillaceum*	—	—
Privet sawfly	*Macrophya punctumalbum*	—	Yes
Pyralid moth	*Hileithia decostalis*	—	—
Red clover seed weevil	*Tychius stephensi*	—	—
Rhizophagid beetle	*Rhizophagus parallelocollis*	—	—
Rove beetle	*Gabrius astutoides*	—	—
Rove beetle	*Sunius melanocephalus*	—	—
Rove beetle	*Oxypoda opaca*	—	—
Rove beetle	*Heterota plumbea*	—	—
Rove beetle	*Coenonica puncticollis*	—	—
Rove beetle	*Staphylinus brunnipes*	—	—
Rove beetle	*Staphylinus similis*	—	—
Rove beetle	*Tachinus rufipes*	—	—
Russian wheat aphid	*Diuraphis noxia*	Introduced outside of U.S. then spread into country	Yes
Sawfly	*Liliacina diversipes*	—	—
Sawfly	*Pristiphora aquilegiae*	—	Yes
Scale predator	*Anthribus nebulosus*	—	—
Seed bug	*Plinthisus brevipennis*	—	—
Seed bug	*Chilacis typhae*	—	—

(continued on next page)

Table 3-1—Continued

Common name	Scientific name	Pathway[a]	Harmful[b]
Shore fly	*Placopsidella grandis*	Stowaway on ship	—
Shore fly	*Brachydeutera longipes*	Stowaway on aquatic plants	—
Siberian elm aphid	*Tinocallis zelkowae*	Stowaway on plants	—
Spider	*Trochosa ruricola*	—	—
Spider	*Lepthyphantes tenuis*	—	—
Spider wasp	*Auplopus carbonarius*	—	—
Spindletree ermine moth	*Yponomeuta cagnagella*	Stowaway on plants	Yes
Spruce bark beetle	*Ips typographys*	Dunnage	Yes
Stink bug	*Pellaea stictica*	—	—
Tatarica honeysuckle aphid	*Hyadaphis tataricae*	Nursery stock	Yes
Thrips	*Thrips palmi*	Stowaway on plants	Yes
Tortoise beetle	*Aspidomorpha transparipennis*	Stowaway on plants	—
Tortoise beetle	*Metriona tuberculata*	Stowaway on plants	—
Tristania psyllid	*Ctenarytaina longicauda*	Stowaway on plants	Yes
Weevil	*Amaurorhinus bewickianus*	—	—
Weevil	*Brachyderes incanus*	Nursery stock	—
Weevil	*Rhinoncus bruchoides*	—	—
Wood-boring wasp	*Xiphydria prolongata*	—	—
Wood-boring wasp	*Urocerus sah*	Stowaway on wood products	—
Wheat bulb maggot	*Delia coarctata*	—	Yes
Waxflower wasp	*Aprostocetus* sp.	Stowaway on plants	—
Whitefly	*Tetraleurodes* new sp.	Stowaway on plants	Yes
—	*Rhagio strigosus*	—	—
—	*Rhagio tringarius*	—	—

(Numerous additional insects and arachnids have been intentionally introduced since 1980 for biological control of pests. None have yet been shown to have harmful effects.)

Fishes (13)

Common name	Scientific name	Pathway	Harmful
Bighead carp	*Hypophthalmichthys nobilis*	Illegal biological control introduction	—
Blue-eyed cichlid	*Cichlasoma spilurum*	Aquarium release	—
European ruffe	*Gymnocephalus cernuus*	Ballast water	Yes
Jaguar guapote	*Cichlasoma manaquense*	Aquarium release	—
Long tom	*Strongylura kreffti*	—	—
Mayan cichlid	*Cichlasoma urophthalmus*	Aquarium release	—
Rainbow krib	*Pelviachromis pulcher*	Aquarium release	—
Redstriped eartheater	*Geophagus surinamensis*	Escape from aquaculture	—
Round goby	*Neogobius melanostomus*	Ballast water	—
Tubenose goby	*Proterorhinus marmoratus*	Ballast water	—
Zebra danio	*Danio rerio*	Aquarium release	—
Yellowbelly cichlid	*Cichlasoma salvini*	Aquarium release	—
—	*Ancistrus* sp.	Aquarium release	—

Mollusks (7)

Common name	Scientific name	Pathway	Harmful
Clam	*Potamocorbula amurensis*	Ballast water	Yes
Clam	*Theora fragilis*	Ballast water	—
Snail	*Alcadia striata*	—	—
Snail	*Potamopyrgus antipodanum*	Contaminant of aquaculture stock that subsequently escaped	Yes
Snail	*Cernuella virgata*	—	Yes
Zebra mussel	*Dreissena polymorpha*	Ballast water	Yes
Zebra mussel	*Dreissena* sp.[g]	Ballast water	Yes

Plant pathogens (9)

Common name	Scientific name	Pathway	Harmful
Blight (on chickpea)	*Aschochyta rabiei*	Stowaway in infected seed	Yes
Citrus canker	*Xanthomonas campestris* pv. *citri*	—	Yes

Common name	Scientific name	Pathway[a]	Harmful[b]
Corn cyst nematode	*Heterodtera zeae*	—	Yes
Needle caste	*Mycospaerella laricina*	Stowaway on infested larch (live or wood?)	Yes
Nematode	*Subanguina picridis*	Biocontrol introduction	—
Potato virus y-necrotic strain (n)	Potyviridae (Potyvirus)	Infected potatoes	Yes
Rust fungus	*Puccinia carduorum*	Biocontrol introduction	—
Rust fungus (on chrysanthemum)	*Puccinia horiana*	Smuggled on infected chrysanthemum	Yes
Smut (on rice)	*Ustilago esculenta*	Smuggled on infected rice	Yes
Other (9)			
Aquatic worm	*Phallodrilus aquaedulcis*	Ballast water	—
Aquatic worm	*Teneridrilus mastix*	Ballast water	—
Asian copepod	*Pseudodiaptomus inopinus*	Ballast water	—
Chinese copepod	*Pseudodiaptomus forbesi*	Ballast water	—
Giant tiger shrimp	*Penaeus monodon*	Escape from research facility	—
Japanese crab	*Hemigrapsus sanguineus*	Ballast water	—
Japanese copepod	*Pseudodiaptomus marinus*	Ballast water	—
Pacific white shrimp	*Penaeus vannamei*	Escape from aquaculture	—
Spiny water flea	*Bythotrephes cederstroemi*	Ballast water	Yes

[a] Listed pathways are according to expert opinions. Often, it is impossible to determine with 100 percent certainty the pathway an NIS followed after the species has become established. A dash in this column indicates that the pathway by which the species entered the United States is unknown.

[b] Known to cause economic, environmental, or other type of harm (see ch. 2). A dash in this column indicates either there are no known harmful effects or they have not yet been well documented.

[c] Where available, common names are those used officially by the Entomological Society of America.

[d] Thought to be a new strain or subspecies of NIS already established in the United States.

[e] The exact origin of the Asian gypsy moth is not yet known; some scientists believe it may be a different species than the established European gypsy moth. The Asian gypsy moth has also been referred to as the "Siberian" gypsy moth in the popular press.

[f] The pointsettia whitefly that recently caused great crop losses in southern California is considered by many to be a new strain of the sweetpotato whitefly which became established in the region several decades ago. Some, however, believe it is a new species.

[g] Recent genetic surveys of Great Lakes zebra mussels suggest a second species of *Dreissena* is also established there; however, its taxonomy remains unclear.

SOURCES: Compiled by the Office of Technology Assessment, 1993 from: J.C. Britton, "Pathways and Consequences of the Introduction of Non-Indigenous Freshwater, Terrestrial, and Estuarine Mollusks in the United States," contractor report prepared for the Office of Technology Assessment, October 1991; J.T. Carlton, "Dispersal of Living Organisms into Aquatic Ecosystems as Mediated by Aquaculture and Fisheries Activities," *Dispersal of Living Organisms into Aquatic Ecosystems*, A. Rosenfield and R. Mann (eds.) (College Park, MD: Maryland Sea Grant, 1992), pp. 13-46; J.T. Carlton, "Marine Species Introductions by Ship's Ballast Water: An Overview," *Introductions and Transfers of Marine Species*, M.R. DeVoe (ed.) (Charleston, SC: South Carolina Sea Grant, 1992), pp. 23-29; J.T. Carlton and J.B. Geller, "Ecological Roulette: The Global Transport of Nonindigenous Marine Organisms," *Science*, vol. 261, July 2, 1993, pp. 78-82; W.R. Courtenay, Jr. "Pathways and Consequences of the Introduction of Non-Indigenous Fishes in the United States," contractor report prepared for the Office of Technology Assessment, September 1991; W.R. Courtenay, Jr., Professor of Zoology, Florida Atlantic University, FAX to E.A. Chornesky, Office of Technology Assessment, Apr. 13, 1993; R.V. Dowell, Entomologist, California Department of Food and Agriculture, FAX to E.A. Chornesky, Office of Technology Assessment, Apr. 12, 1993; R.V. Dowell, Entomologist, California Department of Food and Agriculture, personal communication to E.A. Chornesky, Office of Technology Assessment, May 28, 1993; Entomological Society of America, "Common Names of Insects and Related Organisms, 1989;" D.H. Habeck and F.D. Bennett, "*Cactoblastis cactorum* Berg (Lepidoptera: Pyralidae), a Phycitine New to Florida," Florida Dept. of Agriculture and Consumer Services, Entomology Circular No. 333, August 1990; E.R. Hoebeke and A.G. Wheeler, "Exotic Insects Reported New to Northeastern United States and Eastern Canada Since 1970," *New York Entomological Society*, vol. 91, No. 3, 1983, pp. 193-222; E.R. Hoebeke, "Referenced List of Recently Detected Insects and Arachnids," contractor report prepared for the Office of Technology Assessment, June 22, 1993; E.R. Hoebeke, "*Pityogenes bidentatus* (Herbst), a European Bark Beetle New to North America (Coleoptera: Scolytidae)," *J. New York Entomological Society*, vol. 97, No. 3, 1989, pp. 305-308; E.R. Hoebeke and W.T. Johnson, "A European Privet Sawfly, *Macrophya punctumalbum* (L.): North American Distribution, Host Plants, Seasonal History and Descriptions of Immature Stages (Hymenoptera: Tenthredinidae)," *Proc. Entomol. Soc. Wash.*, vol. 87, No. 1, 1985, pp. 25-33; K.C. Kim and

(continued on next page)

Table 3-1—Continued

A.G. Wheeler, "Pathways and Consequences of the Introduction of Non-Indigenous Insects and Arachnids in the United States," contractor report prepared for the Office of Technology Assessment, December 1991; K.C. Kim, Professor of Entomology, Penn State University, personal communication to E.A. Chornesky, Office of Technology Assessment, May 17, 1993; R.N. Mack, "Additional Information on Non-Indigenous Plants in the United States," contractor report prepared for the Office of Technology Assessment, 1992; R.N. Mack, Professor, Oregon State University, FAX to E.A. Chornesky, Office of Technology Assessment, May 26, 1993; D.R. Miller, Research Leader, Systematic Entomology Laboratory, U.S. Department of Agriculture, Agricultural Research Service, letter to E.A. Chornesky, Office of Technology Assessment, July 1, 1993; J. Morrison, "Cockroaches on the Move,"*Agricultural Research,* vol. 35, No. 2, February 1987, pp. 6-9; B.A. Parfume et al., "Discovery of *Aedes (Howardina) bahamensis* in the United States," *Journal of the American Mosquito Control Association,* vol. 4, No. 3, September 1988, p. 380; M.P. Parrella et al., "Sweet Potato Whitefly: Prospects for Biological Control," *California Agriculture,* vol. 46, No. 1, January-February 1992, pp. 25-26; C.L. Schoulties, "Pathways and Consequences of the Introduction of Non-Indigenous Plant Pathogens in the United States," contractor report prepared for the Office of Technology Assessment, December 1991; U.S. Congress, House Committee on Appropriations, Subcommittee on Agriculture, Rural Development, and Related Agencies, *Hearings on Agriculture, Rural Development, Food and Drug Administration, and Related Agencies Appropriations for 1993, Part 3,* Serial No. 54-888 O, Mar. 18-30, 1992a; A.J. Wheeler, Adjunct Professor, Pennsylvania State University, personal communication to E.A. Chornesky, Office of Technology Assessment, May 6, 1993.

The Application of Decisionmaking Methods | 4

Before the early 1900s, private individuals usually made decisions about whether to introduce non-indigenous species (NIS) with little, if any, government oversight. Even when government was involved, the decision processes were informal and often lenient. Ad hoc judgments and decisions based on precedent predominated. Since then, a trend toward more formal methods has emerged, including risk analysis, legally mandated environmental impact assessment, and economic benefit/cost analysis (table 4-1). Still, these formal approaches rely heavily on judgment and precedent, which in turn are based on the values of the public and its governmental representatives. Whatever the approach, factual gaps and uncertainty complicate the analysis of many existing and potential NIS problems. This chapter examines the prominent decisionmaking methods in use, the role of uncertainty, and the tradeoffs that decisionmakers must face.

Decisions about NIS are made at various levels in Federal and State governments. The flexibility that agency personnel have in making management level decisions depends on their governing statutes, regulations, or policies. A National Park Service (NPS) manager, for example, has very little discretion when deciding whether to introduce a new plant species—in most situations it is prohibited outright by current NPS policies, which seek to preserve the indigenous flora. By contrast, most State and Federal legislation gives broad discretion to managers in dealing with NIS. Agency personnel face two kinds of decisions regarding NIS: which species to allow to be imported and released, and which species to control.

Table 4-1—General Approaches to Making Decisions About Non-Indigenous Species

	Approaches		
	Judgment	Precedent	Formal analysis
Features	Based on relatively undefined procedures Often undocumented	Done according to previous decisions Usually documented	Decisions made according to well-defined procedures Contains explicit documentation
Examples	Judgments by: • General public • Policymakers • Interest groups • Experts	Legal precedent Status quo Tradition	Risk analysis Environmental assessment Economic analysis

SOURCE: P. Kareiva et al., "Risk Analysis as Tool for Making Decisions About the Introduction of Non-Indigenous Species Into the United States," contractor report prepared for the Office of Technology Assessment, July 1991.

WHICH SPECIES ARE IMPORTED AND RELEASED?

Finding:

Most government regulatory approaches to importation and release of NIS use variations of "clean" (allowed) and "dirty" (prohibited) lists of species or groups, with heavy reliance on the dirty list approach. An effective way to reduce risks of harmful invasions is to employ, where practical, a system of both clean and dirty lists, and a "gray" category of unanalyzed species that are prohibited until analyzed and approved.

▌ "Clean" and "Dirty" Lists[1]

The use of "clean" and "dirty" lists reveals a fundamental dichotomy in government decision-making on NIS importation and release. Generally, the clean list approach presumes that all species should be prohibited unless they have been officially listed as allowed, or "clean." The species on the list offer net positive consequences. The dirty list approach presumes that all species may be allowed unless they have been listed as prohibited. Listed species pose net negative consequences. The dirty list method dominates Federal and State decisionmaking,

although several examples of clean lists exist (table 4-2).

Numerous variations of the clean and dirty approaches are employed. These include using a different system for the two phases of introduction, i.e., importation versus release. Also, different methods are used for the major taxonomic groups, e.g., plants, fish, and mammals. Regulators can use a variety of listing criteria, permit requirements, and exemptions; some even adopt total bans on importation or release of major taxonomic groups. Neither clean nor dirty lists per se eliminate the need for inspections and other regulatory compliance measures (25).

Three main factors appear to influence the selection and use of a clean or dirty list approach. These are:

1. **technical feasibility**, that is, whether the potentially threatening NIS in a large taxonomic group, such as non-indigenous plants, are sufficiently limited in number, scientifically understood, and capable of detection so that a comprehensive and accurate clean list can be constructed with reasonable confidence (table 4-3) (25);
2. **requirements for scientific expertise** in fields such as taxonomy, ecology, and risk analysis; these needs are greater to imple-

---

[1] The Federal interagency Aquatic Nuisance Species Task Force has abandoned the terms "clean" and "dirty" due to public objections. Instead, they plan to use the more neutral-sounding "approved," "restricted," and "prohibited." Note that these terms are used by a number of States as well (34).

Table 4-2—Examples of Clean and Dirty Lists in Statutes or Regulations

	Summary
Clean list	
USDA Quarantine 56 (7 CFR 319.56)	Allows import of only listed fruits and vegetables from specified countries
Hawaii Revised Statutes sec. 150A.6	Allows import of only animals and microorganisms on "conditionally approved" list
Dirty list	
Lacey Act	Restricts import of two taxonomic families, 13 genera, and 6 species of fish and wildlife
Federal Noxious Weed Act	Prohibits import of 93 listed weeds

SOURCE: Office of Technology Assessment, 1993.

ment a comprehensive clean list approach and not always available; and

3. **willingness to accept risks** of unanticipated invasions by harmful NIS; a clean list approach can reduce risks, however, decisionmakers may be willing to accept the higher risks of a dirty list approach, especially if control or eradication is feasible.

Several experts have argued for treating NIS under a clean list approach whenever practical; that is, prohibiting all species that are not on a clean list until they have been satisfactorily analyzed and determined to offer net benefits (26,74). This would be comparable to the Food and Drug Administration's general regulatory system for approving a new drug for human use: prohibited until proven net beneficial.

Moving to a clean list approach would require substantial changes in the regulation of *importation* (that is, the act of bringing an NIS across a border into the country or a particular State). Allowing importation only of species on a clean list would place greater restrictions on international trade.

For some groups of organisms, only *release* into a free-living condition has been this strictly

regulated. However, importation of some NIS is likely to lead eventually to their release, whether intentional or by their escape. Imported aquarium fish are a good example. Those that have established free-living populations after being discarded by their owners have often had negative effects, especially in Florida and in the Southwest (11). For such taxonomic groups composed of organisms that readily escape, the regulation of importation in effect *is* the regulation of release. The more restrictive clean list approach would be more effective in preventing harm although this approach is more burdensome in the short run.

Even for those groups in table 4-3 for which clean lists appear technically feasible, the political feasibility of such an approach is questionable. The U.S. Fish and Wildlife Service (FWS) made three politically unsuccessful attempts in the mid-1970s to change the Lacey Act[2] process for regulating importation of "injurious" fish and wildlife from a dirty to a clean list, or to substantially lengthen the dirty list (box 4-A). The available information on environmental and economic consequences of harmful NIS was far less complete than it is today (76,82). Whether the political obstacles remain is unclear.

The Lacey Act was interpreted by FWS to be legally broad enough to allow for a clean list approach without amendment (76). No court has ruled on this interpretation. Apart from this legal issue, the question remains of how to best regulate potentially risky fish and wildlife. One method being considered is a three-part system with an intermediate "gray" category.

■ "Gray" Category

In any given jurisdiction (e.g., country, State, or county) the vast majority of potentially introduced NIS belong to a "gray" category. This consists of all species not already listed as clean or dirty because decisionmakers lack detailed analyses of the likely consequences should they

---

[2] Lacey Act (1900), as amended (16 U.S.C.A. 667, *et seq.*, 18 U.S.C.A. 42 *et seq.*)

Table 4-3—Relative Technical Feasibility of Comprehensive Clean Lists for Regulating Importation of Major Groups of Non-Indigenous Species

Group	Clean list feasibility	Reasons
Fish and other vertebrate animals	High	Well known; fewer species; moderate commercial trade; easily detected
Plants	Medium	Well known; many species; high commercial trade; easily detected
Insects	Low	Poorly known; very many species; low commercial trade; difficult to detect
Other invertebrate animals	Low	Poorly known; very many species; low commercial trade; ease of detection varies
Micro-organisms	Low	Poorly known; very many species; low commercial trade; very difficult to detect

NOTE: These are general ratings. Taxonomic subgroups within each major group may justify different ratings. For example, within the major category of invertebrate animals it would be more feasible to adopt a clean list for the relatively small sub-group of freshwater mollusks.

SOURCE: Office of Technology Assessment, 1993 and R.P. Kahn, letter to P.N. Windle, Office of Technology Assessment, Dec. 2, 1991.

become established. Combining this gray category with the clean and dirty list approaches forms a classification scheme that can be adjusted to suit particular regulatory circumstances (26).

Hawaii, for example, recently amended its laws on importing animals and micro-organisms, creating the most restrictive State laws on the subject (ch. 7). This change responded to the perceived urgency of Hawaii's NIS problems (ch. 8). State law now provides for three lists and a gray category.[3] Species on the *conditionally approved* list require a permit for importation, while those on the *restricted* list require a permit for both importation and possession. Those on the *prohibited* list may not be imported or possessed except in very limited cases. Species not on any list (the gray category) are prohibited without official permission. The State now handles requests for permission as follows (50):

> If the request is for a species that is on an animal or micro-organism list and has received prior approval by BOA [Board of Agriculture] or is a plant that has received such approval, PQ [Plant Quarantine Branch] can issue the permit. If, however, an applicant is requesting a permit for a species that has not received prior BOA approval, PQ will conduct a three-tiered review

process to bring the request before the board.

First, the application is submitted to the BOA's Technical Advisory Subcommittees. The five subcommittees (Land Vertebrates, Invertebrates and Aquatic Biota, Entomology, Micro-organisms and Plants) are composed of researchers, industry representatives and government officials. The subcommittees evaluate the application along technical/scientific lines, particularly for the organism's potential impact. The subcommittees then pass their analyses to the Plant and Animals Advisory Committees which considers the application and the subcommittee findings from a broad perspective, weighing the potential harmful impacts against the potential benefits. BOA then reviews the Advisory Committees' recommendation and issues the final decision on the application.

Much of the rest of this chapter discusses general methods for making the type of listing and approval decisions referred to above, such as how to weigh the potential harmful impacts against the potential benefits.

WHICH SPECIES ARE CONTROLLED OR ERADICATED?

Sometimes greater difficulty can arise in deciding which damaging NIS to control or eradicate,

---

[3] Hawaii Revised Statutes, section 150A-6.

Box 4-A—History of Fish and Wildlife Service Attempts To Implement Clean Lists Under the Lacey Act

The Lacey Act of 1900 and 50 CFR, part 16, enable the Secretary of Interior to restrict fish and wildlife imports beyond those species listed as prohibited in the Act itself. Pursuant to this authority, in December 1973, FWS proposed regulations[1] that concluded all non-indigenous fish and wildlife species had the potential to be injurious and should be prohibited, except for a list of several hundred species and larger taxonomic groups that were believed to pose little risk. FWS prepared this "clean" list after soliciting input from user groups and scientific experts, and it made provisions for future additions.

However, the more than 4,300 comments on the proposal were mostly negative, especially those from people involved with the pet trade, zoos, game ranches, agriculture, and aquaculture. After preparing an environmental impact statement and taking part in a congressional hearing, the agency published a revised proposal to lengthen the clean list, in February 1975.[2] That also received a negative reception, with nearly 1,200 comments. Opponents claimed evidence was insufficient that importation of any particular species would cause harm. The pet industry claimed it would be particularly affected by excluding rare or poorly studied species that were not on the clean list, because they would command the highest prices. After extensive controversy, FWS withdrew the clean list proposal.

As a final effort, in 1977, FWS proposed a rule[3] containing a much longer dirty list. This approach failed as well, with the primary resistance from the hobby fish industry. No major constituency weighed in favoring the concept and further formal attempts to change the regulations were abandoned.

---

[1] 38 *Federal Register* 34970, (Dec. 20, 1973).

[2] 40 *Federal Register* 7935, (Feb. 24, 1975).

[3] 42 *Federal Register* 12972, (Mar. 7, 1977).

SOURCES: R.A. Peoples, Jr., J.A. McCann, and L.B. Starnes, "Introduced Organisms: Policies and Activities of the U.S. Fish and Wildlife Service," *Dispersal of Living Organisms into Aquatic Ecosystems*, A. Rosenfield and R. Mann (eds.) (College Park, MD: Maryland Sea Grant, 1992), pp. 325-352; J.G. Stanley, R.A. Peoples, Jr., and J.A. McCann, "Legislation and Responsibilities Related to Importation of Exotic Fishes and Other Aquatic Organisms," *Canadian Journal of Fisheries and Aquatic Sciences*, vol. 48, suppl. 1, 1991, pp. 162-166.

---

and how to do it, than in deciding which species to allow to be imported or released. If a manager has 10 existing problem species and a control budget that allows elimination of only 3, which ones should he or she choose? Should the goal be complete eradication, or control at some point less than 100 percent eradication? What methods should the manager use?

To complicate matters, eradicating or controlling NIS with chemical pesticides often arouses public opposition. So does killing popular non-indigenous animals, like feral horses (*Equus caballus*), by any method. Both cases involve weighing the potential damage caused by the NIS against other factors. In the pesticide case, the factors are potential human health and environmental impacts; the popular animal case involves mainly ethical values. For both, costs of the available methods may be a major factor. As with decisions about importation and introduction, the formal approaches discussed below may aid these weighing processes.

COMMON DECISIONMAKING APPROACHES

Decisionmakers commonly employ three tools in analyzing NIS: risk analysis, environmental impact assessment, and economic analysis.

■ Risk Analysis

Finding:

Scientists generally cannot make quantitative predictions of the invasiveness or impact

of a new, untested species with high degrees of confidence. Nevertheless, useful qualitative predictions often can be made. Expert judgment based on careful research and diverse input is the most broadly feasible predictive approach. Controlled, realistic-setting experimentation reduces uncertainty but requires more resources.

THE ROLE OF RISK ANALYSIS

A strictly empirical, or after-the-fact, approach to NIS introductions would be clearly inadequate. Always waiting to see if a species causes harm before deciding whether to prohibit it would lead to multiple disasters and huge control costs. Conversely, banning *all* importation and release of NIS would be an effective, but obviously impractical, risk reducer. The most realistic way to prevent human-caused harmful invasions by NIS is to develop better scientific methods to accurately predict them and to act based on these predictions. The field of risk analysis encompasses these predictive methods. Risk analysis looks at the chances that an unwanted event will occur and the consequences if it should occur.

Risk analysis can inform decisionmakers on everything from building nuclear power plants to anticipating oil spills to keeping zebra mussels (*Dreissena polymorpha*) out of the Missouri River. The subfields most relevant here are ''pest risk analysis,'' undertaken to protect agriculture (including forestry) and ''ecological risk analysis,'' which looks at threats to non-agricultural areas and their occupants. The goal is understanding and ordering different degrees of risk, from those as obvious as introducing a mammal that has rabies to those as subtle as introducing an insect that slightly raises the probability that an indigenous insect will go extinct (26).

The ideal risk analysis should specify the likelihood of possible outcomes from a particular activity, estimate the risks associated with the various outcomes, and identify effective means to mitigate the risks. Although much of this follows common sense, as a discipline it forces analytical

accounting for uncertainty, that is, when the data do not permit the ideal analysis. And the process can make the tradeoffs between competing factors clear to the observer.

Clarity regarding tradeoffs in the face of uncertainty is important. A hypothetical example: if current scientific knowledge cannot predict whether a potentially damaging Australian tree fungus will invade valuable redwood stands in northern California, then on what basis can a decision be made to allow Australian logs into northern California? How much would the decisionmaker be willing to spend to reduce that scientific uncertainty? Given the uncertainty, and thus the chance of deciding mistakenly, how does one balance being too restrictive against being too lenient? What numerical chance of being wrong is acceptable? Risk analysis alone does not answer these questions. Nevertheless, a risk analysis process should display the potential tradeoffs clearly, that is, it ''must not cloak what should be societal decisions in the mantle of scientific objectivity when the determinations are not purely scientific'' (39).

Even the best risk analysis methods cannot eliminate all uncertainty. With enough resources, imperfect or incomplete knowledge and human errors—two important sources of uncertainty—can be reduced or eliminated. However, the inherent randomness of the world adds uncertainty that cannot be reduced (71). Also, the ability of NIS and their receiving ecosystems to adapt and evolve means that risk analysis done at the time of introduction may be rapidly obsolete; this adds another source of uncertainty to predictions (70).

In making tradeoffs on the national scale, policymakers must decide the most fundamental question of NIS policy: how much risk of damage will we accept? No formulaic answer exists. Hundreds of harmful NIS are already in the country. Early warnings were available for several recent additions: the zebra mussel, the Asian tiger mosquito (*Aedes albopictus*), and the Asian gypsy moth (*Lymantria dispar*). In each case, a

Controlled scientific studies, such as this study of a biological control organism, can boost the reliability of risk assessment.

fair degree of risk was tolerated. So far, at least, most governmental decisionmakers have not been highly risk averse where potentially damaging NIS were concerned.

THE PROCESS OF RISK ANALYSIS

The first step in risk analysis for *planned* releases is predicting the likelihood that the species to be released will survive and establish one or more self-sustaining populations (27). Then one must assess the probable resulting impacts on the ecosystems and/or agricultural systems involved. The combination of the characteristics of the new organism and the new environment determines the risks associated with the release.

Greater difficulty in prediction arises when one considers *unplanned* introductions. These are NIS that escape from confinement or are unknowingly released. Risk analysis in these cases requires initial determination of the probability that a release will, in fact, occur. The same determination applies to NIS that are knowingly, but illegally released, though some classify these as planned releases (see ch. 3). Probability of release must then be factored into the likelihood of survival, establishment, and environmental impact, as determined for planned releases, also.

The Federal interagency Aquatic Nuisance Species Task Force, formed to respond to the invasion of the zebra mussel and other NIS in the Great Lakes, has adopted a pathways-oriented approach to risk analysis for unplanned releases (75). The Task Force intends to assess all potential pathways for harmful, unintentional releases, ranging from cargo ships dumping their ballast water to pathogens inadvertently transported with fishery stock.

Several models have been developed that generalize about the risks of NIS invasions. Current applications of these models are limited because they do not quantitatively predict with high degrees of confidence either the likelihood that a new species will become established or its impacts (26).

Useful generalities about risks can be drawn, however, some of these lack clear scientific validation. In general, the species most likely to be successful invaders have large natural ranges, a high intrinsic population growth rate, and a large founding population in the new environment (12). The environments most likely to be invaded are those with few species present, a high degree of habitat disturbance, and an absence of species closely related and morphologically similar to the potential invaders (48).

The risk analysis process has relied largely on professional judgments based on "impressionistic syntheses of case studies and anecdotes" (27) rather than rigorous statistical studies or experimental analyses. Formal risk analysis methods for NIS have not been developed or applied (70). This qualitative rather than quantitative approach may be satisfactory in most cases, particularly if a diverse panel of scientists and other experts has input into the analysis. Some expect that more reliable quantitative predictions will be available as data accumulate and computer models are refined (24,57).

The intense commercial interest in risk analysis for the controlled release of new genetically engineered organisms (GEOs) (ch. 9) has helped advance both theoretical and experimental approaches to NIS risks generally (26), as have the research and testing of new biological control agents (ch. 5). The standard paradigm for analyzing risks of these specialized releases relies much more heavily on experimentation, including controlled, small-scale trial releases, than is normally done for other proposed NIS releases.

Recent technological advances have made some experimental releases safer. For certain species, scientists can ensure that released NIS are infertile through sterilization, birth control, or other manipulations such that no more than one generation will survive (ch. 5). Fisheries biologists have used these techniques to assess new introductions of fish and shellfish (51). Some advocate the use of these reproductive control techniques as a precondition for all experimental releases (67).

Experimentation can provide data critical for linking mathematical models to ecosystem behavior, especially for generalized theories of ecosystem response to stress (39). Experimentation also informs the optimal design of monitoring systems and the apportionment of containment or control efforts according to the risks involved. In one facility in England, experiments on invasions are conducted in a large laboratory with 16 connecting microcosm chambers (38). It allows the assembly of a wide variety of plant and animal communities in computer-controlled environments. Still, organisms can behave quite differently in the real world than they do in experimental settings because of untested, often unanticipated, influences. The possibility of chaos in ecological systems suggests that making accurate predictions may be more complex than anticipated (19,60) and not a matter necessarily solved by accumulating more data for better models.

Experimental analyses for NIS (other than GEOs and biological control agents) are not consistently done or required by Federal or State laws. Despite difficulties in interpreting results from small-scale trial releases, experts have called for more use of these and other experimen-

tal approaches as providing better predictions than the largely anecdotal "paper" studies that dominate now (40). An experimental approach would require more personnel, funding, and time.

RISK ANALYSIS BY FEDERAL AGENCIES

Finding:

Within the Animal and Plant Health Inspection Service (APHIS) of the U.S. Department of Agriculture (USDA), there is great variation as far as the stringency of its risk analysis procedures for different types of NIS importation. Internal proposals to improve and standardize risk analysis procedures have not been broadly implemented. Two existing policies hamper the agency's effectiveness at keeping new, harmful NIS from entering the country: its lack of explicit focus on risks to non-agricultural areas, and its general operation under the presumption that unanalyzed imports will be admitted unless risks are proven. Still, APHIS is more analytical than FWS. FWS has implemented very little scientific risk analysis for potentially harmful fish and wildlife.

The primary Federal responsibility for regulating NIS lies with USDA's APHIS and the Department of Interior's FWS (see ch. 6). APHIS can regulate both private and governmental actions that pose risks of introducing agricultural and forestry pests, including weeds. FWS is responsible for "injurious" fish and wildlife under the Lacey Act, which, as applied, primarily means species that threaten interests outside agriculture.

Animal and Plant Health Inspection Service—Much of current APHIS risk analysis consists of preparing a "decision sheet," which often includes only a paragraph or two on the biology of a prospective plant pest (80). Great variation exists within APHIS as far as the stringency of analysis (26). Comprehensive assessments of probabilities and risks are rarely undertaken. The agency is revising a number of its regulatory quarantines and considering adoption of new quarantines, and in the process has sought to improve and standardize its procedures.

The main foundation for this standardization with respect to plants and plant products is the "Generic Pest Risk Assessment Process" developed by the Policy and Program Development office (53). This process has not been finalized yet[4] or broadly adopted within the agency. Once adopted, the process can be tailored to decisions about particular types of proposed new commodity importations, such as cut flowers, nursery stock, and logs (figure 4-1). Since a commodity can carry more than one potential pest, conducting Individual Pest Risk Assessments on each pest will be necessary in addition to the analysis of the risk of the commodity itself (e.g., for its potential weediness). An analyst will make qualitative ratings (low, medium, high) for various factors and assign an uncertainty level. The combination of these will result in an overall Commodity Risk Potential rating and a recommendation by the analyst. APHIS regulatory and operational personnel will make the final decision.

The Agricultural Research Service assists APHIS on risk analysis questions requiring research. ARS conducts experiments on a few potentially serious pests like soybean rust (*Phakopsora pachyrhizi*) (87). This method, in which a small number of samples are imported under controlled conditions and tested in small-scale trials, would be impractical for analyzing risks from all potential pests.

While APHIS has kept thousands of potential agricultural pests from becoming established, it has done little explicit analysis of risks to natural areas. Critics have also pointed to insufficient scientific input, especially from the field of ecology, in its analyses (25,26,36). Long-term risks, such as the potential for pests to evolve more harmful characteristics, are under-analyzed because of lack of input from evolutionary biologists (26).

APHIS lacks sufficient in-house expertise to fully address the questions posed by the regular flow of new potential pests (26). Outside experts are sometimes consulted, but they often lack training or experience in quarantine problems. Further, in the past many risk analyses were not adequately documented to be of use in future decisions (26). The agency is considering several proposals to implement more explicit procedures that are sensitive to natural ecosystems, embrace more diverse input, and provide useful data for the future.

Implementation of these improvements is important. However, a basic policy hampers APHIS's success at keeping out pests—that is, its willingness to allow many types of imports that pose unanalyzed, or incompletely analyzed, risks. Examples of this include virtually all unprocessed wood and wood products, including packing and shipping materials;[5] and potential pests on or in containers and ships that have been in high-risk areas. The agency generally treats unregulated imports under the presumption "that everything is enterable until we [APHIS] determine it should not be" (53). Implicit in this is APHIS's accepting the burden of proving a proposed new import's potential for harm, rather than putting the burden on the importer to demonstrate its safety. This policy relies on inspection at ports-of-entry to interdict potentially harmful organisms despite the fact that many are very difficult to detect or present unknown risks.

---

[4] The final version is anticipated in December, 1993.

[5] APHIS recently published an Advance Notice of Proposed Rulemaking regarding importation of logs, lumber, and certain other wood products, 57 *Federal Register,* 43628-31 (Sept. 22,1992). At this writing it is unclear whether a rule will be issued, or what it will provide, but the Notice indicates that the agency may more proactively address risks from logs and wood products in the future. The Notice did not cover wooden packing or shipping materials.

Figure 4-1—Application of the APHIS Generic Pest Risk Assessment Process

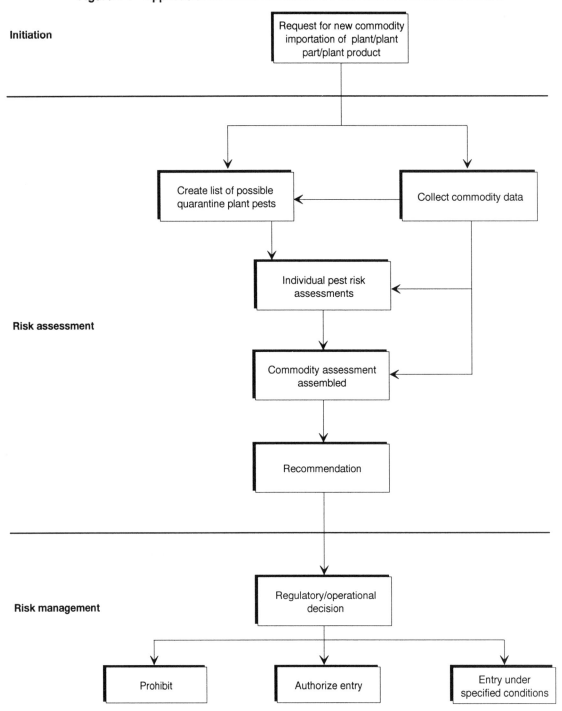

SOURCE: R.L. Orr, Entomologist, and S.D. Cohen, Plant Pathologist, Animal and Plant Health Inspection Service, U.S. Department of Agriculture, "Generic Pest Risk Assessment Process—For Estimating the Pest Risk Associated With Importation of Foreign Plants and Plant Products (draft)," Nov. 20, 1991.

This "presumption of enterability" is not mandated by the Plant Pest Act[6] or by other controlling legislation; it is apparently a policy choice to favor unburdened trade. That choice may itself be the result of weighing the overall risks and benefits of a more restrictive presumption of exclusion. However, OTA has not discovered any evident national weighing of these risks and benefits. The weighing process appears to occur in difficult new cases, one at a time, at high levels of the Department of Agriculture.

> [I]n controversial trade matters, top management outside of APHIS may "weigh" the biological position against the economic or other positions, and the short-term decision made by non-biologists may in some instances prevail regardless of the probability of long-term adverse consequences. (25)

The presumption of enterability has real consequences. In the recently proposed importation of Siberian timber to West Coast sawmills (box 4-B), for example, several critics pointed out that APHIS's starting assumption was that the importation would occur. The agency initially stressed the rights of the importers to proceed rather than the biological issues (7). Indeed, it allowed them to bring in a small shipment of logs, without a formal pest risk analysis or environmental assessment, that was found later to carry pests. It took pressure from academic scientists and members of Congress to stop APHIS from allowing further shipments without a comprehensive risk analysis (14).

For a proposed importation of pine (*Pinus* spp.) wood chips from Honduras into Oregon, APHIS did not require a formal assessment of the potential risk, despite serious warnings from an Oregon State University entomologist (37). The agency would not delay the imports unless risk was first proven; expert opinion was insufficient to overcome the presumption of enterability (66).

The agency's willingness to accept unanalyzed risks is compounded by the low level of effort USDA devotes to researching where risky species are likely to come from and to proactively regulate so as to prevent problems before they arise. The relatively short list of foreign weeds prohibited under the Federal Noxious Weed Act represents one example (ch. 6) (41). Another is the recent Asian gypsy moth infestation in Pacific Northwest ports, which necessitated a $14 million to $20 million emergency eradication program (box 4-B). The moth arrived via cargo ships on which eggs had been laid while in Far East ports. Ships are one of the most obvious pathways for new pest introductions because of their size and frequency of arrival. Yet APHIS had not proactively analyzed the Asian gypsy moth risks nor taken steps to prevent the infestations. In the words of a former California Department of Agriculture official discussing overall U.S. quarantine policy, "ignorance is viewed as a relatively low-level risk compared to the benefits of open trade and other societal needs" (62).

For the items discussed above—unprocessed wood, packing materials, containers from high risk areas, etc.—APHIS lacks specific regulations. The agency assumes the items are suitable for import unless agricultural port inspectors detect a problem. APHIS treats all plants in a similar manner, including nursery stock, seeds, and bulbs, under regulations known as Quarantine 37. Such foreign plants are enterable with a permit if they are *not* listed in these regulations, that is, on the "dirty" list of plants known to carry important pests or diseases in their countries of origin. Quarantine 56, which covers imported fruits and vegetables for consumption, is an exception to APHIS' overall assumption of enterability (25). Under this quarantine, pest risk assessments have judged listed articles "clean" and, thus, able to be imported with a permit.

---

[6] Federal Plant Pest Act (1957), as amended (7 U.S.C.A. 147a *et seq.*)

Box 4-B—Siberian Timber Imports: A Potentially High-Risk Pathway

Siberia has almost half of the world's softwood timber supply. Since the late 1980s a few U.S. timber brokers and lumber companies, short on domestic supplies, have been negotiating for the importation of raw logs from Far East ports to West Coast sawmills. This may create a pathway for non-indigenous forest pests that are adapted to many North American climate zones and tree types. In the past 100 years raw wood or nursery stock imports have provided entry for a number of devastating pathogens, such as chestnut blight (*Cryphonectria parasitica*), Dutch elm disease (*Ceratocystis ulmi*), and white pine blister rust (*Cronartium ribicola*).

In early 1990, the private importers voluntarily notified APHIS and the California Department of Agriculture that they would be shipping two containers of logs representing four Siberian tree species into the northern California port of Eureka. The logs were fumigated, handled, sawn, and disposed of pursuant to agreed upon guidelines. The California officials had sought more time to develop the guidelines before shipment, but were unable to obtain a voluntary delay and lacked regulatory authority to require a delay. According to the program supervisor of the Pest Exclusion Branch, APHIS's California approach to the State's biological concerns was to stress the importers' rights to proceed.

Dead insects were recovered off three of the tree species; the fourth carried a nematode. The agencies concluded that no further shipments should come in until personnel could identify the species and do a pest risk analysis. APHIS arranged a voluntary embargo with the importers. Two of the species were later identified as potentially harmful new pests.

Participation by APHIS in the early phases (April through September 1990) was criticized as "chaotic" by the California official in charge. The agency's Preliminary Pest Risk Analysis was completed in September; it was generally regarded as inadequate, failing to list many *known* Siberian pests and lacking investigation into the many unresearched potential pest species. Worried California and Oregon officials sought independent scientific advice. Several State university professors warned of potentially disastrous consequences from the organisms that were likely to be introduced, even if the logs were fumigated.

Communication among these academics and the State officials in fall 1990 eventually led to congressional pressure in the form of a letter from three members of the Oregon delegation to the Secretary of Agriculture inquiring about APHIS's handling of the matter and requesting a delay pending resolution of the pest issues. At the same time, the importers were negotiating with APHIS to allow large-scale shipments to mills in Humboldt Bay, California. However, "to honor the congressional request," the agency suspended the discussions on December 13. APHIS announced it had imposed a "temporary prohibition" on future imports. Without the congressional pressure, it appears the shipments would have gone ahead without comprehensive analysis.

A joint U.S. Forest Service/APHIS Task Force was convened and worked for almost a year on a detailed risk assessment focusing on larch (*Larix* spp.) from Siberia. The project cost of approximately $500,000 was paid out of a Forest Service contingency fund. APHIS lacked a flexible fund to pay for the unanticipated, unbudgeted work.

The assessment found serious risks posed by several pests. A worst-case scenario examined the economic impacts should they successfully invade Northwest forests. It produced astoundingly high figures for the cumulative potential losses from the Asian gypsy moth (*Lymantria dispar*) and the nun moth (*Lymantria monacha*) between 1990 and 2040—in the range of *$35 billion* to *$58 billion* (net present value in 1991 dollars). Still, the assessment did not resolve all the issues about mitigating the risks. Ultimately, APHIS put the burden back on the importers to propose new pest treatment methods and protocols with "evidence of complete effectiveness". Some experts said the logs would need sawing and kiln-drying to exterminate all risky species, which would probably be prohibitively expensive. The assessment concluded: "If technical efficacy issues can be resolved, APHIS will work with the timber industry to develop operationally feasible import procedures." To date the industry has identified no feasible procedures that APHIS has deemed completely effective.

(continued on next page)

Box 4-B—Continued

A recent discovery may render the timber import risk mitigation efforts moot, at least for the Asian gypsy moth. While APHIS and the Forest Service were looking at the chances it would arrive on logs, the Asian gypsy moth arrived in the Pacific Northwest clinging to grain ships. The risk of this pathway had been overlooked. A $14 million to $20 million program of broadcast biopesticide spraying, trapping, and monitoring has been implemented by Federal and State officials to stop what the Deputy Director of the Washington Department of Agriculture said "has the potential to be the most serious exotic insect ever to enter the U.S." An information program was also initiated to keep shippers that trade in high-risk Far Eastern ports from inadvertently transporting more moths. While officials have found no more Asian gypsy moths in the Pacific Northwest to date, their ultimate success in eradicating this pest remains uncertain.

SOURCES: Associated Press, "Forest Bugaboo—Alarm Over Discovery of Asian Gypsy Moths," *Seattle Times/Post Intelligencer*, Nov. 24, 1991, p. B-8; A. Clark, Program Supervisor, Pest Exclusion Branch, California Department of Agriculture, Sacramento, CA, personal communication to P. Jenkins, Office of Technology Assessment, Feb. 14, 1991; P. DeFazio, U.S. House of Representatives et al., letter to C.K. Yeutter, Secretary, U.S. Department of Agriculture, Washington, DC, Dec. 5, 1990; J.D. Lattin, Professor of Entomology, Oregon State University, personal communication to P. Jenkins, Office of Technology Assessment, Jan. 31, 1991; J.D. Lattin, Professor of Entomology, Oregon State University, memorandum to B. Wright, Administrator, Plant Division, Oregon Department of Agriculture, Salem, OR, Nov. 1, 1990; R. Morris, Division Resources Manager, Louisiana-Pacific Corp., Samoa, CA, internal memorandum to B. Phillips, Dec. 19, 1990; M. Shannon, Chief Operating Officer for Planning and Design, Animal and Plant Health Inspection Service, U.S. Department of Agriculture, Hyattsville, MD, personal communications to P. Jenkins, Office of Technology Assessment, Feb. 5, 1991 and Mar. 2, 1992; U.S. Department of Agriculture, Animal and Plant Health Inspection Service, Hyattsville, MD, "USDA Places Temporary Prohibition on Entry of Siberian Logs Because of Pests," press release, Dec. 20, 1990; U.S. Department of Agriculture, "An Efficacy Review of Control Measures for Potential Pests of Imported Soviet Timber," Miscellaneous Publication No. 1496 (Hyattsville, MD: Animal and Plant Health Inspection Service, September 1991); U.S. Department of Agriculture, Forest Service, "Pest Risk Assessment of the Importation of Larch From Siberia and the Soviet Far East," Miscellaneous Publication No. 1495 (Washington, DC, September 1991); D.L. Wood, Professor of Entomology, and F.W. Cobb, Jr., Professor of Plant Pathology, Univ. of California, Berkeley, letter to Dean Cromwell, California State Board of Forestry et al., Sacramento, CA, Dec. 11, 1990.

Fish and Wildlife Service—FWS does far less than APHIS in analyzing risks from injurious fish and wildlife (26). The current Lacey Act dirty list is short (prohibiting 2 families, 13 genera, and 6 species), and FWS uses no checklist or other standardized procedure to analyze risks from other imported species. While APHIS inspects incoming agricultural livestock for diseases, FWS has no procedure for refusing entry to the remaining unlisted and non-agricultural fish and wildlife.

Service officials acknowledge the need for better evaluation of risks from unlisted NIS: ''it would be desirable to improve internal Service procedures for modifying the list of injurious wildlife ... by establishing listing criteria and procedures'' (54). The Intentional Introductions Policy Review conducted by the Federal interagency Aquatic Nuisance Species Task Force represents one attempt to do so for aquatic species (see ch. 6) (17). Much of the responsibility in this

area rests with State agencies, many of which lack the necessary regulatory authority and/or resources to adequately address these risks (ch. 7).

ANALYSIS OF CONTROL OR ERADICATION EFFORTS

Although risk analysis primarily focuses on preventing harmful invasions, it also assists in setting priorities for control of established, unwanted NIS. In agricultural applications this tactical decisionmaking is part of Integrated Pest Management programs (ch. 5). Farmers use a variety of systems based on factors like pest population size (determined by sampling); weather; and crop stage for efficient allocation of pesticides, cultivation practices, and other control measures. Some systems have been developed for area-wide agriculture and forestry control projects. These systems, in large part computerized, guide responses to important pests such as the European gypsy moth (*Lymantria dispar*).

Outside agriculture and forestry almost no formal systems for pest control decisionmaking existed until recently. Yet, like farmers and foresters, natural area managers must evaluate new NIS and respond if the risks are high, or they may face a major infestation. Recently developed models and ranking systems can help maximize the impact of limited NIS control budgets for natural areas. These models can help a manager determine, for example, whether it is better to first destroy large concentrated populations of an invasive plant or the outlying ''satellite'' populations (usually the latter (47)).

Ronald Hiebert, Chief Scientist with the National Park Service, Midwest Region, developed such a system for ranking control efforts for the more than 250 non-indigenous plant species growing at Indiana Dunes National Lake Shore (23). The system uses a flexible point scale to weigh the current impact of an introduced plant, its potential for harm, control feasibility, and the consequences of delay. The goal is to allow trained ecologists to rank different NIS. New data and theoretical advances may require continual revision of the ranking system. It is undergoing further testing for broader use and has been used by the State of Minnesota Exotic Species Task Force to classify benign, neutral, and threatening plants (46). The Task Force also adapted it to rank animals.

A simpler ranking system using four categories was developed in 1989 for management of 221 species of non-indigenous plants in and around Everglades National Park (85). The National Park Service has also developed a *Handbook for the Removal of Non-Native Animals* which lays out criteria for ranking species for eradication or control projects (15).

■ Environmental Impact Assessment

Environmental impact assessment refers to a governmental decisionmaking process mandated under the National Environmental Policy Act[7] (NEPA) or under analogous State environmental policy acts (SEPAs), adopted in 18 States (ch. 7). The laws generally require assessments for both government-initiated actions (including funding of private actions) and issuing governmental permits for private actions. Using a standardized environmental assessment check list, the responsible agency makes a ''threshold decision'' as to whether a particular action poses potentially significant environmental impacts, which can include impacts on both the natural and the human-built environment. If so, the agency must prepare a detailed environmental impact statement (EIS) analyzing the potential impacts and alternatives to the action before undertaking or permitting it. The laws also provide opportunities for public comment and for legal appeals on the adequacy of these assessments, including the threshold decision.

NEPA and SEPAs generally do not impose the precise methods of analysis required either for the threshold decision or the EIS, but they do provide some standards.[8] Environmental impact assessments tend to be more qualitative than formal risk analyses (26), although some EISs include quantitative risk analysis.

NEPA has received broad recognition for compelling more analytical decisionmaking (although critics say many ways exist to make the information generated more useful (21)). A recent EIS evaluated the introduction of chinook salmon (*Oncorhynchus tshawytscha*) into the Delaware Bay. However, few detailed EISs have been prepared on other decisions related to NIS except

---

[7] National Environmental Policy Act of 1969, as amended (42 U.S.C.A. 4321 *et seq.*)

[8] 42 U.S.C.A. 4332 generally requires Federal agencies to: ''(A) utilize a systematic interdisciplinary approach which will insure the integrated use of the natural and social sciences and the environmental design arts in planning and in decisionmaking . . . ; (B) identify and develop methods and procedures . . . which will insure that presently unquantified environmental amenities and values may be given appropriate consideration in decisionmaking along with economic and technical considerations; . . . [and] (H) initiate and utilize ecological information in the planning and development of resource-oriented projects.''

for control programs involving widespread pesticide spraying. For example, APHIS has never required an EIS for any new plant or wood imports (16). Some observers claim that NEPA is an adequate mechanism to analyze these potential impacts at the Federal level (65). However, existing regulations lack a clear definition of when NEPA should be triggered for government approval of new imports. Thus, neither APHIS nor any other agency has a clear obligation to follow the NEPA process before allowing the increase of agricultural, horticultural, or wood imports from potentially risky sources such as Mexico, South Africa, and Russia.

Various avenues exist to increase consideration of NIS under environmental impact assessment laws. These include:

- Current NEPA regulations do not cover all governmental actions likely to contribute to NIS problems, such as approving major trade agreements like the North American Free Trade Agreement (this is being litigated; see ch. 10).
- Agencies' existing ''categorical exclusions'' —regulations that excuse NEPA compliance for certain activities—can result in unanalyzed importations or releases. An example is the categorical exclusion for the landscaping of Federal highway projects, including those either federally approved or funded, which have historically involved extensive use of non-indigenous plants.[9]
- Detailed questions specific to NIS are not required in the standardized check lists used for preliminary environmental assessments and for making threshold decisions as to whether an EIS is called for (2).
- Most agency regulations and internal policies do not mandate the integration of risk

The potential for wood imports to carry non-indigenous pests has prompted reconsideration of risk and environmental impact assessment procedures.

analysis or other formal decisionmaking tools into the NEPA process.[10]
- The laws vary widely in the 18 States that have SEPA review processes, and 32 States lack them altogether (ch. 7, table 7-5) (18).

The most rigorous application of NEPA and SEPAs would be to require an EIS for all new releases that are not already on a clean list—in other words to declare by law that new, unanalyzed releases are per se potentially significant environmental impacts and require detailed analy-

ANIMAL AND PLANT HEALTH INSPECTION SERVICE

[9] 23 CFR 771.117(7), as amended (Aug. 28, 1987).

[10] To some extent this is happening, however, in analysis of the risks of noxious weeds on Federal lands in accordance with the 1990 Farm Bill's amendment to the Federal Noxious Weed Act, 7 U.S.C.A. sec. 2814; see, Forest Service Manual Interim Directive 2080-92-1, dated Aug. 3, 1992.

sis. Montana already does this for all new fish releases.[11] However, biological control advocates concerned about potential costs and delays caused by NEPA have argued strongly against a proposal to require an EIS for all releases of new biocontrol agents (10).

Some concern exists that NEPA and SEPAs can hinder the responsiveness of NIS regulation and control (63). However, emergency control measures can be excused from environmental impact assessment requirements.[12] For less urgent, broader control measures, such as long-term weed management, Federal and State agencies have already written many EISs. Little support is evident for reducing the role of NEPA and SEPAs in this regard because of the potential health and environmental impacts of the pesticides used.

Environmental impact assessment laws could affect the adoption of new clean and dirty lists for regulating importation and release. FWS prepared the only known EIS for a new listing approach when the agency proposed its clean list regulation under the Lacey Act, in 1974 (box 4-A). The EIS was fairly basic and general, having been prepared in the early years of NEPA. Because FWS withdrew the regulation, the adequacy of that EIS remains untested.

An EIS for adopting a new regulatory clean list of NIS would address the potential impacts of allowing those listed species into the country, or State. Conversely, an EIS for a new dirty list regulation would need to focus on the potential impacts of allowing in the *unlisted* species. Such a task would be quite difficult to do because the number of unlisted, and mostly unanalyzed, species would presumably be quite large.

■ Economic Analysis

Economic analysis of *past* introductions is feasible through careful research, although relatively little has been done and the studies that

exist are of highly uneven quality (see economic consequences section of ch. 2). Even less has been done in the way of *future* projections that attempt to predict economic scenarios with and without a particular introduction. To date no "standard accounting practice" exists for NIS benefits and costs, whether past or projected.

Projecting future economic effects necessarily follows detailed scientific analysis, such as a pest risk analysis or EIS. That is, economists are data hungry—they cannot assess likely effects of a particular NIS until they understand biological baselines and the likely outcomes of an introduction. Projections of future economic effects are available for about a dozen prominent damaging NIS (ch. 10, table 10-2). In these projections uncertainty about biological outcomes compounds the uncertainty about economic outcomes.

Some question the validity of economic analysis as an aid to public policy decisionmaking because of its heavy reliance on market effects—based on things bought and sold in markets—and lesser emphasis on hard-to-quantify non-market effects. Since the mid-1970s, natural resource economists have made major advances in both the theory and methods of valuing non-market effects (56). (Shadow pricing and contingent valuation are the economic terms for this.) Still, a lively debate continues as to whether these methods adequately account for the way people develop and hold different attitudes toward the value of the natural world or its components (58), aspects of which do not seem amenable to quantification (figure 4-2).

Economic projections do not account well for those future events that have a low probability of occurring but will cause high impact if they do occur (9,56). Unfortunately, many potential NIS problems fit this description. Scientific ignorance, long time lags, and cumulative, sometimes irreversible, effects confound the accounting. For example, highly questionable analyses would

---

[11] Montana Code Annotated 87-5-711(2).

[12] 40 CFR 1506.11, as amended (Nov. 28, 1978).

Figure 4-2—Relative Extents to Which Effects of Indigenous and Non-Indigenous Species are Amenable to Economic Quantification

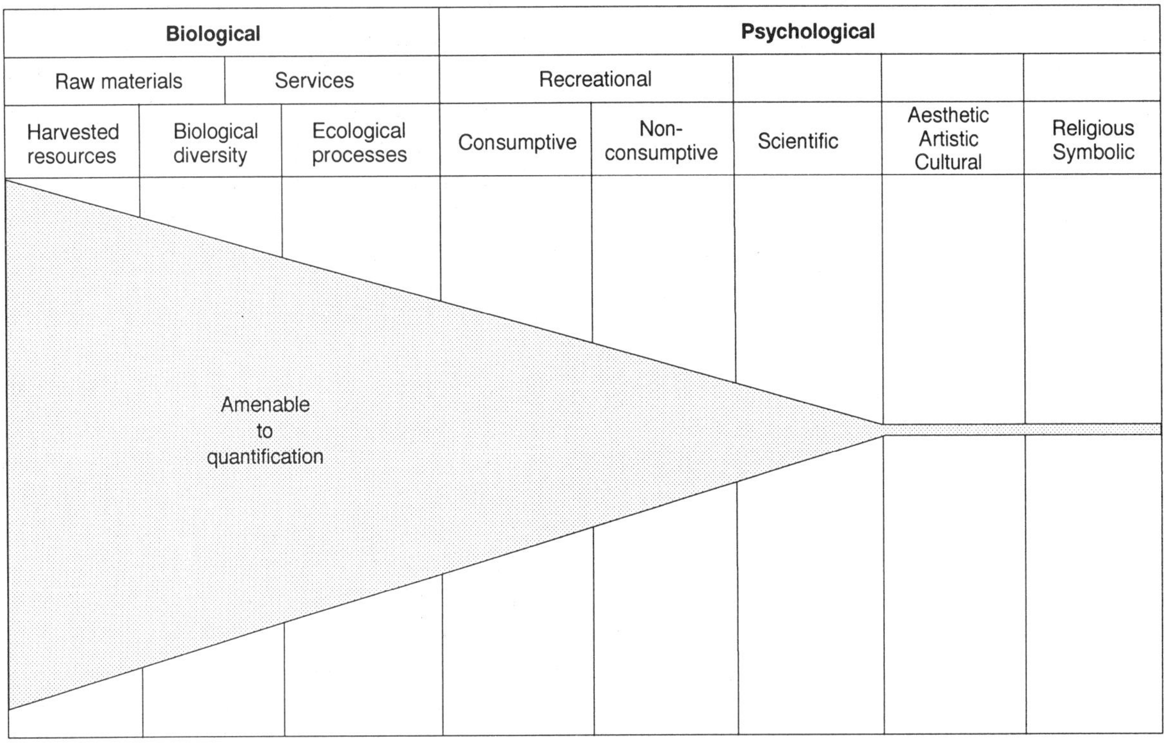

Biological			Psychological				
Raw materials		Services	Recreational				
Harvested resources	Biological diversity	Ecological processes	Consumptive	Non-consumptive	Scientific	Aesthetic Artistic Cultural	Religious Symbolic

Amenable to quantification

SOURCE: C. Prescott-Allen and R. Prescott-Allen, *The First Resource—Wild Species in the North American Economy* (New Haven, CT: Yale Univ. Press, 1986).

derive from estimating the benefits and costs of releasing a sport fish that could, but might not, drive an indigenous, non-harvested fish species to extinction several decades later. Some economists propose assigning rights or entitlements to future generations as an additional way of valuing uncertain future effects (52). However, this "intergenerational equity" has not received wide acceptance in economic accounting to date (56).

Despite these limitations, economic analysis provides a useful rigorous structure to guide decisionmakers who might not otherwise consider all the relevant factors. If the analytical process is accessible to the public and outside experts, it can highlight the areas of debate and uncertainty, making decisionmakers more accountable. This positive effect of economic analysis must be weighed against its costs: personnel,

funding, and time. Incurring these costs may only be justified for cases above a certain threshold of risk that cannot be resolved using other accepted methods.

Economics has utility for broader aspects of NIS decisionmaking than whether a particular NIS should be imported, introduced, or controlled (box 4-C). Well-documented economic analysis can help in designing the most efficient regulatory approaches as well as appropriate incentives (e.g., rewards, bounties) and disincentives (e.g., taxes) to respond to existing problems (56). It can determine effective levels of fines and penalties for violations, that is, disincentives that will keep importers and purchasers of potentially harmful NIS from imposing externalized costs on society.

Economics also serves to ensure that both private and government resources are expended

---

Box 4-C—Macroeconomics and Non-Indigenous Species

Macroeconomics is the study of whole systems and the relationships among different economic sectors. Examination of the increasingly linked global economic system, in which relationships are largely expressed through international trade, illuminates the larger forces behind NIS problems. Some important trends:

- As developing countries pursue export markets for cash crops, traditional agroecosystems are increasingly converted to large monocultures. Global homogenization of crops can reduce biological diversity and increase the crops' vulnerability to pests.
- In the last several years, economic and political changes have resulted in several new significant U.S. trading partners, from Chile to China. These shifts in NIS pathways could lead to new pest problems.
- The North American Free Trade Agreement, if implemented, will increase certain imports from Mexico that pose pest risks, such as fruits and vegetables (see ch. 10).

Economic analysis could also highlight the role NIS play in different sectors of the U.S. national economy and the potential impact of more, or fewer, import restrictions. For example, to what extent do profits of the nursery industry depend on continued infusion of new imported species or varieties? Could an indigenous plant industry substitute for imports in a way that would satisfy consumer preferences and maintain industry profitability? Little analysis of such questions has been done by either government or industry. They represent areas of fruitful inquiry on the relationship between economics and the environment.

SOURCES: R.B. Norgaard, "Economics as Mechanics and the Demise of Biological Diversity," *Ecological Modelling*, vol. 38, 1987, pp. 107-121; T. Dudley, Research Botanist and Project Leader, National Arboretum, personal communication to Office of Technology Assessment, Oct. 4, 1991; C. Regelbrugge, Director of Regulatory Affairs, American Association of Nurserymen, personal communication to Office of Technology Assessment, Oct. 8, 1991.

---

wisely on broad programs. For example, New Zealand's forest industries recently undertook a detailed benefit/cost analysis on conducting forest pest detection surveys at various levels of intensity (6). They found the maximum national net benefit from these surveys resulted at levels that detect 95 percent of new introductions (figure 4-3). The costs of detecting the last 5 percent sharply exceed the marginal benefits. This exemplifies the case that seeking 100 percent success is not always the optimal allocation of resources. However, optimal resource allocation depends entirely on the context, and relatively few detailed studies exist for U.S. NIS programs. In other environmental areas a clear trend exists toward incorporating more economic analysis in designing new policies (13).

BENEFIT/COST ANALYSIS

Where enough is known about the probabilities of future effects from NIS, one can calculate the different expected values of resulting benefits and costs. Benefit/cost analysis (BCA) is a method of weighing particular decisions (box 4-D), such as allowing an NIS to be imported or introduced, or controlling or eradicating it if already present (9). The resulting ratio compares the cumulative potential economic benefits to the costs of the decision, expressing them in 1991 dollars (present value).

Calculating a benefit/cost ratio does not automatically determine a decision. Even when the benefits are greater, the magnitude of the costs may be so high as to make the action unacceptable or unfeasible. Costs and benefits that are unevenly distributed socially, geographically, or generationally can present fairness questions. For example, crop losses from pests can be highly regional—some farmers may lose while others profit from increased market prices (32). Excessive uncertainty or questionable valuation techniques may undercut the analysis. BCA is most useful for ranking a comparable group of desira-

Figure 4-3—National Costs and Benefits of Detecting Forest Pest Introductions in New Zealand

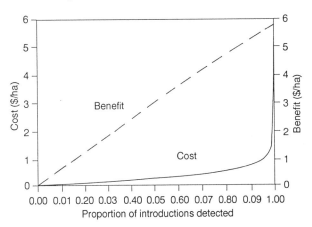

SOURCE: P.C.S. Carter, "Risk Assessment and Pest Detection Surveys for Exotic Pests and Diseases which Threaten Commercial Forestry in New Zealand," *New Zealand Journal of Forestry Science,* vol. 19, Nos. 2/3, 1989, pp. 353-374.

ble actions when budget constraints prevent undertaking them all (9).

In fact, benefit/cost ratios have been calculated for only a few NIS decisions. Most existing studies have focused on the economic justification for eradicating or controlling established infestations. Benefit/cost ratios have been developed for past or potential effects of 12 prominent NIS (table 4-4). In almost all the studies (of highly variable rigor) the ratios are high (median 17.2/1; range 0.23/1 to 1,666/1). That is, the management actions are well justified economically because the overall benefits of eradicating, controlling, or preventing the potential infestations far exceed the costs of the actions. However, these ratios do not give detailed accounting for the uneven distribution of the effects. Also, several of the "potential impacts" represent worst-case scenarios. The analyses did not weigh the likelihood that the worst potential impacts would actually occur. Thus, those resulting ratios are probably too high.

As with risk analysis, future theoretical and technical improvements are likely to make BCA's more comprehensive (56). BCA for NIS will

benefit from the development of standardized practices, such as those proposed in box 4-D and table 4-4, to make results more consistent and comparable. The ability of economists to provide useful analyses will depend to a large extent on whether scientists can estimate probabilities of future effects of NIS in a consistent, comparable way. Economic models provide little assistance, regardless of their sophistication, where they rest on vague or equivocal predictions of biological events ("garbage in, garbage out").

DECISIONMAKING PROTOCOLS

Protocols are written codes used in diplomatic, military, and scientific affairs to guide adherence to a prescribed course of action. In the NIS context, decisionmaking protocols consist of criteria developed by experts to guide the determination of whether a proposed activity involving NIS is appropriate. Some protocols also prescribe precautions to minimize risks. They can be focused narrowly, such as to guide procedures for federally funded research on non-indigenous aquatic species, or broadly on policy-level decisions, such as the model national approach proposed by the International Union for Conservation of Nature and Natural Resources (box 4-E). The broader protocols have the distinctive feature of going beyond scientific or risk-based criteria to encompass value-based considerations and to guide the weighing of benefits and costs.

Protocols lack enforceability except when adopted by law, which has rarely happened (5,84). For example, the American Fisheries Society protocol on new fish introductions has existed for more than 20 years, but no Federal or State laws mandate its use, despite calls for its adoption (33). Few documented cases of its voluntary use exist (11,51). Congress considered, but did not pass, a bill[13] in 1991 requiring agencies to follow a detailed protocol for aquatic introductions (77). Several experts have supported greater use of

---

[13] The Species Introduction and Control Act of 1991, H.R. 5852.

Table 4-4—Documented Benefit/Cost Ratios for Eradication, Control, or Prevention of Selected Non-Indigenous Species

Notes: dollar figures are in millions; totals columns give Net Present Values in 1991 dollars, calculated as indicated in box 4-D to the extent that the information was provided in the original studies; letters after species names refer to references for table 4-4 at end of this table. Note numbers refer to notes at bottom of page. The ratios given compare the benefits to the costs of eradicating, controlling, or preventing the NIS invasion under the circumstances that were studied. (Check index for scientific names.)

| | Direct effects | | Indirect effects | | Costs | | Distribution costs considered | Year of study | 1991 total benefits | 1991 total costs | Benefit/ cost ratio |
	Market goods	Nonmarket goods	Multiplier effects	Related goods	Direct control costs	Opportunity costs					
Past impacts—Plants											
Hydrilla and water hyacinth[a]	0.497			0.016			N	1974	1.260	0.041	31/1
Hydrilla and water hyacinth[a]		0.023			0.100		N	1977	0.047	0.203	0.23/1
Hydrilla and water hyacinth[a]		0.567			0.003		N	1978	1.075	0.006	179/1
Hydrilla and water hyacinth[a]		0.869			0.019		N	1979	1.514	0.033	45.9/1
Hydrilla and water hyacinth[a]		0.468			0.089		N	1982	0.641	0.122	5.25/1
Melaleuca[b]	160[1]	160[1]					Y	1991	160[1]	12.3[1]	13/1
Melaleuca[c]	8.4		15.2		12.3[1]	15.0	Y	1989	182.75	16.259	11.24/1
Leafy spurge[d]				145.0			N	1984			10/1[2]
Pest impacts—Fish											
Sea lamprey[e]		219,748		42.896	8.681		N	1988	296.421	9.797	30.25/1
Sea lamprey[f]		550[3]			40		N	1980	878.588	63.897	13.75/1
Past impacts—Insects											
Alfalfa blotch leafminer[g]	13				1.1	1.6	N	1983	17.129	2.086[4]	
Potential impacts—Plants											
Purple loosestrife[h]	6.54	39.32			0.100		N	1987	53.477	1.982	27/1
Witchweed[i]	389.55				57.4		N	1976	845.6	124.53	6.78/1
Witchweed[i]	997.17				57.4		N	1976	2,163.43	124.53	17.37/1
Witchweed[i]	389.55				52.1		N	1976	845.16	113.03	7.47/1
Witchweed[i]	997.17				52.1		N	1976	2,163.43	113.03	19.1/1
Potential impacts—Insects											
Cotton boll weevil[j]	3.755[5]			-0.84	0.16		Y	1979	5.068	0.279	18.1/1
Cotton boll weevil[j]	5.50[5]			-1.37	0.24		Y	1979	7.193	0.418	17.2/1
Mediterranean fruit fly[k]	1,256[6]				64		Y	1981	1,829.22	93.21	19.62/1
Mediterranean fruit fly[k]	816[6]				64		Y	1981	1,188.41	93.21	12.75/1
Mediterranean fruit fly[l]	3,078				62.76		N	1981	4,482.49	91.40	49/1
Potential impacts—Pathogens											
Foot and mouth disease[m]	11,650[7]				467		N	1976	25,275.51	1,013.19	24.95/1
Foot and mouth disease[m]	11,650[7]				690		N	1976	25,275.51	1,497	16.88/1
Potential impacts—Other Pests of:											
Siberian log imports[n,o]	62,152[8]					37.4	Y	1990	64,704.21	38.94	1,661/1
Siberian log imports[n,o]	35,390.35[8]					37.4	Y	1990	36,843.62	38.94	946/1

NOTES:

1. Direct effects and costs were reported without further classifications, therefore, these figures are listed here under their general headings.
2. Only benefit/cost ratio was reported for this study, without supporting figures.
3. These estimates are the value of all sport and commercial fishers in the Great Lakes. This study used "all or none" valuation technique and hence overstates benefits to sea lamprey control.
4. Costs converted to 1991 dollars by assuming that midpoint of time series was appropriate index year. Assumption was made due to lack of information on the flow of funds through the time series.
5. Two scenarios were examined—the first is for current insect control with boll weevil eradication and the second is for optimum pest management with no government incentives but with a boll weevil eradication program. The analysis is for a 15-year period starting in 1979.
6. High and low cost scenarios were used to estimate the impacts of severe infestations of the Mediterranean fruit fly in California. These were contrasted against only 2 years of current to control costs ($64 million), generating benefit/cost ratios which may be high.
7. High and low control costs were employed as contrasted to the benefits estimated from 1976 to 1990.
8. High and low scenarios for the economic impacts assuming accidental introduction and unmitigated infestations of defoliators (i.e., Asian gypsy moth and Nun moth), nemotodes, larch canker, spruce bark beetles, and annosus root disease resulting from the import of Siberian logs as contrasted to the estimated net welfare gains from the log imports.

REFERENCES FOR TABLE 4-4

a D.E. Colle, J.V. Shireman, W.T. Haller, J.C. Joyce, and D.E. Canfield, Jr., "Influence of Hydrilla on Harvestable Sport-fish Populations, Angler Expenditures at Orange Lake, Florida," *North American Journal of Fisheries Management*, vol. 7, 1987, pp. 410-417.

b *Federal Register*, "Noxious Weeds: Additions to List," vol. 56, No. 201, Oct. 17, 1991, pp. 52,005-52,007.

c C. Diamond, D. Davis, and D.C. Schmitz, "Economic Impact Statement: The Addition of *Melaleuca quinquenervia* to the Florida Prohibited Aquatic Plant List," unpublished paper for the Joint Center for Environmental and Urban Problems, Ft. Lauderdale, Florida.

d U.S. Department of Agriculture, Animal and Plant Health Inspection Service, "Biological Control of Leafy Spurge," program aid #1435, 1988, p. 2.

e W.M. Spaulding, Jr. and R.J. McPhee, "An Analysis of the Economic Contribution of the Great Lakes Sea Lamprey Control Program," *The Report of the Great Lakes Fishery Commission by the Bi-National Evaluation Team*, vol. 2, Nov. 28, 1989, pp. 1-27.

f D.R. Talhelm and R.C. Bishop, "Benefits and Costs of Sea Lamprey (*Petromyzon marinus*) Control in the Great Lakes: Some Preliminary Results," *Canadian Journal of Fishery and Aquatic Science*, vol. 37, 1980, pp. 2,169-2,174.

g J.J. Drea and R.M. Hendrickson, Jr., "Analysis of a Successful Classical Biological Control Project: the Alfalfa Blotch Leafminer (Diptera: Agromyzidae) in the Northeastern United States," *Environmental Entomology*, vol. 15, No. 3, June 1986, pp. 448-453.

h D.Q. Thompson, "Spread, Impact, and Control of Purple Loosestrife (*Lythrum salicaria*) in North America Wetlands," U.S. Fish and Wildlife Service, 1987.

i P.M. Emerson and G.E. Plato, "Social Returns to Disease and Parasite Control in Agriculture: Witchweed in the United States," *Agricultural Economic Research*, vol. 30, No. 1, 1978, pp. 15-22.

j C.R. Taylor et al., "Aggregate Economic Effects of Alternative Boll Weevil Management Strategies," *Agricultural Economics Research*, vol. 35, No. 2, April 1983, pp. 19-28.

k R.K. Conway, "An Economic Perspective of the California Mediterranean Fruit Fly Infestation," National Economics Division, Economic Research Service, U.S. Department of Agriculture, ERS Staff Report #AGES820414, 1982.

l T.T. Vo, "Economic Analysis of the Mediterranean Fruit Fly Program in Guatemala," Animal and Plant Health Inspection Service, U.S. Department of Agriculture, August 1989.

m National Research Council, "Long-Term Planning for Research and Diagnosis to Protect U.S. Agriculture from Foreign Animal Diseases and Ectoparasites," Subcommittee on Research and Diagnosis on Foreign Animal Disease, Committee on Animal Health, Board on Agriculture, NRC paper #PB84-165141, 1984.

n U.S. Department of Agriculture, Forest Service, "Pest Risk Assessment of the Importation of Larch from Siberia and the Soviet Far East," Misc. Publ. No. 1495, September 1991, pp. S-1-K-15.

o U.S. Department of Agriculture, "Economic Benefits from the Importation of Soviet Logs," unpublished draft (1991).

SOURCE: M. Cochran, "Non-Indigenous Species in the United States: Economic Consequences," contractor report prepared for the Office of Technology Assessment, March 1992.

Box 4-D—Outline of Steps for Benefit/Cost Analysis of Non-Indigenous Species

I. Effect estimation
   A. Identify relevant input and output categories
      1. Inputs—(e.g., wetland invasion by non-indigenous melaleuca)
      2. Outputs—(e.g., tourism; honey production)
   B. Define units of measurement for input and output categories
      1. Inputs—(e.g., acres invaded)
      2. Outputs—(e.g., tourist expenditures; quantity of honey sold)
   C. Establish a base of values for input and output categories without the introduction of the NIS
   D. Identify production process relating to introduction of the NIS to a series of outputs, expressed probabilistically
      1. Expected units of invasion—(e.g., acres of distinct environs where NIS would be established and distributed)
   E. Quantify expected magnitude of each output for the relevant magnitudes of each input category
   F. Estimate changes in input and output categories for with introduction versus without introduction scenarios

II. Valuation of direct effects
   A. Market goods
      1. Marginal changes in production
         a. Market price x change in output quantity
      2. Non-marginal change in product in product
         a. Identify market price changes
         b. Measure consumer and producer surplus
   B. Non-market goods
      1. Contingent valuation

III. Calculate indirect effects
   A. Multiplier income and employment effects
      1. Opportunity costs
      2. Unemployed resources
   B. Related goods
      1. Changes in production
      2. Changes in market price
      3. Calculate consumer and producer surplus

IV. Calculate annual benefits and costs

V. Accounting for time
   A. Select appropriate discount rate
      1. Use real (deflated) rate (e.g., riskless rate; Water Resources Council rate)
   B. Convert annual benefits and costs to real terms (e.g., using CPI, GNP Deflator)
   C. Calculate present value
      1. Present value of benefits = $\sum\limits_{n=0}^{N} \dfrac{B_n}{(1+r)^n}$

      2. Present value of costs = $\sum\limits_{n=0}^{N} \dfrac{C_n}{(1+r)^n}$

n = number of the year in time series, N = last year of time series, r = discount rate, B = benefits, C = costs

SOURCE: M. Cochran, "Non-Indigenous Species in the United States: Economic Consequences," contractor report prepared for the Office of Technology Assessment, March 1992.

Box 4-E—The IUCN Position Statement on Translocation of Living Organisms

A broad protocol covering the whole field of NIS releases was developed by the International Union for Conservation of Nature and Natural Resources (IUCN), a body comprised of scientific experts and government officials involved in conservation from around the world. The lengthy IUCN Position Statement on Translocation of Living Organisms, approved in 1987, lays out many questions to answer and steps to follow when considering future releases. In summary it provides that:

- Release of a NIS should be considered only if clear and well-defined benefits to humans or natural communities can be foreseen.
- Releases should be considered only if no indigenous species is suitable.
- No NIS should be deliberately released into any natural area; releases into seminatural areas should not occur absent exceptional reasons.
- Planned releases, including those for biological control, entail three critical phases: rigorous assessment of desirability; controlled experimental release; and extensive release accompanied by careful monitoring and pre-arrangement for control or eradication measures, if necessary.
- Special consideration should be given to eradicating existing introductions in ecologically vulnerable areas.

This approach represents the most broadly applicable model national law on NIS. Indeed, the position statement calls on national governments to provide the "legal authority and administrative support" to implement IUCN's approach. This has not occurred. The statement did substantively influence the initial version of the Convention on Biological Diversity, which was drafted by IUCN's legal branch. However, by the time the convention was opened for signing in Rio de Janeiro the negotiation process had greatly diluted the strong principles summarized above (see ch. 10).

SOURCE: International Union for Conservation of Nature and Natural Resources, Species Survival Commission, "The IUCN Position Statement on Translocation of Living Organisms: Introductions, Reintroductions, and Restocking" (Gland, Switzerland, 1987).

---

protocols; some suggest that they be implemented federally by grafting their use into NEPA when agencies assess potential environment impacts of proposed releases (74).

Adhering to a decisionmaking protocol can require data that are more difficult or expensive to obtain than the information traditionally considered by managers. Even so, protocols often do not eliminate subjectivity and scientific uncertainty— some of the needed data may be unobtainable. Few protocols have been validated by way of follow-up evaluations of decisions based on them (83). Of course, if they are used more broadly greater opportunities for evaluation will exist.

Some prominent decisionmaking protocols do exist or have been proposed (box 4-F). Others could be developed to cover additional NIS groups and situations. Biological control specialists in particular have proposed codifying more comprehensive protocols: 1) to preempt overly restrictive regulations constructed by non-experts and 2) to protect the public from amateur introductions (10). Their emphasis is on flexibility within a reasonable, non-regulatory framework: "the protocols must be dynamic, i.e., capable of being updated in response to ever increasing knowledge and changing conditions" (10). Fisheries specialists have also stressed voluntary compliance with protocols or guidelines, especially combined with education regarding it importance, as a way to avoid the litigation that might accompany overly strict regulations (31).

VALUES IN DECISIONMAKING

Many NIS issues may not be resolvable using risk analysis, environmental impact assessment, or economic analysis, because of lack of necessary information or disagreement over the appro-

---

Box 4-F—Prominent Decisionmaking Protocols

Codes of Practice and Manual of Procedures for Consideration of Introductions and Transfers of Marine and Freshwater Organisms, European Inland Fisheries Advisory Commission, Food and Agriculture Organization, United Nations, Rome, Italy, and International Council for the Exploration of the Sea, Copenhagen, Denmark; revision published in 1988.

Guidelines for Introducing Foreign Organisms into the United States for the Biological Control of Weeds, Working Group on Biological Control of Weeds, joint Weed Committees of the U.S. Departments of Agriculture and Interior; revised in 1980. (The U.S. Department of Agriculture has developed several other guidelines for the importation, interstate movement, and field release of various types of organisms for biological control.)

Guidelines for Re-Introductions—Draft, Re-introduction Specialist Group, Species Survival Commission, International Union for Conservation of Nature and Natural Resources, Gland, Switzerland; proposed in 1992.

IUCN Position Statement on Translocation of Living Organisms, International Union for Conservation of Nature and Natural Resources, Gland, Switzerland; approved in 1987.

Position Statement on Exotic Aquatic Organisms' Introductions, American Fisheries Society, United States; revision adopted in 1986.

Protocol for Translocation of Organisms to Islands, New Zealand; proposed in 1990.

Research Protocol for Handling Nonindigenous Aquatic Species, U.S. Fish and Wildlife Service, National Fisheries Research Center, Gainesville, Florida, adopted by the Federal interagency Aquatic Nuisance Species Task Force in 1992.

The Planned Introduction of Genetically Engineered Organisms: Ecological Considerations and Recommendations, Ecological Society of America; proposed in 1989.

SOURCES: J.T. Carlton, "Man's Role in Changing the Face of the Ocean," *Conservation Biology*, vol. 3, No. 3, September 1989, pp. 270-272; D.L. Klingman and J.R. Coulson, "Guidelines for Introducing Foreign Organisms into the United States for the Biological Control of Weeds," *Bulletin of the Entomological Society of America*, vol. 19, No. 3, 1983, pp. 55-61; J.M. Tiedje et al., "The Planned Introduction of Genetically Engineered Organisms: Ecological Considerations and Recommendations," *Ecology*, vol. 70, No. 2, 1989, pp. 298-315; D.R. Towns et al., "Protocols for Translocation of Organisms to Islands," *Ecological Restoration of New Zealand Islands*, D.R. Towns et al. (eds.) (Wellington, New Zealand: Department of Conservation, 1990).

---

priate method. Decisionmakers may prefer, or be compelled, to decide on the basis of fundamental values. As used in this section, "values" has no monetary connotation, rather, it refers to overarching criteria that people use to make decisions (3). Values, although they are critical, often receive little explicit acknowledgment in studies of decisionmaking because of the focus on science-based models.

For most non-native Americans, being of relatively recent stock in North America and Hawaii, little of their cultural identity revolves around a relationship with indigenous species. Indeed, much pioneer history is the story of clearing the land of threatening or competing indigenous species in favor of tame, familiar, introduced ones. Not surprisingly, preserving indigeneity, both biological and cultural, has only risen as a public value in the last few decades. The Endangered Species Act[14] represents the strongest national law embodying this biological *preservation value*. It is also reflected in native plant societies and similar manifestations of a growing emphasis on using indigenous species for landscaping and other applications (45).

Americans also place strong emphasis on liberty as a value, here encompassing the liberty to sell, purchase, catch, hunt, possess, and use

---

[14] Endangered Species Act of 1973, as amended (16 U.S.C.A. 1531 *et seq.*)

NIS. Most people own pets and/or keep house or garden plants, which are virtually all non-indigenous. This *liberty value* is so strong that at the 1974 congressional hearing on the FWS attempt to implement a clean list decisionmaking approach (box 4-A, above), the successful opponents—largely the pet trade—argued that it usurped their civil rights to import NIS (76). This liberty is not limited to dogs, cats, and poinsettias. Many people want to own novel species because of their novelty (4).

Values can conflict at social or personal levels. The use of non-indigenous fish for recreational fishing, such as hybrid bass (*Morone chrysops x M. saxatilis*), represents a social conflict (59). Anathema to fishing purists, these "put and take" fisheries enjoy broad popularity—some have clubs devoted to their furtherance. Preserving indigeneity in U.S. waters conflicts with the liberty to use the new fish. However, a limited opinion poll (Arizona only) suggests that the public opposes the release of non-indigenous fish that threaten the existence of indigenous fish (22).

No broad public survey data exist on the prevalence of concerns about NIS problems. Surveys do show the public to be very concerned about the health risks of pesticides, however (8). A person who supports preservation of indigenous species may also oppose the use of chemical pesticides because of their health risks. In situations where chemical pesticides offer the only control for NIS that threaten indigenous species, that person has a personal conflict. He or she must decide which carries the most weight, the *preservation* or the *health value*.

Many NIS choices boil down to *humane values*, rooted in basic moral principles. Monkeys may be low-risk invaders, but many people object to their being imported and possessed as pets for ethical reasons. Feral horses and burros (*Equus asinus*) have been successful and often damaging invaders, but vocal citizen groups object to their being killed on ethical grounds. However, few object on ethical grounds to the killing of the less attractive feral hog (*Sus scrofa*)—advocates for

their preservation are the hunters who want to shoot them. (Indeed, a survey has shown that if a decisionmaker is a hunter he or she is more likely to view non-indigenous animals, like feral hogs, as a beneficial resource than if he or she does not hunt (61)). Almost no one objects on ethical grounds to highly deleterious rats or sea lampreys (*Petromyzon marinus*) being killed.

Clearly the attitudes of the public vary with the perceived attractiveness and usefulness of the species involved, indigenous or non-indigenous (28). Nevertheless, most people would probably support the following ethical position: regardless of the species being controlled, if other factors such as costs and risks are equal, managers should use the most humane methods. When applied in the field, though, "humane" methods of control elude easy definition (69).

OTA makes no findings as to which values deserve the greatest weight. Their role in past decisions, however, has tended to lack clarity. Future policy and management decisionmaking would benefit from explicitly separating factual questions from questions of values. Nevertheless, cultural, religious, and historical factors will inevitably color a decisionmaker's perspective.

NEW SYNTHESES OF DIVERSE APPROACHES

Difficulties abound in generalizing about NIS decisionmaking. An approach that holds for one taxonomic group may not hold for another—one size does not fit all. Potential impacts (harmful and beneficial) vary with the species and the environments involved. Different areas of the country often have different interests. A new NIS may favor one group in society and burden another.

Numerous interests can influence NIS decisionmaking (figure 4-4). Each interest is not monolithic; as much contention can occur within an identified group as between them. Not all these interests are brought to bear in all cases nor do all carry equivalent weight. For example, a large or politically influential constituency that favors a particular decision regarding NIS may far outweigh the positions of a small number of expert scientists who caution against the decision (44).

Are methods available to reconcile these diverse interests and to resolve disputes that may otherwise end in expensive and burdensome litigation? If decisionmakers attempt to reconcile these interests, which of the approaches discussed above should they rely on—risk analysis, environmental impact assessment, economic analysis, and/or protocols? None of them alone is currently broadly applied to NIS. And how should diverse values factor in?

Two proposals for synthesis allow incorporation of diverse societal interests and capitalize on the strengths of the various decisionmaking approaches without according any of them trump status. These proposals are outlined here with the caveat that their application in particular contexts may require modifications.

■ Benefit/Cost Analysis Subject to a Safe Minimum Standard

Economist Alan Randall of Ohio State University proposes that current natural resources economics theory justifies this rule: Decide on the basis of maximizing net benefits to society subject to the constraint of a Safe Minimum Standard (56). A "safe minimum standard" is a level of environmental quality that society should not go below, except in extraordinary cases. The rule applies in deciding whether to prevent, support, or take no action on a particular introduction, or whether an existing NIS should be controlled or eradicated. It can be applied to intentional releases and in preparing for or responding to accidental releases. Generic application of the Randall approach by a manager would follow six steps, with the underlying premise that each step involves an open, pluralistic process (56):

Step 1: The manager obtains the judgment of scientists who use risk analysis, experimentation, and/or other methods to predict the likely spread and effects of a particular NIS. They determine likely future scenarios of resulting ecological situations under both baseline conditions, i.e., no introduction or further spread, and "with introduction" (or further spread) conditions. The scientists then determine whether a real possibility exists of a *harmful* invasion. If so, the manager proceeds to Step 2. If no such possibility exists, then the introduction can proceed, providing for further consideration if and when new evidence arises.

Step 2: The manager obtains the judgment of scientists as to whether a possibility exists of ecologically *disastrous*—as opposed to harmful but manageable—consequences. If ecologically disastrous consequences are **not** a real possibility, the manager proceeds to Step 3. If ecologically disastrous consequences **are** a real possibility, the manager omits Steps 3, 4, and 5, and proceeds to Step 2a.

Step 2a: If a real possibility of ecologically disastrous consequences exists, the manager invokes a Safe Minimum Standard rule. This is a presumption based on preservation and other values that actions will not be taken that cause ecologically disastrous consequences even if substantially greater potential benefits are lost. The introduction would be prevented or reversed except for extraordinary cases in which the value of these foregone benefits would be intolerably high. To make that decision, the manager first obtains economic calculations of the foregone benefits, then engages in a public decision process to determine whether these are socially intolerable. If the decision is made to proceed, mitigation of the potentially disastrous consequences would be pursued.

**Figure 4-4—The Major Interests Involved in Shaping
Non-Indigenous Species Policy**

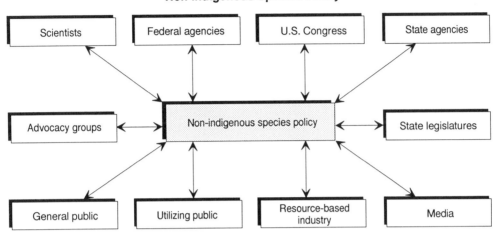

SOURCE: Adapted from S.R. Kellert and T.W. Clark, "The Theory and Application of a Wildlife Policy Framework,"
Public Policy Issues in Wildlife Management, W.R. Mangun (ed.) (New York, NY: Greenwood Press, 1991), pp. 17-38.

Step 3: (from Step 2) Starting with the baseline and "with introduction" scenarios (predicted in Step 1), the manager employs economists to develop accounts of the resulting flows of goods and services. They perform benefit/cost analyses based on these accounts, with appropriate market and non-market valuation methods to measure total value, including use and "existence" values. The manager determines whether the prospective introduction (or spread) is expected to have a net benefit.

Step 4: If Step 3 reveals that the introduction (or spread) will not have a net benefit, the manager develops alternative scenarios to prevent it. If Step 3 reveals positive net benefits, but also significant harmful effects (ecological or economic), alternative scenarios to mitigate the harmful effects are developed. Then the economists perform further benefit/cost analyses based on accountings under these new scenarios that incorporate the prevention or mitigation alternatives. (If positive net benefits result with no significant harmful effects, then no further accounting is needed.)

Step 5: The manager gives full public consideration to the benefits and costs of the alternatives resulting from Step 4. Absent compelling input to the contrary, the alternative with the maximum net benefits is chosen.

The Randall approach represents a compromise between the liberty value and the preservation and humane values discussed in the values section, above. That is, traditional benefit/cost analysis assumes the decisionmaker has the freedom to choose the maximum net benefit alternative, regardless of associated costs, whereas the Safe Minimum Standard (Step 2a) constrains that liberty based on a socially accepted higher good. The constraint also acts as a check on the problem, discussed above, of relying on economic analysis to value effects of low-probability future events that may be irreversible (i.e., disastrous), like extinction.

■ Decision Analysis Combined With Alternative Dispute Resolution

Professor Lynn Maguire of the Duke University School of the Environment proposes a different way to synthesize decisionmaking approaches. It combines decision analysis with alternative dispute resolution. This method has more participation by ''stakeholder'' groups than does the Randall approach, but fewer pre-selected analytical methods (42).

Decision analysis is a framework that ensures that the components common to any decision are recognized and addressed explicitly. Those decision components are: objectives, criteria, alternative actions, sources of uncertainty, and values associated with possible outcomes. The concurrent use of alternative dispute resolution recognizes that leaving difficult decisions to government officials or experts can result in continued conflict among the interest groups involved. The process creates a forum for addressing the decision components, making tradeoffs, recognizing common ground, and making the needed decision. A similar framework has been proposed for decisionmaking for releases of genetically engineered organisms (20). The Maguire approach proceeds through four steps:

Step 1: Identify and convene, in a neutral setting, representatives of stakeholder interest groups in a particular NIS decision (e.g., release, control, eradication, or regulatory changes).

Step 2: Undertake preliminary negotiations to achieve, where possible, joint acceptance of major objectives and sub-objectives, and criteria for judging whether alternative outcomes from the decision to be made meet the objectives (i.e., the ''utility'' of the outcome). To the extent possible, separate technical questions from value-based questions and obtain technical expertise to address the former. When agreed, engage in joint fact-finding efforts.

Step 3: The parties flesh out the sub-components of their views of the probable effects of the alternative outcomes, including factual and value-based effects. These are graphically represented on a ''decision tree'' in which the parties, with expert assistance if needed, assign perceived probabilities to different outcomes (the ''branches'' of the tree), accounting for uncertainties. The ''utility'' (identified in Step 2) of each identified outcome is weighted with the perceived probability of the outcome occurring to calculate the ''expected utility'' of each outcome for each party.

Step 4: All parties identify actions with ''maximum expected utility.'' Other jointly accepted rules, such as minimizing the largest costs, are also possible. Identify and negotiate options to reduce uncertainty by obtaining additional information. If agreed, obtain this additional information. Then discuss creative tradeoff alternatives in view of the maximum expected utilities of all parties or other accepted decision rule. Attempt to negotiate tradeoffs with the aim of achieving a consensus decision.

The Maguire approach, unlike the Randall approach, neither makes presumptions based on values nor prescribes analytical methods. It mandates less input from scientists and economists than the Randall approach. Consequently, the outcomes may reflect less ''good science'' and rely more on the subjective probabilities assigned by the participants. Indeed, the absence of scientific answers may be why the dispute among the stakeholders exists in the first place. The approaches are not mutually exclusive, however. Participants in the Maguire process could ''jointly accept'' that benefit/cost analysis subject to a Safe Minimum Standard embodies the appropriate Step 2 criteria to judge the utility of alternative outcomes. They could choose to obtain more ''good science'' to the extent possible.

OTA finds three common hurdles to implementing these two approaches:

1. Lack of clear guidance as to what should trigger the significant commitment of personnel, expertise, and time necessary to implement formal approaches. Various trigger options exist, however: for preparation of any new clean or dirty list; pursuant to a petition process (similar to listing decisions under the Endangered Species Act); under

NEPA for controversial environmental impact statements (21); and pursuant to the Federal Negotiated Rulemaking Act,[15] discussed below.

2. Lack of convincing treatment of uncertainty, because of their emphases on negotiating, quantifying, or developing scenarios based on unknowns. Admittedly, it is hard to envision *any* convincing treatment of uncertainty in a decisionmaking model.

3. Lack of evaluation of their adaptability to NIS decisionmaking in the real world. Randall's Safe Minimum Standard very roughly resembles the restrained benefit/cost weighing allowed under the Endangered Species Act (55). (The act's Safe Minimum Standard is no further human-caused extinctions unless the ''God Squad'' determines the costs to be intolerably high in a particular case.) The Maguire approach has been utilized successfully in other natural resource contexts, such as reintroducing the endangered grizzly bear (*Ursus arctos horribilis*) in the Northern Rockies, which is comparable in some ways to introducing potentially harmful NIS (43). Obviously, neither model can be evaluated in the NIS context unless a commitment is made to try them.

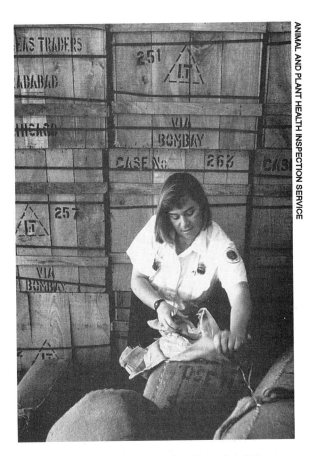

Agencies' implementation of decisions should be evaluated if new decisionmaking methods are tried. Also, the quality of decisions reached must be assessed, i.e., whether new approaches ultimately improve management of harmful NIS.

As far as strengths, both models can incorporate the various decisionmaking approaches discussed in this chapter. In doing so, they organize and structure information from diverse sources but are not overly rigid. Both proposals also call for full documentation of the process. They force methods, assumptions, comparisons, and trade-offs to be explicit, which facilitates their communication, review, and appraisal (20,68).

The question remains how these or comparable decisionmaking approaches could be integrated into a regulatory process. One existing avenue is the Federal Negotiated Rulemaking Act. It provides a process whereby the head of a Federal agency makes a threshold decision about whether an issue would benefit from negotiations. He or she bases this on the need for a new Federal regulation and the feasibility of convening a representative committee likely to achieve consensus. Public notice of the process is required. The agency may hire professional facilitators to run the negotiations. Under the act, the agency commits to using the consensus agreement, if the parties reach one, as the basis for the proposed regulation ''to the maximum extent possible

---

[15] Negotiated Rulemaking Act of 1990 (5 U.S.C.A. section 561 *et seq.*)

consistent with the legal obligations of the agency.''[16] Although it apparently has never been applied before in the NIS context, negotiated rulemaking has successfully resolved disputes in other environmental areas.

Even if these model approaches are used, and consensus achieved, positive improvements in regulation and control of damaging NIS will not necessarily follow. Regular feedback based on monitoring of ultimate results would aid in improving the models. Follow-up evaluation of agency implementation of resulting decisions should be an integral part of any changes in decisionmaking processes (29).

CHAPTER REVIEW

This chapter has examined the means by which decisions about potentially harmful NIS are made: clean and dirty lists, risk analysis, environmental impact assessment, economic analysis,

values, and protocols. This chapter also looked at two methods to synthesize the different approaches. Explicitly addressing three interrelated issues would contribute to clearer decisions in the future: 1) determining the level of risk that is acceptable; 2) setting thresholds of risk at which decisionmakers should invoke formal, more costly, approaches; and 3) clarifying the tradeoffs when deciding in the face of uncertainty. The benefits of taking these issues seriously would be better NIS decisions in many cases or, at least, decisions that take better account of the diverse societal interests involved.

Even under the best of circumstances, some mistaken decisions will be made because of the inherent unpredictability of NIS. Technology provides the means to counter such mistakes. Methods to prevent and control problems due to NIS are the subject of the next chapter.

---

[16] 5 U.S.C.A. 583(a)(7).

Technologies for Preventing and Managing Problems | 5

T his chapter describes technologies and related issues for preventing and managing harmful non-indigenous species (NIS) in the United States. Programs are discussed in the order of their occurrence for dealing with NIS: prevention, followed by eradication, containment, and suppression. Education is a key component within all of these programs.

The adage "an ounce of prevention is worth a pound of cure" holds true for many harmful NIS. However, prevention is not always sufficient. Harmful NIS do enter the country, although it is not possible to predict when or where the next harmful NIS will enter, or what its specific impact will be. Alternative programs are required to prevent establishment of these NIS or to manage them.

Eradication is the first step in such reactive approaches. Destroying a population when it is relatively small or before it spreads can eliminate the need for long-term management programs. Eradication is not always possible, however, or may not be implemented. The next step is containment, or development of a strategy to limit or slow the population's spread. Long-term management using specific control technologies is the final phase. At this point the goal is to suppress the population below acceptable thresholds.

TECHNOLOGIES FOR PREVENTING UNINTENTIONAL AND ILLEGAL INTRODUCTIONS

Finding:

Shortcomings exist in Federal prevention programs. The high volume of people and goods in transit can overwhelm inspectors, limiting thorough surveillance. Confusing regulatory authority can lead to delays in applying known technologies. Lag times often exist between the identification of a harmful NIS and the implementation of an effective prevention technology.

▊ Inspection and Exclusion Activities at U.S. Ports of Entry

Experts often consider prevention the most economical, desirable, and effective management strategy for harmful NIS. The manifestation of this policy is government inspection and exclusion programs for NIS. The main factors involved in successfully preventing the entry of NIS are: the availability and efficacy of technologies for known problems (e.g., fumigation for imported fruits and nuts); the development of applicable technologies and programs for new NIS (e.g., ballast water treatment for zebra mussels, *Dreissena polymorpha*); and applying these technologies effectively (e.g., matching availability of inspectors to volume of passengers from international flights).

Preventing the introduction of harmful NIS involves various Federal and, to a lesser degree, State agencies, often working together. This cooperation may include assuming inspection duties or sharing of resources and information. Chapters 6, 7, and 8 discuss the roles of the different Federal agencies in NIS prevention activities.

TRAVELERS AND BAGGAGE

A recognized pathway for NIS at U.S. ports of entry is the traveling public and their baggage (14). Under normal circumstances, insufficient time and staffing and the numbers of international travelers prevent 100 percent inspection of passengers and baggage. A profile system based on country of origin and passenger descriptions identifies high-risk flights and passengers.

Preferably, selective and efficient inspection technologies are used to reduce NIS introduction.

USDA

Inspections—before imports are shipped, at U.S. ports of entry, and after shipments are treated—are important means of excluding agricultural pests from the country.

The categorization of flights from areas of known NIS of quarantine significance can allow inspectors to most effectively use their limited resources. Human ''rovers'' also play an important role in identifying passengers who might intentionally introduce damaging NIS.

X-ray machines and beagles are important tools in detecting prohibited NIS in baggage. Presently, dogs are used at nine major airports in the United States. X-ray equipment is used at 42 major airports and land-border stations (43). Dogs and xrays have various limitations. For example, they cannot distinguish between permissible and forbidden items of similar type. Their effectiveness also depends on the quantity of goods in a sample and the packaging of the items.

Some innovative approaches to detecting NIS in baggage are being developed; these include carbon dioxide "sniffers" and other electronic or mechanical probes (11).

INTERNATIONAL TRADE: AGRICULTURE AND COMMERCIAL PRODUCTS

International commerce provides another avenue for the introduction of potentially harmful NIS into the United States. Preventing their introduction requires the establishment of regulatory quarantines. Such quarantines can require that a commodity be treated with a specific technology or that live organisms (e.g., large game animals, plant germ plasm, or potential biological control agents) be held in a quarantine facility to test for the presence of restricted pathogens, predators, or parasites.

Commodities (Fruits and Vegetables)— Techniques for preventing unintended introductions of NIS with commodities include treatment schedules and sampling programs. For example, mangoes from Brazil are tested for the presence of Mediterranean fruit fly (*Ceratitis capitata*). Ideally, treatments should provide complete effectiveness (100 percent kill); cause little or no damage to the commodity; cause only minor delays in commercial transit; and have no human health risks (69).

Procedures such as picking fruit and vegetables early to minimize the chance of infestation or using cultivars resistant to specific pests can be implemented before a commodity leaves the originating country. In addition, changing the planting date to avoid pest outbreaks, rotating crops, or using chemical pesticides to establish pest-free zones can reduce the chances of infestation (69).

The goal of a pest-free zone is to remove the pest problem in a specific part of a country. Protocols for establishing such zones include: surveys; required action if the survey detects the target pest within the area; procedures for sampling, marketing, certifying, and safeguarding exported products; and a documented history of pest-free status. The U.S. Department of Agriculture (USDA) has pest-free zone agreements with Mexico, Chile, and other countries (105).

While a commodity is in transit, or after it has arrived at a U.S. port of entry, specific treatments such as the application of chemicals or holding items at specific temperatures for designated time periods are available (table 5-1). Several factors limit the use of temperature or chemicals, including the biology of the NIS, the frailty of the commodity, and the feasibility of application.

Some chemical treatments cause damage or reduce the product's shelf life (29). Temperature treatments are nonchemical alternatives but require strict adherence to protocols for efficacy. For example, a hot water dip for papayas was discontinued because of difficulties in monitoring the process (94).

By combining cultural and physical treatments in the country of origin, some commodities can receive pre-clearance before entering the United States. Pre-cleared commodities are permitted entry without further inspection. For example, inspectors trained by USDA's Animal and Plant Health Inspection Service (APHIS) working in cooperation with local inspectors in Japan, can monitor field production, storage, packaging, and shipment of Satsuma oranges, which are inspected for the presence of citrus canker *(Xanthomonas campestris* pv. *citri)* (72). Pre-clearance programs exist between the United States and 24 other countries, yet, with the exception of Canada, they remain relatively small (43,103).

Subset sampling is part of the pre-clearance inspection for highly perishable commodities or when known NIS potentially infest specific commodities. APHIS has established protocols for subset sampling (93), which involves sampling small portions of an imported commodity to assess whether NIS are present. Limited resources, loading techniques, or large lots can

Table 5-1—Examples of Treatment Technologies for Importing Commodities

Chemical treatment:
Commodities are treated with chemical fumigants at specific atmospheric pressures for specific time periods.
Example: Under normal atmospheric pressure and at 90-96 °F, imported chestnuts are fumigated for 3 hours with methyl bromide for infestations of the chestnut weevil (*Curculio elephas*).

Temperature treatment:
Freezing:
Fruits and vegetables are frozen at subzero temperatures with subsequent storage and transportation handling at temperatures no higher than 20 °F.
Cold treatment:
Commodities are cooled and refrigerated for specific temperatures and days.
Example: Fruit infested with the false codling moth (*Crytophlebia leucotreta*) requires refrigeration for not less than 22 days at or below 31 °F.
Vapor heat:
Commodities are heated in water-saturated air at 110 °F. Condensing moisture gives off latent heat, killing eggs and larvae.
Example: The temperature of grapefruit from Mexico is raised to 110 °F at the center of the fruit in 8 hours and is held at that temperature for 6 hours.
Hot water dip:
Commodities are treated with heated water for specific periods of time.
Example: Mangoes weighing up to 375 grams from Costa Rica are dipped in 115 °F water for 65 minutes.

Combination treatment:
Combination of fumigation and cold treatment.
Example: Fruit infested with Mediterranean fruit fly (*Ceratitis capitata*) is exposed to methyl bromide for 2 hours then refrigerated for 4 days at 33-37 °F.

Irradiation treatments:
Commodities are exposed to irradiation at specific rates and times.
Example: Papayas shipped from Hawaii would be treated with a minimum absorbed ionizing radiation dose of 15 kilorads. (This treatment schedule has USDA approval but is not commercially used at this time.)

SOURCES: 7 CFR Ch. 111 (1-1-91 Ed.) Animal and Plant Health Inspection Service, USDA, Part 319 - Foreign Quarantine Notices, Subpart - Fruits and Vegetables, 319.56; 7 CFR Ch. 111 (1-1-92 Ed.) Animal and Plant Health Inspection Service, USDA, Part 318- Hawaiian and Territorial Quarantine Notices, Subpart - Hawaiian Fruits and Vegetables, 318.13.

reduce the randomness of samples, compromising accuracy (91).

One technology with potential for treating many commodities such as flowers, grain, and fruits is irradiation (e.g., gamma radiation). Irradiation kills organisms directly or indirectly (e.g., causes sterility or other mutations in immature life stages) so that new populations cannot be established. This technology is currently used to increase the shelf life of foods such as strawberries and for treating spices.

To become an effective tool, it is necessary to establish dosage levels for specific pest species and commodities. The doses required to directly kill some non-indigenous pests can damage commodities. For example, some flowers from Hawaii cannot tolerate certain radiation levels (29), but decreasing the doses potentially leaves live (though nonfertile) pests. These present problems for inspectors, because practical methods that distinguish nonfertile from fertile pests are limited.

Public concern over health risks also affects the use of irradiation. Although irradiated products pose no known hazards to consumers, potential occupational health risks exist (63).

Animals (Livestock, Zoos, and the Pet Trade)— NIS such as ''exotic'' game animals are recognized as sources of disease for domesticated and wild indigenous animals (47). Therefore, various non-indigenous animals being imported are temporarily held at quarantine stations, where they are examined for general clinical signs of disease, ectoparasites, and specific diseases based on the species and country of origin. Categories of vertebrate animals quarantined include domestic livestock and swine, poultry, pet birds, and various "exotic" game animals. Other categories of vertebrates have no or few restrictions. For example, no Federal quarantine requirements exist for non-indigenous fish, and few exist for non-indigenous reptiles.

Animals are held either in USDA Veterinary Services quarantine stations or in various private facilities approved by the USDA at or near ports of entry. Veterinary Services maintains quarantine stations in Newburg, New York; Miami, Florida; and Honolulu, Hawaii. In addition, the Harry S. Truman Animal Import Center at Fleming Key, Florida, quarantines imported animals when highly contagious diseases (e.g., foot-and-mouth disease) are a risk or where high security is required.

Animal quarantine does not completely prevent the introduction of animal disease or disease vectors, however. Some non-indigenous animals circumvent quarantine when they are shipped to approved zoos. While these animals are technically held in a permanent quarantine (i.e., the zoo), the potential exists for diseases to escape via other vectors such as insects. Importation of animals such as red deer (*Cervus elaphus*) for game and ostriches (*Struthio camelus*) for commercial purposes also provides a potential pathway for NIS. A gap in prevention occurs because it is difficult to recognize diseases or their vectors carried on these novel imports and to develop appropriate tests quickly.

Plant Germ Plasm—High-risk plant germ plasm is quarantined to check for the presence of pests or pathogens such as viruses, bacteria, insects and mites, or fungi. The National Plant Germplasm Center in Beltsville, Maryland, con-

ducts tests for detection methods. Present facilities and staffing are inadequate to process expected future volumes of incoming material (65), and the Center is in the process of expansion. Ongoing construction activities may extend into 1997 (92).

Some standard techniques for detecting pathogens in germ plasm include visually looking for signs and symptoms of disease, and checking for transmission to healthy plants (79). More specific techniques involving electron microscopy, immunosorbent assays (ELISA, EIA), molecular probes, and other tools have been developed or improved for particular pathogens (38). These tools, used alone or in combination, allow faster and more precise pathogen detection, although they also have limitations to their use. Research is needed to detect other pathogens of quarantine significance and to make these technologies more practical at inspection stations (38).

Biological Control Agents—Certain groups of non-indigenous biological control agents (e.g., insects and pathogens) are also quarantined upon importation. The quarantine may screen for non-target effects of control agents, for hyperparasites, or for purity to guard against the inadvertent introduction of additional NIS (43).

Biological control quarantine facilities exist in Federal, State, and university laboratories. The USDA provides guidelines for their development and sets standards for features such as air intake systems, drains, escape-proof containers, and greenhouses. These standards vary depending on the type of organisms being held. Quarantine facilities in Frederick, Maryland, for example, are designed to prevent plant pathogens from escaping (58).

■ Education at Ports of Entry

A portion of travelers carrying prohibited NIS are unaware of Federal restrictions or have made honest mistakes about possessing prohibited items. These travelers would more likely comply with restrictions if they were aware of the reasons for

Attempts to educate travelers regarding the dangers of importing non-indigenous species have relied on posters and other written materials, with mixed success.

regulatory actions, and the environmental and economic risks involved (38). A well-organized, active public education campaign could disseminate such information.

One example of a public education campaign for travelers was a USDA program begun in the early 1960s. It used the media to build general awareness in order to deter entry of prohibited products (54). The program included printed information, radio and television advertisements, films, foreign language fliers, and the development of the symbol ''Pestina'' (akin to the U.S. Forest Service's Smoky the Bear).

The program had mixed results. No formal evaluation attempted to determine the program's effectiveness (52). The program did illustrate a lack of cooperation and coordination between Federal agencies and the private sector, as airlines, travel agencies, and port authorities were indifferent about giving full support to the USDA's programs (54,91).

Although public education is considered an essential element of prevention programs, OTA could not identify a formal national education program directed against NIS importation. Limited public education at ports of entry depends primarily on printed materials (e.g., posters and pamphlets). Showing videos on airplanes is an interesting approach. Hawaiian, Northwest, and Continental Airlines are sporadically involved in such a program on flights to Hawaii.

Where, when, and how to educate the public about NIS policy are important questions. Education before travelers depart (allowing them to leave prohibited items behind) offers perhaps the best way to prevent introductions. Educating after departure but before arrival also is beneficial, acting not so much as a safeguard for the existing trip, but as a method for building awareness for future trips (54).

■ Evaluation of Prevention Programs and Methods

Assessing the effectiveness of inspection and quarantine programs is difficult. For example, the number of reported interceptions at a port of entry only provides the quantity and types of regulated NIS discovered. This information provides little data on the effectiveness of the prevention system because it does not estimate the total pest entries. OTA was only able to identify ad hoc programs that evaluate the effectiveness of prevention programs.

THE "BLITZ"

One approach to understanding how many prohibited items enter the country is through ''blitzes,'' or brief 100 percent inspection. During one week in May 1990, USDA/APHIS, the California Department of Food and Agriculture, and some southern California counties conducted a blitz at Los Angeles International Airport. Out of a total of 490 flights, 100 percent of the baggage of 153 targeted flights (from high-risk countries of origin) and several non-targeted flights was inspected. The remainder underwent standard USDA inspection.

The blitz showed that passenger baggage on foreign flights is an important pathway for plant and animal pests (7). Inspection involving 16,997 passengers (i.e., passengers *and* their baggage) from the targeted flights intercepted 667 lots of prohibited fruits and vegetables and 140 animal products (equaling 2,828 pounds). Another 690 lots of prohibited fruits and vegetables and 185 of animal products (2,969 pounds) were intercepted from non-targeted flights. The results also demonstrated that at this airport considerable illegal importation occurs. A study of the blitz concluded that more resources are needed to close this pathway and to more strongly deter common illegal activity (8).

"Shutting the Door"—Blitzes can evaluate the effectiveness of prevention programs already underway. Assessing when and how new programs are established is another important issue. Lag times often occur between the identification of new pathways (and new NIS) and the implementation of new prevention programs (table 5-2). Eliminating such lags could help prevent the establishment of new harmful NIS.

Both political and technical limitations cause delays. For example, effective methods such as xrays and dogs exist for identifying domestic first-class mail containing prohibited agricultural products. But postal laws and lack of departmental interest have limited the control of this pathway (7). And while many techniques are available to treat ballast water, few are practical for large-scale use (97).

Even when programs are established, gaps in their implementation may continue to allow the entry of NIS. The protocols to prevent introductions via ballast water apply only for the Great Lakes (97). Ships entering other U.S. ports can still introduce non-indigenous aquatic organisms. The development of a domestic first-class mail inspection program between Hawaii and California

does not address the potential movement of harmful NIS between Puerto Rico and California (77).

TECHNOLOGIES FOR MANAGING ESTABLISHED HARMFUL NON-INDIGENOUS SPECIES

Prevention programs are less than perfect at keeping potentially damaging NIS out of the United States. Programs to manage already introduced species are essential and use additional technologies.

Finding:

Accurate and timely species-level identification is essential at all levels of a NIS management program. Applications of computer technologies provide new approaches to NIS monitoring and information acquisition. However, these technologies are only tools. Their information output is only as good as what is put in.

■ Species Identification and Detection

As illustrated in chapter 3, information concerning the identity and number of NIS in the United States is incomplete. Correct identification is vital for distinguishing NIS from indigenous ones and for establishing management programs. For example, some scientists now believe that the 1991 infestations of the sweet potato whitefly *(Bemisia tabaci)* in California were in fact a different species (2). If true, the search for control methods would require a different focus because many technologies are species specific (e.g., pheromone traps, classical biological control). Improper species identification can lead to the failure of these species-specific management programs.

COLLECTIONS AND STAFFING

National, State, and university taxonomic collections provide reference material for comparing and identifying species. They maintain records of known species and their historical and present-day distribution. Plant and animal collections of USDA and the Fish and Wildlife Service are held

Table 5-2—Lag Times Between Identification of Species' Pathway and Implementation of Prevention Program.

Species	Pathway	Date pathway identified	Date prevention program implemented	Remaining gaps
Mediterranean fruit fly (*Ceratitis capitata*)	Fruit shipped through first-class domestic mail from Hawaii	mid 1930s	1990, mail traveling from Hawaii to California inspected	First-class mail from elsewhere or other potential pathways (e.g., Puerto Rico to California)
Aquatic vertebrates, invertebrates, and algae	Ship ballast water	1981	1992, Coast Guard proposes guidelines for treating ballast water into the Great Lakes	International shipping into other U.S. ports; ship ballast water from domestic ports
Asian tiger mosquito (*Aedes albopictus*)	Imported used tires	1986	1988, protocols established for imported used tires	Interstate used tire transport
Forest pests	Unprocessed wood (including dunnage, logs, wood chips, etc.)	1985	1991, first restrictions imposed on log imports from Siberia	Wood imports other than from Siberia

SOURCES: Bio-environmental Services Ltd., *The Presence and Implication of Foreign Organisms in Ship Ballast Waters Discharged into the Great Lakes*, vol I, March 1981; C.G. Moore, D.B. Francy, D.A. Eliason, and T.P. Monath, "*Aedes albopictus* in the United States: Rapid Spread of a Potential Disease Vector," *Journal of the American Mosquito Control Association*, vol. 4, No. 3, September 1988, pp. 356-361; I.A. Siddiqui, Assistant Director, California Department of Food and Agriculture, Sacramento, CA, testimony at hearings before the Senate Committee on Governmental Affairs, Subcommittee on Federal Services, Post Offices, and Civil Services, *Postal Implementation of the Agricultural Quarantine Enforcement Act*, June 5, 1991; United States Department of Agriculture, Animal and Plant Health Inspection Service, "Wood and Wood Product Risk Assessment," draft, 1985.

at the Smithsonian Institution's National Museum of Natural History, the National Arboretum, and taxonomic laboratories of the USDA Agricultural Research Service. In addition, the American Type Culture Collection, a non-profit, privately held organization, maintains reference and research material on microorganisms.

Some groups of organisms are better known and easier to identify than others. Indigenous birds and mammals are thoroughly inventoried, but experts believe more than half of the indigenous insects and arachnids in the United States are unidentified (40). The lack of information on indigenous species hampers the identification of some NIS in the United States. The Clinton Administration's proposed national biological survey, slated by the Department of Interior to begin in October 1993, is an attempt to bolster information on U.S. biological diversity (81).

Taxonomists (people who describe, identify, and classify species) work at field locations, museums, and universities across the country. A shortage of trained taxonomists at all levels in the United States (40,102) impedes rapid and accurate identification of intercepted species and the collection of scientific information on NIS (40).

MOLECULAR BIOLOGY TECHNIQUES

Traditionally, taxonomists study variations in anatomy, physiology, and morphology to distinguish between different species. For many NIS, identification is hampered by the species' small size or because of taxonomic complexity or ambiguity. Alternatively, methods of molecular biology can provide effective options. Tools such as gel electrophoresis can reveal enough genetic variation to separate species (60). Molecular biology methods can identify genetic strains, or distinguish between hybrids and natural populations (27,36).

Molecular techniques may also provide faster identifications, which is important for NIS like the African honey bee. European (*Apis mellifera*) and African (*A.m. scutellata*) honey bees can exist at the same location, and quick identification of the African type is important for management programs. The morphological approach to identification measures variation of specific body parts, while mitochondrial DNA testing works faster and is more accurate (15).

Aside from species identification, molecular testing is useful for determining geographic origin of a NIS (56). For example, molecular markers may in the future help identify the origin of Californian populations of the Mediterranean fruit fly (ch. 8, box 8-A). Understanding a species' origin can help identify routes of invasion or spread and aid in developing appropriate prevention or management programs (39,74).

■ Species Surveys and Population Monitoring

Planned detection systems are useful for identifying early infestations of NIS, monitoring populations after they are established, and documenting effects. For example, monitoring water systems for young zebra mussels can provide early warnings of an invasion (55).

DETECTION TECHNOLOGIES

Visual surveys, traps, and physical inspection can locate infestations of NIS. Visual surveys are used for such species as weeds, birds, and mammals. Trapping locates organisms that are more difficult to see, such as insects or aquatic invertebrates. Physical inspection is especially useful for diseases associated with livestock.

Surveys for known harmful NIS occur at the local level, as part of pest management programs; at the State level, as part of domestic quarantine programs; and at regional or national levels. Surveys to detect new introductions are generally conducted by the Federal Government (California is an exception), in part because surveys generally have little or no immediate economic value and can have significant long-term costs.

Traps can provide information on the presence and geographical distribution of NIS. Further information, such as the host, geographic origin, age, and sex of a NIS are potentially obtainable

JACK DYKINGA

Fast and accurate species identification is essential for designing detection methods and management plans but distinguishing some species, e.g., European and African honey bees, requires expertise that is in short supply.

(9). The basic components of a monitoring system are the attractant, the trap itself, and information about the species' biology (100). Desirable attributes of trapping systems are low cost, ready availability, easy servicing and inspection, and provision of specimens in good condition for taxonomic identification (13).

Commercially available traps incorporating behavior-modifying compounds (biorationals) such as sex pheromones or other attractants are relatively inexpensive and effective tools for surveying NIS in certain situations. Most research involving pheromones and other attractants in traps is aimed at non-indigenous insects that are agricultural pests. Such traps are potentially useful with other NIS (e.g., terrestrial vertebrates) (25). (For more on the use of pheromones see ''Tools of the Control Trade'' below.)

Limitations to the broader use of pheromone monitoring programs include the high cost of the active ingredients, inadequacies in synthetic pheromone formulation technologies, the lack of commercial development, and shortcomings in technology transfer to the marketplace (78).

REMOTE SENSING

Remote sensing shows promise in NIS detection programs. Remote sensing of habitats with video and still-camera equipment can provide information on the distribution and spread of certain NIS, especially plants. Helicopters, planes, and even satellites gather information using infrared or near-infrared photography. Image-processing software creates a digital mosaic in which dominant species can sometimes be distinguished on a regional basis.

Federal and State agencies are conducting research into and applying remote sensing technology. The data collected are important for identifying new infestations of damaging NIS and developing management plans. For example, the Agricultural Research Service used Landsat imagery in a bollworm (*Helicoverpa zea*) control program for cotton in Texas (32). Remote sensing data are also often suitable for use in geographical information systems.

GIS TECHNOLOGY

Geographical information systems (GIS) store, manipulate, analyze, and display spatial data. The combination and display of variables such as topography, vegetation types, and climate has recently been enhanced by the merging of GIS with online satellite data. By sorting and filing vast amounts of information, GIS can rapidly correlate and map such variables. Limiting factors in GIS technology are the high cost of data acquisition and a lack of data linking NIS to geographical variables (39).

Federal and State agencies and universities use GIS technology for various natural areas' issues, e.g., to study wildlife migration patterns and rates of wetlands loss. Such tools are also applicable for monitoring NIS. The National Fisheries Research Center in Gainesville, Florida, now uses GIS to analyze non-indigenous fish and certain mollusks (84). The National Park Service determines resources vulnerable to fire or gypsy moths (*Lymantria dispar*) (85).

The applications of GIS vary with the availability of suitable NIS data. Detailed knowledge of a NIS allows the prediction of high-risk areas for unplanned invasions or expansion. Conversely, monitoring planned or known introductions can generate NIS data by identifying habitat correlations. Hypotheses can rapidly be tested, for example, relating invasions to habitat disturbance or identifying particular corridors that invasions are likely to follow (39).

■ Information Collection and Dissemination

The development of tools to collect information about NIS quickly and easily is important, as are mechanisms to disseminate the information. Methods to distribute information about NIS presence and distribution should be timely and reliable. The range of potential mechanisms varies from printed books, journals, newsletters, and abstracts to electronic computer storage, CD-ROM (Compact Disk-Read Only Memory), and expert systems.

Few programs for disseminating information strictly about NIS are available within the United States. As one example, the New York Sea Grant Marine Advisory Service operates the Zebra Mussel Information Clearinghouse in Brockport, New York, to provide information on zebra mussel distribution, impacts, research, and other issues (84).

Potentially, computer technologies could help develop national or even global centralized NIS databases. The function of such databases would be not only to provide information on available management technologies, but also to warn of possible harmful NIS. No single organization is likely to develop such programs, as the creation and maintenance of the databases is expensive (33).

Technologies such as computerized databases could aid information management related to NIS. For example, the BIOCAT database records the results of nearly 5,000 introductions of biological control agents in about 200 countries since 1880 (28).

An interest at the Federal level (especially within the USDA) exists for increased use of computerized databases (17,88). Within the USDA, however, OTA has found sharp contrasts between the start-up and long-term support of databases involving NIS. NAPIS (the National Agricultural Pest Information System) and DATAPEST (the National Historical Pest Database) under CAPS (the Cooperative Agricultural Pest Survey), WHAID (Western Hemisphere Immigrant Pest Database), NAIAD (North American Immigrant Arthropod Database), ROBO (Releases of Beneficial Organisms), and PINET (Pest Information Network) are among some of the USDA databases that have been recently developed. However, few of these databases are properly functioning (17, 40). For example, critics find that NAPIS suffers from poor data (43); ROBO only was published in 1988, with information collected in 1981 (17).

Advances in computer technologies provide relatively inexpensive approaches for quick dissemination of information on NIS. Various Federal agencies have begun to apply these technologies to NIS problems.

CD-ROM first appeared in 1985 and has developed into an easy-to-use, well-standardized technology (48). By applying indexing techniques, CD-ROM is commercially suitable for building both general and specialized databases (e.g., the National Agricultural Library's AGRICOLA database, which indexes agricultural papers). Information specific to NIS could be gathered in this format.

Electronic mail or computer-based message systems are used by various agencies to transfer NIS information. For example, information on plant pests is collected and electronically sent to the NAPIS. The rapid transmittance and minimal costs of information via electronic mail can allow for better and more timely decisionmaking (48).

Expert systems may also have use for NIS concerns. An outgrowth of artificial intelligence research, expert systems are computer programs that make inferences and draw conclusions from statements supplied by a user. These systems have begun to find commercial application in the last few years (48). For example, a prototype system was recently developed to assist in European gypsy moth management.

■ Eradication

Finding:

Feasible eradication technologies do exist for many NIS, but public opinion and cost often prohibit implementation of a fully effective program. Three issues that complicate a successful eradication program include: the difficulty in identifying the zero-population level, diminishing returns as the population approaches zero, and the potential for reinfestation from surrounding areas. Although eradication of a NIS can have high short-term costs, the alternative is often a long-term management program with far greater cumulative costs.

It is important to distinguish between eradication and control. Both strategies use the same technologies (e.g., chemical pesticides or biologically based methods), but they have different goals. The goal of eradication is to remove the entire population of a species from a specific area. The alternative is to keep the population below a defined threshold through containment or suppression. Eradication programs for NIS (especially terrestrial vertebrates) are often long, costly, frustrating, and controversial (73), yet the failure to fully eradicate a harmful NIS can lead to long-term management programs, with continual yearly investments of time and money.

APPLICATION OF ERADICATION

Both governmental (State and Federal) and non-governmental organizations (NGOs) conduct NIS eradication programs. The reasons for eradication vary. For example, a Federal program to eradicate witchweed (*Striga asiatica*) in North and South Carolina is based on the potential economic effects that would result if the weed were to spread to the Midwest. Localized eradication programs for Asian tiger mosquito (*Aedes albopictus*) infestations occur because they are vectors for human diseases. Eradication programs for feral goats (*Capra hircus*) in Hawaii Volcanoes National Park were implemented because of the goats' impact on the natural resources of the area.

Studies assessing different eradication programs indicate that several factors influence the ease of eradicating NIS (19,42). Some of the most important include:

- adequate monitoring and early detection,
- quick implementation after detection,
- sensitive enough tools to detect low population densities,
- effective control technologies, and
- public perception and cooperation.

Eradication programs also require adequate planning and a commitment of sufficient resources (19,98). These two elements in particular affected the outcomes of eradication programs for imported fire ants (*Solenopsis invicta, S. richteri*) and boll weevil (*Anthonomis grandis*) (box 5-A).

THE ROLE OF THE PUBLIC

Public interaction can play a significant role in eradication programs for both governmental and non-governmental organizations. Favorable public opinion can lead to help and cooperation during a program while opposition can lead to legal actions aimed at ending a specific program. Perceived risk from control technologies, outrage from involuntary quarantine restrictions, or moral issues of animal rights may charge public opinion against an eradication program. The desire for humane treatment of NIS can restrict or prohibit the use of specific control technologies or eradication generally. Programs to eradicate damaging NIS (like feral horses (*Equus caballus*) and donkeys (*Equus asinus*)) have evoked such public opposition (23).

In some instances, negative reaction can simply stem from a lack of accurate information (73). Implementing education programs around the use of specific technologies and the reasons for removing particular NIS can help alleviate public fears.

■ Domestic Quarantine and Containment

The goals of domestic quarantine and containment are to prevent or limit the spread of potentially harmful NIS. Domestic quarantine provides a regulatory means to prevent or slow down the spread of a NIS within the United States, often during control or eradication programs. Plants, animals, and diseases have all been subject to domestic quarantine. Containment more often applies to non-indigenous animals. Some containment of cultivated game and other non-indigenous animals is required, for example, to prevent their spread into natural areas.

DOMESTIC QUARANTINE

Domestic quarantine attempts to slow or limit the spread of a harmful NIS within or to a State or region of the United States. Generally, domestic quarantines exist for pests that threaten agriculture, horticulture, or forestry. All States have some type of domestic quarantines (68).

Two important factors for a successful domestic quarantine program, like that for witchweed (71), are an effective certification process for pest-free commodities and other items within the quarantine area, and the cooperation of the general public (71).

Unfortunately, not all domestic quarantines work as well. The domestic quarantine of the imported fire ant has not prevented it from spreading. Movement reportedly has occurred in association with nursery material (1).

Domestic quarantines cannot slow or prevent NIS from moving by natural means; they can only hinder NIS from spreading through human-assisted mechanisms such as interstate shipments of nursery stock or household goods. Their

Box 5-A—Failure and Success: Lessons From the Fire Ant and Boll Weevil Eradication Programs

Imported Fire Ant Eradication:

Two species of imported fire ants are assumed to have entered at Mobile, Alabama, in dry ship ballast: *Solenopsis richteri* in 1918 and, around 1940, *Solenopsis invicta*. The ants became a public health problem and had significant negative effects on commerce, recreation, and agriculture in the States where they were found. In late 1957, a cooperative Federal-State eradication program began. It exemplifies what can go wrong with an eradication program.

Funding was provided to study the fire ants, but information on the biology of the species was lacking, and the ant populations increased and spread. Various chemicals (heptachlor and mirex) were used to control and eradicate the ants over a 30-year period. Although they did kill the ants, the chemicals caused more ecological harm than good. Their widespread application, often by airplane, destroyed many non-target organisms, including fire ants' predators and competitors, leaving habitats suitable for recolonization by the ants.

The chemicals eventually lost registration by the Environmental Protection Agency, leaving few alternatives available. In the 5 years after 1957, fire ant infestations increased from 90 million to 120 million acres.

Boll Weevil Eradication:

The boll weevil, *Anthonomus grandis*, a pest of cotton, naturally spread into Texas, near Brownsville, from Mexico, in the early 1890s and crossed the Mississippi River in 1907. By 1922, it infested the remainder of the southeastern cotton area. Unlike the imported fire ant eradication program, boll weevil eradication does not rely solely on chemicals.

The eradication program centers around the weevil's life cycle and uses many different techniques. Part of the boll weevil population spends the winter in cotton fields. Insecticides are used to suppress this late season population. In spring and early summer, pheromone bait traps and chemical pesticides reduce populations before they have a chance to reproduce. Still other control technologies (e.g., sterile male release or insect growth regulators) limit the development of a new generation of boll weevils.

Boll weevil eradication trials were conducted from 1971-1973 (in southern Mississippi, Alabama, and Louisiana) and from 1978-1980 (in North Carolina and Virginia). Although results of the trials were mixed, cotton producers in the Carolinas voted in 1983 to support the boll weevil eradication program in their area and to provide 70 percent of the funding. The USDA Animal and Plant Health Inspection Service was charged with overall management of the program.

By the mid-1980s, the boll weevil was eradicated from North Carolina and Virginia. This 1978-1987 eradication program achieved a very high rate of return, mainly from increased cotton yields and lower chemical pesticide spending and use. In 1986, pesticide cost savings, additions to land value, and yield increases amounted to a benefit of $76.65 per acre. The benefit was $78.32 per acre for the expansion area in southern North Carolina and South Carolina.

SOURCES: G.A. Carlson, G. Sappie, and M. Hamming, "Economic Returns to Boll Weevil Eradication," U.S. Department of Agriculture, Economic Research Service, September 1989, p. 31; W. Klassen, "Eradication of Introduced Arthropod Pests: Theory and Historical Practice," Entomological Society of America, Miscellaneous Publications, No. 73, November 1989; E.P. Lloyd, "The Boll Weevil: Recent Research Developments and Progress Towards Eradication in the USA," *Management and Control of Invertebrate Crop Pests*, G.E. Russell (ed.) (Andover, Hampshire, England: Intercept, 1989), pp. 1-19; and C.S. Lofgran, W.A. Banks, and B.M. Glancey, "Biology and Control of Imported Fire Ants," *Annual Review of Entomology* vol. 30, 1975, pp. 1-30.

effectiveness is based on enforcement by government agencies and the education of the general public to prevent inadvertent spread.

State border station systems are one mechanism to enforce domestic quarantines. Presently they are used in California and Florida to inspect agricultural commodities for the presence of State quarantined pests (68). The effectiveness of State border inspection is illustrated by California's enforcement of the Federal domestic gypsy moth program. Stricter enforcement raised compliance with quarantine restrictions from about 20 percent in 1985 to approximately 80 percent in 1990 (7).

CONTAINMENT OF LARGE GAME AND FISH

Non-indigenous animals are kept as pets, for food production, sport, and as part of conservation programs. The escape of a NIS can introduce disease or parasites to wild populations, alter habitats, and lead to competition for limited resources or hybridization with wild populations. The scenarios that follow illustrate where deleterious effects might occur or have occurred.

Large-Game Ranching—Ranchers have kept large game in the United States for at least 40 years. Non-indigenous animals such as African ungulates are raised for sport, show, food, and for their aesthetic value. Interest in species preservation has also increased the numbers of large game in the United States. The first documented escape of contained non-indigenous mammals occurred approximately 45 years ago, from private ranches in Texas, California, and New Mexico (47; see ch. 7).

For most large mammals, no official national minimum containment standards exist. States such as California and Florida have established guidelines, but they are far from uniform (75). The USDA has asked the American Association of Zoological Parks and Aquariums to develop minimum standards for mammal containment, but these are still under development (75).

Big game animals are most commonly contained with standard-grade sheep or goat fencing, often electrified. The reasons and means of escape vary, but they usually include poor fence maintenance or design, weather damage, or vandalism (47). Further, when startled or upset, many mammals are capable of escaping either over or through fences.

CHARLES E. CICHRA

Triploid grass carp (Ctenopharyngodon idella) are tested for sterility before their release as biological control agents for aquatic weeds.

Aquaculture—In aquaculture, NIS are propagated for food (e.g., salmon, crayfish, and oysters), biological control (e.g., grass carp—*Ctenopharyngodon idella*), and for the pet trade (e.g., tropical fish). Improvements in production systems and new developments in genetics and biotechnology are expanding the size of the industry. Fish have escaped from commercial and experimental culture facilities (12), raising concern about the containment of NIS as aquaculture markets expand.

Scientists have created guidelines for the containment of transgenic or non-indigenous fish for research purposes (35, 96). These guidelines aim to prevent the escape of NIS from containment facilities. They have little application to commercial aquaculture, however, because they often involve small, indoor buildings. Many States, such as Florida, have minimum containment standards for commercial aquaculture. In general, no national standards exist for commercial aquaculture.

Outdoor facilities for containing NIS for aquaculture include ponds, pools, raceways, canals, tanks, and floating pen nets. Escapes can be prevented by constructing levees, placing ponds above 100-year flood lines, or using fences or nets. Escapes from tanks or pools can be prevented with the use of closed circulatory systems and filtered drainage systems. Floating pen nets are generally anchored to prevent drifting and covered with nets to prevent escape or removal of animals.

The production of sterile or single-sex populations can prevent establishment of reproducing populations if escape occurs. Single-sex fish populations are created by hybridization and sex reversals. Sex reversal in fish is possible in the early developmental period by administering hormones in the diet or in slow-release implants. These methods are not 100 percent effective, however (35).

Reproductive sterilization is accomplished with radiation, chemicals, or hybridization. Reproductive sterilization is perhaps the most secure approach for the biological containment of NIS. Currently, the use of triploid sterility[1] has the greatest potential (35). Although the sterilization techniques are not 100 percent effective, some NIS can be tested for triploidy. For example, tests to guarantee grass carp and Pacific oyster (*Crassostrea gigas*) sterility are available.

▮ Tools of the Control Trade

Finding:

No "silver bullets" exist for NIS control. Alternatives to chemical pesticides are being developed, but these new pesticides must provide advantages (cost, efficacy, environmental stability) before they can replace chemicals. Biotechnological improvements may overcome some of the limitations of biological control agents. As with chemicals, the potential for pest resistance exists.

The final stage in the management of a NIS is the development of a long-term control to suppress the population below specific thresholds. Three major groups of control technologies exist: physical controls, including manual, mechanical, and cultural methods; chemical pesticides, including synthetic and organic chemicals; and biologically based technologies, including natural or modified organisms, genes, or gene products and related techniques (table 5-3). The broad array of NIS in the United States requires an assortment of controls for use in agriculture, urban and suburban habitats, and natural areas. Whether to eradicate an NIS, contain it, or limit its economic damage to a crop, no control technology is optimal for all species, or in all settings.

PHYSICAL CONTROL

Physical controls may be mechanical (e.g., mowing), manual (e.g., hand pulling), or cultural (e.g., burning) (table 5-3). Physical controls are often applied to small populations of NIS because of the time (and therefore cost) associated with controlling larger populations. Physical controls may also be used where other control technologies are infeasible (e.g., a control program for an aquatic plant occurring close to a municipal water supply).

Use of physical controls may be limited by their low efficacy and other environmental factors. Hand pulling or cutting may leave roots, vegetative fragments, or seeds to resprout or germinate, leading to the establishment of new populations. Similarly, small populations of non-indigenous animals (e.g., goats) can repopulate an area if hunting or trapping does not remove all reproductive pairs.

Physical techniques may also lead to high levels of disturbance. The disturbance involved in the removal of non-indigenous plants, for example, may encourage invasion by other, nearby weedy non-indigenous plants and the germination of weed seeds already present.

---

[1] Triploid organisms have 3, instead of 2, sets of chromosomes. For the most part, these organisms cannot reproduce. This third set of chromosomes arises from altering the earliest stages of development. Techniques to induce triploidy include temperature, chemical, and pressure treatments.

Table 5-3—Examples of Control Technologies for Non-Indigenous Species

	Physical control	Chemical control	Biological control
Aquatic plants	Cutting or harvesting for temporary control of Eurasian watermilfoil (*Myriophyllum spicatum*) in waters	Various glyphosate herbicides (Rodeo is one brand registered for use in aquatic sites) for controlling purple loosestrife (*Lythrum salicaria*)	Imported Klamathweed beetle (*Agasicles hygrophila*) and a moth (*Vogtia malloi*) to control alligator weed (*Alternanthera philoxeroides*) in southeastern United States
Terrestrial plants	Fire and cutting to manage populations of garlic mustard (*Alliaria petiolata*) in natural areas	Paraquat for the control of witchweed (*Striga asiatica*) in corn fields	Introduction of a seed head weevil (*Rhinocyllus conicus*) to control musk thistle (*Carduus nutans*)
Fish	Fencing used as a barrier along with electroshock to control non-indigenous fish in streams	Application of the natural chemical rotenone to control various non-indigenous fish	Stocking predatory fish such as northern pike (*Esox lucius*) and walleye (*Stizostedion vitreum*) to control populations of the ruffe (*Gymnocephalus cernuus*)
Terrestrial vertebrates	Fencing and hunting to control feral pigs (*Sus scrofa*) in natural areas	Baiting with diphacinone to control the Indian mongoose (*Herpestes auropunctatus*)	Vaccinating female feral horses (*Equus caballus*) with the contraceptive PZP (porcine zona pellucida) to limit population growth
Aquatic invertebrates	Washing boats with hot water or soap to control the spread of zebra mussels (*Dreissena polymorpha*) from infested waters	In industrial settings, chlorinated water treatments to kill attached zebra mussels	No known examples of successful biological control of non-indigenous aquatic invertebrates (Target specificity is a major concern)
Insects/mites	Various agricultural practices, including crop rotation, alternation of planting dates, and field sanitation practices	Mathathion bait-sprays for control of the Mediterranean fruit fly (*Ceratitis capitatis*)	A parasitic wasp (*Encarsia partenopea*) and a beetle (*Clitostethus arcuatus*) to control ash whitefly (*Siphoninus phillyreae*)

SOURCE: Office of Technology Assessment, 1993.

CHEMICAL CONTROL

When used properly, chemical pesticides are an effective tool for controlling pests. Their greatest application has occurred within agriculture. In 1989, U.S. users spent approximately $7.6 billion for conventional pesticides, with agriculture accounting for more than two-thirds (4). The use of chemical pesticides for NIS control is limited based on availability and application to specific environments.

Quick and effective control technologies are often desirable to limit the impact of a NIS, and chemical pesticides can be applied and take effect within a short period of time. For example, in natural areas, systemic herbicides applied to a non-indigenous plant population can suppress it before it has a chance to produce seeds and thereby prevent future populations.

Although chemical pesticides are effective for many NIS, problems do exist in using many of them in control programs. For non-indigenous aquatic plants, effective chemical pesticides may be available, but are not registered for use in aquatic settings. Public concern can also limit the use of chemical pesticides by government agencies. For example, Utah's decision to use the biopesticide *Bacillus thuringiensis* instead of chemical pesticides to control the European gypsy moth was influenced by the general public and environmental groups (44).

An important issue related to the use of chemical pesticides is their future availability. Methyl bromide, a widely used chemical pesticide, may soon become unavailable because of its effect on the atmosphere (63). In addition, the 1988 amendments to the Federal Insecticide, Fungicide, and Rodenticide Act[2] may also limit the availability of many chemical pesticides for NIS (see the following section, "EPA Reregistration and Minor Use Pesticides").

BIOLOGICAL CONTROLS

Alternatives to chemical pesticides are often desirable for either economic or ecological reasons. Biological control has been in use in the United States and elsewhere for more than 100 years, although the development of synthetic chemicals in the 1940s shifted focus away from biological control (61). Attention has recently focused again on the development and use of biological control. Biological control attributable to natural enemies (i.e., classical biological control) is distinguished here from controls involving other biologically based methods (e.g., genetic control, hormones and pheromones, and contraceptives) (70). Both forms are important alternatives to chemicals for NIS control.

Biological Control With Natural Enemies— The standard definition of biological control is the use of natural enemies—parasites, predators, or pathogens—to reduce populations of target species and thereby reduce their damage to tolerable levels (16). Applying biological control involves research in many branches of biology—behavior, development, physiology, genetics, reproduction, systematics, biogeography, population biology, and ecology.

Biological control is divided into three broad categories: *importation* (or classical), involving the establishment of a NIS as a natural enemy in a new habitat; *augmentation* (often called the biopesticide approach), involving direct manipulation of established populations of natural enemies through mass production or colonization; and *conservation*, involving habitat manipulations to encourage populations of natural enemies. To date, importation is considered the most successful of these approaches (16).

Classical Biological Control—In theory, classical biological control re-establishes natural control by predators or parasites for foreign NIS that were introduced without their natural enemies. The goal of classical biological control is not to eradicate a NIS, but to lower the population level to economically or aesthetically acceptable levels.

Classical biological control has several advantages over other types of control technologies. When successful, reasonably permanent management of the target species results. Control agents are self perpetuating, will increase and decrease with populations of the pest, and are self disseminating. Costs are non-recurrent and benefit/cost ratios are high relative to other types of control (20,101). The average benefit/cost ratio for successful biological control projects is about 30:1, although the ratio varies widely among various projects (83).

Historically, however, most biological control projects have not been successful (59). The worldwide rate of establishment of introduced beneficial predators and parasites is about 30 percent; approximately 36 percent of these established agents successfully reduced or completely controlled their targeted pests—a proportion that is probably estimated too high (28). According to another author, the introduction of natural enemies sufficiently reduced host densities to replace

---

2 Federal Insecticide, Fungicide, and Rodenticide Act of 1947 (7 U.S.C.A. 135 *et seq.*); 1988 amendments, Public Law 100-532.

chemical control only in approximately 16 percent of 600 projects (59).

Constraints to implementing biological control stem from uncoordinated efforts among agencies, inadequate funding for overseas and domestic research, as well as the lack of a theoretical framework for determining what species or combinations of species will likely control a target pest in a given situation (20). Classical biological control does not work well in certain agricultural settings (e.g., annual crops where control must be rapid). It does show great promise for controlling NIS in natural areas or rangelands. For example, an Australian weevil is the first natural enemy imported for use against melaleuca (*Melaleuca quinquenervia*) in the Everglades (3).

Microbial Pesticides—Microbial pesticides (or biopesticides) include the use of fungi, viruses, bacteria, protozoa, and nematodes to control targeted species. Microbially derived herbicides and insect pathogens are commercially available in the United States (table 5-4, table 6-5). Microbial pesticides represent only a small portion of the pesticide market. The biggest obstacles in their development and commercialization involve host specificity, production technologies, lack of virulence, and the time frame needed to suppress the pest populations. The prospects for developing additional microbial pesticides, naturally or through genetic modification, are considered good (83).

The research and development costs of biopesticides are significantly less than those for chemical pesticides. The estimated cost for developing and deploying a biopesticide is between $1 million and $2 million, involving 11 to 13 scientist-years, whereas a chemical pesticide takes at least $10 million (10). Although biopesticides will not completely replace chemicals in the foreseeable future, they will complement chemicals and allow the development of improved integrated control measures (37). Market size is an important criterion in the development of these control technologies because lead times are long and the

Table 5-4—Examples[a] of Registered Microbial Biological Control Agents

Fungi
Phytohthora palmivora controls citrus strangler vine (*Morrenia odorata*)
Lagenidium gigantium controls various mosquito larvae
Viruses
Heliothis nuclear polyhedrosis virus (NPV) controls the cotton bollworm (*Helicoverpa zea*)
Gypsy moth NPV controls European gypsy moth larvae (*Lymantria dispar*)
Bacteria
Bacillus popilliae controls Japanese beetle larvae (*Popillia japonica*)
Bacillus thuringiensis controls various moth larvae
Protozoa
Nosema locustae controls various grasshoppers

[a] See table 6-5 for a complete list.

SOURCE: F. Betz, Acting Chief, Science Analysis and Coordination Staff, U.S. Environmental Protection Agency, letter to E.A. Chornesky, Office of Technology Assessment, Apr. 10, 1992.

development and registration costs for new products are high.

Other Biologically Based Methods—Several types of other biologically based methods have become available for NIS control.

Sterile Male Release (genetic control)—The release of sterile male insects was first successfully used in the United States in 1953 to control the new world screwworm (*Cochliomyia hominivorax*). Since then, it has been attempted with a large variety of insects, such as the Mediterranean fruit fly and the boll weevil, with varying success (51).

Sterile males released in large numbers mate with females, leading to the production of unfertilized eggs. Difficulties in implementing this technology exist, especially with mass rearing. Not only are appropriate facilities necessary to breed large populations of a given species, but adequate information about dietary needs and biology are vital. Accurate sterilization techniques are also required, as is knowledge about the effects of sterilization on species behavior.

Vertebrate Contraceptives—Contraceptives provide reversible fertility control for captive and free-roaming non-indigenous animals. Their use is seen as a humane alternative to hunting or other management practices. Use of contraceptive methods requires continual monitoring and repeat applications.

New research is centering on the use of immuno-contraception (relying on an animal's immune system) instead of hormone levels to interfere with a part of the reproductive process. Other research has focused on the use of commercially available contraceptives such as Norplant and in identifying antisperm antigens for male animals (41). These controls are still in the research and development stages for most NIS.

Semiochemicals—Semiochemicals are a group of compounds (e.g., sex pheromones) that can modify behavior. The compounds, either natural forms or synthetic copies, are useful for large-scale trapping or to disrupt mating behavior (78).

Semiochemicals are presently useful only against insects (46). Their use has been inhibited by high development and registration costs and low use in specialized markets. The Environmental Protection Agency (EPA) considers pheromones pesticides, requiring toxicity and residue testing under FIFRA. Such species-specific technologies are often more expensive than more traditional techniques such as chemical pesticides. In agricultural settings, this generally makes the use of semiochemicals economical only on high-value crops (46).

Host Plant Resistance—Enhanced host plant resistance is the artificial selection and breeding of plants to produce specific physical traits (e.g., very hard or hairy leaves) or biochemical traits (e.g., production of specific chemicals) that deter pest damage (16). It is useful in agricultural and horticultural settings.

Resistance is developed against non-indigenous plant diseases and plant-eating insects. It is useful in situations where no registered chemicals exist or when alternative controls are unavailable (16). Host plant resistance is compatible with other control measures.

Development of host plant resistance requires large-scale support. A lack of specific information about plant genetics can limit the use of this technology. Long production times mean it has little application as a quick fix against new harmful NIS (16).

Biotechnology—Many new biological control technologies currently in the research stage depend on biotechnology to increase the virulence and efficacy of controls. This approach, involving recombinant DNA, so far has been applied only to microorganisms. Limited knowledge curtails the genetic manipulation of more complex organisms, such as insects used for biological control.

The long-term goals of biotechnology research include increasing the shelf life of microbial pesticides and their persistence in the field. For example, the bacterium *Bacillus thuringiensis* (Bt) releases an insecticidal toxic crystal along with its reproductive spores. Researchers have inserted the toxin gene into another bacterium that produces the toxin during the non-reproductive phase. After the bacterium is killed chemically, the dead cell wall protectively coats the crystal and increases its stability. This process also eliminates the release of viable spores, an area of environmental concern.

The importance of biotechnology for biological control will likely increase in the future, although more economic research into biotechnology methods is needed (83). One application of biotechnology that will have a significant impact, especially in agriculture, is the development of transgenic plants, an alternative approach to chemical or classical biological control that involves genetically engineering crops to express insecticidal or antifeedant proteins.

The first successful application of transgenic technology occurred within the past 5 years (57). Most of the work has focused on inserting genes from various Bt strains into plants, which then produce the insecticidal toxins. The Bt toxin is considered safe (specific to certain groups of species) and is relatively simple to work with (57). Research has so far focused on cotton, tomato, and potato. Private companies hope to have transgenic tomato and cotton plants on the market by the mid-1990s (45).

Concerns exist that pests, especially insects, will develop resistance to transgenic plants. Recently, resistance to Bt has been documented in both laboratory and field settings (45). Efforts to prevent resistance counter-intuitively seek to maintain the susceptible population, thus delaying complete population resistance. Possible techniques for maintaining susceptible populations include rotating Bt toxins with other toxins, establishing nontoxic plant refuges, spatially alternating toxic and nontoxic plants, and expressing toxicity only in specific plant parts (53).

Scientists are just beginning to study the effectiveness of these techniques in preventing pest resistance. Some feel government legislation to coordinate use by farmers will be required for the proper application of this technology (50). Other issues surrounding the used of transgenic organisms are discussed in chapter 9.

Integrated Pest Management—Integrated Pest Management (IPM) is used in agricultural and natural areas for the control of NIS. IPM is defined as a management system that uses all suitable techniques in an economical and ecologically sound manner to reduce pest populations and maintain them at levels that do not have an economic impact while minimizing danger to humans and the environment (90).

IPM may combine biological control, pest resistance, autocidal, cultural, and mechanical and physical control technologies with limited use of chemical pesticides (64). IPM uses monitoring and other decisionmaking tools to gauge the health of the ecosystem, and consequently requires an understanding of the biology and ecology of the resource, the pest, and the pest's natural enemies.

Research establishes the needed economic thresholds and natural suppression factors. An understanding of the effectiveness of the control technologies and damage caused by different stages of pests is important. Because IPM does

USDA

The boll weevil (Anthonomis grandis) *eradication program integrates a variety of control measures: chemical pesticides, releases of sterile males, pheromone bait traps, and insect growth regulators.*

not necessarily rely on chemical pesticides, quick, simple, inexpensive but accurate tools are needed to monitor the environment and implement programs before a pest becomes an economic problem.

■ Education and Management

The need for greater public awareness regarding harmful NIS and for educating various specialized groups was cited repeatedly in recommendations by OTA's expert contractors (39,43,49,82) and its advisory panelists. Also, this theme surfaced frequently in recommendations by non-governmental groups (39). For example, successful education campaigns have been identified by many experts as a key mechanism for gaining public support of NIS management programs (18,31,39).

To assess the breadth of current NIS education programs, OTA asked the North American Association for Environmental Education to conduct a survey of government and non-governmental organizations (NGO) involved in educational programs relating to NIS. Federal and State agencies and NGOs conduct many activities related to NIS education. The survey of NIS education programs found:

- Education programs are typically small: funding averages less than 10 percent of agencies' budgets.
- Predicted funding outlays over the next 3 years varied depending on the organization.
- NGOs generally devote a larger share of their budgets to NIS issues as compared with Federal and State agencies.
- The need for increased funding for NIS education was often voiced.
- Little coordination of educational efforts among agencies and organizations exists.
- Information exchange is hampered by a lack of networks and materials to exchange.
- The success of the education programs is rarely evaluated.
- Programs that are evaluated rely on assessing subjective factors (76).

THE SCOPE AND METHODS OF EDUCATION PROGRAMS

Some environmental education programs tackle overarching environmental issues while others focus on NIS in particular. Groups in Hawaii are among the leaders in environmental education. Generally, they have taken a broad approach, linking NIS to endangered species, land development, park protection, and agriculture. For example, the formal school-based Ohia project educates children about the biology of the Hawaiian islands (ch. 8). Part of the project deals with the effects of NIS on Hawaii's ecology.

On the other hand, numerous groups have created focused educational materials on single NIS such as zebra mussels, gypsy moths, or purple loosestrife (*Lythrum salicaria*), sometimes for specific user groups. For example, APHIS has produced pamphlets and small fliers to educate people leaving the quarantine zone for the European gypsy moth. They provide information about how to identify, inspect, and treat for moths on firewood, vehicles, and outdoor household

items. Vermont's Department of Environmental Conservation began with a program focused on stopping the movement of Eurasian watermilfoil (*Myriophyllum spicatum*). It is moving now to a broader, regional watershed approach (76). Sometimes the selection of a narrow approach relates to a program's enabling legislation and funding rather than its educational merits.

Few formal national programs exist to identify and distribute information concerning harmful NIS. Minnesota's Department of Natural Resources has compiled this kind of information at the State level in its "Exotic Species Handbook" (62). The Handbook provides basic information on organizing citizen-level awareness programs and contains reference materials on various NIS in Minnesota. Information on obtaining educational material and a directory to the many agencies and organizations involved are included. The USDA's Cooperative Extension Service has been cited as a good Federal model for relaying information about invasive NIS to the public (76). The Extension Service does some technical training now, e.g., for pesticide applicators. And the Extension Service, in combination with Land Grant and Sea Grant universities, is doing the most comprehensive and innovative public education regarding zebra mussels (76).

Media and methods used in education about NIS mirror the larger field of environmental education in both scope and type. Techniques and media vary considerably and include almost any device or activity commonly used in education and informational efforts (76). For example, Federal and State organizations and NGOs have relied on a wide variety of channels to inform people about zebra mussel problems (table 5-5).

RELATED ISSUES

■ Ecological Restoration

Finding:

Ecological restoration is a relatively new practice that shows some promise in prevent-

Table 5-5—Examples of Technologies Used in Zebra Mussel (*Dreissena polymorpha*) Education Programs

Technique	Organization	Description or title
Booklet, brochure, or leaflet	Ohio Department of Natural Resources	"Zebra Mussels in Ohio"
Fact sheet	Illinois-Indiana Sea Grant Program Ohio Sea Grant Program	Information on how to report a sighting Information on zebra mussels in the Great Lakes
Newsletter, magazine	Minnesota Department of Natural Resources Vermont Department of Environmental Conservation	"On the LOOSE" "Out of The Blue"
Poster or sign	Ohio Department of Natural Resources	Boater's advisory on zebra mussels
Report	Zebra mussel Task Force Report to the Michigan Legislation	Zebra mussel control in Michigan
Workshops/lectures	Indiana Academy of Sciences	Presentation on zebra mussels, Conference on Biological Pollution: the Control and Impact of Invasive Exotic Species, October 1991
Video or slide show	Ohio Department of Natural Resources	Zebra mussel slide series Zebra mussel video
Classroom kits	Illinois Department of Conservation	"Lakes in My World" K-8 Workbook

SOURCE: Office of Technology Assessment, 1993.

ing NIS introductions and controlling reintroductions of NIS. The goal of ecological restoration, when applied to NIS control or eradication, is to modify those biotic and abiotic conditions that make the habitat suitable for NIS.

Ecological restoration is a branch of applied ecology that became visible as a management tool in the 1980s. It is the intentional return of an ecosystem to a close approximation of its condition before human disturbance (66). The goal is re-creation of whole, healthy, self-maintaining ecosystems in which natural ecological processes, such as nutrient cycling and succession, can operate without continual intervention by resource managers or reliance on synthetic engineered structures (5). Generalizations about ecological restoration's effectiveness are difficult, mainly because of the time it takes to see a project through to completion.

Ecological restoration is almost invariably a sequel rather than a preventive prelude to NIS invasion. Reestablishing prairie burns (i.e., fire as a restoration tool) is an exception to this statement. To date, ecological restoration has not been widely used to control harmful NIS (5) and its importance varies. At one extreme, the success of a restoration project may rest entirely on the removal of NIS. In other cases, control of a NIS may occur only after other phases of restoration have been completed (i.e., in which the restoration itself may eliminate the introduced species).

Existing data suggest ecological restoration is useful for NIS control, as it has been in part of Everglades National Park, Florida, for example (box 5-B). Limitations of ecological restoration in the management of NIS do exist, however. It will not repel an invader that is genetically or behaviorally very similar to a desired indigenous species. Ecological restoration also does not seem effective in managing NIS capable of invading ecosystems in pristine condition. For example, the non-indigenous garlic mustard (*Alliaria petiolata*) is capable of invading relatively stable forests in Illinois (5).

The genetic make-up of species used in restoration projects has recently become an important issue. Locally adapted germ plasm is important for assessing ecosystem performance, avoiding restoration failure, and assuring long-term genetic conservation (5).

Box 5-B—Ecological Restoration in the Hole-in-the-Donut, Everglades National Park, Florida

Work in the "Hole-in-the-Donut," 4,000 hectares of former agricultural land in Everglades National Park, Florida, is testing ecological restoration's ability to manage a damaging non-indigenous species and prevent its reintroduction. Chemical and fire techniques were used to rid the site of Brazilian pepper (Schinus terebinthifolius). Neither method was successful. In 1989, attempts were made to alter the environmental factors favoring NIS over indigenous species and to restore the site to pre-agricultural conditions.

In the 1950s, approximately half of the site was rock plowed, i.e., the limestone substrate was crushed to produce soil better suited for crops. The area remained in cultivation for 25 years. The changes in the soil—from primarily low-nutrient, anaerobic conditions to higher nutrient, aerobic conditions—were more favorable to Brazilian pepper and other non-indigenous plants.

In 1975, Everglades National Park acquired the land. With the end of agriculture, the vegetation began to change. The nonrock-plowed land returned, for the most part, to indigenous species. The 2,000 hectares of rock-plowed land were invaded and eventually dominated by Brazilian pepper. Between 1979 and 1985, fire was used to control Brazilian pepper, but monitoring of the burned sites indicated that repeated burning did not retard or reduce its growth. Studies on the economic feasibility of Brazilian pepper control with chemicals concluded that killing female trees was not an effective control strategy.

In 1989, a study on a 24.3-hectare site in the Hole-in-the-Donut attempted to determine the feasibility of ecological restoration on this former agricultural land. The idea was to remove the present vegetation and soil down to the limestone bedrock, establishing pre-agricultural conditions. Since 1989, recolonization by Brazilian pepper has been significantly reduced. The experimental site is still being monitored to determine the extent of the indigenous flora's return.

SOURCES: R.F. Doren and L.D. Whiteaker, "Comparison of Economic Feasibility of Chemical Control Strategies on Differing Age and Density Classes Schinus terebinthifolius," Natural Areas Journal vol. 10, No. 1, 1990, pp. 28-34; R.F. Doren and L.D. Whiteaker, "Effects of Fire on Different Size Individuals of Schinus terebinthifolius," Natural Areas Journal vol. 10, No. 3, 1990, pp. 107-113; F.J. Webb, Jr. (ed.), Proceedings of the Seventeenth Annual Conference on Wetlands Restoration and Creation, Hillsborough Community College, Tampa Florida, 1990, pp. 35-50.

A common recommendation is to use germ plasm adapted to the restoration site, preferably from the original gene pool. The notion that the germ plasm source might be important to restoration success is too new to have been tested rigorously. The reason locally adapted germ plasm is not used in plant restoration programs may be because of a lack of available seed (5).

■ Environmental Impacts of Control Technologies

Finding:

Adverse environmental impacts associated with chemical pesticides have been documented. Host specificity, residual effects, and human toxicity also need to be taken into consideration when biologically based meth- ods are used. Classical biological control should also receive careful consideration before application, as it becomes very difficult to remove an agent from the environment once it is established.

CHEMICAL CONTROL

Since the 1940s, the chemical industry has produced an array of chemical pesticides to control damaging NIS. Many pesticides are effective against more than one species (i.e., broad spectrum), and their application can pose significant environmental or human health risks when used in natural or agricultural settings.

One consequence of chemical pesticide control of NIS is the occurrence of secondary pest outbreaks. Chemical pesticides may kill not only the target pest, but also the natural enemies that keep different pests under control. For example, both indigenous and non-indigenous pest outbreaks are associated with malathion used for Mediterranean fruit fly eradication in California in 1980 (21,22).

Beginning with the 1972 amendment of FIFRA, EPA has been reviewing chemical pesticides used in the United States for their toxic effects on nontarget organisms, including humans.

The issue of human toxicity, either through accidental poisoning in the field or in residues on food, is a large and complex issue. Because chemical pesticides will continue to play an important role in NIS management, support is needed for EPA to finish its assessment of chemical pesticide risk.

In addition, the development of resistance to chemical pesticides by NIS threatens management of problem species. At least 500 insect species are resistant to at least one synthetic insecticide, and many are resistant to several (45).

In agricultural settings, chemical resistance can lead to additional pest problems. For example, numerous new plant viruses are reported associated with the emergence of a more aggressive, pesticide-resistant, sweet potato whitefly (72). Similarly, the tomato spotted wilt virus may become an important disease outside its present range if its insecticide-resistant vector, the western flower thrips (Frankliniella occidentalis), spreads (72).

BIOLOGICAL CONTROL

Biological control is often considered a safer, cleaner, and environmentally friendly alternative to chemical pesticides for the control of NIS. As with chemical pesticides, the risks associated with a biological control agent must be considered before it is released into the environment. Some scientists believe that, like chemical pesticides, biological control agents may disrupt existing or future control programs (34). This concern often focuses on introduced predators. For example, an introduced predator could attack a pest's existing natural enemies. Secondary pest

outbreaks could result if previously controlled pests flourish. Also, newly introduced and previously established biological control agents could compete, lowering the efficacy of one or both. This topic is hotly debated among the many scientists who study and apply biological control.

Recognition of such potential environmental effects is important, since it is normally impossible to eliminate a biological control agent from the environment once it is established (30,34). Comprehensive study before and after release of a control agent would establish baseline data on the environmental effects of such agents and could limit future adverse effects.

Many species have been found to be harmful as biological control agents. Vertebrates, in particular, are poor choices for effective, host-specific control. The mosquito fish (Gambusia spp.), the Indian mongoose (Herpestes auropunctatus), and the cane toad (Bufo marinus), for example, were introduced for biological control and had extremely harmful non-target impacts (34). The selection of species that have relatively narrow host preferences, such as some predatory insects or microbial organisms, provides greater likelihood of minimizing the impacts on non-target organisms.

Environmental impacts of microbial pesticides also require evaluation. Although microbial pesticides are considered safer than chemical pesticides, risks and uncertainties exist. Indirect effects often are not recognized because of a lack of general research (99), although studies are beginning to assess the impacts of microbial pesticides. The use of Bt can seriously affect indigenous butterflies and moths (6,67). The effects of insect pathogens (e.g., nematodes) on species closely related to the target are not well known (34).

■ EPA Reregistration and Minor Use Pesticides

Finding:

During the present EPA reregistration process, many old chemicals will become unavaila-

---

Box 5-C—The Loss of Chemical Pesticides: A Real Example

The loss of minor use chemical pesticides and the lack of alternative technologies pose a significant problem for NIS control. The loss of chemical pesticides used to control the sea lamprey (*Petromyzon marinus*) in the Great Lakes illustrates the importance of the problem. The Great Lakes Fishery Commission relies on two chemicals, TFM and Bayer 73 for the control of sea lampreys. TFM is a selective chemical that kills sea lamprey larvae. Bayer 73 is an additive to TFM. These two chemicals must be reregistered under FIFRA 88. Because of high reregistration costs and low revenue, the sole manufacturer of the two chemicals does not plan to reregister them. The scenario is complicated by the lack of effective alternatives. The two chemical lampricides are the only effective control. New, feasible technologies are not yet available. For example, a program based on sterile male release needs at least 10 more years of research before its effectiveness will be known (86).

The Great Lakes Fishery Commission is the only user of TFM in the world, and it has been unsuccessful in identifying additional suppliers. In order to maintain use of these pesticides, the Commission is faced with assuming reregistration costs, estimated to be $8 million over 4 years (86). The Commission has not begun incorporating the cost for reregistration into future budget proposals (89). However, FIFRA allows emergency use of unregistered pesticides for pests new to the country.

SOURCES: U.S. Congress, House Committee on Merchant Marine and Fisheries, "Status of Efforts to Control Sea Lamprey Populations in the Great Lakes," Sept. 17, 1991, U.S. Congress, General Accounting Office, *Great Lakes Fishery Commission: Actions Needed to Support an Expanded Program*, March 1992, and *Pesticides: 30 Years Since Silent Spring*, July 23, 1992.

---

ble, and fewer chemicals will receive registration. Concern exists that over the next 10 years, new or alternative technologies to replace chemicals will not be available for large-scale use.

Chemical pesticide use will continue to be essential for control of a significant number of NIS through the next decade, especially in agricultural settings (80). The 1988 amendments to FIFRA established reregistration guidelines for active ingredients in pesticides first registered before November 1, 1984. This reregistration process uses tightened standards for human health and environmental risk, and is scheduled for completion by December 1997.

The cost for developing and marketing a conventional chemical pesticide is more than $10 million (10). Although less expensive, reregistration also costs millions of dollars. FIFRA 88 will have its biggest impact on minor use chemical pesticides. Minor use is defined as low volume use that is not sufficient to justify the cost to a pesticide manufacturer to obtain federal registration (95).

In agricultural areas this includes chemical pesticides used on most vegetables, fruits and nuts, herbs, commercially grown ornamentals, trees, and turf. In non-agricultural areas, minor use chemical pesticides are used on aquatic plants, terrestrial vertebrates, fish, and aquatic invertebrates.

Many minor use chemicals are expected to become unavailable under FIFRA 88 (24). For example, the loss of herbicide registrations for aquatic weeds will leave a void in control programs because effective, economical substitutes are not now available (26). Chemical registration for vertebrate control has similar problems (box 5-C). It is estimated that about 1,000 minor use pesticides' registrations, having priority uses, will lose sponsorship during the reregistration process (104).

A potential model for the reregistration of minor use chemical pesticides for NIS is the Interregional Research Project No. 4 (IR-4), a USDA Cooperative State Research Service program organized in 1963 to obtain residue tolerances for minor use pesticides on food and feed crops. Since 1963, IR-4 has expanded to include registration information for pesticides used on nursery and floral crops, forestry seedlings, and turfgrass; animal health drugs, antibiotics, and antihelminthics; and for the further development and registration of microbial and specific biochemical materials used in pest management systems (95).

The IR-4 program is heavily burdened. It is estimated that 3,600 new uses and chemical reregistrations will try to pass through the IR-4 program by 1997 (95). Under the present funding schedule and timetable it is unlikely that the IR-4 program will complete the research and analysis necessary by the 1997 deadline (87,95). At best, the IR-4 program provides a model for the reregistration of minor use chemical pesticides for NIS.

CHAPTER REVIEW

This chapter examined the technologies to prevent the entry of harmful NIS and to control or eradicate those that slip through. These include a wide array of useful chemical, biological, physical, educational, and regulatory methods. Several related circumstances raise concern whether as many effective controls will be available in the future. Some important chemical pesticides probably will not be reregistered under FIFRA and so will go out of use. The environmental impacts of microbial, biological, or bioengineered substitutes are not yet clear. And efforts to make habitats less suitable for NIS in the long-term, via ecological restoration, are not now possible on a wide scale. For all of these reasons, continued research and development remain essential.

Effective management of harmful NIS involves institutional, as well as technical, issues. In the next 3 chapters, OTA examines the efforts of Federal and State institutions.

A Primer on Federal Policy | 6

T his chapter presents an overview of the Federal Government's activities related to non-indigenous species (NIS). It examines both the prevention and control of harmful NIS and the intentional introduction and use of desirable NIS. The reason for this dual focus is that, in the past, some presumably beneficial NIS introduced or promoted by Federal agencies have subsequently caused great economic or environmental harm.

OTA has drawn from this analysis a number of significant conclusions that cross agency jurisdictions and undergird several policy options presented earlier (ch. 1). The chapter begins with these conclusions, followed by a discussion of existing national policies on NIS. The remainder of chapter 6 presents a detailed reference to Federal programs, broken down along agency lines (box 6-A).

LESSONS FROM THE PRIMER

Finding:

The current Federal framework is a largely uncoordinated patchwork of laws, regulations, policies, and programs. Some focus on narrowly drawn problems. Many others peripherally address NIS. In general, present Federal efforts only partially match the problems at hand.

■ Keeping Harmful Species Out of the United States

The Federal Government currently plays a much larger role in preventing the entry of agricultural pests than in excluding other potentially harmful NIS. The Animal and Plant Health Inspection Service's (APHIS) fiscal year 1992 budget for agricultural

Box 6-A—A Locator for Federal Agencies Discussed in Chapter 6

Agency	Page/s
Department of Agriculture	
Animal and Plant Health Inspection Service (APHIS)	170-177
Agricultural Marketing Service (AMS)	177
Foreign Agricultural Service(FAS)	177
Forest Service (USFS)	177-179
Agricultural Research Service (ARS)	179-181
Soil Conservation Service (SCS)	181-183
Agricultural Stabilization and Conservation Service (ASCS)	183-184
Cooperative State Research Service (CSRS)	184
Department of the Interior	
Fish and Wildlife Service (FWS)	184-188
National Park Service (NPS)	188-189
Bureau of Land Management (BLM)	189-193
Bureau of Indian Affairs (BIA)	193
Bureau of Reclamation (BOR)	193
Department of Commerce	
National Oceanic and Atmospheric Administration (NOAA)	194-195
Department of Defense (DOD)	195-196
Environmental Protection Agency (EPA)	196-199
Department of Health and Human Services	
Public Health Service (PHS)	199
Department of the Treasury	
Customs Service	199
Department of Transportation	
Coast Guard (USCG)	199-200
Department of Energy (DOE)	200
Department of Justice	
Drug Enforcement Agency (DEA)	200

SOURCE: Office of Technology Assessment, 1993.

quarantine and port inspection was at least $100 million, compared with the $3 million for port inspections of fish and wildlife requested by the Fish and Wildlife Service (FWS) (97,100,170). The hundreds of agricultural pests restricted from entry by Federal regulations form the largest category of excluded NIS.[1] Current FWS and Public Health Service (PHS) regulations covering injurious fish and wildlife and potential human disease vectors restrict entry of far fewer NIS (by an order of magnitude). Certain categories of harmful NIS are not restricted from entry at all, such as many potentially affecting only natural areas.

Direct assessment of the effectiveness of Federal efforts to exclude harmful NIS is not possible because both APHIS and FWS lack performance standards for their port inspection activities or routine evaluations of their programs. The continuing entry of harmful species even in regulated categories (ch. 3) suggests that the agencies are not entirely successful.

Current Federal efforts may fail to exclude a significant number of harmful NIS because entry of many is prohibited only after they have become established or caused damage in the United States. Under certain laws, such as the Lacey Act[2] and the Federal Noxious Weed Act,[3] harmful species can continue to be imported legally until added by regulation to a published list. However, adding species to these lists is often difficult and time consuming (40,83,140).

Delays in preventing entry of harmful NIS also sometimes occur when new pathways emerge with no regulatory history. Recent examples include the slow reaction of PHS to the entry of

[1] CFR vols. 7,9.

[2] Lacey Act (1900), as amended (16 U.S.C.A. 667 *et seq.*, 18 U.S.C.A. 42 *et seq.*)

[3] Federal Noxious Weed Act of 1974, as amended (7 U.S.C.A. 2801 *et seq.*)

the Asian tiger mosquito (*Aedes albopictus*) in used tire imports, and of APHIS to the potential entry of forest pests and pathogens with proposed timber imports from Siberia (see also boxes 3-A and 4-B) (22,25). APHIS's efforts to take a more proactive approach for certain categories of agricultural pests have had varying success in part because of erratic support of the databases necessary for worldwide monitoring and anticipation of potential pest threats (54).

■ Dealing With Harmful NIS Already Here

The Federal Government devotes significant resources to managing and preventing interstate movement of many NIS that are agricultural pests. However, insufficient impetus or authority exists for Federal agencies to impose emergency quarantines on other highly damaging species. Noxious weeds, for example, despite explicit authorization under the Federal Noxious Weed Act,[4] receive little attention from APHIS. Interstate transport of injurious fish and wildlife listed under the Lacey Act, such as the zebra mussel (*Dreissena polymorpha*), is not prohibited by Federal law (30).

No coordinated control efforts exist to prevent the spread of large categories of harmful NIS, such as the many that damage only natural areas or are vectors of human diseases. Current Federal efforts to control non-indigenous fish and wildlife developed piecemeal and are noncomprehensive. The Nonindigenous Aquatic Nuisance Prevention and Control Act[5] authorized a coordinated program that might go far toward correcting this shortcoming in the future. Lack of appropriations has impeded implementation of the Act thus far (31).

■ Federal Land and Resource Management

Federal agencies manage about 30 percent of the nation's lands and play a major role in

NATIONAL PARK SERVICE

The National Park Service has strict policies to exclude or eradicate non-indigenous species. Still, control of harmful species is not adequate in Everglades National Park and many others.

determining the distributions and population sizes of NIS in the United States. Their policies regarding NIS vary from rigorous to nonexistent. The National Park Service (NPS) has the most stringent policies designed to conserve indigenous species and exclude or eradicate NIS. Nevertheless, even this agency does not adequately control harmful NIS.

Most other Federal land management agencies have general policies favoring the use of indigenous species or already established NIS in planned introductions or stocking of fish and wildlife. Few have similar policies regarding plant introductions. Routine planting of NIS for

---

[4] 7 U.S.C.A. 2804

[5] Nonindigenous Aquatic Nuisance Prevention and Control Act of 1990, as amended (16 U.S.C.A. 4701 *et seq.*, 18 U.S.C.A. 42)

landscaping, soil conservation, and to provide vegetation for wildlife occurs on many Federal lands, including FWS's National Wildlife Refuges and other reserves (4).

Grazing by non-indigenous livestock, feral horses (*Equus caballus*), and burros (*Equus asinus*) is specifically allowed by law on vast areas of Federal land. In some places overgrazing in the past has contributed to rangeland degradation and domination by noxious weeds (134). Many Federal land managers consider the currently widespread and growing distribution of noxious weeds to be a significant management concern (136). Noxious weed control programs generally are small and underfunded, however. Widespread interest exists in the use of biological control agents to control noxious weeds, but few agencies have clearly defined policies for evaluating their safety before release.

Federal policies also affect millions of privately owned acres through the Conservation Reserve Program of the Agricultural Stabilization and Conservation Service. There are no requirements for planting indigenous species or controlling non-indigenous insect pests and noxious weeds on lands enrolled in this program.

■ Evaluating NIS Before Introduction

Federal agencies vary in how rigorously they assess potential environmental effects before recommending NIS for technical applications or introducing them through Federal or federally funded activities. Neither the Soil Conservation Service nor the Agricultural Research Service systematically evaluates plant invasiveness before releasing species for use in soil conservation or horticulture. FWS's Federal Aid Program makes it the responsibility of State applicants to ensure any proposed introductions comply with the National Environmental Policy Act[6] and Executive Order 11987[7] (138,139).

■ NIS in Commerce

Historically, seed purity laws significantly reduced the entry and spread of non-indigenous weeds by requiring accurate labeling and by setting standards for purity of agricultural seed. Many other categories of NIS are commercially distributed today with varying degrees of equivalent coverage. The significance of contamination of transported goods as a potential pathway for harmful introductions is uncertain for these other NIS. Nevertheless, areas with expanding production and markets pose the greatest concern. For example, Federal regulations specifying labeling requirements and standards for product purity are lacking for horticultural seeds (including wildflowers) and certain biological control agents (including insects and nematodes).

CURRENT NATIONAL POLICY

Finding:

No clear national policy presently exists on NIS. President Carter issued a far-reaching executive order on NIS in 1977; in practice it has been ignored by most Federal agencies. Moreover, the U.S. Fish and Wildlife Service has yet to implement the order in regulations although specifically directed to do so.

■ President Carter's Executive Order

President Jimmy Carter issued an executive order in 1977 that could have created a national policy on NIS if it had been broadly implemented (box 6-B). It instructed executive agencies to restrict introductions of "exotic" species into U.S. ecosystems, to encourage State and local governments and private citizens to prevent introductions, and to restrict the export of indigenous species for introduction into ecosystems outside of the United States. While the order's definition of "exotic" is usually interpreted to be those species not yet established in the United

---

[6] National Environmental Policy Act of 1969 (42 U.S.C.A. 4321 *et seq.*)

[7] Executive Order No. 11987, Exotic Organisms, 42 FR 26949, May 24, 1977

Box 6-B—Executive Order 11987—May 24, 1977, Exotic Organisms

By virtue of the authority vested in me by the Constitution and statutes of the United States of America, and as President of the United States of America, in furtherance of the purposes and policies of the Lacey Act (18 U.S.C. 42) and the National Environmental Policy Act of 1969, as amended (42 U.S.C. 4321 et seq.) it is hereby ordered as follows:

Section 1. As used in this Order:

(a) "United States" means all of the several States, the District of Columbia, the Commonwealth of Puerto Rico, American Samoa, the Virgin Islands, Guam, and the Trust Territory of the Pacific Islands.

(b) "Introduction" means the release, escape, or establishment of an exotic species into a natural ecosystem.

(c) "Exotic species" means all species of plants and animals not naturally occurring, either presently or historically, in any ecosystem of the United States.

(d) "Native species" means all species of plants and animals naturally occurring, either presently or historically, in any ecosystem of the United States.

Section 2. (a) Executive agencies shall, to the extent permitted by law, restrict the introduction of exotic species into the natural ecosystems on lands and waters which they own, lease, or hold for purposes of administration; and, shall encourage the States, local governments, and private citizens to prevent the introduction of exotic species into natural ecosystems of the United States.

(b) Executive agencies, to the extent they have been authorized by statute to restrict the importation of exotic species, shall restrict the introduction of exotic species into any natural ecosystem of the United States.

(c) Executive agencies shall, to the extent permitted by law, restrict the use of Federal funds, programs, or authorities used to export native species for the purpose of introducing such species into ecosystems outside the United States where they do not naturally occur.

(d) This Order does not apply to the introduction of any exotic species, or the export of any native species, if the Secretary of Agriculture or the Secretary of the Interior finds that such introduction or exportation will not have an adverse effect on natural ecosystems.

Section 3. The Secretary of the Interior, in consultation with the Secretary of Agriculture and the heads of other appropriate agencies, shall develop and implement, by rule or regulation, a system to standardize and simplify the requirements, procedures and other activities appropriate for implementing the provisions of this Order. The Secretary of the Interior shall ensure that such rules or regulations are in accord with the performance by other agencies of those functions vested by law, including this Order, in such agencies.

JIMMY CARTER

SOURCE: Executive Order No. 11987, 42 *Federal Register* 26949 (May 24, 1977).

States, the wording is sufficiently vague to allow a species presently in one U.S. ecosystem to be "exotic" in other U.S. ecosystems (30).

The Secretary of the Interior was instructed to implement the order in regulations. Attempts by FWS to develop regulations in 1978 met with strong opposition from agriculture, the pet trade, and other interest groups (see ch. 4, box 4-A). To date, FWS has not succeeded in issuing regulations under the order, although the earlier draft regulations continue as internal guidelines for the agency (37).

No direct evidence exists that other executive agencies changed internal guidelines or agency policies in response to the Executive Order. No Federal agency contacted by OTA, other than FWS and NPS, provided any explicit policy statement on NIS, although officials from several were aware of the Carter order. Considerable variation exists among Federal agencies in how they define and treat NIS. This sometimes makes coordination among them difficult. Given its minor effects, Executive Order 11987 did not generate a consistent national policy on NIS.

Interest in implementing the Carter order continues in some parts of FWS and other agencies. However, executive orders are an inherently weak mechanism for establishing new national policy. Executive Order 11987 has not been fully implemented for 16 years. Consequently, its future significance is questionable.

■ Recent Related Efforts

Two acts of Congress in 1990 have recently focused Federal attention on specific groups of harmful NIS.

AQUATIC NUISANCE SPECIES TASK FORCE

The Nonindigenous Aquatic Nuisance Prevention and Control Act created an interagency task force to deal with harmful aquatic NIS in response to the spread of zebra mussels in the Great Lakes. The Act's goals go beyond control of this single species and include significant anticipatory functions for preventing and controlling future invasions of other harmful aquatic NIS.

The Task Force is cochaired by FWS and the National Oceanic and Atmospheric Administration (NOAA) and draws additional members from five other Federal agencies. The Act set out a number of assignments for the Task Force, including many having required completion dates (table 6-1). The delivery of most has been delayed considerably on account of several factors (31).

First, little funding has been appropriated for the program and policy development that is authorized and necessary for fulfilling the Task Force's responsibilities (31). For most staff on working groups, Task Force functions were simply added to their existing responsibilities. A lack of funds has also seriously hampered initia-

tion of the required ballast exchange and biological studies (table 6-1). The related appropriations that have been forthcoming in fiscal years 1991 and 1992 went primarily to zebra mussel control programs and research (91).

In addition, the Task Force has a broad membership with differing missions and goals. It has taken time for member agencies to air their differences, negotiate priorities, and set consensus goals. Had a national policy on NIS already been incorporated into the internal policies of all agencies, this process probably would have been more rapid. Nevertheless, the Task Force's development of common policies and approaches may lay the foundation for future efforts in this area.

Finally, administrative details related to the mandated structure and function of the Task Force have also slowed its progress. Early on, attorneys for several member agencies decided the Task Force needed to be chartered.[8] Further, the charter was deemed a prerequisite for the memorandum of understanding required under the Act and for allowing non-Federal entities to participate in Task Force meetings (31).

A key to future prevention and control efforts will be the development and implementation of an "Aquatic Nuisance Species Program."[9] The Act does not set out details of this program. Instead, it instructs the Task Force to develop the program, describe the responsibilities of individual agencies, and recommend funding levels. A draft of the program was released for public comment in November 1992. Although the draft sets out general areas of potential agency activity, it does not clearly assign agency duties or provide guidance to Congress on future funding. Member agencies have hesitated to take on new responsibilities unmatched by new appropriations.

Should the prevention and control provisions of the Nonindigenous Aquatic Nuisance Prevention and Control Act eventually be funded and implemented, they could have a significant role in

---

[8] as required by the Federal Advisory Committee Act (1972), as amended (5 Ap 2 U.S.C.A. 1 *et seq.*)

[9] 16 U.S.C.A. 4722

Table 6-1—Delivery of Requirements Under the Nonindigenous Aquatic Nuisance Prevention and Control Act

Responsibility assigned to:	Task:	Required by:	Delivered by:
Task Force	Request the Great Lakes Commission convene a coordination meeting	Feb. 29, 1990	Nov. 26, 1991
Task Force	Issue protocols for research on aquatic nuisance species	Feb. 29, 1991	Sept. 24, 1992 (draft)
USCG	Issue voluntary guidelines for ballast exchange	May 29, 1991	Mar. 15, 1991
Task Force	Sign memorandum of understanding on roles of agencies in the task force	May 29, 1991	Apr. 17, 1992
USCG	Issue education and technical assistance programs to assist in compliance with ballast exchange guidelines	Nov. 29, 1991	Dec. 1991
Task Force	Report to Congress on a program to prevent and control aquatic nuisance species ("Aquatic Nuisance Species Program")	Nov. 29, 1991 (annual reports thereafter)	Nov. 18, 1992 (draft)
Task Force	Report to Congress on intentional introductions policy review	Nov. 29, 1991	anticipated mid-1993
USCG	Report to Congress on needs for controls on vessels other than those entering the great lakes ("Shipping Study")	May 29, 1992	Dec. 1992
Task Force	Report to Congress on effects of aquatic nuisance species on the ecology and economic use of U.S. waters other than the Great Lakes ("Biological Study")	May 29, 1992	anticipated mid-1995
Task Force	Report to Congress on the environmental effects of ballast exchange ("Ballast Exchange Study")	May 29, 1992	anticipated mid-1994
USCG	Issue regulations on ballast exchange	Nov. 29, 1992	Apr. 8, 1993

SOURCES: Nonindigenous Aquatic Nuisance Prevention and Control Act of 1990 (16 U.S.C.A. 4701-4751; 18 U.S.C.A. 42); G.B. Edwards and D. Cottingham, Cochairs, Aquatic Nuisance Species Task Force, letter to E.A. Chornesky, Office of Technology Assessment, Nov. 25, 1992; 58 Federal Register 18330 (April 8, 1993).

preventing the unintentional entry and dissemination of harmful aquatic species. However, since the draft program requires detailed and time-consuming analyses of requests for funds, this probably will not result in a rapid-response control program for new infestations (91). The absence of any mechanism to disperse funds for emergency control was a significant concern in State reviews of the draft program (17,49). The Act's implementation also will not address the escape of aquatic NIS from aquaculture facilities: the Task Force has interpreted all introductions related to aquaculture as intentional, and therefore not under the general purview of the Act (91).

UNDESIRABLE PLANT MANAGEMENT ON FEDERAL LANDS

The 1990 Farm Bill contained an amendment to the Federal Noxious Weed Act requiring agencies to control "undesirable plants," including "exotic"[10] species, on Federal lands. It requires each agency to develop, staff, and support a program for undesirable plant management. Implementation has been patchy thus far. The U.S. Department of Agriculture (USDA) issued a department-wide policy on noxious weeds in 1990 to more fully integrate its existing programs and activities (103). Several agencies, such as the Bureau of Land Management, Forest Service, and Bureau of Indian Affairs, have

[10] The amendment does not define "exotic." Instead it specifies "undesirable" as those plants classified "undesirable, noxious, exotic, injurious, or poisonous, pursuant to State or Federal law." (7 U.S.C.A. 2814)

noxious weed programs in place, although these tend to be a small component of overall land management activities, and the level of effort varies among sites. NPS has a long-standing program for management of non-indigenous plants, some of which are noxious weeds. Several other agencies have not yet developed noxious weed management programs, including FWS and the Department of Energy.

Representatives of several Federal land management agencies met in September 1992 to discuss future efforts to control noxious weeds. There was general consensus that the problems are severe and growing, programs are generally underfunded and understaffed, and needs exist for greater coordination among agencies. Such interest could presage greater efforts in this area.

POLICIES AND PROGRAMS OF FEDERAL AGENCIES

Finding:

Of the 21 Federal agencies engaged in NIS activities, APHIS has the largest role, with a sizable staff performing its responsibilities to prevent the importation and dissemination of agricultural pest species. FWS, although its programs are smaller, also has an important role in regulating the importation of fish and wildlife. Other relevant Federal activities are scattered among agencies and primarily relate to other uses or management of NIS or research.

■ Areas of Federal Activity

Federal activities related to NIS occur in several areas (table 6-2):

- Movement of species into the United States. This involves restricting entry of harmful NIS by regulation, inspection, and quarantine or enhancing entry by intentional importation of desirable species or by importation of materials that unintentionally harbor harmful NIS.

- Movement of species within the United States across State lines. This involves restricting movement of harmful NIS by regulation, inspection, and quarantine or enhancing movement of desirable NIS by intentional transfers and of harmful NIS by transporting materials that unintentionally harbor NIS.
- Regulating product content or labeling. This involves restricting entry or interstate movement of harmful NIS by regulating contamination or mislabeling of NIS in commerce.
- Controlling or eradicating harmful NIS.
- Introducing desirable NIS.
- Federal land management. This involves preventing, eradicating, or controlling harmful NIS on Federal lands and introducing or maintaining desirable NIS on Federal lands.
- NIS research. This addresses prevention, control, and eradication of harmful NIS and beneficial uses of NIS.

The following section examines the roles and responsibilities of 21 Federal agencies (box 6-A) in each area of activity. Included are several specific topics, such as control of noxious weeds; development or application of aquaculture and biological control (both often are based on the transfer or cultivation of species in areas where they did not formerly occur); and management of livestock, wild horses, and burros—all of which are NIS. These same domestic activities of the various Federal agencies are shown for different groups of organisms in table 6-3.

■ Department of Agriculture

At least eight separate agencies of the U.S. Department of Agriculture have responsibilities related to NIS. Their roles are diverse and include most categories shown in tables 6-2 and 6-3.

ANIMAL AND PLANT HEALTH INSPECTION SERVICE

The Animal and Plant Health Inspection Service has broad assignments related to the importa-

Table 6-2—Areas of Federal Agency Activity Related to NIS

Agency[a]	Movement into U.S. Restrict	Movement into U.S. Enhance	Interstate movement within U.S. Restrict	Interstate movement within U.S. Enhance	Regulate product content or labeling	Control or eradication programs	Fund or do introductions	Federal land management Prevent eradication or control	Federal land management Introduce or maintain	Fund or do research Prevention control eradication	Fund or do research Uses of species	Aquaculture development	Biocontrol development
APHIS	✓		✓		✓	✓	✓	✓	✓	✓	✓		✓
AMS			✓		✓								
FAS	b												
USFS	✓					✓	✓	✓		✓	✓		
ARS	✓	✓		✓		✓	✓			✓	✓	✓	✓
SCS	✓	✓		✓						✓	✓		
ASCS							✓						
CSRS							✓				✓	✓	✓
FWS	✓		✓		✓	✓	✓	✓	✓	✓	✓	✓	✓
NPS						✓		✓	✓	✓	✓		✓
BLM						✓		✓	✓	✓			✓
BIA						✓		✓					
BOR						✓	✓	✓	✓				
NOAA	✓					✓		✓		✓	✓	✓	
DOD	✓	✓		✓		✓	✓		✓			✓	✓
EPA			c		✓					✓	d		
PHS	✓							✓					
Customs	✓												
USCG	✓							✓					
DOE								e	e				
DEA	✓												

a For acronyms of Federal agencies see box 6-A.
b Monitors animal diseases abroad.
c Monitors spread of human disease vectors within the United States.
d Regulates experimental releases of microbial pesticides.
e DOE lacks policies on NIS.

SOURCE: Office of Technology Assessment, 1993.

Table 6-3—Federal Coverage of Different Groups of Organisms[a]

	Movement into U.S. Restrict	Movement into U.S. Enhance	Interstate movement within U.S. Restrict	Interstate movement within U.S. Enhance	Regulate product content or labeling	Control or eradication programs	Fund or do introductions	Federal land management — Prevent eradication or control	Federal land management — Introduce or maintain	Fund or do research — Prevention control eradication	Fund or do research — Uses of species	Fund or do research — Assist industry uses
Plants	APHIS DOD Customs DEA	ARS[c] SCS[c]	APHIS AMS	ARS SCS DOD[b]	APHIS AMS	APHIS FWS BIA BOR NOAA DOD	ARS[c] ASCS[c]	USFS FWS NPS BLM DOD	FWS NPS DOD	APHIS ARS SCS CSRS FWS NPS BLM BOR DOD	USFS[c] ARS[c] SCS[c]	ARS[c] SCS[c]
Terrestrial vertebrates	APHIS FWS DOD PHS Customs	DOD[b]	APHIS FWS		FWS	APHIS FWS	FWS	FWS NPS	USFS FWS NPS BLM DOD	APHIS FWS NPS		
Insects (and arachnids)	APHIS FAS ARS DOD PHS Customs	ARS[d] DOD[b]	APHIS	ARS[d] DOD[b]		APHIS USFS	ARS[d] USFS[d] DOD[d]	USFS NPS BLM	USFS[d] NPS[d] BLM[d]	APHIS USFS ARS CSRS NPS PHS	APHIS[d] ARS[d] ARS NPS[d] DOD[d]	ARS[d] CSRS[d]
Fish	FWS Customs USCG		FWS	DOD[b]	FWS	FWS BOR	FWS BOR[d]	NPS BLM	USFS FWS NPS BLM DOD	FWS NPS NOAA EPA USCG	ARS[e] CSRS[e] FWS[e] NOAA[e]	ARS[e] CSRS[e] FWS[c][e] NOAA[e]
Invertebrates (non-insect)	APHIS ARS FWS DOD PHS Customs USCG	ARS[d] DOD[b]	APHIS FWS	DOD[b]	FWS	APHIS				FWS NPS NOAA EPA USCG	ARS[c] NOAA[e] CSRS[e]	ARS[e] CSRS[e] DOD[e]
Microbes	APHIS FAS ARS FWS NOAA DOD EPA PHS Customs USCG	ARS[d] DOD[b]	APHIS		EPA	APHIS USFS FWS	ARS[d] USFS[d]	USFS NPS	USFS[d] NPS[d]	APHIS USFS ARS CSRS FWS NPS NOAA USCG	ARS[d] CSRS[d] NPS[d]	ARS[d]

a For acronyms of Federal agencies see box 6-A.
b Pests move unintentionally with equipment or due to construction.
c Plants for agriculture, horticulture, or soil conservation.
d Biological control agents.
e Aquaculture.

SOURCE: Office of Technology Assessment, 1993.

tion, interstate movement, and management of NIS under the Federal Plant Pest Act,[11] the Plant Quarantine Act,[12] and several related statutes. The agency's primary concern is species that pose a threat to agriculture, including plant pests and pathogens, animal pests and pathogens, and noxious weeds. APHIS, for the most part, does not deal with species capable of harming natural ecosystems or creating a human nuisance, unless they also affect agriculture or forestry. Exceptions include its responsibilities to control vertebrate pests and to prevent importations of noxious weeds. In addition, APHIS is a member agency of the Aquatic Nuisance Species Task Force.

Movement of Species Into the United States— APHIS restricts the movement of agricultural pests and pathogens into the country by inspecting, prohibiting, or requiring permits for the entry of agricultural products, seeds, live plants and animals, and other articles that might either be or carry pests and pathogens. In fiscal year 1992, actual expenditures for agricultural quarantine and inspection were $105,787,000, with 1,929 full-time employees (170). APHIS's task of controlling movement of NIS into the country continues to expand because of increased international travel and trade (table 6-4). Pest exclusion activities are projected to double between 1991 and the year 2000 (42).

Most import restrictions relate to the relative risk that an item will be or will carry agricultural pests or pathogens. Past risk assessments were informal and based on review of the scientific literature, previous experience, and expert judgment (ch. 4). Development of more formalized risk assessment procedures is under way.

A shortcoming of current pest exclusion is that potential pests are not always restricted from entry in a timely fashion. In 1990 APHIS did not scrutinize the potential movement of forest pests

and pathogens with proposed imports of timber from Siberia until substantial congressional concern surfaced (25). Delays also occur in excluding noxious weeds from entry, which requires formal listing of species by agency regulation under the Federal Noxious Weed Act.[13] The listing approach is difficult and time consuming, allowing species fulfilling the criteria of a noxious weed to be legally imported until added to the list (40,83).

The overall success of APHIS's efforts to exclude pests is difficult to evaluate. Complete exclusion probably is infeasible. However, it is unclear what level of exclusion APHIS aims for or routinely attains, since the agency lacks performance standards for its port inspection activities or routine evaluation of its programs.

APHIS ''pre-clears'' some commodities before they are shipped to the United States by inspecting or treating commodities to eliminate pests or by inspecting growing areas, processing facilities, or handling and shipping facilities (55). Approved countries sometimes provide staff for these functions. Pre-cleared materials can enter the United States without further inspection, although they are subject to random examination at the point of entry (55). Thus far, APHIS's pre-clearance programs are small, with inspections of fruits, vegetables, and plant material occurring in 24 countries (170).

Most of APHIS's pest exclusion activities occur at ports of entry, where inspection of incoming passenger baggage and cargo and assignment to quarantine take place. Thirty-seven million passengers arrived in fiscal year 1990. That year APHIS found 1,303,000 baggage violations and assessed $723,345 in penalties for 23,676 of these (42). APHIS forwards certain plants, animals, and commodities from ports of entry to quarantine facilities within the country for detection and treatment of any pests or pathogens they might carry.

---

[11] Federal Plant Pest Act (1957), as amended (7 U.S.C.A. 147a *et seq.*)

[12] Nursery Stock Quarantine Act (1912), as amended (7 U.S.C.A. 151 *et seq.*; 46 U.S.C.A. 103 *et seq.*)

[13] 7 U.S.C.A. 2809

Table 6-4—APHIS's Pest and Disease Exclusion Activities

Recent increases in inspections, incoming passengers, and commodities (thousands)					
	1977	1984	1989	1990	Percentage increase
Total inspections	—[a]	18,917	—[a]	390,278	2000%
Inspections of animals	—[a]	1,690	—[a]	2,965	75%
Interceptions of prohibited material	—[a]	1,250	—[a]	1,858	49%
Plant importations	155,000	—[a]	318,000	—[a]	105%
Trade in commercial birds	313	—[a]	368	—[a]	18%
Passenger traffic	—[a]	26,000	34,000	—[a]	31%

Numbers of agricultural quarantine inspections		
	1990	1991
Airplanes inspected	364,000	356,915
Vessels inspected	54,000	52,119
Railroad cars inspected..	156,838	151,988
Mail packages inspected	237,024	256,964
Regulated and misc. cargo inspections	1,054,000	1,109,175
Animal/plant import inspections	2,965,000	—[a]
Personally owned pet birds inspected	2,130	1,612
Commercial birds inspected	361,373	180,706
Poultry inspected (chicks and poults)	7,121,000	5,440,976
Seed samples processed	12,923	5,099

Numbers of interceptions of unauthorized material		
	1990	1991
Unauthorized plant material	1,652,000	1,527,922
Unauthorized animal products, by-products	206,000	221,174
Noxious weeds: total interceptions (sent for inspection)	3,219	3,065
Noxious weeds: number of taxa	27	30
Mail containing unauthorized material.....................................	8,900	10,785
Baggage containing unauthorized material.................................	1,303,000	1,149,508

[a] Data not obtained.

SOURCES: U.S. Department of Agriculture, Animal and Plant Health Inspection Service, "WADS Information: October 1991," Information Fact Sheet, October 1991; U.S. Congress, House Committee on Appropriations, Subcommittee on Agriculture, Rural Development, and Related Agencies, Hearings on Agriculture, Rural Development and Related Agencies Appropriations for 1992: Part 4, Serial No. 43-171 O, May 2, 1991; D. Barnett, Staff Officer, USDA Animal and Plant Health Inspection Service, FAX letter to E.A. Chornesky, Office of Technology Assessment, Nov. 19, 1992.

Movement of Species Within the United States—APHIS restricts interstate movement of agricultural plant pests or pathogens by imposing domestic quarantines and regulations. Affected States usually adopt parallel measures to restrict intrastate movement (55).

Domestic quarantines exist for 14 non-indigenous plant pests.[14] Such quarantines re-strict interstate transport of items that might carry a pest, such as firewood and recreational vehicles for the gypsy moth (*Lymantria dispar*). APHIS also regulates the interstate transport of livestock, animal products, hay, manure, and other items that could spread animal pathogens, as well as nursery stock, soil, and soil-moving equipment that could spread plant pathogens listed in domes-

[14] 7 CFR 301

tic quarantines (55). Some domestic quarantines restrict interstate transport of imported commodities. For example, Japanese Unshiu oranges (*Citrus reticulata* var. *unshiu*) can carry citrus canker (*Xanthomonas campestris* pv. *citri*). APHIS allows their importation, but restricts their transport within the country to non-citrus growing areas.

Restricting the movement of non-indigenous pests with high natural rates of spread is difficult. Consequently, APHIS does not attempt eradication, containment, or suppression of pests like the Russian wheat aphid (*Diuraphis noxia*) (55). APHIS also does not regulate some areas where the States are active, unless problems occur requiring a national approach. For example, although regulation of the honey bee (*Apis mellifera*) industry has been a State function, introduction of varroa mites (*Varroa jacobsoni*) prompted APHIS to consider developing regulations on interstate movement of honey bees in 1991 (42).

APHIS's current authority requires a warrant for inspection of first-class mail between States, although this can be an important pathway for pest spread. The shipment of agricultural products and associated pests, such as the Mediterranean fruit fly (*Ceratitis capitata*), between Hawaii and the mainland has been a growing concern. APHIS confiscated 4,228 pounds of prohibited plant material and imposed 85 civil penalties during the first five months of a trial inspection program conducted with the U.S. Postal Service in 1990. Fruit fly larvae occurred in 45 inspected packages; other important agricultural pests were found in 177 packages (42). APHIS supported formalization of first-class mail inspection either in Postal Service regulations or in additional legislation in 1991 (42). By 1992, the agency was no longer seeking an easing of the warrant system, because the interdiction program, coupled with extensive public education, had

USDA

*Witchweed (*Striga asiatica*) is the only noxious weed that USDA's Animal and Plant Health Inspection Service has attempted to quarantine.*

reduced attempted quarantine violations by 80 percent (64).

APHIS narrowly interprets its authority under the Federal Noxious Weed Act to restrict interstate transport of noxious weeds. The agency only regulates interstate transport if a quarantine is in place, and imposes a quarantine only if a control or eradication program exists (41). Few control or eradication programs exist for noxious weeds, and the agency has imposed only one domestic quarantine—witchweed (*Striga asiatica*).[15] Consequently, although all 93 designated noxious weeds are prohibited from entry to the United States, 9 of these presently are sold in interstate commerce (55).

Monitoring—APHIS conducts several monitoring programs abroad and in the United States to track non-indigenous pests and pathogens. International pest detection surveys focus on approximately 100 non-indigenous fruit fly species, khapra beetle (*Trogoderma granarium*), citrus canker, and Karnal bunt fungus (*Tilletia indica*)—primarily in Mexico, the Caribbean, or Latin America (42). While monitoring of worldwide animal disease agents is relatively successful, widespread criticism exists of programs for plant pests. Many observers consider current systems to be inadequate for providing predictive information of use to regulators (54). This may, in part, be due to the inherent difficulty of developing plant pest databases (see ch. 4) (12). However, it also reflects erratic support.

The agency has domestic survey programs for at least 23 non-indigenous insect pests (42). APHIS also participates in the National Animal Health Monitoring Program, a cooperative Federal-State-Industry monitoring system that provides information on the geographic scope of infectious pathogens threatening livestock, poultry, and related industries.

Control and Eradication—APHIS's management plans often combine regulatory actions with monitoring, eradication, or control programs. The choice among these options depends on feasibility and the existence of appropriate technologies. Many management plans are in cooperation with State agencies.

APHIS eradicates or controls certain species that are newly introduced or present in confined areas. Its advanced planning includes "action plans" for eradicating pests not yet in the United States, but which previously have been intercepted at U.S. borders (32). Once a pest is widely established, however, control responsibilities often shift to other Federal, State, and private agencies. For example, APHIS attempted to eradicate early swarms of the African honey bee (*Apis mellifera scutellata*) along the Texas border, but switched its strategy to technology transfer and advice to the States when eradication no longer seemed feasible (42).

APHIS does have some eradication campaigns to eliminate or suppress widespread pests that are under domestic quarantines, such as the boll weevil (*Anthonomus grandis*), the bluetongue virus, several equine pathogens, golden nematode (*Globodera rostochiensis*), and witchweed (55).

More often, however, the goal is to eliminate isolated infestations of pests, like the gypsy moth or imported fire ants (*Solenopsis* spp.).

Suppression of noxious weeds is a minor component of APHIS's eradication and control efforts. Small control programs exist for only 8 of the 45 listed noxious weeds that are known or thought to occur in the United States (164). APHIS spent an estimated $725,000 in fiscal year 1992 for control of noxious weeds. As perspective, the agency's budget for domestic quarantine and control totaled at least $42 million (98). The budget request for noxious weed control in fiscal year 1993 was even smaller, $412,000 (98). Among other things, the agency plans to discontinue control efforts for common crupina (*Crupina vulgaris*) (98), even though, according to experts, this harmful weed of rangelands infests about 60,000 acres in the United States and is spreading (87).

APHIS is increasingly involved in biological control (55). Biocontrol programs exist for several pests, including the European corn borer (*Ostrinia nubilalis*), diffuse and spotted knapweed (*Centaurea diffusa* and *C. maculosa*), leafy spurge (*Euphorbia esula*), and Russian wheat aphid (98). In 1990, the National Biological Control Institute was created within APHIS to "promote, facilitate, and provide leadership for biological control" (106). Planned functions include increasing the visibility of biological control within APHIS, developing related regulations, and performing liaison with other Federal and State agencies that use biological control (106).

APHIS's Animal Damage Control Program (ADC) controls or eradicates both indigenous and non-indigenous wildlife that conflict with agriculture[16] (15). It also is responsible for controlling the brown tree snake (*Boiga irregularis*), under the Nonindigenous Aquatic Nuisance Prevention and Control Act. ADC is working on methods to prevent snake transfers in cargo and toxicants to

---

[16] 7 U.S.C.A. 426a.

reduce snake populations. It has begun to develop a cooperative program with Guam, with control efforts expected to begin in 1993 (16).

Under the Organic Act of 1944,[17] APHIS conducts eradication programs in countries adjacent to or near the United States. For example, a suppression program exists for the Mexfly (*Anastrepha ludens*), a pest of more than 40 fruits, in the northwestern region of Mexico to prevent its migration into the United States (98).

Research—Research at APHIS focuses on methods to support the agency's regulatory activities. Current areas include techniques to detect noxious weeds at ports of entry, treatments to eliminate pests from commodities, pest identification and control methods, and biological control (1,97). APHIS had research under way on control methods for at least nine non-indigenous pests in fiscal year 1992 (98). The agency sometimes works with industry and other government agencies to evaluate promising control agents (97). APHIS also funds some related research by the Agricultural Research Service.

AGRICULTURAL MARKETING SERVICE

The Federal Seed Act[18] authorizes USDA to regulate the labeling and content of agricultural and vegetable seed imported to the United States or shipped in interstate commerce. Historically, implementation of this Act significantly reduced the movement of non-indigenous plants into the United States and between the States by setting standards for seed purity and requiring that seed packages accurately identify their contents (60). The Act does not cover seeds of flowers or ornamental plants (104). The Agricultural Marketing Service (AMS) originally was responsible for regulating both seed importations and movement of seeds in interstate commerce. However,

APHIS assumed responsibility for importation in 1982 (75).

AMS works closely with States in regulating interstate seed shipments. About 500 State seed inspectors inspect seed subject to interstate provisions of the Federal Seed Act (98). Regulations require accurate labeling, including specification of all seed in excess of 5 percent, and designation of "weeds" and "noxious weeds" conforming to those of the State into which the seed is transported or offered for sale.[19] It is illegal to transport seeds containing weeds or noxious weeds into a State in excess of specified tolerances. When inspectors detect infractions, AMS usually resolves the case administratively, rather than by prosecution (98). In fiscal year 1991, AMS tested 934 seed samples in connection with interstate shipments and collected $76,075 in penalties under the Act (98). The fiscal year 1991 budget for Federal Seed Act functions was about $1.1 million (98).

FOREIGN AGRICULTURAL SERVICE

The Foreign Agricultural Service (FAS) is the lead agency in all USDA foreign activities (75). It maintains agricultural counselors, attachés, and trade officers in 74 offices, embassies and consulates covering about 110 countries (95). FAS staff periodically report on plant or animal health issues that might affect expected importations, and the agency sometimes alerts U.S. Customs, APHIS, or other agencies of developing problems (75). FAS also facilitates the overseas activities of APHIS staff supervising pre-clearance or monitoring foreign pest and pathogen conditions (75).

FOREST SERVICE

Primary responsibilities of the Forest Service (USFS) relate to its management of the National Forest System and research on forest pests and pathogens.

---

[17] Department of Agriculture Organic Act of 1956, as amended (7 U.S.C.A. 428a *et seq.*)

[18] Federal Seed Act (1939), as amended (7 U.S.C.A. 1551 *et seq.*)

[19] 7 CFR 201, as amended (Jan. 4, 1940).

Gypsy moth (Lymantria dispar*) research is the U.S. Forest Service's responsibility while the Forest Service and USDA's Animal and Plant Health Inspection Service share obligations for controlling the pest.*

Land and Resource Management—The 191 million-acre National Forest System is distributed in 43 States (74) and makes up roughly 8 percent of the U.S. land area. Congress has designated 32.5 million of these acres, or 17 percent, as wilderness (92). Policies regarding NIS are more restrictive in wilderness areas; for example, stocking of "exotic"[20] fish is prohibited, restoration of disturbed vegetation must incorporate only indigenous species, and wildlife may be controlled when they harm indigenous species (75).

In general, however, the National Forest System is managed for multiple uses,[21] including timber production, outdoor recreation, rangeland grazing, watershed preservation, and fish and wildlife habitat (94). Thus, aside from constraints on wilderness areas, the Forest Service manages its lands for purposes that sometimes include the introduction of NIS.

Grazing—In 1989, a total of 1,147,916 cattle (*Bos taurus*), horses, and burros and 944,843 sheep (*Ovis aries*) and goats (*Capra hircus*)—all of them non-indigenous—grazed on lands of the National Forest System (163). The Forest Service has inventoried approximately 50 million acres as suitable for grazing (93). According to a recent Forest Service internal survey, 24 percent of the grazing allotments in six Western Regions had problems with vegetation or soil and water resources caused either by improper livestock grazing or by grazing occurring where it conflicts with other valued resources such as wildlife or recreation (92).

Introductions of Fish and Wildlife—As a general policy, when stocking or introducing fish or wildlife, the Forest Service favors "native"[22] or "desirable" non-native species (108). Introductions of new NIS desired by the public may be allowed (108). The Forest Service considers management of fish and wildlife in the National Forests primarily a State responsibility. Releases of NIS at new sites involve joint agreements with State fish and wildlife agencies and coordination with FWS (108,163). In evaluating such introductions, the Forest Service and States consider probable effects on adjoining private and other public lands, as well as compatibility with multiple-use management (108,109). More careful consideration is given to introductions of new NIS than to repeated stocking of species introduced in the past, such as the chukar partridge (*Alectoris chukar*). The latter do not require an environmental analysis unless they are controversial (108).

Control of Noxious Weeds—The Forest Service has an active program to control noxious weeds. The current emphasis is on use of integrated management systems, and the Forest

---

[20] "Exotic" is defined in the FS manual as "Species not originally occurring in the United States and introduced from a foreign country. Exotic species that have become naturalized, such as the ring-neck pheasant [*Phasianus colchicus*], are considered the same as native species" (111).

[21] under the Multiple-Use Sustained Yield Act of 1960, as amended (16 U.S.C.A. 528 *et seq.*)

[22] According to the Forest Service Manual, "native" refers to species indigenous to the United States (111).

Service has a strong interest in using biological control agents (112,168).

A recently issued interim directive on noxious weeds includes several notable components (112). Where possible, forage and browse seed for planting and feed, hay, or straw brought onto Forest Service lands must be certified free of noxious weed seed (112). The directive further encourages the use of desirable plant species that out-compete noxious weeds and requires where appropriate that equipment brought onto Forest Service lands by contractors or permittees be free of noxious weed seeds (112). Forest Supervisors are specifically instructed to assess the risks of introducing noxious weeds in projects that disturb plant communities (112).

Control of Forest Pests and Pathogens—The Forest Service has responsibility for detecting, identifying, surveying, and controlling forest pests affecting forested lands in the United States under the Cooperative Forestry Assistance Act.[23] While the Forest Service directly manages species affecting the National Forest System, management elsewhere is through cooperative agreements with other Federal and State agencies using funds specifically appropriated to the Forest Service for this use (162).

Most of this program does not deal with NIS, since the majority of significant pests and pathogens affecting the nation's forests are indigenous (110). Nevertheless, it does address several well-established non-indigenous pests, including gypsy moth, white pine blister rust (*Cronartium ribicola*), balsam woolly adelgid (*Adelges piceae*), and Port-Orford cedar root disease (*Phytophthora lateralis*) (163). Gypsy moth, considered the most damaging of these, is controlled cooperatively by the Forest Service and APHIS (163). The Forest Service manages larger infested areas, and it shares eradication responsibilities with APHIS for isolated outbreaks (163). The Forest Service expended an average of at least $10 million annually for gypsy moth suppression and eradica-

tion on Federal, State, and private lands from 1987 to 1991 (163). Non-indigenous insects and pathogens could become an even more significant component of forest pest management if species from Siberia ever become established in the Pacific Northwest—some localized infestations have already occurred (26).

Research—Forest Service research on timber management includes the selection, testing, and distribution of plant materials to improve forests. The United States is rich in indigenous woody species, and only a few NIS have been developed and distributed for specialized applications, such as windbreaks in treeless areas, urban plantings, and Christmas trees (56).

Forest Service research on forest insects and pathogens previously had large programs on introduced pathogens such as white pine blister rust, Dutch elm disease (*Ceratocystis ulmi*), and chestnut blight (*Cryphonectria parasitica*) (162). It currently has a large program (funded at $3,849,000 in fiscal year 1992) on the gypsy moth at the agency's Northeastern Forest Experiment Station (163).

AGRICULTURAL RESEARCH SERVICE

The Agricultural Research Service (ARS) is the research branch of USDA. Its functions include the evaluation of agricultural NIS, which later are disseminated throughout the country by the commercial sector. ARS also conducts research on the prevention, control, or eradication of harmful NIS, often in cooperation with APHIS.

Development of New Varieties—The National Plant Germplasm System (NPGS) is an important repository of seeds and other plant materials (germ plasm) for plant breeding in the United States (53,166). ARS plays a pivotal role in coordinating, funding, and staffing NPGS, although the system is actually a network of cooperating Federal, State, and private institutions (77). ARS's functions in the NPGS include

---

[23] Cooperative Forestry Assistance Act of 1978, as amended (7 U.S.C.A. 2651-2654; 16 U.S.C.A. 564 *et seq.*)

foreign exploration to bring back new plant varieties of potential use to breeders and the inspection and quarantine of imported plant materials, which it conducts in cooperation with APHIS (77). In addition, some of the U.S. plant germ plasm collection is stored by ARS (105).

An annual average of 8,503 accessions were incorporated into NPGS between 1985 and 1989 (165). About 90 percent of these were of foreign origin (165). Screening of this plant material for pathogens or contamination by other species is generally successful. Only one introduced pest, the peanut stripe virus has been traced to the National Plant Germplasm Program during the past 25 years (165).

Non-indigenous plant species and varieties are not evaluated for potential invasiveness or other harmful ecological qualities before being placed in NPGS. Many are cultivated plants posing few ecological risks (75). However, the collection does contain some harmful plants that are sources of useful genes for plant breeders (e.g., noxious weeds like wild oats (*Avena fatua*)) (166). Individuals receiving noxious weed seed from the collection must obtain Federal and State permits (166).

ARS's National Arboretum is part of the National Plant Germplasm System. Its functions include overseas plant exploration and importation, although the Arboretum's main focus is on plants for ornamental horticulture (24). The Arboretum imported a total of 2,371 species between 1986 and 1988 (165). In addition, scientists at the Arboretum develop ornamental plants and then release them to researchers or to the commercial sector for multiplication, distribution, and sale. Plants are evaluated for hardiness, pest and disease resistance, and other desirable characteristics before release. The Arboretum does not systematically evaluate plants for invasiveness. Some ARS botanists, however,

may be sensitive to such concerns and incorporate them into plant assessments (27).

ARS presently is developing the National Genetic Resources Program required by the 1990 Farm Bill.[24] This program will eventually subsume work currently in the NPGS (75). Its functions include the collection, classification, preservation, and dissemination of genetic material of importance to U.S. agriculture. Its biological breadth is greater than that of NPGS, encompassing genetic resources of animals, aquatic species, insects, and microbes in addition to those of plants. The National Genetic Resources Program may thus eventually expand ARS's role in foreign exploration and importation to include a greater variety of organisms.

Aquaculture—An additional research area involving the use of NIS is aquaculture. ARS projects include culture techniques and disease diagnosis and control (99). Total expenditures in this area were at least $7 million in fiscal year 1992 (99).

Biological Control and Other Uses of Beneficial Insects—ARS considers biological control to be one of the most important pest control tactics and has a sizable program for locating, importing, and evaluating insects and other organisms (5,99). The budget request for this program was about $9.5 million for fiscal year 1993 (99). The agency operates several laboratories abroad where researchers locate and study new biological control agents and ship them to the United States. The recently closed laboratory in Italy shipped a total of 80,175 individuals of 28 biological control agents to the United States in 1990 (34). (The laboratory's functions shifted to a new facility in Montepellier, France.) Some of ARS's research on biological control is in cooperation with other Federal agencies. For example, ARS and the Army Corps of Engineers cooperate extensively on control of aquatic weeds in the

---

[24] Food, Agriculture, Conservation, and Trade Act of 1990 (7 U.S.C.A. 5841 *et seq.*)

NOAH PORITZ

The Agricultural Research Service is studying methods to control or eradicate the African honey bee (Apis mellifera scutellata), the tracheal mites (Acarapis woodi) that infect European honey bees (A. mellifera), and other agricultural pests.

southeastern United States, and the Bureau of Reclamation contributes funding to ARS work on biological control of salt cedar (*Tamarix* spp.).

ARS researchers follow protocols for the importation and release of non-indigenous biological control agents, in addition to fulfilling APHIS's requirements for import and interstate transport permits (18). General provisions include adherence to applicable Federal and State laws, quarantine, detailed documentation, and evaluation of potential environmental and safety effects (18). These protocols provide guidance for ARS workers, but are largely voluntary for other researchers in academia and industry (13). Detailed requirements for evaluation of environmental effects before release have not yet been developed by ARS for all categories of biocontrol agents (18).

ARS also imports non-indigenous bees for research on crop pollination (88). APHIS requires permits for importation and release of bees to prevent entry of bee pathogens, parasites, predators, or harmful germ plasm (58).

Prevention, Control, or Eradication Methods— In addition to its biological control program, ARS has research aimed at the control or eradication of several non-indigenous agricultural pests, such as the Russian wheat aphid; sweet potato whitefly (*Bemisia tabaci*); Mediterranean fruit fly; African honey bee; pear thrips (*Taeniothrips inconsequens*); and tracheal mites (*Acarapis woodi*), which infect honey bees (99).

The agency spent almost $9 million for research on these six NIS in fiscal year 1992 (99). Another relevant research area is plant disease resistance, which aims to prevent infections by non-indigenous plant pathogens. ARS also studies animal pathogens not yet present in the United States at four specialized laboratories in the United States (75).

Some funds for ARS research come from State or local governments. These are for research on the control of NIS of great local concern. For example, in 1991 Florida provided at least $200,000 to ARS for work on biocontrol of melaleuca (*Melaleuca quinquenervia*) and aquatic weeds (99).

SOIL CONSERVATION SERVICE

The Soil Conservation Act of 1935[25] established the Soil Conservation Service (SCS). Its central mission continues to be the protection of land and related resources against soil erosion.[26] SCS gives technical advice to nearly all public agencies and many private entities in the United States on grasses, forages, trees, and shrubs suitable for erosion control (75). The agency devotes a significant part of its efforts to the development and dissemination of new plant materials for conservation.

Some plants released and recommended by the SCS are non-indigenous to the United States (79). Others are species of U.S. origin spread beyond their natural ranges through soil conservation applications. SCS uses NIS at least in part

---

[25] Soil Conservation and Domestic Allotment Act (1935) (16 U.S.C.A. 590a *et seq.*)

[26] 7 CFR 600, as amended (April 6, 1982).

because indigenous species sometimes may not satisfy all soil conservation needs, especially for plants that grow rapidly in disturbed, contaminated, or polluted habitats (75).

Movement of Species Into and Within the United States—SCS operates 20 plant materials centers throughout the United States, and an additional 6 are operated either jointly with other agencies or by State agencies with SCS assistance (116). These centers assemble, test, release, and provide for the commercial production and use of plant materials. Plants evaluated for any given application may come from collections of indigenous vegetation, foreign plant introductions, strains from plant breeders, or commercial seed (114). SCS has a small, informal program to locate new species abroad (69). However, the principal source of foreign plant materials is ARS, which provides an estimated 90 percent of the NIS evaluated by SCS (80).

At any given time, the plant materials centers collectively may be evaluating as many as 20,000 plant types (117). Of these, about 25 percent are non-indigenous to the United States[27] (117). From 1981 through 1990, the plant materials centers formally released for public use a total of 75 species or cultivars (varieties); 29 percent had origins outside the United States, including Turkey, China, and Africa (113). Once into commercial production, plants developed by SCS can have wide distribution. For example, in 1989, 200 SCS cultivars were in production, resulting in 24.8 million pounds of seed and 27.1 million plants, with a retail value of $78.3 million (117).

Within the SCS, no explicit agency-wide policy governs the use of indigenous versus NIS, although SCS officials state that priority is generally given to indigenous[28] species (69,80). The SCS does provide general guidance to the plant materials centers regarding testing for potential weediness. Specifically, it requires de-

termination of whether a plant "has any toxic qualities or has a potential for becoming a pest." Should the plant have these qualities, "control methods are to be developed and hazards are to be carefully assessed before the plant is considered for release" (114). Annually about 10 percent of species under evaluation are discarded because of their potential to become weeds (80).

Within those general national guidelines, the review process and species choice occurs at the individual plant material centers (69). Procedures for evaluating plants are not standardized and can vary among centers and even among individual researchers (79,80). In the past, SCS has recommended some plants that have become notable pests, such as multiflora rose (*Rosa multiflora*), Russian olive (*Elaeagnus angustifolia*), and salt cedar (75). SCS staff believe that many, if not all, of these harmful species would not pass the plant review process today (75,80).

Nevertheless, present review processes may fail to adequately screen out potential pests, especially those that only become pests in forests and other natural areas. According to one expert, at least 7 of the 22 non-indigenous cultivars released between 1980 and 1990 have the potential to become invasive in natural areas (61). In addition, even U.S. species spread beyond their natural ranges by soil conservation applications might cause problems: the Illinois Department of Conservation recently expressed concern over the release of Elsmo lacebark elm (*Ulmus parvifolia*) by the Missouri plant material center for use in windbreaks and ornamental and conservation areas (76).

Control and Eradication—SCS does not control or eradicate species it has released when they become pests (80). However, SCS is involved in an effort to replace noxious weeds on grazed lands with other palatable plants that outcompete the weeds (80,115). Current and

---

[27] SCS specifies that 75 percent are "native," presumably meaning indigenous to the United States (117).

[28] SCS staff use the term "native."

planned work includes grazing management studies, development of methods to encourage re-invasion by long-lived indigenous plants, and the collection and screening of new grassland plants (115). The collection and screening may itself involve new introductions, since SCS is considering "importing plants that have been under centuries of intensive grazing in Inner Mongolia because they have evolved to withstand abusive and intensive grazing" (79).

Providing Indigenous Germ Plasm for Restoration—Since 1990, SCS has collaborated with the National Park Service to propagate indigenous plants for revegetation following park road construction (149). SCS expanded this program to include providing plants for general park maintenance in 1992 and adopted it as an agency plan (81). A unique aspect of this effort is the use of genetic strains that are indigenous[29] to individual parks. The program provides mutual benefits to the participating agencies. SCS obtains plant materials for potential use in soil conservation. Park managers receive indigenous plants that otherwise are difficult to obtain (80).

A SCS draft strategic plan suggested this program and other SCS work could contribute to the development of banks of indigenous[30] species with known ecological zones for future needs (117). The plan recommended an expanded role in the preservation of indigenous germ plasm, including the establishment and operation of an indigenous germ plasm center (117). Whether and how this center would coordinate with the National Genetic Resources Center under development by ARS is unclear. In any case, a repository of indigenous plant material might decrease SCS reliance on potentially harmful NIS for conservation.

AGRICULTURAL STABILIZATION AND CONSERVATION SERVICE

The Agricultural Stabilization and Conservation Service (ASCS) administers the Conservation Reserve Program (CRP), created under the Food Security Act of 1985.[31] CRP's primary objective is to help reduce water and wind erosion on highly erodable croplands (19,95). Farmers enroll eligible acreage, and then plant soil-conserving plants for a 10-year contract period (19). In exchange, participants receive annual rental payments and a one-time payment for half of the eligible costs of establishing the plant cover (95). The 1990 Farm Bill broadened the program to include wetland preservation and other conservation practices (75).

CRP is set at a maximum of 44 million acres (95). As of 1990, 33,922,565 acres were enrolled (19), or roughly 8 percent of U.S. cropland and 1 percent of the total U.S. land area. In 1990, 58 percent of CRP lands were planted with grasses non-indigenous to the United States, while only 24 percent were planted with indigenous grasses[32] (19). The difference probably relates to per acre planting costs of $37.39 for NIS versus $44.95 for indigenous species (19).

CRP lands may inadvertently provide habitats for non-indigenous weeds, such as tumbleweed (*Salsola iberica*), kochia (*Kochia scoparia*), and leafy spurge (19). Plants on CRP lands can also provide habitats for non-indigenous crop pests during periods when crop hosts are not available; for example, the Russian wheat aphid persists on several grasses recommended for western sites (10,19).

Between 1986 and 1987, CRP acreage jumped by approximately 17 million acres (107). This unanticipated rapid rate of enrollment caused the demand for grass seed to exceed supply and resulted in large legal importations from abroad

---

[29] Text uses term "native" (149).

[30] Text uses term "native," referring to species indigenous to the United States (117).

[31] Food Security Act of 1985, Public Law 99-198, Title XII.

[32] Text uses "introduced" and "native" for non-indigenous and indigenous to the United States, respectively (19).

and widespread use of uncertified seed (75). While ASCS is not aware of any resulting weed problems (75), such conditions provide a ripe opportunity for unintentional importation and distribution of non-indigenous weeds.

COOPERATIVE STATE RESEARCH SERVICE

The Cooperative State Research Service (CSRS) funds research on agricultural pest control and aquaculture through State agricultural experiment stations, forestry schools, land-grant colleges, the Tuskegee Institute, and veterinary colleges. CSRS awarded grants for research on the management and control, including biological control, of non-indigenous pests totaling at least $450,000 in 1990 and $550,000 in 1991 (96). These included leafy spurge, gypsy moth, imported fire ants, Eastern filbert blight (*Anisogramma anomala*), and Russian wheat aphid. CSRS also provides funds for the use of NIS in technical applications such as biological control or aquaculture. In 1990, $338,900 was awarded to develop facilities for biocontrol of Japanese beetle (*Popillia japonica*) (96). CSRS funds five regional aquaculture centers. At these and other locations, research is under way on the detection and prevention of diseases in aquaculture species and the development of species for aquaculture applications.

■ Department of the Interior

At least five agencies within the Department of the Interior have responsibilities related to NIS. Of these, the U.S. Fish and Wildlife Service (FWS) has the most diverse role. Collectively, management policies of the department's agencies affect the distributions and impacts of NIS on at least 20 percent of the U.S. land area.

FISH AND WILDLIFE SERVICE

FWS simultaneously engages in both controlling and intentionally introducing or stocking NIS. The agency has responsibilities to prevent and control injurious fish and wildlife and to protect threatened and endangered species. At the same time, FWS promotes recreational fisheries,

Educational efforts, such as this brochure for travellers, are part of the U.S. Fish and Wildlife Service's work to prevent and control injurious introductions and protect endangered species.

hunting, and aquaculture that involve NIS. Although FWS uses regulations drafted under Executive Order 11987 as an internal policy to discourage introductions of NIS, the policy has not been uniformly adopted throughout the agency (30). Conflicting goals sometimes occur between different programs, and even between different parts of individual programs.

FWS's participation as co-chair of the Aquatic Nuisance Species Task Force has required some synthesis and internal evaluation of the agency's role in NIS issues. While the ultimate effects of this effort are presently unknown, it potentially will generate increased communication and coordination among the currently disparate programs within FWS.

Movement of Species Into the United States— FWS has responsibility for regulating the importation of injurious fish and wildlife under the Lacey Act. Current regulations prohibit or restrict entry to the United States of two families of fishes; 18 genera or species of mammals, birds, reptiles, and shellfish; and two fish pathogens.[33] FWS also restricts the importation of hundreds of threatened and endangered species from abroad under the Convention on International Trade in Endangered Species (CITES).

The FWS port inspection program is relatively small, especially in comparison with agricultural inspection. The budget request for fiscal year 1992 included $3,294,000 for 65 wildlife inspectors and an additional $500,000 for an automated import clearance system (100). In 1990, FWS port inspectors inspected 22 percent (a total of 17,562 inspections) of the wildlife shipments at international ports of entry (100).

The potential exists for FWS to play an increased role in regulating fish and wildlife imports, but current shortcomings of the FWS law enforcement division might compromise expanded efforts. A recent advisory commission found the division seriously understaffed and underfunded

and lacking clear priorities, adequate staff supervision, or sufficient technical expertise to identify species (145). Unfunded needs for law enforcement identified by FWS regional offices totaled at least $7 million for fiscal year 1992 (67).

Movement of Species Within the United States—Under the Lacey Act, interstate transport of *federally* listed species is legal. Thus, intentional movements within the country of harmful fish and wildlife such as zebra mussels face no Federal prohibition. In contrast, amendments to the Lacey Act in 1981 made the interstate movement of *State*-listed injurious fish and wildlife a Federal offense, potentially subject to FWS enforcement (70,90). No interceptions of such interstate shipments were listed among the 1990 accomplishments of FWS enforcement, suggesting this is not a high priority within the agency (100). Future implementation of the Nonindigenous Aquatic Nuisance Prevention and Control Act might increase the FWS role in preventing interstate transfers of harmful aquatic species.

Federally Funded Introductions—The FWS Federal Aid Program allows States to recover up to 75 percent of acceptable costs for various projects related to fish and wildlife restoration. Funds come from Federal excise taxes on sales of firearms and hunting and fishing equipment and supplies. The receipts have grown steadily over the past few years (figure 6-1), and payments to States totaled more than $320 million in fiscal year 1991.

The program frequently is criticized for its historical role in supporting numerous introductions of non-indigenous fish and wildlife species (20,141). Determining the exact number of introductions funded is difficult, however, since few project titles include species names or the words ''exotic'' or ''non-indigenous'' (63).

The Federal Aid Program now discourages introductions of NIS not yet established in an area. It requires States to assess the environmental

---

[33] 50 CFR 16, as amended (Jan. 4, 1974).

impacts of any introductions they propose (4,138,139). Although proposals for introductions presently are uncommon, they do continue (142). Most involve introductions of U.S. species into areas where they are not indigenous, such as the recent proposal by the New Jersey Division of Fish, Game and Wildlife to introduce chinook salmon (*Oncorhynchus tshawytscha*) from the Pacific coast to the Delaware Bay (159). Such introductions have become controversial only recently (4), and the Federal Aid Program lacks a clear policy regarding their eligibility for funds. Additional concerns are that proposals for introductions are closely scrutinized only when they engender vocal public controversy, and that State agencies sometimes inadequately fulfill requirements for assessing environmental effects of introductions. Further, States can avoid scrutiny by using State funds for the initial introduction of a species; once the species is established, funding can be sought from the Federal Aid Program for stocking without any requirement for environmental assessment.

Control and Eradication—FWS has no centralized, comprehensive program for the control and eradication of harmful NIS. Instead, control programs have variable goals, such as control of individual species, recovery of endangered species, and control of fish diseases affecting aquaculture. The most notable control program is for the sea lamprey (*Petromyzon marinus*) in the Great Lakes, conducted by the North Central Regional Office in Minnesota in cooperation with other regional entities. Under the Great Lakes Fish and Wildlife Restoration Act,[34] FWS plans to expand sea lamprey control as part of a Great Lakes initiative (100).

FWS had reported NIS as a factor contributing to the decline of approximately 30 percent of species listed as threatened or endangered as of June 1991 (see table 2-3) (4). Control of NIS is a component of the recovery plans of many listed

Figure 6-1—Account Receipts of the FWS Federal Aid to States Program

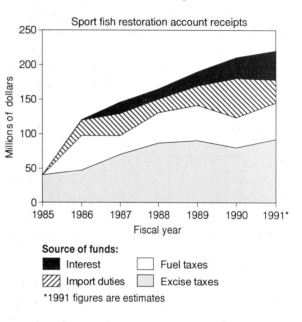

Source of funds:
- Interest
- Import duties
- Fuel taxes
- Excise taxes

*1991 figures are estimates

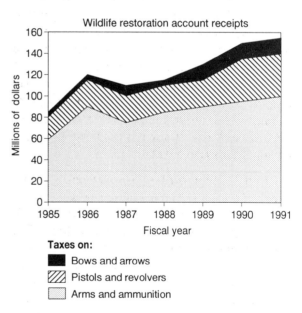

Taxes on:
- Bows and arrows
- Pistols and revolvers
- Arms and ammunition

SOURCE: U.S. Congress, House Committee on Appropriations, Subcommittee on the Department of the Interior and Related Agencies, "Hearings on Department of the Interior and Related Agencies Appropriations for 1992," Serial No. 43-294 O, (Washington, DC: U.S. Government Printing Office, 1991), pp. 1091, 1099-1100, 1111, 1117-1118.

---

[34] Great Lakes Fish and Wildlife Restoration Act of 1990 (16 U.S.C.A. 941 *et seq*)

species (4). Examples include control of feral animals and non-indigenous vegetation in Hawaii and reduction of non-indigenous fish populations in the upper Colorado Basin (100). Implementation of many recovery plans has been poor, however (4, 152). Endangered species recovery plans consequently contribute little to the control of NIS at this time.

Fisheries Enhancement and Aquaculture— FWS produces fish for stocking waterways at 77 National Fish Hatcheries throughout the country (147). While much of this effort goes to culturing indigenous fishes, it also produces NIS commonly stocked in U.S. waters. Rainbow trout (*Oncorhynchus mykiss*) and striped bass (*Morone saxatilis*), for example, are widely stocked beyond their natural ranges.

FWS created an office to coordinate aquaculture within the agency and with other Federal agencies under the National Aquaculture Act of 1980 (70). The office's primary activity is providing technical assistance related to natural resource issues and fish diseases to State agencies and the private sector. FWS helps control the spread of fish pathogens by promoting a National Fish Health Strategy and by providing voluntary diagnosis and inspection to the private sector through technical centers associated with the National Fish Hatcheries.

Land Management—FWS manages approximately 91 million acres, about 4 percent of the U.S. land area, mostly within the National Wildlife Refuge System. This system includes 500 national wildlife refuges, 166 waterfowl production areas, and 51 wildlife coordination areas (46). General goals include the preservation of natural diversity, although various units were established under different authorities and for varying purposes (4). Sometimes these even include preservation of NIS—for example, management of longhorn cattle (*Bos taurus*) at the Wichita Mountains National Wildlife Refuge.

The National Wildlife Refuge System Administration Act[35] only allows land uses that are compatible with the refuges' original purposes. In practice, this results in inconsistent NIS policies. Some NIS may be purposely introduced—for example, planting non-indigenous grass mixtures (i.e., wheatgrass (*Agropyron* spp.), alfalfa (*Medicago sativa*), and sweet clover (*Melilotus* spp.)) to enhance waterfowl production and stocking non-indigenous fish to achieve management objectives (4). Other NIS are controlled when they interfere with refuge management goals (72,147). Approximately 12 percent of the wildlife refuges experienced problems with NIS in 1991 (72).

Research—FWS has ongoing NIS research in the following areas: the distribution, biology, and control of aquatic nuisance species; the identification and treatment of fish pathogens; control of wildlife diseases; control of the brown tree snake; effects of non-indigenous vegetation on nongame migratory birds; biological control of purple loosestrife (*Lythrum salicaria*); and aquaculture techniques (72,85,100). Much of the work on aquatic species is conducted at the National Fisheries Research Centers in Gainesville, Florida; Ann Arbor, Michigan; and LaCrosse, Wisconsin.

The Gainesville center sometimes is referred to as the "Exotic Species Laboratory." One of its missions is to identify the distribution, status, and impacts of non-indigenous fish (85). The center has a database to monitor the spread of non-indigenous fishes in the United States and is developing a geographic information system (ch. 5) for monitoring non-indigenous aquatic species in general. The center's prominent role in research and information exchange has been due to the intense efforts of a small, experienced staff. However, recent staff turnover coupled with the ambiguous status of NIS among the center's various responsibilities makes its future unclear.

---

[35] National Wildlife Refuge System Administration Act of 1966, as amended (16 U.S.C.A. 668dd *et seq.*)

The Federal Aid Program of FWS funds some State research on uses, impacts, and management of non-indigenous fish and wildlife. For example, from 1989 to 1990, $100,036 went to research on the brown trout (*Salmo trutta*) and $24,671 to research on feral dogs (*Canis familiaris*) and pigs (*Sus scrofa*) (143,144). Such projects are a small part of the total research funded by this program.

Certification of Sterile Grass Carp—The FWS has operated an inspection service to certify that grass carp (*Ctenopharyngodon idella*) are triploid since 1979 (146). Presently, this is done at the Warm Springs Regional Fisheries Center in Georgia. Grass carp are non-indigenous fish that have wide application as biocontrol agents for aquatic weeds. However, they can also spread and cause environmental harm if reproductive populations become established in the wild. The triploid grass carp are sterile, and can be released without risk of establishing self-sustaining field populations.

NATIONAL PARK SERVICE

Although the law that created the National Park Service (NPS) says nothing about NIS, it does set out a general goal to "conserve the scenery and the natural and historic objects and the wild life therein and to provide for the enjoyment of the same in such a manner and by such means as will leave them unimpaired for the enjoyment of future generations."[36] This responsibility is the basis for NPS's policies promoting the eradication and control of NIS and prohibiting introductions except under very limited circumstances (4). As early as 1933, NPS had explicit policies regarding the need to control "exotic" species on park lands (52).

When the National Park System was created, preservation of U.S. ecosystems could be accomplished largely by leaving things alone. Increasingly, however, intervention has become essential to control the ecological disruption caused by

harmful NIS. This changing need has not been met by an adequate shift in management priorities, funding, and staffing within the NPS.

A rough estimate is that NPS allocates less than 1 percent of its annual budget to research, management, and control of NIS. Natural resource issues in general receive low priority within NPS. In fiscal year 1990, only 6 percent of the NPS budget went to management of natural resources (66).

Growing recognition exists that NPS will need to shift its funding priorities if it is to address the degradation of natural resources, including that related to NIS, resulting from human encroachment around park boundaries (86).

Land Management—NPS manages approximately 80 million acres divided into 10 geographic regions, or about 3 percent of the U.S. land area (2). The system is made up of about 364 units having 22 different designations such as parks, monuments, recreation areas, historic sites, and battlefields (2). Reflecting this diversity, NPS lands are divided into natural, cultural, park development, and special use management zones (148). NPS's strictest policies related to NIS are for natural zones (148).

A survey done in 1986 and 1987 on natural resource conditions in the parks found control of harmful NIS to be a significant management concern throughout NPS (47). Respondents cited non-indigenous plants as the most common threat to park natural resources. Non-indigenous animals were the fourth most commonly reported threat. Parks negatively affected by NIS occur in all 10 NPS regions (47).

Most decisions regarding control and management of NIS are made by individual parks during development of resource management plans. Within any given park, the priority given to NIS projects depends on the park's goals and present condition. NIS projects have relatively high priority among natural resource concerns within

---

[36] National Park Service Organic Act (1916), as amended (16 U.S.C.A. 1 *et seq.*)

NPS; according to NPS officials, 42 percent of NIS projects were either funded (39 percent) or ranked as highest priority among unfunded projects (3 percent) for the period from 1991 to 1995, compared with only 36 percent for all other resource management projects (51). National Parks with especially pressing problems with NIS include Haleakala and Volcanoes in Hawaii, Everglades in Florida, Great Smoky Mountains in Tennessee, and the Indiana Dunes National Lakeshore. Even smaller parks like Rock Creek Park in the District of Columbia have numerous pressing problems with non-indigenous plants.

NPS generally seeks to perpetuate indigenous plants and animals, and its policy is to manage or eradicate NIS that threaten park resources or public health whenever prudent and feasible. NIS introductions are generally prohibited by agency regulation.[37] To further prevent introductions, some parks, such as Yosemite, have park-specific regulations requiring feed materials transported into the park be certified weed free or requiring use of pelletized feeds in the backcountry (52). Notwithstanding these various bans, intentional introductions are tolerated to varying degrees in NPS's four management zones (box 6-C) (148).

Still, NPS differs from other Federal land management agencies in having strict guidelines for introductions. Plants and animals must be from populations closely related genetically and ecologically to park populations, except when the goal is to correct losses of the gene pool caused by human activities (148). In natural zones, revegetation efforts are to use plant materials not only of indigenous species, but of indigenous gene pools as well (148).

NPS Control of Activities Outside the National Parks—NPS officials increasingly see park resources affected by land use practices in surrounding areas (151). The potential impact of NIS is clear, since live organisms can move freely on and off park lands and few other public or private land managers are as restrictive as NPS. However, few parks actually do control NIS on neighboring lands, even though the 1991 NPS Natural Resources Management Guidelines list this as an appropriate approach when surrounding land owners are cooperative (59).

Research—NPS conducts research to provide "an accurate scientific basis for planning, development, and management decisions" (148). Research in the national parks is conducted by both NPS staff and researchers from outside institutions. NPS provided about $2 million for over 200 research projects related to NIS in fiscal year 1990. Research topics included evaluating environmental effects, monitoring, management, eradication methods, and restoration following species removal (150,151). NPS both conducts research on the potential use of biological control to control NIS and participates in related cooperative projects with State agencies (36).

BUREAU OF LAND MANAGEMENT

The Bureau of Land Management (BLM) manages about 270 million acres, or 11 percent of the total U.S. land area, mostly located west of the Mississippi River (2). The Federal Land Policy and Management Act of 1976 (FLPMA) directs BLM to manage lands under its jurisdiction for a mix of uses including grazing, mining, timber harvest, recreation, and wildlife conservation.[38] FLPMA thus authorizes certain uses that facilitate the spread and establishment of NIS (4).

Grazing—Grazing is one of the most common and widespread uses of BLM lands (4). It also has been a factor in the transformation and degradation of rangeland vegetation, including the spread and establishment of many non-indigenous weeds (39,134). The agency annually authorizes grazing by 4.3 million cattle, sheep, goats, and horses on

---

[37] 36 CFR 2.1 (June 30, 1983).

[38] Federal Land Policy and Management Act of 1976, as amended (43 U.S.C.A. 1701, 1702).

Box 6-C—Introduction of Non-Indigenouse Species in the National Parks

NPS divides its holdings into four management categories. Natural zones are managed to protect natural resources. Cultural zones are managed to preserve and foster appreciation of cultural resources. Park development zones are managed and maintained for intensive visitor use. And special use zones are managed for uses not appropriate in other zones, such as commercial use, mineral exploration and mining, grazing, forest use, and reservoirs. NPS policies on introductions of NIS differ among the four zones.

In natural zones, non-indigenous plants and animals may be introduced only rarely. Allowed introductions include: nearest relatives of extirpated indigenous species; improved varieties of indigenous species when the local variety cannot survive current environmental conditions; and agents used to control established NIS. Introductions to natural zones are also permitted when there is explicit direction by law or legislative intent; for example, the enabling legislation for Great Basin National Park allows for the perpetuation of free-ranging livestock within the park. The emphasis of natural zone management is on maintaining fundamental ecological processes, rather than individual species *per se*. Thus, ring-necked pheasants (*Phasianus colchicus*) and chukars (*Alectoris chukar*), introduced long ago to Haleakala National Park, are tolerated because they may satisfy ecological roles previously filled by now-extinct Hawaiian birds. Also, biological control agents have been introduced into natural zones of several national parks to control harmful NIS.

NIS may be introduced in cultural zones when they are a desirable, and historically authentic, part of the historical landscape. Such introductions are permitted only if the plant or animal is controlled so that it cannot spread. In park development zones, all of the above uses are allowed, as well as introductions to satisfy management needs that cannot be met by indigenous species. Again, such introductions are only permitted if the NIS will not spread, become a pest, or harm indigenous plants and animals.

Stocking of waterways with non-indigenous fish may occur only in special use zones, either in altered waterways that are inhospitable to indigenous species or in rivers and streams where non-indigenous fish are already established. Similarly, stocking non-indigenous game species may be allowed in national recreation areas and preserves where they are already established. When stocking fish and game, NPS gives precedence to indigenous species wherever possible, and stocking is contingent on evidence that the species cannot spread or do harm to indigenous species.

SOURCES: M.J. Bean, "The Role of the U.S. Department of the Interior in Non-Indigenous Species Issues," contractor report prepared for the Office of Technology Assessment, November 1991; D.E. Gardner, U.S. Department of the Interior, National Park Service, "Role of Biological Control as a Management Tool in National Parks and Other Natural Areas," technical report NPS/NRUH/NRTR-90/01; G.H. Johnston, Chief of Wildlife and Vegetation Division, Natural Resources Program Branch, National Park Service, personal communication to E.A. Chornesky, Office of Technology Assessment, July 10, 1991, Mar. 13, 1992; L. Loope, U.S. Department of the Interior, National Park Service, "Public Outreach in Controlling Alien Species in Haleakala National Park," talk presented at the National Park Service Headquarters, Aug. 21, 1991; U.S. Department of the Interior, National Park Service, "Management Policies," Washington, DC, 1988.

---

about 164 million acres, or 61 percent, of the BLM lands (100).

Additional grazing on BLM lands occurs under the Wild Free-Roaming Horses and Burros Act.[39] This law explicitly perpetuates NIS by protecting wild horses and burros and preserving them as a living reminder of the history of the American West. An estimated 50,000 free-roaming horses and burros occurred on BLM lands at the start of 1991 (100).

Control of Non-Indigenous Weeds—Non-indigenous weeds are widespread on BLM lands within the contiguous 48 States (figure 6-2) (132, 161). They degrade rangelands because many are unsuitable for forage. Although some emphasis is already being placed on weed management in

---

[39] Wild Free-Roaming Horses and Burros Act (1971) (16 U.S.C.A. 1331 *et seq*).

Figure 6-2—Growing Distributions of Three Noxious Weeds in the Northwest

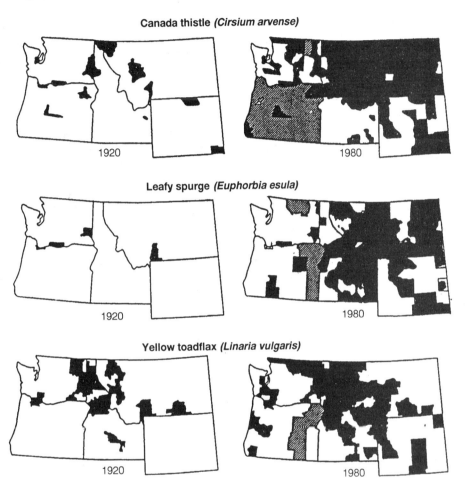

Canada thistle *(Cirsium arvense)*

1920   1980

Leafy spurge *(Euphorbia esula)*

1920   1980

Yellow toadflax *(Linaria vulgaris)*

1920   1980

Many noxious weeds are widespread on BLM lands. These maps show how three species spread in five States over a 60-year period.

SOURCE: U.S. Department of the Interior, Bureau of Land Management, "Northwest Area Noxious Weed Control Program: Final Environmental Impact Statement," December 1985.

BLM, much more is needed (136). Weed management is a small component of rangeland management, receiving only about $1.2 million annually (100,136). A 1991 internal evaluation concluded that even though noxious weed problems are widespread and growing, their control program is seriously underfunded and lacks adequate staff (136). Moreover, existing staff lack technical training or an awareness of noxious weed problems (136). Documenting the extent and severity

of noxious weed infestations on BLM lands is almost impossible because of inadequate monitoring and inventory (136).

Cooperative weed control efforts exist among BLM and other Federal, State, and county agencies, and BLM's funding provides for control on about 225 sites within 8 States (100). BLM also is involved in the management of noxious weeds in the greater Yellowstone area, in a coordinated effort with several Federal and State agencies

(44). Recent draft policies on weed management include requirements for anticipating and addressing factors that facilitate the spread and establishment of noxious weeds (136), although such long-term strategies have not yet been implemented. Examples include requiring contractors to clean equipment before entering BLM lands and using only seed, hay, mulch, or feed that is free of noxious weed seed.

The 1990 Amendment to the Federal Noxious Weed Act[40] gave Federal land managers explicit authority to develop programs for control of undesirable plants. BLM's internal evaluation cited a need for increased coordination and cooperation with State agencies (136), and the agency has instructed its State Directors to develop cooperative agreements with State agencies and review their programs to ensure full compliance (71).

Introduction of Biological Control Agents— BLM encourages introductions of biological control agents as part of an integrated management of weeds (161). The agency differs from other Federal land managers in having developed specific guidelines for the release of biological control agents. BLM requires compliance and coordination with State and Federal authorities, including evaluation of an agent's potential environmental effects before its release in an environmental assessment prepared by APHIS (135). BLM contributes funding to the Agricultural Research Service for the development and release of biological control agents. ARS also operates several small, 1-acre laboratories on BLM lands to propagate insects for biological control; in return ARS makes these agents available to BLM (161).

Introductions and Control of Fish and Wildlife—BLM manages more fish and wildlife habitat than any other Federal or State agency (100,130). The agency's long-standing policy is to give top priority to protecting, maintaining, and

LEWIS WATERS

*The Bureau of Land Management is beginning a program to manage weeds—like dyer's woad (*Isatis tinctoria*)—on public lands.*

enhancing indigenous fauna and flora (131). Requirements for introducing fish and wildlife include prior assessment of environmental effects, creation of a buffer zone around the introduction area, and a trial release of at least 2 years (131). In addition, animals must be quarantined to prevent pathogen or parasite introductions. Except under limited circumstances, current policy prohibits introductions into wilderness areas, into areas with threatened and endangered species, or of species that can hybridize with indigenous fauna (131). A unique feature of BLM policy is a provision that "individuals or organizations may be held liable for damages and responsible for expenses incurred in control of unauthorized exotic wildlife introductions" (131). However, no related regulation or law specifies such liability (4).

---

[40] 7 U.S.C.A. 2814

The current BLM manual lacks any statement concerning harmful NIS already established on BLM lands (4). A 1986 draft revision of the fish and wildlife section did promote control of feral species adversely affecting indigenous species, and it would have permitted the persistence of NIS that had become "naturalized" prior to passage of the 1976 Federal Land Policy and Management Act (133). However, this draft was never finalized, and BLM lacks any explicit policy regarding whether and under what circumstances established non-indigenous fish and wildlife should be controlled or eliminated (4).

BLM is indirectly involved in the control of non-indigenous fish through a new joint initiative with the National Fish and Wildlife Foundation. The "Return of the Natives" project was begun in 1991 and is cooperatively funded by public and private sources. Its goal is to restore indigenous fisheries in western streams, primarily through habitat restoration (68).

BUREAU OF INDIAN AFFAIRS

The Bureau of Indian Affairs (BIA) is now in the fourth year of a 10-year program for management of noxious weeds, which agency staff estimate infest 726,000—or 12 percent—of the approximately 56 million acres found on Indian reservations (129). The plan's objective is to eliminate approximately 90 percent of the weed infestation by the end of fiscal year 1999. According to BIA, the most serious problems with noxious weeds occur in North and South Dakota and Montana (65). The management plan provides funds on a 50 percent cost-share basis for control of noxious weeds on reservations to States, counties, and individual farmers. Control programs must last a minimum of three years. BIA requested $1,974,000 for fiscal year 1993 to fund control on approximately 80,000 acres (101).

BUREAU OF RECLAMATION

Congress created the Bureau of Reclamation (BOR) in 1902 to reclaim arid lands in the West for development. Much of its efforts have been to construct dams and irrigation systems for water management, although the agency's objectives have expanded to include development of recreational waterways and other goals. Systems built by the Bureau altered wetland habitats, and some agency programs have begun to address resulting changes in the resident plant and animal populations by controlling NIS. These projects are not part of a coordinated program, but instead have arisen according to need through the Bureau's regional offices (89).

Salt cedar now constitutes, in single or mixed-species stands, 83 percent of riverside vegetation along the Lower Colorado River (137). It provides poor habitat for most wildlife and consumes water more rapidly than indigenous vegetation. BOR currently is developing a long-term program for the management and eradication of salt cedar (137). As part of this effort, BOR is funding research by ARS on biological control. BOR presently spends between $250,000 and $400,000 annually to remove salt cedar mechanically (89).

In the Columbia River Basin Project, problems occur with Eurasian watermilfoil (*Myriophyllum spicatum*) and purple loosestrife—the latter infests about 20,000 wetland acres in the area (89). Non-indigenous aquatic weeds, like hydrilla (*Hydrilla verticillata*) and water hyacinth (*Eichhornia crassipes*), now clog waterways and reservoirs in Texas and California. BOR is working with Federal, State, and private agencies in control programs, which have included introductions of triploid grass carp into irrigation systems as well as the development of chemical control methods for aquatic plants (89).

One by-product of BOR's water management programs has been the creation of habitats more suitable for non-indigenous rather than for indigenous fish, with indigenous species becoming threatened or endangered in some cases (89). BOR currently has several projects designed to control non-indigenous fishes and protect threatened and indigenous ones.

■ Department of Commerce—National Oceanic and Atmospheric Administration

The National Oceanic and Atmospheric Administration's (NOAA) involvement with NIS originates from its role in the management of the Great Lakes and coastal resources. NOAA has conducted much of the Federal research and funded much of the outside research on the zebra mussel. The agency also co-chairs the Aquatic Nuisance Species Task Force.

MOVEMENT OF SPECIES INTO THE UNITED STATES

The National Marine Fisheries Service (NMFS) of NOAA inspects imported shellfish to prevent the introduction of non-indigenous parasites and pathogens. NMFS has cooperative inspection agreements with Chile and Australia. Venezuela has requested a similar cooperative agreement, although it is not yet in place because of a lack of funds (167).

ERADICATION AND CONTROL

NOAA awards annual matching grants to the States for coastal zone management as authorized by the Coastal Zone Management Act.[41] States use some of these funds for the eradication or control of harmful NIS. For example, Pennsylvania received a grant in fiscal year 1991 for eradication of four non-indigenous plants in Presque State Park to aid in restoration of wetland and dune communities (14). Additional funds used for species eradication and control may sometimes be allocated as a component of other general management categories, such as "marsh management" (160).

LAND AND RESOURCE MANAGEMENT

NOAA cooperates with States in managing the National Estuarine Research Reserve System, also under authority of the Coastal Zone Manage-ment Act. The agency provides 50 percent in matching funds for States to acquire, develop, and operate estuarine areas as natural field laboratories. As of 1990, there were 18 reserves, or a total of 267,000 acres of estuarine lands and waters, in the system (120). Multiple uses can occur in the reserves as long as they are consistent with the program's goals, including maintenance of a stable environment through protection of estuarine resources, and the uses do not "compromise the representative character and integrity of a reserve." The regulations allow, but do not require, restoration activities to improve the representative character and integrity of a reserve, including removal of NIS.[42]

RESEARCH

NOAA funds both in-house and outside research on NIS through Sea Grant, the Great Lakes Environmental Research Laboratory, the National Estuarine Research Reserve System, and the National Marine Fisheries Service. Research topics include the ecology and control of harmful species as well as the use of NIS in aquaculture.

Sea Grant's competitive grants program funded 15 projects on the zebra mussel in fiscal year 1991, totaling about $1.5 million (45). Sea Grant also funds aquaculture research, some of which deals with NIS (119,121).

NIS have become a major research priority at NOAA's Great Lakes Environmental Research Laboratory (GLERL) since invasion of the zebra mussel (102). The Laboratory was conducting six projects on zebra mussels and one on the newly introduced spiny water flea (*Bythotrephes cederstroemi*) in fiscal year 1991 (118). Funding included $1.2 million, with a similar amount provided for fiscal year 1992 (9).

NOAA funds some research projects on NIS in its estuarine reserves. Six projects related to NIS were supported from 1985 through 1991 (23).

---

[41] Coastal Zone Management Act of 1972, as amended (16 U.S.C.A. 1451 *et seq.*).

[42] Reserve regulations refer to "intentional/unintentional species changes—introduced or exotic species" as a factor that may diminish "the representative character and integrity of a site" (15 CFR 921).

Plans for 1995 to 1996 are to increase the focus on restoring habitats in the reserves; in many cases this may be to correct problems caused by NIS (23).

NOAA's National Marine Fisheries Service also conducts research on NIS. The NMFS Laboratory in Oxford, Maryland, studies the detection and diagnosis of non-indigenous pathogens and parasites of aquatic species (167). Much of the $270,000 (fiscal year 1992) program on oyster research involves studies of non-indigenous parasites and pathogens (91). NMFS also conducts research on aquaculture.

■ Department of Defense

The Department of Defense (DOD) has diverse activities related to NIS. These generally relate to its movements of personnel and cargo, management of land holdings, and maintenance of navigable waterways.

MOVEMENT OF SPECIES INTO THE UNITED STATES

The Armed Forces move large shipments of equipment, supplies, and personnel into the United States from around the world. These usually are not inspected by APHIS. Instead, each branch of DOD conducts its own inspections using military customs inspectors trained by APHIS and the Public Health Service (124).

Although APHIS officials express confidence in the capability of military customs inspection (33), concerns exist that it lacks sufficient rigor, especially during periods of enhanced military activity. Insect pests were found within material cleared for entry by U.S. Army inspectors during Operation Desert Storm, and shipped equipment sometimes carried excessive dirt or sand (3). While APHIS considered these problems minor (12), subsequent internal review by DOD suggested some Army inspectors may not be adequately trained and that careful inspection suffers under the pressure to move materials rapidly (3). Similar problems may affect other branches of the military.

DON SCHMITZ

*The Army Corps of Engineers helps States control aquatic weeds such as water hyacinth (*Eichornia crassipes*) and also conducts specialized research on control methods.*

The potential spread of NIS through military movements was graphically illustrated by discoveries of the brown tree snake at military airports and in naval cargo on Pacific islands where this noxious pest is not yet established (35). DOD now conducts special pre- and post-flight inspections of military planes flying from Guam to Hawaii to ensure they do not carry brown tree snakes. The program has been commended by experts in Hawaii (84).

MOVEMENT OF SPECIES WITHIN THE UNITED STATES

Movement of military equipment within the United States can also spread non-indigenous insect pests, like the European gypsy moth (62), and noxious weed seeds. A specific objective of the Army pest management program is to prevent the spread of economic pests throughout the United States by controlling them at Army installations (127).

The Army Corps of Engineers sometimes is indirectly involved in interstate transfers of species through its efforts to develop aquaculture and build wetlands (11). For example, during wetlands construction the Corps will use NIS from nearby areas when indigenous species are not available

(11). In addition, Corps construction of dams, reservoirs, and channels can create new habitats or pathways for the spread of aquatic NIS.

CONTROL AND ERADICATION

The Aquatic Plant Control Program of COE controls aquatic weeds in cooperation with State and local agencies by providing about 50 percent of the funds for approved projects. The program has supported control efforts in 10 States, the District of Columbia, and Puerto Rico. Appropriations for fiscal year 1992 were $5 million (91). In addition, the COE is a member of the Aquatic Nuisance Species Task Force.

LAND AND RESOURCE MANAGEMENT

DOD is the fifth largest land manager in the Federal Government, owning at least 25 million acres and managing another 15 million through agreements with other Federal or State agencies (82). DOD manages natural resources for multiple uses, including hunting, fishing, forestry, grazing, and agriculture (122). NIS are routinely introduced to DOD lands as livestock, agricultural crops, landscaping plants, and vegetation for wildlife. Management plans exist for all DOD lands, and they must include control of noxious weeds[43] (122). Cooperative agreements involving DOD, FWS, and host State agencies are the vehicle for DOD management of fish and wildlife, and new species introductions only occur when consistent with such an agreement (122). Draft Army regulations for resources management further require an environmental assessment to determine the impact of introductions on existing flora and fauna (126). These constraints are not comprehensive, however: the Air Force, like the Forest Service, excludes "certain game birds that have become established, such as pheasants" from its definition of "exotic" species (125).

DOD established the Legacy Resource Management Program in 1991 to "inventory, protect, and manage biological, cultural, and geophysical resources on lands owned or used by DOD" in cooperation with other Federal, State, and nongovernmental agencies and organizations (123). The Legacy program funded two projects for control of non-indigenous plants in Ohio and California in fiscal year 1991 (123).

RESEARCH

The COE conducts research on the biological and chemical control of aquatic weeds at its facility in Vicksburg, Mississippi, an effort related to its Aquatic Weed Control Program. The research presently focuses on hydrilla and Eurasian watermilfoil. Research efforts are coordinated with other Federal and State agencies. The appropriation for fiscal year 1992 was $4 million (91).

■ Environmental Protection Agency

The Environmental Protection Agency (EPA) deals with NIS in two general areas. First, it regulates the entry and dissemination of various microorganisms. Second, it conducts research on aquatic nuisance species.

MOVEMENT OF SPECIES INTO AND THROUGH THE UNITED STATES

EPA regulates the movement of certain nonindigenous microbes into and through the United States under the Federal Insecticide, Fungicide, and Rodenticide Act[44] (FIFRA) and the Toxic Substances Control Act[45] (TSCA). Since both statutes address the development, distribution, and sale of commercial products, they generally do not apply to the importation or distribution of microbes for research uses before product development. EPA regulates pesticidal microbes, like the bacterium Bt (*Bacillus thuringiensis*), under FIFRA. Microorganisms that are neither agricul-

---

[43] DOD defines noxious weeds as including both federally and State-listed species.

[44] Federal Insecticide, Fungicide, and Rodenticide Act (1947), as amended (7 U.S.C.A. 135 *et seq.*)

[45] Toxic Substances Control Act, as amended (15 U.S.C.A. 2601 *et seq.*)

tural pests nor pesticides—for example, nitrogen-fixing fungi—are regulated under TSCA. Any microorganism falling under regulation by FIFRA or TSCA that is also either a potential agricultural pest or a human pathogen would also be regulated by APHIS or the Public Health Service.

Pesticidal Microbes—FIFRA authorizes EPA to regulate importation, environmental release, and commercial distribution and sale of pesticides. Living microorganisms used as pesticides include bacteria, fungi, protozoa, and viruses (8). Manufacturers must register such microbial pesticides with EPA before commercial distribution and sale. Reporting requirements for registration are quite extensive and include detailed analyses of effects on organisms other than the target pest and of the eventual fate of the microbe following release to the environment (155). In addition, FIFRA requires explicit labeling of microbial pesticides (155). Violations of this or other provisions of the Act can result in civil or criminal penalties.[46]

Only registered microbial pesticides may be imported into the United States for commercial distribution and sale (43). Unregistered pesticides may be denied entry by U.S. Customs. As of March 1992, 2 of the 23 microbes registered as pesticides in the United States were non-indigenous (table 6-5) (7). Origins of an additional 11 are unknown, since EPA did not require reporting of this information until 1984 (7). Under FIFRA, EPA considers only those microbes from continents other than North America to be non-indigenous to the United States (6).

During pesticide research and development, EPA requires manufacturers to provide notification before small-scale tests of non-indigenous microbial pesticides. EPA may then require additional information, or application for an experimental use permit. Such permits are re-quired for large-scale tests. Permit applications include information on microbe identity, origin, host range, mode of action, intended application, and potential effects on nontarget organisms and the environment.[47] Similar notification and application for an experimental use permit is not required for small scale tests of indigenous microbes. EPA currently is considering whether it should continue to require notification for small scale tests of NIS, since APHIS and the Public Health Service require permits for tests involving potential agricultural pests or human health threats (6).

Non-Pest, Non-Pesticidal Microbes—Under TSCA, EPA could regulate certain non-indigenous microbes that fall outside of other regulatory authorities, such as nitrogen-fixing bacteria and fungi. Thus far EPA has regulated only genetically engineered microbes under TSCA (38). TSCA regulations do not explicitly distinguish between indigenous and non-indigenous microbes, except in the requirement for EPA notification when microbes are imported for commercial purposes or into commerce. TSCA's applicability is further restricted to only those microbes having an identified risk to human health or the environment, since naturally occurring microorganisms are considered to be ''in commerce'' and therefore implicitly on the TSCA inventory of unregulated substances (38). Nevertheless, should a risk be shown, EPA could potentially ban, limit production of, or remove from sale the non-indigenous microbes that fall under TSCA.[48]

MONITORING

The goal of EPA's Environmental Monitoring and Assessment Program (EMAP) is to monitor the condition of the Nation's ecological resources (156). EPA began developing EMAP in 1987, and the program is still in the preliminary phases of

---

[46] 7 U.S.C.A. 136.

[47] 40 CFR 172.4 (May 11, 1981).

[48] Toxic Substances Control Act, as amended (15 USC 2601).

Table 6-5—Microbial Pesticides Registered by EPA

Microorganism	Year registered	Origin	Pest controlled
Bacteria	1948	*	Japanese beetle larvae (*Popillia japonica*)
Bacillus popilliae + B. lentimorbus			
B. thuringiensis "Berliner"	1961	*	Lepidopteran larvae
Agrobacterium radiobacter	1979	*	crown gall disease (*Agrobacterium tumefaciens*)
B. thuringiensis israeliensis	1981	Israel	Dipteran larvae
B. thuringiensis aizawai	1981	*	wax moth larvae (*Galleria mellonella*)
Pseudomonas fluorescens	1988	U.S.	*Pythium, Rhizoctonia*
B. thuringiensis San Diego	1988	U.S.	Coleopteran larvae
B. thuringiensis tenebrionis	1988	Germany	Coleopteran larvae
B. thuringiensis EG2348	1989	U.S.	Lepidopteran larvae
B. thuringiensis EG2371	1989	U.S.	Lepidopteran larvae
B. thuringiensis EG2424	1990	U.S.	Lepidopteran/Coleopteran larvae
B. sphaericus	1991	U.S.	Dipteran larvae
Viruses			
Heliothis nuclear polyhedrosis virus (NPV)	1975	*	cotton bollworm (*Helicoverpa zea*), budworm (*Choristoneura* spp.)
Tussock moth NPV	1976	*	Douglas fir tussock moth larvae (*Orgyia pseudotsugata*)
Gypsy moth NPV	1978	*	Gypsy moth larvae (*Lymantria dispar*)
Pine sawfly NPV	1983	*	Pine sawfly larvae (*Neodiprion* spp.)
Fungi			
Phytophthora palmivora	1981	*	citrus stangler vine (*Morrenia odorata*)
Colletotrichum gloeosporioides	1982	*	northern joint vetch (*Aeschynomene virginia*)
Trichoderma harziarum ATCC20476 +			
T. polysporum ATCC20475	1990	U.S.	wood rot
Gliocladium virens GL21	1990	U.S.	*Pythium, Rhizoctonia*
Trichoderma harzianum KRLAG2	1990	U.S.	*Pythium*
Lagenidium gigantium	1991	U.S.	mosquito larvae
Protozoa			
Nosema locustae	1980	*	grasshoppers

* Reporting of the origin of registered microbes was not required before 1984 so their origins are unknown.

SOURCE: F. Betz, Acting Chief, Science Analysis and Coordination Staff, U.S. Environmental Protection Agency, letter to E.A. Chornesky, Office of Technology Assessment, Apr. 10, 1992.

design and small-scale application. However, EMAP's planners expect the program eventually will involve the accumulation and analysis of information on the plants, animals, and physical environment throughout the country. Although EMAP could conceivably be used to monitor NIS in the United States, that is not one of its goals, and its current design would not provide suitable information for this purpose (50,57).

RESEARCH

EPA's most direct involvement with NIS is through its Office of Research and Development. Staff from this office represent EPA on the Aquatic Nuisance Species Task Force. EPA's Environmental Research Laboratory in Duluth, Minnesota, conducts in-house research on the environmental effects and control of zebra mussels and the ruffe (*Gymnocephalus cernuus*), and participates in collaborative projects with NOAA and the Coast Guard on zebra mussel monitoring (48). In 1992, the laboratory also funded related research at several other institutions. EPA appropriations related to harmful aquatic NIS totaled $1.65 million in fiscal year 1992 (91).

Department of Health and Human Services—Public Health Service

The Public Health Service (PHS) regulates entry of living organisms that might carry or cause human diseases.[49] Current PHS regulations restrict, require inspection of, or require permits for the importation of all cats, dogs, monkeys, turtles, and bats, as well as certain snails, insects, and microbes.[50] PHS does not perform primary inspection at ports of entry. Instead, it provides training for Customs and USDA inspectors who directly examine people, baggage, and cargo and make referrals to PHS when problems arise (158).

PHS has only small efforts abroad to identify species and commodities that might serve as human disease vectors, and it generally develops regulations only after a potential route of human disease entry has been demonstrated. For example, PHS developed regulations requiring fumigation of used tire imports at least 2 years after evidence demonstrated that the tires were a major pathway by which the Asian tiger mosquito—a vector of several human diseases—entered the country (see box 3-A). For certain human health threats, like the African honey bee, PHS has taken a minimal role. In this case, primary responsibility for devising a response has fallen to APHIS; however, since APHIS is not a public health agency, it has not fully addressed the public health issues (78).

PHS does not impose quarantines or regulations to prevent the interstate spread of human disease vectors once they become established in the country (73). For such organisms, the agency does, however, monitor spread and conduct research on their potential to transmit indigenous diseases. PHS research also examines general techniques for tracking and controlling organisms that can transmit human diseases (157).

Department of the Treasury— U.S. Customs Service

The U.S. Customs Service (Customs) has a major operational role in restricting the entry of harmful NIS. Customs personnel inspect passengers, baggage, and cargo at U.S. ports of entry to enforce the regulations of other Federal agencies (154). They inform interested agencies when a possible violation is detected and then usually detain the suspected passenger or commodity for inspection by agency staff. APHIS, FWS, and PHS each has a cooperative agreement with Customs and provides specialized training to Customs inspectors. Customs inspects only some incoming passengers, baggage, and cargo, aiming to examine higher risk categories established by country of origin and other criteria (153). APHIS has established its own high risk categories for agricultural port inspection using different criteria (12).

Department of Transportation— U.S. Coast Guard

The U.S. Coast Guard (USCG) was given certain responsibilities related to preventing introductions of harmful aquatic species by the Nonindigenous Aquatic Nuisance Prevention and Control Act and is a member of the Aquatic Nuisance Species Task Force. USCG issued voluntary ballast management guidelines for ships entering the Great Lakes in March 1991. Mandatory ballast management regulations went into effect May 10, 1993 to prevent further

---

[49] under the Public Health Service Act (1944), as amended (42 U.S.C.A. 201 *et seq.*)

[50] 42 CFR 71,72, as amended (Jan. 11, 1985).

introductions of aquatic species into the Great Lakes.[51] These regulations require ships to exchange ballast water at sea, to retain ballast water on the vessel, or to use an alternative approved method.

USCG is also researching methods of ship design that might prevent the survival and transport of NIS in ballast water (91).

■ Department of Energy

Approximately 2.4 million acres, or 0.1 percent of the U.S. land area, fall under the management of the Department of Energy (128). These holdings include research laboratories, electric utilities, and petroleum reserves (29). DOE has no general policies regarding the control of NIS, including noxious weeds, on its lands. The agency plans to issue a programmatic Environmental Impact Statement in 1993 that should help establish consistent land use policies (169).

DOE conducts restoration in some areas. Although the primary goal now is removal or containment of nuclear or toxic wastes, DOE is beginning to restore ecological communities of plants and animals at a few sites (28). DOE lacks a general policy regarding the use of indigenous versus non-indigenous organisms in restoration, presently relying on State policies for guidance.

■ Department of Justice—Drug Enforcement Agency

The Drug Enforcement Agency (DEA) restricts importation of a few non-indigenous plants and fungi because they contain narcotic substances. Importation of NIS such as coca (*Erythroxylum coca*), marijuana (*Cannabis sativa*), and opium poppy (*Papaver somniferum*) is only allowed with a permit from DEA.

CHAPTER REVIEW

This chapter described the large number of Federal agencies and programs responsible for different aspects of managing harmful NIS or introducing desirable ones. Clearly, much is being done. However, OTA's analysis shows that the U.S. system for dealing with harmful NIS falls short in a number of important areas. An overall assessment requires looking beyond the Federal Government, however. For example, when the Asian tiger mosquito became established in the country, control was left to State public health authorities; they simply were unable to respond effectively (21). In the next chapter, OTA looks more closely at such interactions between Federal and State efforts.

---

[51] 58 *Federal Register* 18330 (April 8, 1993).

State and Local Approaches From a National Perspective | 7

This chapter picks up from the last, adding how State and local efforts affect the management of non-indigenous species (NIS). Here, OTA discusses Federal and State relations and relationships among States. The chapter's centerpiece is an analysis of the States' 50 distinct approaches to regulating importation and release of "fish and wildlife"— mammals, fish, birds, reptiles, and amphibians.[1] In some cases, States have pioneered exemplary approaches and these are highlighted. The chapter examines how States treat non-indigenous invertebrates and plants also. Various proposed model State laws and local approaches conclude the chapter.

THE RELATIONSHIP BETWEEN THE FEDERAL GOVERNMENT AND THE STATES

Generalities come with difficulty regarding Federal-State relationships. The authority of the Federal and State Governments varies not only with the type of organism regulated, but also depending on the particular Federal and State laws and agencies involved. Mainly, however, States control the entry of NIS across State borders and release of NIS within the State. Often these are pests, of either foreign or U.S. origin, that are already established elsewhere in the country.

For fish and wildlife, States retain almost unlimited power, notwithstanding the Federal Lacey Act,[2] to make decisions about

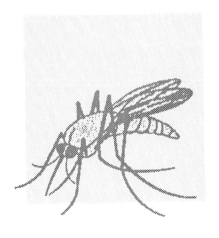

---

[1] Some State and Federal laws include all, or certain groups of, invertebrate animals under their definitions of "fish and wildlife." For example, the Lacey Act covers invertebrates like snails and crayfish. Occasionally, "wildlife" is defined to include all fauna and flora, as in Illinois. The term, as used here, refers only to vertebrates, but it does include domesticated or cultured species.

[2] For full citations of this and other Federal laws see footnotes to ch. 6.

Box 7-A—*Maine* v. *Taylor:* A Key Constitutional Decision

The Commerce Clause of the U.S. Constitution grants to Congress the power to regulate international and interstate trade. This grant puts limits on, but does not eliminate, the power of States to ban imports of NIS. The limits were outlined by the U.S. Supreme Court in a 1986 ruling on the constitutionality of a Maine law that prohibited importation into the State of "any live fish, including smelts, which are commonly used for bait fishing in inland waters." The case of *Maine* v. *Taylor* upheld the law even though it clearly discriminated against out-of-state bait fish dealers. The Supreme Court applied a two-part test for validity under the Commerce Clause: "the statute must serve a legitimate local purpose, and the purpose must be one that cannot be served as well by available nondiscriminatory means." The Supreme Court approved a lower court's findings that both parts of the test had been met:

First, the lower court found that Maine "clearly has a legitimate and substantial purpose in prohibiting the importation of live baitfish," because "substantial uncertainties" surrounded the effect that baitfish parasites would have on the State's unique population of wild fish, and the consequences of introducing nonnative species were similarly unpredictable . . . . Second, the court concluded that less discriminatory means of protecting against these threats were currently unavailable, and that, in particular, testing procedures for baitfish parasites had not yet been devised . . . . "[T]he constitutional principles underlying the commerce clause cannot be read as requiring the State of Maine to sit idly by and wait until potentially irreversible environmental damage has occurred or until the scientific community agrees on what disease organisms are or are not dangerous before it acts to avoid such consequences."

The Supreme Court has long upheld State quarantine laws that, notwithstanding the Commerce Clause, ban importation of pests of *known* significance. Importantly, the *Maine* v. *Taylor* ruling upholds a ban based on threats whose significance involved "substantial uncertainties." This gives States leeway in drafting laws on NIS importation in the face of such uncertainties so long as they do not needlessly discriminate against out-of-State interests.

SOURCES: 12 Me. Rev. Stat. Ann. sec. 7613; *Maine* v. *Taylor,* 477 U.S. 131 (1986).

---

which species are imported and/or released. Congressional incursions on this traditional State control over fish and wildlife have been limited and controversial (16). In contrast, several major Federal laws—such as the Federal Plant Pest Act and the Federal Noxious Weed Act—set national policy for weeds and other plant pests.

Where Federal programs miss significant problems, States, in effect, determine the success of nationwide efforts to manage harmful NIS. There are important limits to the States' capacities, however.

The Constitution vests the power to regulate international and interstate commerce in Congress.[3] Therefore, States cannot unnecessarily restrict such commerce. The key Supreme Court

case is *Maine* v. *Taylor* (box 7-A). As a result of the Commerce Clause, States lack the power to stop the importation and release of a potentially invasive NIS in a neighboring State.

A few States, e.g., Hawaii and Alaska, have geographical barriers against the interstate spread of NIS. A small number of States, like California, have border inspection stations to interdict pests in transit. Without these kinds of barriers, a State cannot do much to slow the influx of State-prohibited plants or seeds that were acquired legally in another State or country (53). Nor can a State effectively stop mail-order sales of plants or seeds it prohibits, as policing the mails is a Federal function.

---

[3] U.S. Constitution, Article I, section 8, clause 3.

Also, States cannot legislate in direct conflict with Federal law. Nor can they directly regulate activities on Federal lands, absent a cooperative agreement. Occasionally, Federal laws explicitly preempt State involvement.

▮ Federal Preemption of State Law

Finding:

Federal preemption of State law varies among categories of NIS. It is more common in agricultural laws than in those related to fish and wildlife. Cooperative programs are a more feasible way for the Federal Government to influence State actions.

A key issue in the relationship between Federal and State authorities is whether an applicable Federal law preempts State laws, keeping States from legislating in the area. This occurs when the Federal law explicitly or implicitly provides for preemption, or regulates an area so comprehensively as to leave no practical State role.

Federal preemption is more common in agricultural laws than in those pertaining to fish and wildlife—traditionally an area of State prerogatives. The Lacey Act required that a list of "injurious" species or groups be created and it preempts States from allowing foreign importation of the 23 "injurious" taxonomic categories of fish, wildlife, and fish pathogens on that list. The Lacey Act does not, however, forbid more restrictive State laws.[4] Similarly, no State may permit foreign importation of a weed species prohibited and listed under the Federal Noxious Weed Act, although it does not otherwise preempt State weed laws.[5] The Federal Plant Quarantine Act also allows States to be more restrictive under certain circumstances, but it imposes a strong Federal presence. For example, the Federal Government can quarantine an entire State under the Act.[6] The Federal Plant Pest Act similarly provides strong emergency authority to override State laws.[7]

The Federal power to preempt does not mean that the Federal approach is always the best. Some State laws regulate more comprehensively than parallel Federal laws and their implementation is more effective (see below). Such States are, in effect, laboratories where different approaches are tested; their successes can spawn Federal imitation. Nevertheless, when States adopt widely varying laws, the regulated industries may support federally imposed uniformity to facilitate commerce.

Using Federal preemptive powers to implement a national approach is fraught with political difficulties—especially for fish and wildlife—and usually engenders resistance from the States. Thus, the trend is toward programs administered cooperatively by State and Federal officials. In these the Federal Government provides incentives to pull, and sanctions to push, the States toward certain general goals or national minimum standards. Several points made in a 1987 U.S. Fish and Wildlife Service discussion paper on aquatic introductions appear applicable to NIS introductions in general:

> Introduced aquatic organism issues are inherently interjurisdictional and, thus, clearly national, indeed international in scope. Despite this Federal interest, however, emergence of a fully effective program for avoiding undesirable introductions of aquatic organisms requires that involvement by the Federal Government not preempt State authority. Rather, the Federal Government should function as a catalyst/facilitator establishing incentives for action by the States and the other co-managers of the Nation's fishery resources. However, it will also be imper-

---

[4] Lacey Act (1900) (16 U.S.C.A. 3378(a)).

[5] Federal Noxious Weed Act of 1974 (7 U.S.C.A. 2812).

[6] Federal Plant Quarantine Act (1912) (7 U.S.C.A. 161).

[7] Federal Plant Pest Act (1957) (7 U.S.C.A. sec. 150dd(b)(1)).

ative to ensure universal applicability of any action. Although it must be exercised as a last resort, a credible threat of Federal sanctions against non-complying jurisdictions is essential to ensure uniform and, therefore, fair application of any corrective strategy. (66)

Congress has previously recognized circumstances that justify overriding State management of NIS when it conflicted with Federal goals. Congress restricted State control of feral horses (*Equus caballus*) and burros (*Equus asinus*) through the Wild Free-Roaming Horses and Burros Act. State officials may not kill them, or allow their killing, even if they stray off Federal lands.[8]

A major extension of Federal authority resulted from litigation over the palila (*Loxioides bailleui*), a rare bird found only in Hawaii.[9] The State's Department of Land and Natural Resources had been managing feral goats (*Capra hircus*) and introduced mouflon sheep (*Ovis* spp.) for the benefit of sport hunters but to the detriment of the palila and its habitat. A Federal court ruled that Hawaii's action amounted to an illegal "taking" of the palila under the Endangered Species Act and ordered the State to remove the non-indigenous goats and sheep (6). Under this reasoning, other States could be compelled to manage NIS to prevent conflicts with threatened or endangered species.[10] Thus, precedents exist for Federal preemption even in the traditionally State-dominated area of fish and wildlife management.[11]

New emergency powers to override State control were added to the Federal Plant Pest Act after the 1980-1982 medfly (*Ceratitis capitata*) crisis in California.[12] Delays occurred in developing a coordinated Federal-State response because of many factors including California's unwillingness to spray chemical insecticides over cities. These helped drive the eventual costs to the highest ever for a single eradication project—at least $100 million (17). Although they have not yet been invoked to preempt State authority, these powers represent a potent assertion of Federal prerogatives, but only under defined circumstances. They provide sufficient leverage such that actually invoking them may never be necessary. They also provide a potential model for preempting State control efforts if they are found lacking for other NIS (box 7-B).

Federal preemption can engender controversy when applied to new areas, even in agricultural regulation where preemption has a long history. In 1993, Federal officials asserted their authority to preempt more restrictive State laws regarding releases of genetically engineered organisms, raising concerns among some State officials (see ch. 9).

■ Federal-State Cooperation

Cooperative programs serve several key functions in Federal and State efforts. Many provide a means for developing consistent strategies in areas of common concern. Federal and State agricultural officials, for example, collaborate in the regulation of NIS importation, interstate commerce, and control. Postentry quarantine of certain federally restricted plants is a joint program, in which private importers keep the plants in quarantine, usually subject to State inspection (50). The National Plant Board, and four regional

---

[8] Wild Free-Roaming Horses and Burros Act (1971) (16 U.S.C.A. sec. 1334).

[9] *Palila v. Hawaii Department of Land and Natural Resources*, 471 F.Supp. 985 (D. Ha. 1979), *aff'd*, 639 F.2d 495 (9th Cir. 1981).

[10] The Endangered Species Act does not provide the same protection against "takings" of endangered or threatened plants as it does for fish and wildlife, 16 U.S.C.A. 1538(2).

[11] In certain narrow cases, Federal laws regulating States may be unconstitutional under the Tenth Amendment. *New York* v. *United States*, 112 S.Ct. 2408 (1992). Federal laws may set up powerful incentives for State action, or may impose preemptive Federal standards; however, they may not compel State legislatures to enact federally desired legislation.

[12] 7 U.S.C.A. 150dd(b)(1).

**Box 7-B—When Federal and State Interests Collide:
Control of Harmful NIS In and Around Protected Lands**

Where Federal- and State-regulated lands adjoin, conflicts can arise over differing management goals. Some national parks and other natural areas provide safe havens for non-indigenous pests of agriculture that are controlled elsewhere. However, harmful NIS also invade Federal reserves from lands under State jurisdiction. The lack of comprehensive State regulation and control exposes the reserves to these species' impacts when they are introduced nearby and then spread.

Federal agencies can be stymied in trying to address problems attributable to State-supported NIS with multiple impacts. An example occurs in and around the Great Smoky Mountains National Park, where park and Forest Service managers were compelled to cooperate with North Carolina in a trapping plan for introduced hogs (*Sus scrofa*). The plan limits control efforts in lower elevations of the park, despite the widespread ecological damage the hogs have caused. The park engages in time-consuming and costly transfer of live-trapped hogs, which could otherwise be killed, so that they can be released on State lands. The reason: North Carolina's wildlife agency wants to maintain hogs in the area for hunters and it had support in dealing with the Park Service from the State's congressional delegation.

The Park's hog management budget dropped drastically from FY 1992 to FY 1993 -from $197,000 to $65,000. The hog numbers will likely increase as will their negative effects.

Federal managers sometimes must commit resources to control or eradicate threatening NIS in areas *outside* their boundaries and their jurisdiction. A clear Federal interest lies in improving this situation by providing an unambiguous mechanism for Federal managers to act beyond their boundaries, but only if compelling circumstances exist. While cooperative, negotiated agreements are always preferable, unresolved NIS threats may justify Federal preemption of State management to protect Federal reserves.

SOURCES: R. Joseph Abrell, Chief, Resource Management and Science Division, Great Smoky National park, personal communication to P.T. Jenkins, Office of Technology Assessment, Dec. 15, 1992; F.C. Craighead and R.F. Dasmann, Bureau of Land Management, "Exotic Big Game on Public Lands," September 1964; E.F. Hester, "The U.S. National Park Experience with Exotic Species," *Natural Areas Journal*, vol. 11, No. 3, 1991, pp. 127-28; L. Loope, Research Scientist, Haleakala National Park, personal communication to P.T. Jenkins, Office of Technology Assessment, Aug. 21, 1991.

---

plant boards, composed of officials from State departments of agriculture, help coordinate Federal and State regulations (50).

Certain programs aim for consistent goals in the management and control of harmful NIS across a geographic region; it does little good for an invasive NIS to be controlled in one area but not in adjacent areas from which it can reinvade. The 1990 amendment to the Noxious Weed Act acknowledged this by requiring Federal land managers to control State-prohibited weeds.[13] Several other cooperative programs for non-indigenous weeds are voluntary. For example, representatives of Federal, State, and local jurisdictions with holdings in the area surrounding Yellowstone National Park signed a memorandum of understanding to control noxious weeds. The agreement included adoption of comprehensive management guidelines (3). In Hawaii, Federal and State officials have an interagency agreement to research the biological control of forest weeds (ch. 8). Similarly, the Western Weed Coordinating Committee, with members from western Federal and State agencies, enhances cooperation in weed management (44). Florida's Exotic Pest Plant Council (EPPC) fills this role for primarily non-agricultural weeds; agency officials, botanists, and others from private groups in California recently created their own EPPC using Florida's model.

---

[13] 7 U.S.C.A. 2814.

Some programs allow targeting of Federal funds or technical assistance to the States for actions serving both national and State needs. Both APHIS and the U.S. Forest Service cooperate extensively with States in the suppression of forests pests such as the European and Asian strains of the gypsy moth (*Lymantria dispar*). The Forest Service trains State personnel in the management of forest insects and diseases (65). Funding for pests surveys and control is on a cost-sharing basis, with States providing 50 percent or more of the funds for some activities (65). According to the Forest Service, such coordinated approaches have greater effectiveness and lower overall costs than separate efforts (65). The U.S. Army Corps of Engineers also oversees a program for the control of aquatic weeds in which State or local governments can partially recover costs for weed control in navigable waterways (64). The Fish and Wildlife Service provides information and expertise on diseases affecting aquaculture, an area where no comprehensive Federal program currently exists (47).

In some areas, the Federal Government assists or provides funds to address State needs. Sometimes these programs rely on Federal powers, for example, the program to help California prevent entry of agricultural pests via first class mail from Hawaii (58). Also, Federal inspectors at ports of entry in a particular State may help interdict species prohibited by that State, even if they are not federally listed (19).

Federal assistance for local problems makes sense if, over the long run, they may become national ones (e.g., a rapidly spreading NIS) or if local problems are so common they become a national concern. The Nonindigenous Aquatic Nuisance Prevention and Control Act of 1990 provides for State submission of comprehensive aquatic nuisance species management plans. States with approved plans may receive Federal matching grants for implementation. No Federal funds have yet been budgeted for these grants (64).

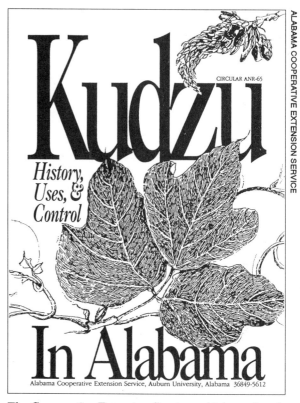

The Cooperative Extension Service, which produced this booket on kudzu (Puereria lobata) in Alabama, is one of several means by which Federal and State efforts are joined.

Federal and State agencies cooperate extensively in the prevention, quarantine, and control of agricultural pests, but several problems exist. Federal agencies do not always inform States of foreign pest threats in a timely fashion. For example, although the Animal and Plant Health Inspection Service (APHIS) was aware of the apple ermine moth (*Yponomeuta malinellus*), a serious orchard pest, in British Columbia in 1981, it did not advise Washington State officials until 1985. Shortly thereafter, the pest spread into the State. According to a Washington State agriculture official, it ''just fell between the cracks''; in other words, Federal officials lacked a good system for communicating about potential threats (1).

The balance between Federal and State efforts sometimes shifts too quickly to adequately address potential problems. After APHIS removed Federal quarantine restrictions on the movement of nursery stock from Japanese beetle-infested areas (*Popillia japonica*), a number of States, but not all, promulgated quarantine regulations of their own. The resulting patchwork of State regulations led to the inadvertent movement of infested nursery stock to States both with and without their own quarantines (49). In another case, black stem rust (*Puccinia graminis*), APHIS has maintained a Federal quarantine, but has delegated nearly all responsibility to the States. Inconsistent enforcement by the States has increased the possibility that barberry (*Berberis vulgaris*) varieties susceptible to black stem rust will be shipped to areas protected by the quarantine (49).

Some observers maintain that the balance of responsibility for eradicating agricultural pests has tilted to the States since roughly 1980. This was forcefully argued by a Florida official in 1991, after seven frustrating years of trying to eradicate citrus canker (*Xanthomonas campestris* pv. *citri*):

> The concept of dual responsibility, a partnership, if you will, between States and the USDA has never fallen into greater disrepair or erosion than it has over the last decade or so. Simply put, USDA/APHIS has become less and less responsive to domestic and exotic pest eradication programs. (2)

The official further complained that the State had been forced to carry out quarantines of several well-known, damaging NIS like the varroa mite (*Varroa jacobsoni*) and Caribbean fruit fly (*Anastrepha suspensa*), because APHIS considered them local pest problems of little economic significance (2).

RELATIONSHIPS AMONG STATES

Finding:

Conflicts, particularly regarding aquatic releases, arise among States because of their differing ecological, economic, and policy contexts. Regional approaches provide opportunities for States to resolve their differences and influence the actions of neighboring States. Such approaches have been used most frequently for evaluating aquatic releases. Expanding the use of regional approaches for other types of releases appears promising, but is limited by their voluntary nature.

States lack the power to stop the importation and release of a potentially invasive NIS in a neighboring State. Since few Federal laws compel States to cooperate with each other, and States have differing priorities, conflicts can and do occur. A recent conflict between Virginia and Maryland over the proposed introduction of the Pacific oyster (*Crassostrea gigas*) to the Chesapeake Bay has largely economic origins and is partly rooted in different patterns of public versus private ownership of oyster beds (10,30). Harvests of the indigenous Atlantic oyster (*Crassostrea virginica*) have declined to a historic low, especially on the Virginia side of the Bay (34). Virginia has a greater economic incentive to promote the introduction than Maryland, which still maintains a viable oyster fishery based on the indigenous species. Virginia approved an experimental release of sterile Pacific oysters in 1992, but later reversed this decision.

The experimental release by North Dakota of a new sport fish, the European zander (*Stizostedion lucioperca*), demonstrated how a State can introduce NIS notwithstanding concerns of adjacent States. Minnesota had objections to the release because of ecological and disease risks. (Federal and provincial Canadian governments also disputed North Dakota's action; see Scarratt and Drinnan (51) for a description of Canadian fisheries policies). Still, Minnesota officials supported the principle of paramount State sovereignty over natural resources (71). States themselves are unlikely to be advocates for less State sovereignty.

Several councils or commissions exist to coordinate introduction policies across a particular

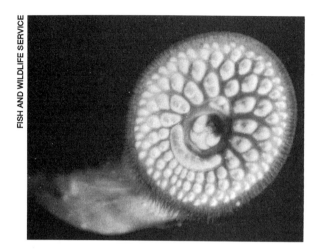

The Great Lakes Fishery Commission, like its 4 counterparts elsewhere in the United States, coodinates introduction policies across the region; controlling the damaging sea lamprey has been a major focus in the Great Lakes.

region. For fish and wildlife, these include the Great Lakes Fishery Commission, the Colorado River Fish and Wildlife Council, and the three Marine Fisheries Commissions (Atlantic States, Gulf States, and Pacific). They provide venues for State officials to agree on guidelines for releases, inspections, and permits. For example, 5 western States and the province of British Columbia signed a cooperative agreement in 1980 for the interstate transfer of shellfish under the auspices of the Pacific Marine Fisheries Commission (28). The U.S. Fish and Wildlife Service provides technical and research assistance to the various regional groups.

The National and four regional Plant Boards, composed of State plant health officials, fill a similar role for agricultural pests, i.e., facilitating coordination of quarantines. They have commissioned a compilation of all State laws on weeds and pests with the goal of improving communication and reducing inadvertent violations. These boards move slowly, however, because of limited funding and spotty State participation.

Sometimes no mechanism exists for resolving conflicts between States short of a Federal lawsuit. The regional organizations that exist, however, provide important forums for proactively addressing potential differences. Indeed, many States require approval by the regional council or commission as a prerequisite for certain NIS introductions (52). Most of these regional organizations currently deal with aquatic releases, although similar structures could be useful for nonaquatic NIS issues. Regional organizations are limited in that they are essentially voluntary and not all States are members. Moreover, they have no independent regulatory authority. Robson Collins (11), a California official, notes the clear need for interstate cooperation but also that the members of the Pacific Marine Fisheries Commission have largely gone their own ways since the efforts of the 1970s and early 1980s.

STATE LAWS REGULATING FISH AND WILDLIFE IMPORTATION AND RELEASE

Findings:

- States prohibit importation and/or release of a median of only eight potentially harmful fish and wildlife species or groups. In a survey of State fish and wildlife agency officials, about one-third responded that their lists of prohibited species are too short.

- About one-quarter of the States lack legal authority over importation and/or release of one or more of the five major vertebrate groups (mammals, birds, fish, reptiles, and amphibians). Also, about 40 percent of State agencies would like to receive additional regulatory authority from their State legislatures.

- Among those States that do have decision-making standards for approval of importation and/or release of non-indigenous fish and wildlife, none legally requires adherence to a scientific protocol when considering a proposal. A few States mandate scientific studies for certain proposals. About half the States require a general determination of potential impacts, defined broadly enough to include all ecological impacts. The rest lack rigorous decisionmaking standards.

- Most State agencies rate their own implementation and enforcement resources (staff, funding, or others) as "less" or "much less" than adequate; on average, they would like increases of resources of about 50 percent to meet their responsibilities.

- Several States present exemplary approaches to managing non-indigenous fish and wildlife. On the other hand, many States are under-regulating in several important respects. Overall, States are not adequately addressing non-indigenous fish and wildlife concerns.

■ Overview of State Laws

OTA researched the laws[14] of all 50 States to answer the following questions regarding fish and wildlife importation and release: What regulatory approaches are used? Are large groups of clearly harmful NIS not being regulated? What decision-making standards are agency officials required to meet? The aim of this undertaking is to determine which laws are exemplary, providing potential models for national approaches. However, drawing conclusions from State-to-State comparisons requires caution because each State has an unique ecological, agricultural, and institutional setting.

No efficient way exists to find and compare State laws and OTA's process was time consuming and expensive. States' key provisions diverge broadly, use different terminology, are often scattered within their codes, and some rules and regulations are unpublished. No group has the job of maintaining a comprehensive, up-to-date compilation. The last private compilation was based on 1983 laws; it rapidly became obsolete (29). Any future oversight of State efforts will require updating the information summarized in this chapter.[15] In order to supplement this legal research, OTA also surveyed the heads of the responsible State agencies for their opinions about their own laws as implemented (box 7-C).

---

[14] "Laws" here means State statutes and formal rules and regulations adopted by the executive agencies. Table 7-6 cites the key provisions. OTA's initial legal research was sent for review, correction, and updating to the 50 relevant State agencies in fall 1992. Thirty-six States responded and their information was used for the analysis throughout this chapter. Another two States responded too late to be incorporated into the full analysis but their corrections are included in table 7-6. Respondents are listed in App. B.

[15] A research project is under way at the University of New Mexico Law School's Center for Wildlife Law to collect all State wildlife laws and regulations (not just those affecting NIS) in an accessible, standard-format collection, which eventually may be computerized (45).

Box 7-C—Views From the State Fish and Wildlife Agencies

States responding: *36 (72%)*—AK, AL, AR, AZ, CO, FL, GA, HI, IA, ID, IL, IN, KS, LA, MA, MD, ME, MN, MO, MS, MT, NC, ND, NE, NJ, NY, OH, OK, PA, RI, TN, UT, VT, WI, WV, WY

Not responding: *14 (28%)*—CA, CT, DE, KY, MI, NV, NH, NM, OR, SC, SD, TX, VA, WA

NOTE: The OTA survey was conducted by mail in fall 1992. Percentages below are for the respondents listed in appendix B. Explanations provided with the answers are not included here. South Dakota and New Hampshire's responses were received too late to be included in the analysis, although their corrections for table 7-6 are tabulated.

Question 1: Beyond your existing authority, are there additional areas of legal authority that your agency would like to receive from your State legislature to regulate the importation, possession, or introduction of non-indigenous (exotic) fish and wildlife?

Yes: *15 (42%)*—AK, AL, GA, HI, ID, MA, MD, MO, MS, MT, ND, NE, NY, WI, WY

No: *18 (50%)*—AR, AZ, CO, FL, IA, IL, IN, KS, LA, ME, MN, NC, NJ, OK, PA, TN, UT, WV

No answer/other: *3 (8%)* — OH, RI, VT

Question 2: Evaluate the numbers of non-indigenous species that are prohibited outright (disregarding minor exemptions such as for research) from importation, possession, or introduction into your State.

List is too short: *13 (36%)*—AL, FL, KS, LA, MD, MN, MO, MS, MT, ND, NE, RI, WI

List is about right: *17 (47%)*—AK, AR, CO, GA, HI, IA, IL, IN, MA, ME, NC, NJ, OH, OK, UT, VT, WY

List is too long: *0*

Not sure: *2 (6%)*—PA, WV

No answer/other: *4 (11%)*—AZ, ID, NY, TN

BASIC LEGAL APPROACHES

The States employ several basic legal approaches (table 7-1).[16] The most restrictive approach is to prohibit all NIS except those individually evaluated and listed as allowed, that is, a "clean" list. Hawaii is the only State with laws that require this for both importation and release of all major fish and wildlife groups. A few other States have adopted clean lists for particular actions, most commonly for fish releases.

More than half the States have "dirty" list approaches, in which certain listed NIS are prohibited from importation and/or release because of their economic, ecological, or health effects. A smaller proportion of States have neither clean nor dirty lists, that is, they have *no* species prohibited by statute or regulation. For importation this is true for 11 States regarding all major vertebrate groups and for 7 States regarding some groups. For release, 12 States prohibit no

---

[16] Some important preliminary qualifications and observations: 1) The information summarized represents the *main* provisions of the State laws that directly govern whether or not importation and release of NIS is allowed in particular cases. This narrow scope of inquiry excludes minor provisions, limited exemptions, and a myriad of veterinary, commercial, endangered species, humane, and other provisions that may incidentally affect NIS importation and release. 2) Some definitional differences exist regarding what is included when States regulate "non-indigenous" or "exotic" species. Generally, the legal definitions refer to any species not naturally found within the State; a small number, such as Delaware, include only species not indigenous to the United States as a whole. A few States define these terms ecologically, similar to OTA's definition of "indigenous" (ch. 2), so as to potentially cover intrastate movements. 3) The agencies responsible for carrying out the laws vary. Many States divide responsibility for different taxonomic groups among different agencies, which can lead to inconsistencies and even conflict within a State (54).

Question 3: Has your agency undertaken internal or external evaluations of your programs in this area?
Yes: *11 (31%)*—FL, HI, KS, MA, ME, MT, OH, RI, VT, WI, WY
No: *23 (63%)*—AK, AL, AR, CO, GA, IA, ID, IL, IN, LA, MD, MN, MO, MS, NC, ND, NE, NY, OK, PA, TN, UT, WV
No answer/other: *2 (6%)*—AZ, NJ

Question 4: How closely do your agency's resources (staff, funding or others) match your current and anticipated responsibilities in enforcing your State's existing laws regulating the importation, possession, or introduction of non-indigenous fish and wildlife?
More than adequate: *0*
Adequate: *7 (19%)*—IA, LA, MD, MO, NY, OH, OK
Less than adequate: *20 (56%)*—AL, AR, CO, GA, HI, ID, IL, KS, MA, ME, MN, MS, NC, ND, NJ, PA, RI, UT, WV, WY
Much less than adequate: *7 (19%)*—AK, AZ, FL, IN, MT, VT, WI
Not sure: *1 (3%)*—TN
No answer/other: *1 (3%)*—NE

Question 5: In future regulation of the importation, possession, or introduction of non-indigenous fish and wildlife, how would your agency prefer to see the Federal role in relation to the role of the States?
Increased: *23 (63%)*—CO, FL, GA, HI, IN, KS, LA, MA, MD, ME, MN, MT, NC, ND, NE, NJ, NY, PA, RI, TN, VT, WV, WY
Decreased: *1 (3%)*—WI
About the same: *8 (23%)*—AL, AR, IA, IL, MO, OH, OK, UT
Not sure: *1 (3%)*—MS
No answer/other: *3 (8%)* — AK, AZ, ID

SOURCES: Office of Technology Assessment, 1993 and Center for Wildlife Law, University of New Mexico Law School, "Selected Research and Analysis of State Laws on Vertebrate Animal Importation and Introduction, " contractor report prepared for the Office of Technology Assessment, Washington, DC, April 1992.

species in any fish or wildlife group and 9 States prohibit none in some groups. State's that do treat vertebrate groups differently usually treat fish apart from the other wildlife groups.

A species or group that is not prohibited may be allowed in one of two ways: formal agency permission is required, which the agency may grant or deny, or no formal permission is required, except possibly to comply with incidental veterinary, commercial, or other laws. Many States use a combination of these two. They may have a list of species for which permits are required and allow any unlisted species to be imported or released without government oversight. Others use the opposite, and stricter, approach of only listing the permit-*exempt* species, such as common pets, and requiring permits for all others.

Wide variety exists both in the structure of statutory approaches and the detail of implementing regulations, even within the basic categories of table 7-1. For example, California lists no prohibited species but requires a permit for importation of dozens of listed groups— including whole orders, families, and genera.[17] The total of individual species requiring a permit is probably well into the thousands. Unlisted species and groups do not require a permit for importation, but all species do for release.[18] By

---

[17] Cal. Fish and Game Code sec. 2118.

[18] 14 Cal. Code Reg. sec. 671.

Table 7-1—Basic Legal Approaches Used by States for Fish and Wildlife Importation and Release

Basic approach	Importation[a][b]		Release	
	Number	States	Number	States
All species are prohibited unless on allowed ("clean") list(s).	2 + 1pt[c]	HI,IDpt, VT[d]	1 + 5pt	AKpt, FLpt, GApt, HI, IDpt, KYpt
All species may be allowed except those on prohibited ("dirty") list(s).				
Prohibited list(s) have 5 or more identified species or groups.	20 + 3pt	AL, AR, CO, CT, FL, IL, KS, KY, MI, MN, MTpt, NC, NE, NY, OH, PA, SCpt, SD, TN, TXpt, UT, WA, WY	14 + 6pt	AL, AR, CO, CT, FLpt, GApt, IL, KS, KYpt, MN, NE, NY, OHpt, PA, SCpt, TN, TXpt, UT, WA, WY
Prohibited list(s) have fewer than 5 identified species or groups.	11 + 3pt	AK, DE, IN, LApt, MD, ME, MS, NH, NV, NJ, ORpt, RI, VA, WVpt	11 + 6pt	AKpt, IN, LApt, NC, NDpt, NJ, MD, MN, MS, NH, NV, OR, RIpt, SD, VA, VTpt, WVpt
All species may be allowed; there is no prohibited list.	11 + 7pt	AZ, CA, GA, IDpt, IA, LApt, MA, MO, MTpt, ND, NH, NM, OK, ORpt, SCpt, TXpt, WI, WVpt	12 + 9pt	AZ, CA, DE, IDpt, IA, LApt, MA, ME, MI, MO, MT, NDpt, NM, OHpt, OK, RIpt, SCpt, TXpt, VTpt, WI, WVpt

[a] State regulation of "possession" of a group or groups is considered here as regulation of both "importation" and "release," since neither act can be done without having possession. For the few States that specifically regulate "importation with intention to release (or introduce)," it is not treated here as comprehensive regulation of "release" because it covers only acts of importation done with a specific intent.

[b] Many States that regulate importation of particular groups exempt mere transportation through the State. These are not distinguished here.

[c] Some States treat different groups of vertebrates differently. This is designated, where applicable, by using the abbreviation "pt" after the State initial to indicate the entry covers only "part" of the vertebrates regulated. They are totaled separately.

[d] The summary classifications are general; in many States there are limited exemptions, such as for scientific research, and other minor provisions which are not covered here. The extensive State regulation of falconry is excluded.

SOURCES: Office of Technology Assessment, 1993 and Center for Wildlife Law, University of New Mexico Law School, "Selected Research and Analysis of State Laws on Vertebrate Animal Importation and Introduction," contractor report prepared for the Office of Technology Assessment, Washington, DC, April 1992.

contrast, Texas prohibits 50 fish species or groups outright, and it requires a permit for release of all but two fish species and for importation of many others.[19] However, Texas lacks a permit system to regulate importation and release of non-indigenous reptiles, amphibians, birds, or mammals, except for 15 mammal species that are public safety risks such as lions (*Panthera leo*).

Analyzing the numbers of groups a State prohibits outright presents an attractively quantitative, but problematic, measure of the State's attentiveness to potentially harmful NIS. Comparing the totals is difficult for some States that list by taxonomic categories larger than single species. A few list large indeterminate categories (which are only counted as one listing here), such as Alaska's prohibition against importing or releasing "venomous reptiles."[20] States with few or no species prohibited outright may still be restrictive in their review of permit applications, so that in practice they prohibit more species than do States with a larger number of species prohibited outright but lower decisionmaking standards. And, of course, States vary in their ecological vulnerability to NIS invasions such that they would not be expected to all have the same number of prohibited species.

---

[19] Vernon's Tex. Code Annot. sec. 134.020.

[20] AK. Stat. 16.05.920.

Table 7-2—Numbers of Species or Groups Prohibited From Importation and/or Release by States

Number NIS prohibited:	0	1-4	5-9	10-19	20-29	30-39	40-49	50-99	100+[a]
Number States:	9	10	8	7	1	1	1	2	11

[a] 100+category includes those States that generally prohibit importation or release of one or more of the five vertebrate groups as a whole, e.g., all non-indigenous fish.

SOURCES: Office of Technology Assessment, 1993 and Center for Wildlife Law, University of New Mexico Law School, "Selected Research and Analysis of State Laws on Vertebrate Animal Importation and Introduction," contractor report prepared for the Office of Technology Assessment, Washington, DC, April 1992.

Given these limitations, breaking down the numbers of prohibited species does provide a rough sense of the variability (table 7-2). A total of 34 States prohibit fewer than 20 species or groups, and 19 of those prohibit fewer than 5; the median number prohibited is 8.

The species most commonly prohibited include piranhas, walking catfish (*Clarias batrachus*), grass carp (*Ctenopharyngodon idella*), European (also called San Juan) rabbit (*Oryctolagus cuniculus*), nutria (*Myocaster coypus*), and coyote (*Canis latrans*)—the latter by the eastern States into which it is expanding its range because of human activities (21). The processes States use in listing species vary extensively, with some based on expert input and others of unclear origin. State lists of prohibited fish, in particular, have been criticized for lack of scientific input (13, 26).

At least one-third of the State fish and wildlife officials surveyed rated their own lists of prohibited species as "too short" (box 7-C, question 2). North Dakota's self-evaluation typifies the comments of this group:

There are presently no non-indigenous species of animals other than fish that are prohibited from importation, possession, or introduction into North Dakota. Given the documented problems that other states have had with the introduction and escape of non-indigenous species, this is obviously an unacceptable state of affairs.

No States rated their prohibited species list as "too long." Slightly less than half rated their list as "about right."

GAPS IN LEGAL AUTHORITY

The least restrictive approach would be to have *no* laws regulating importation or release for any groups. No States fit this description, although a few come close. Several either omit or only partially cover major taxonomic categories of fish and wildlife (table 7-3).[21] OTA's listing of gaps is limited to those States in which no legal authority exists to regulate a particular group comprehensively; it does not include those in which the laws do give such authority, but the agencies have, for whatever reason, chosen not to exercise it. Thus, table 7-3 gives a conservative picture.

Thirteen States lack legal authority over *importation* of one or more of the major vertebrate groups. Twelve States lack legal authority over *release* of one or more of the groups. Fish are least likely to be left uncovered. The only State without authority over fish releases is Mississippi, which lacks authority over all releases except birds.

Almost half of the State officials who responded to OTA's survey wanted additional legal authority from their legislatures (box 7-C, question 1). They typically commented that their existing authority left potentially harmful activities, such as NIS importation for game farming,

---

[21] Most of the gaps are complete omissions where the entire vertebrate group is unregulated. A few gaps are due to partial coverage of a group; for example, Connecticut's law only regulates mammals that are "quadrapeds" (Conn. Gen. Stat. Annot. 26 sec. 55). This covers most potentially harmful non-indigenous mammals, but it does omit authority over several taxa such as pinnipeds (e.g., seals), primates, and bats.

Table 7-3—Gaps in Legal Authority

Vertebrate group	Legal authority over *importation* omits, or only partially covers, the group		Legal authority over *release* omits, or only partially covers, the group	
	Number	States	Number	States
Mammals	9	CT, IA, LA, ND, OR, SC, TX, WI, WV	10	CT, MI, MS, ND, OH, RI, SC, TX, VT, WV
Birds	8	IA, LA, ND, OR, SC, TX, WI, WV	8	MI, ND, OH, RI, SC, TX, VT, WV
Fish	4	IA, ND, NJ, WI	1	MS
Reptiles	10	IA, LA, MI, ND, OR, PA, SC, TX, WI, WV	9	MI, MS, ND, OH, RI, SC, TX, VT, WV
Amphibians	10	AK, IA, LA, MI, ND, OR, PA, SC, TX, WI	9	AK, MI, MS, ND, OH, RI, SC, TX, VT

SOURCES: Office of Technology Assessment, 1993 and Center for Wildlife Law, University of New Mexico Law School, "Selected Research and Analysis of State Laws on Vertebrate Animal Importation and Introduction," contractor report prepared for the Office of Technology Assessment, Washington, DC, April 1992.

uncovered. Those States with authority gaps might try to keep harmful NIS out under their general laws, but they could be legally challenged in disputed cases.

DECISIONMAKING STANDARDS

How are State agencies required to exercise their discretion in cases where they do have legal authority? "Decisionmaking standards" refers to the legal criteria imposed on, or adopted by, the agencies to guide this discretion. With respect to NIS, these criteria typically address potential ecological impacts of the proposed action. States have more restrictive standards for releases than for importation, but overall few States require careful studies, even for releases (table 7-4).

The most restrictive standard, of course, is where the legislature prohibits entire groups of NIS outright, eliminating agency discretion. Florida's statute prohibiting any marine releases is an example.[22] But predeterminations are rare—agencies commonly have broad discretion when permitting or denying NIS proposals.

For allowing NIS importation, 17 States lack standards for all vertebrate groups and 3 States lack them for some groups; for NIS releases, 15 States wholly lack standards, and 6 in part.[23] In these States the discretion of the responsible agency may still be generally guided by the statute(s) that grants the agency's general powers. Nevertheless, having no defined, legally enforceable standards, and thus less accountability, increases the likelihood of widely varying decisions. Political and citizen pressure, personal preferences or values of agency officials, and other unpredictable factors will more likely be influential, especially as this regulatory area is relatively volatile and fast changing (13).

Among the States that *do* have express decisionmaking standards for allowing importation and/or release of NIS, none legally requires that a scientifically based protocol, such as that developed by the American Fisheries Society, be followed. Such protocols are designed by experts to provide formal guides for examining all potential risks and benefits of a proposal (see protocols section in ch. 4). Three States—Florida,

---

[22] 28 Fla. Stat. Annot. sec. 370.081(4).

[23] These numbers include the States previously identified in table 7-3 as lacking legal authority to regulate in these areas; plainly, if a State's laws provide no authority to make a decision, neither do they provide decisionmaking standards.

Table 7-4—Decisionmaking Standards Used by States

Decisionmaking standard[a]	For *importation* permission		For *release* permission	
	Number	States	Number	States
Agency has no discretion; action prohibited	1pt[b]	VTpt	6pt	AKpt, FLpt, GApt, KYpt, **MDpt**, **WApt**[c]
Mandated study of potential ecological impacts	1pt	FLpt	3pt	FLpt, **HIpt**, **MTpt**
Determination of potential impacts, defined broadly enough to include all ecological impacts	18 + 5pt	AL, **CApt**, CO, **CT, DE, FLpt, GA, HI**, ILpt, **IN**, KY, **MD**, ME, **MN, NC**, NE, NH, **NY**, SCpt, TN, **UT**, VTpt, **WA**	15 +12pt	AL, AZpt, CO, **CT**, GApt, DE, **HIpt**, ILpt, **IN**, IA, KYpt, **MDpt**, ME, **MN**, MSpt, **MTpt, NC**, NE, NH, **NY**, SCpt, TN, **TXpt, UT**, VApt, **WApt, WI**
Determination of potential impacts, not defined broadly enough to include all ecological impacts	8 + 4pt	AZ, AKpt, **CApt**, ID, ILpt, **MT, NJ**, NM, NV, PA, RIpt, **VA**	4 + 6pt	AZpt, ID, ILpt, **NJ**, NV, OKpt, ORpt, PA, **VApt, WApt**
No specific decisionmaking standards	17 + 3pt	AKpt, AR, IA, KS, LA, **MA, MI**, MO, MS, ND, OH, OK, OR, RIpt, SCpt, **SD, TX, WI**, WV, WY	15 + 6pt	AKpt, AR, **CA**, KS, LA, **MA, MI**, MSpt, MO, ND, NM, OH, OKpt, ORpt, RI, SCpt, **SD**, TXpt, VT, WV, WY

[a] "Decisionmaking standards" refers to the requirements legally imposed on, or adopted by, the permitting agencies when they exercise discretion.

[b] Some States treat different groups of vertebrates differently. This is designated, where applicable, by using the abbreviation "pt" after the State initial to indicate the entry covers only "part" of the vertebrates regulated. They are totaled separately.

[c] The 18 States indicated in ***bold italics*** have general environmental policy statutes, regulations or executive orders that may overlay NIS permitting and require higher decision-making standards with regard to environmental impacts than the standard indicated (18). They are: *CA, CT, HI, IN, MD, MA, MI, MN, MT, NJ, NY, NC, SD, TX, UT, VA, WA, WI.*

SOURCES: Office of Technology Assessment, 1993 and Center for Wildlife Law, University of New Mexico Law School, "Selected Research and Analysis of State Laws on Vertebrate Animal Importation and Introduction," contractor report prepared for the Office of Technology Assessment, Washington, DC, April 1992.

Hawaii, and Montana—mandate ''studies'' for certain groups to investigate the potential ecological effects a proposed species will have if released.

The main drawbacks to mandating scientific protocols or detailed studies are the costs to applicants and agencies (52). This is reflected, for example, by Maine's explicit decision not to require rigorous scientific studies as a precondition for marine NIS releases, on the grounds that ''[existing] regulations require substantial pre-introduction screening and review processes that are the most appropriate safeguard and the most efficient utilization of scarce resources''

(12). Some States require that NIS be scientifically studied and evaluated *after* release, e.g., Washington.[24]

Many States require some determination—but not detailed scientific studies—of the potential impacts, and they define this broadly enough to include all ecological impacts. Eighteen States require such determinations for *importation* of all vertebrate groups and five require them for some groups. Fifteen States require determinations of impacts for *release* of all vertebrate groups and 12 require them for some groups. These standards vary remarkably in their attention to detail.[25] A few States set out long and complex permitting

[24] Wash. Admin. Code 232-12-271(2)(a).

[25] The classification by OTA in table 7-4 is liberal as to whether the laws provide for consideration of all ecological impacts, even when such impacts are not mentioned specifically. Thus, Alabama's standard of ''best interests of the State'' is treated as potentially including all ecological impacts.

criteria, such as Maine's regulatory standards for wildlife imports.[26] By contrast, Alabama's standard governing the Commissioner of Conservation and Natural Resources' decision to prohibit a species is simply "the best interests of the State."[27] It is difficult to hold decisionmakers accountable for their actions regarding NIS when legal standards are vague.

Several States require determination of potential impacts of the decision but do not define these broadly enough to include all ecological impacts. For example, Oregon's standard for denying a fish release permit is "if the [Fish and Wildlife] Commission finds that the release of the fish into a body of water would adversely affect existing fish populations."[28] That standard does not require consideration of the other organisms potentially affected by a fish release, such as plants, insects, and non-fish predators, nor of the overall condition of the ecosystem.

Adding the number of States in table 7-4 with *no* decisionmaking standards to the number of States with standards that are not broad enough to include all ecological impacts gives the following totals: For importation, 25 States have no or narrow standards for all vertebrate groups and 7 States have such standards for some groups. For release, 19 States have no or narrow standards for all vertebrate groups and 12 States have such standards for some groups. These are the "States without comprehensive decisionmaking standards in their NIS laws" (category (a) in table 7-5).

However, 18 States have a superimposed layer of decisionmaking standards in the form of State environmental policy acts (SEPAs) (table 7-4 in italics). The application of SEPAs varies widely, and they appear to have had little effect in State NIS decisionmaking.[29] However, they can provide general protection against ill-considered

Table 7-5—Non-Indigenous Species Decisionmaking Standards in Relation to State Environmental Policy Acts

	For *importation* permission	For *release* permission
(a) Number of States without comprehensive decision making standards in their NIS laws	25 + 7pt	19 + 12pt
(b) Number of States in category (a) that have adopted general environmental policy acts	8 + 1pt	6 + 3pt
(c) Remainder of States lacking comprehensive decisionmaking standards (a minus b)	17 + 6pt	13 + 9pt

NOTE: Some States treat different groups of vertebrates differently. This is designated, where applicable, by using the abbreviation "pt" after the State initial to indicate the entry covers only "part" of the vertebrates regulated. They are totaled separately.

SOURCES: Office of Technology Assessment, 1993 and Center for Wildlife Law, University of New Mexico Law School, "Selected Research and Analysis of State Laws on Vertebrate Animal Importation and Introduction," contractor report prepared for the Office of Technology Assessment, Washington, DC, April 1992.

decisions by requiring formal environmental review of both agency-permitting and agency-initiated actions (18). For example, Montana requires a detailed environmental impact statement under its SEPA for all new releases of non-indigenous fish, the only State to do so explicitly.[30]

These SEPAs could make the decisionmaking processes more rigorous in the States that lack comprehensive standards written directly into their NIS laws. But in how many States do SEPAs make up for their low (or no) standards? The pattern of adoption of SEPAs answers this question (table 7-5). Even after considering those

---

[26] 402 Code Me. Rules, part IV, sec. 7.60.

[27] Code Al. 9-2-13.

[28] Or. Rev. Stat. 498.228.

[29] State releases supported by Federal funds may require environmental review under the National Environmental Policy Act (ch. 6).

[30] Rev. Code Mont. 87-5-711(2).

States that have SEPAs, approximately one-third of the States have agencies that permit NIS importation and release with no legal requirement that they give comprehensive consideration to the potential ecological impacts of their decisions.

■ Emerging Fish and Wildlife Issues

With a general decline in hunting opportunities on public and open private lands, numerous States face new proposals for releases of non-indigenous mammals and birds on private hunting preserves (22). A trend also exists toward use of "exotics" such as red deer (*Cervus elaphus*) for livestock. When the Wyoming Game and Fish Commission was confronted with a proposal for a large ranch using several hundred animals from 15 non-indigenous species, officials surveyed 13 other western States and four Canadian Provinces that had experience with these ranches. They found a good deal of variation, including quarantine and fencing requirements and responsibility for escapees (32). The key finding: "As they have become more experienced with the problems of disease, competition, and hybridization with exotics and game farms, regulations governing exotics and game farms in 7 States and 3 provinces have become more restrictive for biological reasons." Four of the States and Provinces either lacked legal authority or did not respond to the survey; only one State (Arizona) indicated it had become less restrictive in certain circumstances.

As additional confirmation of the greater State concern in this area, in 1991 and 1992 Montana and Washington imposed emergency moratoriums on various game farm activities, including NIS importation. They cited mainly disease and hybridization risks.

Another emerging area of State concern is the release of non-indigenous fish stocks. (Stocks are sub-species or recognized strains.) The concern focuses on genetic dilution resulting from releases within the larger species' range, but outside the particular stock's range. The most prominent genetic dilution problems occur in the Northwest where massive intentional releases of non-indigenous stocks of hatchery salmon have diluted several wild stocks, contributing to their endangered status (67).

All States but Mississippi have general legal authority to regulate non-indigenous fish releases (table 7-3). A 1990 survey found that 26 of the 39 responding States had some restrictions on interstate and intrastate fish movements based on genetics (70). But, 19 of the 26 States restricted movements of only one or a few species. Usually these were popular sport fish. Only 7 of the 26 had policies applicable to *all* non-indigenous stock releases.

The growth of aquaculture, with the potential for accidental releases, compounds the risks of genetic dilution.

■ Lessons From State Fish and Wildlife Laws

The above comparison of State wildlife laws yields several lessons about exemplary approaches, areas of under-regulation, and problems regarding enforcement.

EXEMPLARY APPROACHES

Which States' approaches represent good examples for other States and the Federal Government? OTA's broad answer, based on overall comprehensiveness and attention to detail in existing statutes and regulations, is that exemplary States include (in alphabetical order): Florida, Georgia, Hawaii, Montana, and Utah. They leave no major authority gaps, they have detailed laws, and they require decisionmakers to observe rigorous standards. This does not mean that their approaches cannot be improved or that OTA endorses decisions these States have made in particular cases.

Also, a number of States' individual legal provisions stand out. The States listed below are not necessarily the only ones with the provisions discussed. The wide variety of these exemplary provisions illustrates the strength of the U.S.

system, in which 50 different regulatory approaches can be developed and tested.

- *Burdens of Proof*: Georgia strongly asserts that importation and release of NIS are a "privilege" to be granted only upon a "clear demonstration" that the review criteria are satisfied (Ga. Game and Fish Code 27-5-1).
- *Expert Input*: Illinois created an Aquaculture Advisory Committee, which makes recommendations regarding importation and possession of NIS for aquaculture (17 Ill. Admin. Code sec. 870.10(e)). The regulation provides for participation by experts from universities, government, and private industry.
- *Funding*: In the past, State fish and wildlife agencies focused mostly on providing fishing and hunting opportunities. Many still rely for operating funds on license fees and taxes on purchases by hunters and anglers. Understandably, these agencies balk at meeting the costs of additional department responsibilities, like new NIS regulations, out of traditional revenues. Tennessee addressed this problem directly by mandating that "costs of administration" of NIS laws come from either NIS permit fees or the general fund (Tenn. Code Annot. 70-4-417). New Jersey authorizes its Commissioner of Environmental Protection to charge user fees adequate to cover the costs of NIS inspections and other necessary governmental services (N.J. Stat. Annot. 23:2A-5).
- *Control of Escapees*: Louisiana's regulation of non-indigenous game breeders is clear. Applicants must submit a written plan for recapture of an escaped animal that includes: equipment, personnel, recovery techniques, and the method of payment for any damages caused (La. Wildlife and Fisheries Reg. sec. 107.11.D.).
- *Compensation for Damages*: Many States hold private owners of NIS responsible for damages caused both to the State and to private claimants if their animals escape. Vermont goes further than most by assessing *treble* damages against importers of illegal NIS for expenses

Illegal releases of fish and wildlife, such as the introduced wild boar (Sus scrofa), are a major concern to States. Hogs and other animals that become feral are seldom brought under State law.

incurred (10 Vt. Stat. Annot. sec. 4709). Nevada created a compensation fund for private property damage and crop loss caused by "game mammals not native to this State" (Nev. Rev. Stat. 504.165). Georgia requires a major insurance policy to cover potential damages caused by certain "inherently dangerous" animals, such as lions (Ga. Game and Fish Code 27-5-4(f)).

- *Emergency Powers*: Legal authority to respond quickly to newly perceived threats can cut off problems before they become widespread. Montana imposed a 4-month moratorium in 1991 on importation of certain non-indigenous game species on the basis of disease concerns, using emergency rule-making powers (Mont. Admin. Register 2-1/30/92).
- *Hybrids and Ferals*: Although non-naturally occurring hybrid animals are non-indigenous, few States explicitly bring them under their laws. Wisconsin spells out coverage of hybrids (Wisc. Admin. Code NR 19.05). Almost all States exempt domesticated species from wildlife laws, leaving their authority over feral domestic animals ambiguous. However, Alaska specifically defines regulated "game" so as to include ferals (Ak. Stat. sec. 16.05.940(17)).

- *Bait Fish*: The importation of live bait fish, followed by its release during or after sport fishing trips, can cause NIS infestations (37, 43). Some States have specific laws regulating live bait; Maine flatly bans all importation of live bait fish commonly used in inland waters (12 Me. Rev. Stat. Ann. 7613) (see box 7-A, on the constitutionality of this ban).

- *Sanctions*: A Vermonter's hunting or fishing license may have "points" assessed against it for violation of animal import laws, in addition to a fine and/or imprisonment (10 Vt. Stat. Ann. sec 4502(b)(2)(L)). This is similar to points assessed against auto drivers convicted of traffic offenses—a certain number results in license suspension. In Montana, a conviction for violation of NIS laws can lead to loss of hunting, fishing, or trapping privileges for 2 years (Rev. Code Mont. 87-1-102).

- *Compliance Incentives*: Hawaii recently amended its laws to provide some of the most severe fines for violations of its importation permit laws—up to $10,000 for a first offense and up to $25,000 for a subsequent offense within 5 years of the prior offense (Ha. Rev. Stat. sec. 150A-14[31]). However, the same statute provides a strong compliance incentive by granting amnesty to any violator who "voluntarily surrenders any prohibited plant, animal, or microorganism or any restricted plant, animal, or microorganism without a permit issued by the department [of Agriculture], prior to the initiation of any seizure action by the department."

- *Comprehensive Planning*: Many States have uncoordinated patchworks of NIS provisions. Minnesota recognized this in its own laws and directed a public-private task force to prepare a major report on NIS threats (41). Based on this, the Commissioner of Natural Resources was to develop a comprehensive management plan for "ecologically harmful exotic species" by January 1993.

UNDER-REGULATION

The comparison of State non-indigenous fish and wildlife laws also reveals areas of under-regulation of clearly harmful NIS by some States. Five States (listed alphabetically) represent those lacking complete regulatory authority, lacking detailed implementing regulations, and/or not legally requiring careful decisionmaking for proposed NIS: Mississippi, North Dakota, Ohio, Texas, and West Virginia. (This does not mean that OTA disagrees with particular decisions these States have made.) Many others also under-regulate in one or more respects—a conclusion supported by the survey of State officials, 42 percent of whom wanted additional regulatory authority.

The most important areas of NIS regulation in which many States fall short are:

- prohibiting harmful species or groups,
- adopting legal authority covering all major fish and wildlife groups and harmful activities,
- following rigorous decisionmaking standards that look at all ecological impacts,
- requiring scientific study of potential significant impacts,
- defining "non-indigenous" so as to potentially include both interstate and intrastate releases,
- regulating all releases of fish stocks to protect genetic diversity,
- covering hybrids and ferals unambiguously,
- making comprehensive rules for containment and other ownership duties,
- clarifying liability for escapes and damages they may cause,
- mandating post-release monitoring and evaluation, and
- obtaining expert input to aid in decisionmaking.

---

[31] Amendments enacted in Hawaii House of Representatives Bill No. 2597, effective on June 17, 1992.

OBSERVATIONS REGARDING ENFORCEMENT

As with the laws themselves, great variability exists in legal enforcement regarding NIS (24). The following admission from Michigan's Department of Natural Resources probably applies to many States:

> [Michigan's] laws and regulations have developed over many years and now exist in a somewhat complex and fragmented manner. These laws and regulations should be reviewed, consolidated, and publicized. Most people in the State are probably not aware of the existing regulations, and the impacts of ignoring these regulations. Moreover, these regulations are often not vigorously enforced. (40)

A major enforcement difficulty is that States generally lack effective ways to monitor imports from within the United States, except for Hawaii and Alaska. Few real geographic checkpoints exist; State borders only provide meaningful enforcement points in the rare States, like California, with inspection stations. A popular or wide-ranging species imported or released into one unrestrictive State can soon spread on its own or be taken into others.

Illegal releases are a major concern of State managers, especially of sport fish (52). Fisheries agencies repeatedly eradicate illegal releases. California recently spent about $2 million to clear white bass (*Morone chrysops*) out of a Central Valley reservoir, where they were threatening native salmonids, only to find them introduced again in a neighboring reservoir (43). Indeed, in some States, thwarting illegal private fish releases is an impetus for officials to undertake their own, more carefully managed, releases (52). Nevertheless, legal releases intended for one watershed can be illegally transplanted by citizens into other watersheds (72).

Illegal releases of animals for sport hunting also occur occasionally, particularly of wild boar (*Sus scrofa*) (35,36). Several other NIS have escaped from game farms, especially in Texas. In Montana, on March 2, 1992, the Wildlife Division conducted a statewide inspection and enforcement blitz of the 107 licensed game farms in the State, looking for illegal or negligent practices (42). They uncovered a number of serious violations, falling into 22 different categories. Five categories involved escape or other opportunities for NIS, such as red deer, to come into contact with indigenous wildlife. As a result of the blitz, the Division pursued legal action against 12 of the farms' operators (42).

These types of enforcement operations are relatively new for many States' fish and wildlife agencies.[32] Their traditional focus on fishing and hunting still holds. In many cases, their budgets depend almost exclusively on dollars generated by hunters and anglers. For example, Utah's State Division of Wildlife Resources receives only 6 percent of its budget from the State legislature (52). They have a strong incentive to introduce popular, harvestable NIS. However, non-game concerns, including NIS regulation, have risen dramatically in the last 15 years or so (52). Internal and external evaluations are important ways to assess whether an agency is meeting its obligations, especially at times when its clients are rapidly changing. Still, only 11 (31 percent) of the agencies that responded to OTA's survey had undertaken prior evaluations of their NIS programs (box 7-A, question 3).

Also, a majority of responding State agencies —20 of 36 (56 percent)—rated their own implementation and enforcement resources (staff, funding, etc.) as ''less than adequate'' (box 7-A, question 4).

In the opinions of several commentators, the States' limited mandates, authority, laws, policies, and resources, when taken as a whole, have led States to do relatively little to slow the establishment or spread of harmful non-indigenous

---

[32] In a few States, agriculture departments have primary enforcement responsibility for non-indigenous fish (especially aquaculture) and wildlife.

fish and wildlife (table 7-6) (9,13,31,62). OTA's analysis supports these opinions. On the positive side, OTA's research revealed that many States have recently taken steps to upgrade their laws and programs, particularly in the West where threats from non-indigenous fish and wildlife have caused significant concern.

STATE LAWS ON NON-INDIGENOUS PLANTS, INSECTS, AND OTHER INVERTEBRATE ANIMALS

Finding:

State laws governing agricultural pests are relatively comprehensive. However, for non-indigenous invertebrates and plants that do not affect agriculture, State laws provide only spotty coverage.

■ Overview of State Laws

The Federal Government dominates the regulation of foreign plants and invertebrate agricultural pests—much more than for fish and wildlife. Nevertheless, States play a major role in quarantining interstate and intrastate movements of weeds and pests of both foreign and U.S. origin.

No government agency maintains a compilation of State laws regulating plants and invertebrates. The National Plant Board, composed of State and Federal agriculture officials, has commissioned a new compilation of nursery regulations and plant quarantines, available in June, 1993. Regional compilations are also underway. For example, the Southern Plant Board had compiled restrictions for 10 of the region's 12 States as of December 1992 (25). These included a ''quick reference'' to each State's full regulations and lists of: definitions; shipping and additional permit requirements; fees; regulated professions or industries; State noxious weeds; applicable Federal and State quarantines; and apiary and miscellaneous information. A similar, standardized format for the national compilation is planned.

State seed laws are compiled annually by Seed World magazine (57). However, a State's restrictions on seeds do not necessarily mean that corresponding restrictions exist against importing or planting whole plants of the same species. Also, limited tolerances of most noxious weed seeds are allowed per unit weight of imported seed. In other words, State seed laws primarily protect seed consumers (farmers) rather than the environment.

As with fish and wildlife, variability exists in State approaches to non-indigenous plants and invertebrates (68). However, all States have agricultural pest prevention programs and certification programs for pest-free nursery stock (68). Most States inspect nursery stock before commercial interstate shipments (50). These programs have been successful in eliminating the occurrence of certain pests in some States (27). Many States also have interior quarantines designed to limit infestations to certain counties.

WEEDS

Almost all States list some prohibited agricultural weeds beyond those listed under the Federal Noxious Weed and Seed Acts. In these cases, State prohibitions may reduce interstate spread of some harmful non-indigenous weeds otherwise allowed by Federal laws and regulations. Relatively few States, however, have natural area weed laws, that is, plant prohibitions separate from agricultural quarantines. The lack of such prohibitions in most States has left them unable to address some harmful NIS, such as the wetland invader purple loosestrife (*Lythrum salicaria*) (63). A trend exists to adopt non-agricultural weed prohibitions, especially to protect aquatic or wetland areas. Washington, for example, has recently adopted detailed regulations on natural area weeds (box 7-D).

Since no national compilation of State plant laws exists yet, OTA commissioned a case study on the adequacy of the weed and seed laws for five contiguous western States: Idaho, Oregon, Utah, Washington, and Wyoming. An expert on

Table 7-6—References to Key State Statutes and Regulations on Importation and Release of Fish and Wildlife

State	Statutory authority	Authority in regulations
Alabama	9-2-13	220-2-.26, -.93
Alaska	16.05.251, -.255(8), -.920, -.940(10), 20-(17)	5 AAC 41.005, -.030, -.070, -92.029
Arizona	3-2901; 17-306	R12-4-401, -405, -406, -410, -412, -413
Arkansas	15-46-101	Game and Fish Comm'n's Code Book §§04.07; 18.12; 32.12-.16; 42.05, -.09
California	Fish and Game Code 2118, -2150	Fish and Game Comm'n regs §§171-171.5; 236; 670.7; 671.1-671.5
Colorado	33-6-112, -114, -114.5	Art. VII.007, -.008, -.009
Connecticut	26-40a, -55, -56	26-55-1, -2
Delaware	3 §7201, 7 §741, -772	Dep't of Nat. Res. and Env't'l Control, Div. of Fish and Wildlife regs. 10, 14
Florida	370.081; 372.26, -.265, -.922; -.98, -.981	Vol. 14, 39-4.005; 39-6; 39-12.004, -.011; 39-23.006-.008; 39-23.088
Georgia	27-5-1, -2, -4, -5, -7	391-4-2-.06; 391-4-3-.12
Hawaii	142-94, 150A-6, -7, -8; 197-3	Title 4, chs. 18, 71; Title 13, ch. 124
Idaho	36-104(6), -701	13K 1, 5.4, 7; 13L 3
Illinois	8 §240; 56 §10-100, -105; 61 §2.2, 2.3	17 IAC 630.10, -870.10, -870.80
Indiana	14-2-7-20, -21	310 IAC 3.1-6.7, -10-1, -10-11
Iowa	109.20, -.47, -.83	none
Kansas	32-956, -1004	23-16-1; 115-20-3
Kentucky	150.180	301 KAR 1:115; -:120; -:122; 1:171; 2:040; 2:080
Louisiana	56:20; 56:319, -:319.1	Title 76, §107
Maine	7 §1809, 12 §6071, -7202, -7204, -7237, -7237a, -7239, -7240, -7613	Tab 402, Pt. IV, §7.60; Dep't Marine Res. regs. Ch. 24
Maryland	Agric. Code 5-601; Health-Gen. 18-219, 24-109; Nat. Res. 4-11A-02, 10-903	08.02.14.05, -.07; 08.03.09.0; 08.02.11.05K
Massachusetts	131 §§19, -19A, -23	321 CMR 2.12, -9.00-.9.02
Michigan	300.253 §3.(1), -(8), 300.257; 300.258(m); 304.2 §2(a); 305.9; 308.115a; 317.81	Wildlife Conservation Act Comm'n Order update #92, 9/17/91: §§4.2, 5.2, 5.5
Minnesota	17.45, -.497; 84.967, -.968, -.9691; 97C.515, -.521	Dep't of Nat. Res. Comm'r's Order No. 2450 published in June 22, 1992 State Register, Chs. 6216, 6250
Mississippi	75-40-113; 79-22-9, -11	Dep't of Wildlife Conservation Public Notice No.s 1405, 2768
Missouri	252.190; 578.023	3 CSR 10-4.110, -.134
Montana	75-1-201; 87-3-105, -210, -221; 87-4-424; 87-5-701 et seq.	12.7.602, -.701; 12.6.1506, -.1507, -.1512, -.1514, -.1515

State	Statutory authority	Authority in regulations
Nebraska	37-713, -719	Title 163, ch. 2, §§002, 004.03, 008.08
Nevada	503.597; 504.295	503.110, -.140
New Hampshire	207:14; 211.62-(e) I and II (previous provisions as reenacted in HB 1183, ch. 171 of 1992 Laws), 211:64; 212:25 and 467:3	FIS ch. 800
New Jersey	23:4-50; 23:4-63.1, -63.2, -63.3, -63.4; 23:5-30, -33.1	7:25-4.1 et seq., -5.1 et seq., -10.1 et seq.
New Mexico	17-3-32; 77-18-1	Reg. 677, Ch. 5, Art. 3, §A
New York	Ag. and Markets Law §74-9; Env't'l Cons. Law §11-0507, -0511, -0917, -1703, -1709, -1728	Title 6, part 174; part 180, §180.1
North Carolina	113-158, -160, -274, -291, -291.3, -292	T02:52B.0212; T15A:03B.0108; T15A:10B.0100; T15A:10C.0211
North Dakota	20.1-01-02; 20.1-02-05.14; 20.1-04-03	29-04-04-01, -03; 30-04-04
Ohio	1533.31	1501.31-19-01
Oklahoma	29 §§5-103, 6-504, 7-801	800.25-25
Oregon	498.052, -.222.b, -.242; 609.309	635-07-515, -522, -523, -527, -585, -600, -615, -620
Pennsylvania	30 §2102; 34 §102, -2163, -2961, -2962, -2963	58 §§71.1-71.6, 73.1-73.2, 77.7, 137.1
Rhode Island	4-11-2; 4-18-3, -5; 20-1-12; 20-10-12; 20-17-9	Dep't of Env't'l Management, Div. of Fish and Wildlife, Rules and Regs. no.s 61-63; Dep't of Health, Rules and Regs., R4-18-IWA, §§2.0, 3.0, 4.0
South Carolina	50-11-1760; 50-13-1630; 50-16-20, -40, -60	none
South Dakota	41-2-18, -3-13, -13-1.1, -13-3	41:07:01:11; 41:00:01:02, 41:09:02:02; 41:09:02:06.01; 41:09:08; 41:14:01
Tennessee	70-2-212; 70-4-401, -403, -412	Rules of Tenn. Wildlife Resources Agency, ch. 1660-1-18-.01(5), -.02(2), -.02(5), -.03(1), -.03(4), -.03(5)
Texas	Ag. Code §134.020; Parks and Wildlife Code §§12.015, 66.007	31 TAC 52.202-.401, 55.201 et seq., 57.111 et seq., 57.251 et seq.
Utah	23-13-5, -14	R657-3-1 et seq., -16-1 et seq.
Vermont	10 §4605, -4709	Fish and Wildlife Regs. Governing Importation of Wild Birds and Animals
Virginia	28.1-183.2; 29.1-521, -531, -542, -545	325-01-1. sec. 5, 325-01-2. secs 1-4; 325-02-27 §§12, 13; 325-03-1 §§5, 6
Washington	75.08.295; 77.12.020, -030, -.040; 77.16.150	220-20-039, -040; 232-12-017, -271
West Virginia	20-1-2; 20-2-13	none
Wisconsin	29.47(6), -.51, -.535	NR 19.05; 150.03
Wyoming	23-1-302; 23-3-301; 23-4-101	Game and Fish Comm'n regs. Chap. X.

This mailing package was designed to complement a State-produced videotape on the dangers of zebra mussels (Dreissena polymorpha) *in Illinois. Generally, State laws on importation and release of these and other aquatic mollusks are less comprehensive than for agricultural pests.*

non-indigenous plants of that region, Richard Mack of Washington State University, assessed the adequacy of the protection afforded by the restrictions under the States' noxious weed and seed lists (also considering the species restricted under the Federal Noxious Weed Act and Federal Seed Act) (33). He based his assessment on the likelihood of *un*listed weeds causing economic or ecological problems. His conclusions:

Idaho—list of 47 weeds (species or larger taxonomic groups) provides adequate protection but omits at least 6 well-known threats.[33]

Oregon—list of 67 provides more than adequate protection, although a few additions would be appropriate.

Utah—list of 23 does not provide adequate protection, omitting at least 11 threatening species.

Washington—list of 75 provides more than adequate protection (box 7-D), although a few additions would be appropriate.

Wyoming—list of 34 provides barely adequate protection, omitting at least 11 threatening species.

Thus, the adequacy of the case-study States' lists of prohibited weeds varies considerably, but only Utah's was rated as inadequate. Also, some State lists include inaccurate or misspelled scientific names, raising questions about the lists' technical validity (33).

---

[33] A partial list of the weeds most commonly found unlisted by these States that nevertheless present economic or ecological threats includes: poison hemlock (*Conium maculatum*), kochia (*Kochia scoparia*), Russian thistle (*Salsola kali*), silver-leaf nightshade (*Solanum elaeagnifolium*), tamarisk (*Tamarix gallica*), tansy ragwort (*Senecio jacobaea*), and yellow nutsedge (*Cyperus esculentus*).

Many western States have implemented a promising approach to protect both agriculture and natural areas through certification of noxious weed-free forage (feed, hay, straw, or mulch) (4). Forage is grown, marketed, and transported throughout the West and is often taken into natural areas to feed pack animals. The certification program reduces the pathways for the spread of noxious weeds and protects consumers who want to purchase pure feed.

INSECTS AND OTHER INVERTEBRATE ANIMALS

In many States, the same laws governing importation and release of vertebrate animals govern those invertebrate animals not otherwise covered by agricultural pest quarantines. Nonindigenous aquatic invertebrates that can be cultured, like oysters, are commonly covered by specific laws regulating aquaculture. Most States also have specific laws on bee culture. But in many States, other non-agricultural pest invertebrates are simply left unregulated, including, for example, aquatic mollusks—one of the most potentially invasive animal groups—imported for use in home aquariums (7).

As of 1992, only three States had adopted regulations specifically on biological control agents. They are California, Florida, and North Carolina (39). However, a later survey identified seven States with laws encouraging the development and application of biological control (see ch. 1).

ENFORCEMENT

State pest and weed programs lack the personnel to undertake comprehensive enforcement against illegal importations. Almost all States lack border inspection stations. Existing programs also have been weakened in recent years by two major outside factors: widespread budget crises affecting State Governments, and demographic changes favoring urban areas, with rural

interests losing their former dominance in many legislatures (56).

Weed prevention and control programs are highly underfunded (44,48), perhaps more than other pest programs. Montana has addressed the funding problem by creating an innovative Noxious Weed Trust Fund.[34] Funded by a 1-percent surcharge on retail herbicide sales and a "vehicle weed fee" imposed through automobile registration, it provides $1.2 million per year for grants for weed control, with one-fourth earmarked for "research and development of non-chemical methods of weed management" (44). Another avenue Montana has pursued that lessens the need for government expenditures is imposing greater legal responsibility on private landowners to prevent the spread of weeds from their property. Designated noxious weeds are treated as common nuisances, and it is illegal to "permit any noxious weed to propagate or go to seed" unless the landowner is in adherence with a local weed management plan.[35]

The leading agricultural production State, California, is the most well equipped to address importation of weeds and pests. The California Department of Food and Agriculture (CDFA) has 16 border agricultural inspection stations to check the almost 30 million incoming vehicles annually, and it carries out cooperative inspection programs with USDA at ports and airports (15). CDFA also inspects parcel post. It carries out intensive insect detection trapping (over 100,000 traps per year), as well as active pest eradication programs. Public education and involvement receive high priority. In 1990, CDFA began an apparently unique enforcement program called "We Tip," with its own toll-free hotline. It offers rewards of up to $10,000 (from funds donated by private growers) for information leading to convictions of people who smuggle in quarantined fruit (8). Yet even with such programs, three agriculturally significant new NIS were detected

---

[34] Mont. Code Ann. 80-7-801 et seq.

[35] Mont. Code Ann. 7-22-2115, -2116.

Box 7-D—Washington State's New Quarantines on Natural Area Weeds

In response to concerns about natural area degradation, in 1992 the Washington Department of Agriculture promulgated sweeping regulations prohibiting all transactions that could lead tdo the spread of seeds or whole plants of 39 invasive plants not indigenous to the State. Previously, the only non-agricultural weed under quarantine was purple loosestrife (*Lythrum salicaria* and *L. virgatum*). The new listings are:

Scientific name	Common name
Amorpha fruticosa	indigobush, lead plant
Anchusa officinalis	common bugloss, alkanet, anchusa
Anthriscus sylvestris	wild chervil
Carduus acanthoides	plumeless thistle
Carduus nutans	musk thistle, nodding thistle
Centaurea diffusa	diffuse knapweed
Centaurea jacea	brown knapweed, rayed knapweed, brown centaury, horse-knobs, hardheads
Centaurea maculosa	spotted knapweed
Centaurea macrocephala	bighead knapweed
Centaurea nigra	black knapweed
Centaurea nigrescens	Vochin knapweed
Chaenorrhinum minus	dwarf snapdragon
Chrysanthemum leucanthemum	oxeye daisy, white daisy, whiteweed, field daisy, marguerite, poorland flower
Cytisus scoparius	Scotch broom
Daucus carota	wild carrot, Queen Anne's lace
Echium vulgare	blueweed, blue thistle, blue devil, viper's bugloss, snake flower
Heracleum mantegazzianum	giant hogweed, giant cow parsnip
Hibiscus trionum	Venice mallow, flower-of-an-hour, bladder ketmia, modesty, shoo-fly
Hieracium aurantiacum	orange hawkweed, orange paintbrush, red daisy, flameweed, devil's weed, grim-the-collier
Hieracium pratense	yellow hawkweed, yellow paintbrush, devil's paint-brush, yellow devil, field hawkweed, king devil
Hypericum perforatum	common St. Johnswort, goatweed, St. Johnswort
Isatis tinctoria	dyers' woad
Kochia scoparia	kochia, summer-cyprus, burning-bush, fireball, Mexican fireweed
Linaria genistifolia dalmatica	Dalmatian toadflax
Lepidium latifolium	perennial pepperweed
Mirabilis nyctaginea	wild four o'clock, umbrella-wort
Onopordum acanthium	Scotch thistle
Proboscidea louisianica	unicorn-plant
Salvia aethiopsis	Mediterranean sage
Silybum marianum	milk thistle
Torilis arvensis	hedgeparsley
Ulex europaeus	gorse, furze
Zygophyllum fabago	Syrian bean-caper

Wetland and Aquatic Plants

Scientific name	Common name
Myriophyllum spicatum	Eurasian watermilfoil
Hydrilla verticillata	hydrilla
Spartina patens	salt meadow cordgrass
Spartina anglica	common cordgrass
Spartina alterniflora	smooth cordgrass
Myriophyllum aquaticum	Parrot's-feather, parrotfeather or waterfeather
Egeria densa or *Elodea densa*	Brazilian elodea or egeria

SOURCE: Washington State Department of Agriculture, Plant Services Division, *Plant Quarantine Manual*, Seattle, WA, 1992.

---

in 1990—one weed (jointed vetch—*Aeschynomene rudis*), a fungal plant disease (a smut—*Ustilago esculenta*), and one nematode (*Hirschmanniella* spp.) (8). This is further evidence that completely preventing entry of harmful nonindigenous species is not possible.

California's park system is also active in NIS issues. Its policies support replacing NIS, such as eucalyptus (*Eucalyptus* spp.), with indigenous species; however, the expense is high and opposition occasionally comes from members of the public who prefer the NIS (ch. 2) (69).

A few other States have begun to emphasize the use of indigenous plants for soil conservation, wildlife habitat, landscaping, and other public purposes, which have traditionally depended heavily on NIS. Illinois has blazed a trail in this change (box 7-E).

PROPOSED MODEL STATE LAWS

Model State laws have been developed by experts outside the legislative process to help legislators improve, and achieve consistency in, States' statutes and regulations. Legislatures have adopted them, sometimes wholly but usually in part, in a wide range of contexts. Model State laws have been directed to a wide range of topics, e.g., controlling narcotics, enforcing child support obligations, and facilitating interstate business (the Uniform Commercial Code).

A model law can be a preferred alternative to a superimposed, preemptive Federal uniform law from the perspective of preserving State sovereignty (see Federal/State section of ch. 8). Robert McDowell, Director of New Jersey's Division of Fish, Game, and Wildlife, expressed this in testimony against a proposed congressional House of Representatives bill that would have imposed greater Federal control over State fish and wildlife releases (38). He supported, as an alternative to Federal control, a ''model law that states could adopt to control undesirable impacts of introductions''; adopting the model law ''would be a requirement in order to have, for example, . . . lack of Federal intervention in the issue'' or possibly as a condition for obtaining related Federal funding (38).

Three proposed model laws address NIS issues. The first, and by far most detailed, is for fish and wildlife.

■ "Model for State Regulations Pertaining to Captive Wild and Exotic Animals"

In 1985, the Animal Health Association, a national veterinary group, resolved to develop a model law for upgrading State laws on NIS introduction and related subjects, an effort led by the Southeast Cooperative Wildlife Disease Study Center (SCWDSC) at the University of Georgia's College of Veterinary Medicine. The Center proposed a broad regulatory system for animal importation that addressed veterinary, humane, public safety, ecological, and other concerns (46). After extensive external review and revisions, SCWDSC sent the model out to all appropriate State agencies in late 1988 (60).

Box 7-E—Illinois Shifts to Indigenous Plants

In the early 1980s, the Illinois Department of Conservation took a hard look at the benefits and costs of its heavy reliance on non-indigenous species (NIS) in the two State-run nurseries. These produce plants for such uses as landscaping of State property, erosion control, and for wildlife habitat and feed. Department officials recognized the risks of degrading natural ecosystems and even endangering indigenous plants through competition and hybridization with NIS. They found no evidence that NIS were better food or habitat for wildlife. In 1983 they decided to phase out NIS. It took roughly 5 years to change over.

First, they proved that indigenous plants could be grown in nurseries using existing techniques and equipment. Then, they collected seeds from State parks and began producing plant materials on a commercial scale. Currently, they grow 67 species of indigenous trees and shrubs, 61 species of prairie wildflowers and grasses, plus 13 woodland herbs. In 1990, the two nurseries filled 2, 517 orders with 4.5 million plants.

Also, they adopted a general policy restricting the use of NIS on Department lands. Harmful NIS are to be controlled or eradicated from Department-owned or managed land "as time, manpower, and funds allow." Officials rewrote several manuals and public information pieces, such as "Landscaping for Wildlife, " to emphasize indigenous species. The Department of Conservation, USDA's Soil Conservation Service, and the Cooperative Extension Service at the University of Illinois jointly prepared a manual for all agencies and organizations planning and designing wind and snow breaks in the State. It specifies 31 indigenous trees and shrubs and just three well-tested, non-invasive NIS (blue spruce (*Picea pungens*), Norway spruce (*Picea abies*), and Douglas-fir (*Pseudotsuga menziesii*)).

Despite the Department's trailblazing efforts against the use of potentially harmful NIS, it was stymied in the Illinois Legislature by commercial nursery and agricultural interests when it sought to add more prohibited species to the State's Exotic Weed Act. The act, designed to protect natural areas, prohibits only three species, each of which is already extensively present—purple loosestrife (*Lythrum salicaria*), multifora rose (*Rosa multiflora*), and Japanese honeysuckle (*Lonicera japonica*). To put this number in perspective, at least 811 non-indigenous plant species grow in a free-living condition in Illinois, representing 29 percent of its total plant species. About 37 of these 811 are considered to be damaging invaders of natural communities, yet Illinois law allows most of them to be planted.

SOURCES:M. Bolin, R. Oliver, S. Brady, and F.M. Harty, *Illinois Windbreak Manual* (Springfield, IL: Illinois Department of Conservation, 1987); F. M. Harty, "How Illinois Kicked the Exotic Habit, " conference on Biological Pollution: the Control and Impact of Invasive Exotic Species, Indiana Academy of Sciences, Indianapolis, IN, Oct. 25-26, 1991; R.D. Henry and A.R. Scott, "Time of Introduction of the Alien Component of the Spontaneous Illinois Vascular Flora, " *American Midland Naturalist*, vol. 106, No. 2, 1981, pp. 318-324; J. Schwegman, Botany Program Manager, Illinois Department of Conservation, personal communication to P.T. Jenkins, Office of Technology Assessment, Aug. 20, 1992.

The entire model law runs to 45 pages. It gives States an optional resource to fill gaps in their laws. Key features include:

- A permit requirement to "own, possess, transfer, transport, exhibit, or release" non-exempt and non-"established exotic wild animals."
- A list of 30 common domestic animals that should be exempt from the model's regulatory requirements.
- A list of "established exotic wild animals" that have "become widespread and are generally considered native wild animals." These "will vary from State to State and the species listed below are a partial list offered for consideration." It consists of ring-necked pheasant (*Phasianus colchicus*), chukar (*Alectoris chukar*), Hungarian partridge (*Perdix perdix*), European starling (*Sturnus vulgaris*), English or house sparrow (*Passer domesticus*), Muscovy duck (*Cairina moschata*), mute swan (*Cygnus olor*), European carp (*Cyprinus carpio*), brown trout (*Salmo trutta*), and nutria.

- Criteria for deciding on ''environmentally injurious animals'' and a list of animals that meet the criteria ''offered for consideration.'' The list includes all 18 vertebrate species or groups already prohibited under the Federal Lacey Act plus 36 other species or groups—28 more than the median number of State-prohibited species. The list was designed to be tailored to each State's particular circumstances.
- A Technical Advisory Committee to provide advice regarding regulations and exemptions, consisting of 12 members representing scientific, commercial, humane, and other interests.

No States have adopted the model wholly, but some, such as Missouri, have used different parts. Utah recently adopted the most detailed non-indigenous animal regulations of any State; it considered the SCWDS model, but chose their own approach instead (20). No further revisions of the model are planned, nor has it been formally evaluated.

▌ Model Honey Bee Certification Plan

In response to the impending invasion by the African honey bee (*Apis mellifera scutellata*), State and Federal officials and private beekeepers developed a Model Honey Bee Certification Plan. In 1991, they offered it to the States for adoption or modification. It sets out methods to certify that queen bees are the desired European type, rather than African types, and it recommends steps for quarantining areas in which the African bee appears. It also prescribes beekeeping practices to reduce ''Africanization.'' Texas, the first State affected by the new bee, has adopted most of the plan; other States are considering it (23). However, some experts question the plan's technical assumptions and probable effectiveness, particularly in light of the very limited enforcement personnel States commit to bee inspection and certification (61).

▌ Outline of a Model Law for Non-Indigenous Weeds in Natural Communities

John Schwegman, the Botany Program Manager of the Illinois Department of Conservation, has outlined the only known model State law approach to combating weeds in natural areas (55):

States should enact laws that:

1. Allow for designation of State exotic weeds by a flexible administrative procedure under control of conservation interests as opposed to agricultural interests.
2. Prohibit the sale, offering for sale, or planting of plants or seed of designated exotic weeds.
3. Designate plants and seeds of exotic weeds offered by dealers as contraband subject to seizure by the State in addition to imposition of fines.
4. Do not force landowners to remove or control exotic weeds growing naturally on their lands (based on the idea that doing so would rouse intolerable public opposition).
5. Set policy on removal and control of exotic weeds on all State owned and managed lands.
6. Require testing or other proof of safety from escape to natural communities of new potential problem plants proposed for marketing in the State.

Nonregulatory components of the model program include supporting research into control methods, providing adequate management staff, supporting Federal efforts, and public education. Schwegman's suggested approach has not been widely adopted, even in his own State (box 7-E). Indeed, few States have comparable programs, although some have made steps toward them, e.g., Washington (box 7-D).

LOCAL APPROACHES

Some local governments have ordinances covering harmful NIS. Generally, local authority has not included imposing quarantines or prohibiting importation of particular NIS except in public health and safety matters. (However, particular

Eurasian watermilfoil (Myriophyllum spicatum) *is among the non-indigenous weeds of natural areas newly targeted by State and local efforts.*

counties are routinely quarantined by State authorities to stop the intrastate spread of pests.) Local governments most commonly address localized problems, such as the capturing of dangerous escaped exotic pets by animal control officers.

Local authority has predominated in the control of agricultural weeds in many western States, in the form of weed control districts. These districts generally develop a county-wide management plan and provide enforcement mechanisms. In the event a landowner fails to comply with the plan by allowing designated weeds to flourish, the districts often have authority to take control measures directly and charge the landowner for its costs. Operations are typically funded through local property assessments with some State sup-

port. Funding can vary greatly from county to county, depending on local economies and property values (44). In regions without such districts—most of the East and South—weed control, other than private efforts, is a State Government function. The historical reasons for this split relate to the greater roles of county governments in the West, the greater size of western States, and their relatively severe weed problems.

Another key area of local authority is the regulation of land development and use. As development involves alterations to vegetation, the local permit process affords an opportunity to require the elimination of existing weeds. Ordinances can also require that certain areas be kept in indigenous vegetation or prohibit the planting of certain NIS. However, the nursery and landscaping industries, already concerned with 50 disparate State approaches, view increasing local regulation with alarm (5). They would prefer not to have to adjust their activities to a variety of ordinances adopted by hundreds of sub-units within a State.

The only "model local law" addressing NIS combines weed control with regulation of land development. In 1985, the South Florida Exotic Pest Plant Council, an association of government and private individuals concerned with non-agricultural weeds, drafted a "Model Exotic Species Ordinance for Municipalities and Counties" (59). Below OTA summarizes, with their titles, the ordinance's main provisions:

- "Model Ordinance Prohibiting the Importation, Transportation, Sale, Propagation and Planting of Harmful Exotic Vegetation"—an outright prohibition is imposed on the listed activities for particular designated harmful species.
- "Requiring Removal of Harmful Exotic Vegetation Prior to Development of Land or When Such Vegetation Constitutes a Nuisance"—before development, the landowner must remove all of the designated species, subject to the plant removal standards; also,

owners of land that is not being developed can be ordered to remove the designated species within 1 year if their property lies within given distances of defined environmentally sensitive areas.

- "Providing Property Tax Reductions for Removal of Harmful Exotic Vegetation"—landowners who have been ordered to conduct removal efforts to protect sensitive areas under the previous provision are entitled to a 1-year 25 percent property tax reduction for the portions of their land from which the vegetation was removed.
- "Establishing Standards for Exotic Vegetation Removal"—specifies removal techniques and precautionary measures.
- "Establishing Standards for Acceptance of Covenants for the Protection and Management of Environmentally Sensitive Lands"—lays out a procedure encouraging the long-term protection of ecologically important areas, with an emphasis on maintaining them free of harmful vegetation.

At least seven South Florida counties and two cities have adopted parts of the model ordinance

(14). Clearly, South Florida's non-indigenous plant problems are among the worst in the country (ch. 8). The model ordinance offers a useful example for other regions with similar, but perhaps currently less severe, problems.

CHAPTER REVIEW

This chapter surveyed State and Federal relationships and State laws regulating fish and wildlife, insects, other invertebrate animals, and weeds. States' approaches vary widely, some tend to under-regulate certain types of potentially damaging NIS, and their enforcement of existing standards is often inadequate. Other States' show exemplary approaches. More successful management of harmful NIS depends upon addressing the deficiencies, disseminating noteworthy State approaches, and ensuring that Federal and State efforts are mutually supportive. This chapter, along with chapter 6, suggests that much more can be done by both Federal and State Governments. In the next chapter, OTA takes a closer look at the situation in two States where severe NIS-related problems have prompted special concern: Hawaii and Florida.

Two Case Studies: Non-Indigeneous Species in Hawaii and Florida | 8

I n this chapter, OTA focuses on the status, problems, and policies regarding nonmarine, non-indigenous organisms in two particular States: Hawaii and Florida. These two States have large numbers of non-indigenous species (NIS) because of their particular geography, climate, and history. Each has experienced considerable problems as a result. And each area has developed interesting policy responses in the attempt to solve these problems. Their efforts are worth attention in their own right and also because they may provide lessons for other parts of the United States.

Several common themes appear in both States. Invasive NIS threaten the uniqueness of certain areas. In Hawaii, this threat is to the remaining indigenous species, most of which occur nowhere else in the United States or the world. In both States, the greatest threat of NIS is to unusual natural areas as a whole. Both States are transportation hubs and tourist destinations. Therefore, entry and establishment of non-indigenous pests in either State provide a route for further spread into other parts of the United States.

Of course, Hawaii and Florida are very different from each other. Hawaii is the only State subject to a Federal agricultural quarantine that includes comprehensive Federal inspection activities. Many policies affecting Hawaii would be different if California, with its massive agricultural sector, were not nearby. No other State receives as much U.S. military traffic and, thus, needs to pay as much attention to this pathway. Florida is the center for U.S. production of tropical aquarium fish, and few other States have engaged in environmental manipulation on the large scale Florida has.

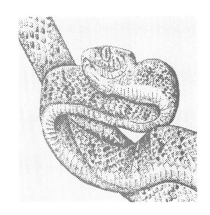

These two States have learned certain lessons in dealing with harmful NIS:

- Federal and State approaches need to be coordinated;
- seldom do those who are the source of NIS problems also bear their cost;
- agriculture and natural areas bear a high cost for introductions, whatever their source; and
- public education is vital to preventing new species entry and spread.

These lessons are worth the attention of other States, perhaps with less severe problems right now. Also, these lessons are worth the attention of Federal policymakers. The Federal Government has both helped and hindered these States in their efforts to deal with harmful NIS. Better integration of Federal and State policies and programs in the future would benefit both the Nation and the States.

NON-INDIGENOUS SPECIES IN HAWAII

Finding:

Hawaii has a unique indigenous biota, the result of its remote location, topography, and climate. Many of its species, however, are already lost, and at least one-half of the wild species in Hawaii today are non-indigenous. New species have played a significant role in the extinction of indigenous species in the past and continue to do so. Hawaii, the Nation, and the world lose something valuable as the indigenous flora and fauna decline.

■ The Nature of the Problem

By many measures, the Hawaiian Islands represent the worst-case example of the Nation's NIS problem. No other area in the United States receives as many new species annually, nor has as great a proportion of NIS established in the wild. At the same time, Hawaii, the Nation's so-called extinction capital, has the greatest concentration of threatened and endangered species in the United States and the greatest number of extinct species as well. While habitat destruction has been and continues to be a main factor in the demise of the indigenous biota, NIS[1] have been identified as an important, if not the most important, current threat (27,85,86,128).

In addition, Hawaii may be the State most visibly transformed by NIS. Most of the coastal areas and lowlands of the mountainous islands appear to be the proverbial paradise—green, often lush, replete with birds and flowers. But except in a few pockets, most of the trees, foliage, flowers, and birds are non-indigenous. Only at higher elevations can one find any appreciable expanse of the globally unique flora and fauna.

Non-indigenous species have had a distinctive impact in Hawaii for several reasons.

- **The island ecology.** The Hawaiian Islands are the most remote land mass in the world, separated from the continents by a 2,500-mile-wide ocean moat. As a result, only a relatively few kinds of plants, insects, birds, and other organisms managed to colonize the islands before human settlement (see "original immigrants" in table 8-1). The original several hundred species that arrived by ocean or air currents evolved into many thousands of species, more than 90 percent of which are endemic (unique) to Hawaii.

  Missing from this assemblage were many of the predators, grazers, pathogens, and other organisms that have shaped the ecology of the continents. Birds, plants, brightly colored snails, and insects dominated the original Hawaiian landscape. Yet there were no ants, mosquitoes, or cockroaches, nor any snakes or other reptiles. The only mammals were a small insect-eating bat and a marine mammal, the Hawaiian monk seal (*Monarchus schauinslandi*).

---

[1] In Hawaii, alien species is the preferred term.

Table 8-1—Past and Present Status of Nonmarine Species in Hawaii

Group	Original immigrants (number)	Indigenous species (number)	Endemic species (no./%)	Extinct species (no./%)	Threatened/ endangered[a] (no./%)	Established NIS[b] (no./%)
Plants[c]	407	≈1,400	≈1,200/86%	/≈10%	/≈30%	≈900/45%
Birds[c]	21	≈100	92/≈92%	60/≈60	30[a]/≈70%	38/≈48%
Mammal	1	1	0	0	1/100%	19/95%
Reptiles	0	NA	NA	NA	NA	13/100%
Amphibians	0	NA	NA	NA	NA	4/100%
Freshwater fish		5	5/100%	0	0	29/84%
Mollusks[c]	22-24	≈1,060	/≈99%	/≈50%	/100%	≈30/6%
Insects[c]	350-400	≈8,000	/≈98%		/≈30%	≈2,500/≈25%

[a] Percentage of remaining species, for most cases representing unofficial estimates. As of December 1992, 104 plant species (all but one as endangered) and 30 bird (marine and nonmarine) species and subspecies (all but one as endangered) were on the U.S. Endangered Species List. Another 61 plant species were proposed for listing (all but one as endangered). A total of 189 plant species were slated to be listed by 1993 under a Federal court settlement (Civil No. 89-953 ACK).

[b] Refers to species non-indigenous to Hawaii. This includes many species originating in the continental United States.

[c] Numbers for plants, birds, mollusks (mostly land snails), and insects in most categories are rounded estimates based on species lists, other published reports, and expert opinion.

NA = not applicable.

SOURCES: Adapted by the Office of Technology Assessment from W.L. Wagner, D.R. Herbst, and S.H. Sohmer, *Manual of the Flowering Plants of Hawaii* (Honolulu, HI: University of Hawaii Press, Bishop Museum Press, 1990); L.L. Loope, O. Hamann, and C.P. Stone, "Comparative Conservation Biology of Oceanic Archipelagoes," *BioScience*, vol. 38, No. 4, April 1988, pp. 272-282; G.M. Nishida (ed.) *Hawaiian Terrestrial Arthropod Checklist* (Honolulu, HI: Bishop Museum Press, 1992); and personal communications from H.F. James, ornithologist, National Museum of Natural History, Smithsonian Institution, Jan. 23, 1992; W. Devick, aquatic resources specialist, Hawaii Department of Land and Natural Resources, Jan. 7, 1992; M. Hadfield, zoologist, University of Hawaii, Honolulu, Jan. 6, 1992; and F.G. Howarth, entomologist, Bishop Museum, January 1992.

Because they evolved in the absence of any large herbivorous animals like deer, many of the plants lost their physical and chemical defenses against such animals (17). Hawaii's indigenous raspberries (*Rubus hawaiensis*) do not have the sharp thorns of related species. The 50 species of indigenous mints lack the herbivore-deterring aromatic scent of sage (*Salvia officinalis*), basil (*Ocimum basilicum*), and other continental mints. Similarly, more than a dozen species of flightless, ground-dwelling birds (88) evolved on the islands, as did several unusual flightless moths, flies, and other insects (55).

This isolated evolution is seen as the prime reason why Hawaii, and oceanic islands in general, are especially vulnerable to ecological invasions (70). In addition, most indigenous species in Hawaii are not adapted to fire, which has increased considerably with human settlement. This now common physical disturbance

not only eliminates indigenous species, particularly rare and threatened or endangered plants, it provides an inroad to invasions by better adapted NIS (109). Trampling by large non-indigenous animals also facilitates invasions.

• **The tropical climate.** Hawaii's average temperatures vary little between winter and summer, at sea level ranging from about 72 to 78 degrees F. In contrast, rainfall, delivered to the islands by trade winds from the northeast, varies tremendously. Windward mountain slopes can receive 300 to 400 inches per year, while leeward coasts receive as few as 10 to 20 inches.

The variation in rainfall, along with the diverse, volcano-created terrain, accounts for Hawaii's large variety of habitats, which in turn accounts at least in part for the diversity of recently arrived organisms that have successfully colonized the islands (69). And the lack of a killing frost except at high elevations means that Hawaii is subject to invasion by many species that would not be a threat to the largely temperate continental United States.

- **The transportation hub.** Lying close to the middle of the Pacific Ocean, Hawaii is a portal between Asia and North America. Traffic through the islands has been increasing dramatically, given the rising economic importance of the Pacific Rim nations and the increasing popularity of Hawaii as a vacation spot. With a 50-percent increase in traffic during the 1980s, Honolulu's airport was 15th busiest in the United States in 1990, according to the Federal Aviation Administration. Equally important is the military traffic through Hawaii, the Pacific center for U.S. defense (see below).

The large volume and variety of traffic is responsible for the great number of NIS that arrive in the State. In addition to stowaways on transport equipment or cargo, plants and animals are brought in, intentionally or unintentionally, by the increasing number of travelers, both residents and tourists.

RATES OF INTRODUCTIONS

The rate of NIS introductions in Hawaii increased dramatically with the start of regular air service to the islands in the 1930s. But Hawaii's transformation by NIS began 1,500 or more years ago, with the arrival of sea-faring Polynesians.

Polynesians intentionally introduced about 30 kinds of plants for cultivation—including sugar cane (*Saccharum officinarum*) and coconut (*Cocos nucifera*), two images closely allied with Hawaiian culture today—and accidentally brought along several weeds. They also brought a few domesticated animals (pigs, dogs, chickens) and stowaways like rats, lizards, and probably several insects. The rate of species becoming established in the islands thus changed from the natural rate of one new species every 50,000 years to three or four new species every 100 years (70).

Hawaii began to absorb a new wave of species with the arrival of Europeans in 1778, when the rate of successful introductions jumped to hun-

dreds of thousands of times the natural rate. Among the most significant and persistent introductions were the goats (*Capra hircus*), sheep (*Ovis aries*), European pigs (*Sus scrofa*), and cattle (*Bos taurus*) released by explorer James Cook and other early ship captains as gifts or to create herds to feed their crews. Feral European pigs and goats in particular remain serious pests of natural areas (and to some extent agriculture) today.

In the subsequent two centuries of European and Asian settlement, horses, deer, and more rodents have also been introduced. More non-indigenous bird species (including 15 game species) have become established in Hawaii than anywhere else (64). More than 4,600 non-indigenous plant species have been introduced, primarily for cultivation. Of these, almost 900 have become established, so that Hawaii's wild non-indigenous plant species today are approaching the number of indigenous species (129). Non-indigenous freshwater fish, most of which were intentionally introduced for sport, food, or other reasons (71), far outnumber the relatively few indigenous freshwater species. In the case of insects, NIS make up perhaps 25 percent (table 8-1). Many of Hawaii's NIS are indigenous to the continental United States; according to the Hawaii Department of Agriculture, about one-quarter of Hawaii's non-indigenous pests are mainland species (47).

Like goats and pigs, many other present-day pest species were deliberate, well-intentioned introductions in the past (table 8-2). Several plants originally brought in for agricultural or ornamental purposes have become extremely invasive, as in the case of strawberry guava (*Psidium cattleianum*) or banana poka (*Passiflora mollissima*). Some animals brought in to control other pests became problems themselves. The Indian mongoose (*Herpestes auropunctatus*), introduced via Jamaica in the 1880s, was supposed to control rats in sugar cane fields, but has come to prey on birds, including the Hawaiian goose (nene, the State bird) (*Branta sandvicensis*), and

Table 8-2—Significant Non-Indigenous Pest Species in Hawaii

Species	Origin	Date introduced	Reason	Impacts
Pig (*Sus scrofa*)	Europe	1778	Gift, food	Damages crops; degrades natural habitats by foraging, trampling; spreads alien plants; causes erosion, harming watersheds
Goat (*Capra hircus*)	Europe	1778	Gift, food	Degrades natural habitat by foraging, trampling; spreads alien plants; causes erosion, harming watersheds
Myna bird (*Acridotheres tristis*)	India	1865	Control armyworm in pastures	Spreads alien plants; damages crops; spreads avifaunal diseases
Cattle egret (*Bubulcus ibis*)	Southern Eurasia, Africa	1959	Control insect pests on cattle	Damages crops, aquaculture; airport hazard; preys on indigenous waterbird chicks
"Trifly"	Widespread		Accidental	$300 million in lost produce markets; $3.5 million in damaged produce; $1 million in postharvest treatment in 1989
Melon fly (*Dacus cucurbitae*)		1895		
Mediterranean fruit fly (*Ceratitis capitata*)		1907		
Oriental fruit fly (*D. dorsalis*)		1945		
Strawberry guava (*Psidium cattleianum*)	Brazil	1825	Cultivated for fruit	Forms a thicket shading out indigenous plants; fruit attracts pigs; crowds out cattle forage; serves as primary host to oriental fruit fly
Koster's curse (*Clidemia hirta*)	Tropical America	pre-1941	Possibly for erosion control	Highly invasive, forming a thicket in forest understory; 80,000 acres affected
Banana poka (*Passiflora mollissima*)	Andes	pre-1921	Ornamental	Heavy vines damage indigenous trees; alters forest understory; 100,000 acres affected
Fountain grass (*Pennisetum setaceum*)	Africa	early 1900s	Ornamental	Invades bare lava flows, natural areas, rangelands; provides fuel for damaging wildfires and is spread by fire
Fire tree (*Myrica fava*)	Azores, Canary Islands	pre-1900	Ornamental, or for fruit (wine) or firewood	Invades natural areas to form a dense stand, obliterating indigenous ground cover; upsets nitrogen balance in soils, encouraging other weeds; attracts pigs

SOURCE: Office of Technology Assessment, 1993.

at least seven other endangered species. The rosy snail (*Euglandina rosea*) from Florida was introduced in 1955 to prey on a non-indigenous pest, the African giant snail (*Achatina fulica*), but is widely believed to have also hunted many of the endemic snails to extinction (55).

Today organisms brought in for biological control are more rigorously screened to avoid nontarget effects; "no purposely introduced species, approved for release in the past 21 years, has been recorded to attack any native or other desirable species" in Hawaii (40). Other scientists, however, question whether monitoring adequately assesses other important impacts, such as competition with indigenous species (55). Still, most new problem species today are believed to be the result of accidental or smuggled introductions.

The rate of NIS establishment nevertheless remains high. About five new plant species per year have become established during the 20th century (133). For the 50-year period from 1937 to 1987, Hawaii received an average of 18 new insect and other arthropod species annually (6, 48)—more than a million times the natural rate and almost twice the number absorbed each year by North America (77). Since the mid-1940s, the annual rate for this fairly well-documented group has been highly variable (see also ch. 3)—ranging from at least 35 new species in 1945 and 1977 to 10 or fewer in 1957 and the beginning of the 1990s (86). It has been suggested that some of the upsurges may be related to wartime activities at the ends of World War II and the Vietnam War (6). Annually about three of Hawaii's new arthropod species turn out to be economic pests (7).

STATE OF INDIGENEOUS SPECIES

The impact of the high rate of biological invasions in Hawaii is partly reflected in the extreme numbers of its extinct and threatened or endangered indigenous species (table 8-1). Some of the best evidence of extinction by NIS comes from Hawaii, as in the case of the rosy snail (ch. 2). Although habitat destruction was probably the greater force behind extinctions in the past, today NIS, through predation and competition, are often considered to be the main threat because they can invade parks and other natural areas protected from development (128).

Hawaii has been described as the 50th State but first in terms of biological imperilment. It occupies only 0.2 percent of U.S. land area—the fourth smallest State—but takes up disproportionate space on the Federal Endangered Species List: about a third of the plants and birds listed or being considered for listing belong to Hawaii.

Much of the unique plant and animal life is already gone. Of all the plants and birds known to have gone extinct in the United States, two-thirds are from Hawaii (128).

Hawaii's spectacular bird life has been the most visibly diminished. Half of the original bird species, including all of the flightless birds, are known only from skeletal remains. Polynesians and their animals probably hunted the birds to extinction, or ensured their demise by clearing their habitat. About a dozen additional species are thought to have gone extinct since Cook's arrival. Most of the remaining birds are either threatened or endangered (table 8-1), accounting for the greatest known concentration of endangered birds in the world.

At least a tenth of Hawaii's plant species are already extinct, and about 30 percent of the remaining species are considered threatened or endangered (129); some botanists say as many as half may be at risk. The indigenous insects and other life forms are too poorly known to allow an assessment of their status, but experts believe they have been similarly affected (table 8-1). At least half of Hawaii's distinctive land snails, for example, are thought to be extinct, while the remaining species are probably all threatened or endangered, in large part because of the imported rosy snail (43,54).

Because islands are especially vulnerable to biological invasions, many of their indigenous species—Hawaii's in particular—were once thought to be doomed to extinction. But recent work in ecological restoration in Hawaii has been promising, and some biologists and conservationists now express optimism that some habitats can recover when browsing animals, for example, are removed (55,70).

■ Causes and Consequences

Findings:

As a set of islands, Hawaii is unique among the 50 States in its vulnerability to the sometimes devastating ecological impacts of NIS. On the other hand its geographic isolation limits the pathways for introductions and presents unique opportunities for the design of prevention strategies.

Hawaii's natural areas and agriculture bear the brunt of new species' harmful impacts. However, agriculture, including horticulture and forestry, also has been a source of problem introductions.

Few economic or noneconomic activities in Hawaii are unaffected by or uninvolved in the influx of NIS to the State. Specific costs incurred because of harmful NIS, however, are available in only some cases. (The State does not maintain records of crop damages from pests.) Many of the consequences of invasions, especially in natural areas, are unquantified.

NATURAL AREAS

In Hawaii, harmful NIS have taken their greatest toll on natural areas. Although they produce no commodities like timber in substantial amounts, they are of value for their unique biological diversity, for maintaining the islands' freshwater supply, for providing scenery and some recreation in a tourist-dependent economy, and as a scientific laboratory.

Hawaii is considered an unparalled site for the study of evolution (see special issues of *Bio-Science*, April 1988; *Trends in Ecology and Evolution*, July 1987; *Natural History*, December 1982). The diverse indigenous species all evolved from a small number of colonizers (table 8-1) and

Harmful non-indigenous species have taken their greatest toll on Hawaii's natural areas, including Haleakala National Park.

as such have been important for understanding how new species arise. One of the world's most dramatic examples of this process is Hawaii's 600 or more species of fruit flies, a quarter of the world's species, all the evolutionary descendants of one colonizing species. Similarly, a single colonizing finch species gave rise to 40 remarkably varied species of honeycreepers.

This evolutionary proliferation of species has endowed Hawaii with the most biological diversity per unit area in the United States (68); as such it is a potential source of useful new biological materials for research and development (123). Hawaii's endemic cotton plant (*Gossypium tomentosum*), for example, lacks the nectar-producing glands of other cotton species and has been used by plant breeders to create a commercial strain that is less attractive to insect pests. A marine coral produces a promising antitumor compound. Only a fraction of Hawaii's unique species, however, have been screened for such properties.

Many indigenous species—perhaps one-third or more of the insects, for example—have not even been described, prompting calls for a thorough inventory of the remaining species and important baseline population studies. The recently signed Hawaii Tropical Forest Recovery Act[2] specifies development of "actions to encourage and accelerate the identification and classification of unidentified plant and animal species" (sec. 605) and baseline studies (sec. 607) in Hawaii forests. The legislation also authorizes grants for NIS control (sec. 610). The 1992 Hawaii legislature also took action[3] to establish a biological survey of the islands' indigenous and NIS.

Natural areas that still support indigenous species in relatively intact habitat make up about 25 percent of Hawaii (114). These areas are protected by the Federal Government (56 percent), the State (41 percent), and others, primarily the Nature Conservancy of Hawaii (3 percent).

The State forest reserves were established at the beginning of this century in recognition of the forests' importance as watersheds (27). Early management involved large-scale plantings of non-indigenous trees, as well as fencing and removal of feral goats, pigs, and other ungulates. By rooting, browsing, and trampling, these animals destroy the vegetation that holds soil in place, especially on steep terrains, resulting in run-off into rivers and streams. Communities have spent millions of dollars for water filtration systems to deal with the contamination, siltation, and discoloration (41).

Damage by feral ungulates is still one of two main non-indigenous threats to forests and other natural areas. Control of feral ungulates has been best achieved in parts of two national parks, but at considerable cost. Areas must be fenced off then cleared of animals by shooting. At Haleakala National Park (HALE), for example, 45 miles of fencing were installed around two important areas—including a rainforest of exceptional biological diversity—at a cost of $2.4 million, provided by the National Park Service's Natural Resource Preservation Program. Maintenance of fences—because of damage from storms, humid-

ity, tree falls, and the like—costs an estimated $130,000 per year (67). Fencing is also under way at Hawaii Volcanoes National Park at a comparable cost.

Weeds constitute the second main non-indigenous threat to natural areas. About 90 of the estimated 900 established non-indigenous plant species in Hawaii are serious pests (109), capable of invading undisturbed natural areas. Hawaii's national parks have a much greater proportion of non-indigenous plant species than do other U.S. national parks (65). At Hawaii Volcanoes National Park, the non-indigenous plant problem is especially severe: 30 of the worst plant pest species are present, 24 of which are widespread (26). Out of 900 total plant species in the park, two-thirds are non-indigenous. Control by hand clearing, chemicals, or in some cases biological agents is concentrated on portions of the park that are especially sensitive; parkwide control is considered impossible.

Non-indigenous insects also threaten natural areas, by competing with or preying on indigenous species and altering pollination patterns, although the extent of their impact is less understood and has received less attention. Perhaps the worst of the insect pests are the predatory Argentine ant (*Iridomyrmex humilis*) and western yellow jacket (*Vespula pensylvanica*), which are the subject of monitoring and control research in the national parks.

For all natural areas, the control and management of harmful NIS consume the vast bulk of their resource management budgets. In the case of the two national parks, which have the most aggressive management programs, the 1987 resource management budget was $1.8 million (114); the 1991 budget was $1.2 million (86) prompting concerns among managers regarding shrinking and inconsistent funding. (Resource management represents 40 percent of the total park budget at HALE (66). By contrast, in the

---

[2] Hawaii Tropical Forest Recovery Act (1992), Public Law 102-574

[3] H.B. 3660

Table 8-3—Non-Indigenous Species in Hawaii: Roles of Federal and State Agencies

Federal Agencies

Treasury Department
  Customs Service—inspects cargo and passengers from foreign points of origin; directs cases to USDA or FWS

Interior Department
  Fish and Wildlife Service—manages 14 wildlife refuges, includes NIS control
  • Law Enforcement Division—inspects wildlife imported into United States to enforce CITES, ESA, and Lacey Act
  National Park Service—manages 2 nature parks, includes NIS control and research

Agriculture Department
  Agricultural Research Service—research on pest control and eradication

  Animal and Plant and Health Inspection Service
  • Animal Damage Control—works to reduce feral animal problems
  • Plant Protection and Quarantine—inspects foreign arrivals and domestic departures for U.S. mainland to prevent movement of agricultural pests
  • Veterinary Service—quarantines animals for rabies and other diseases

Forest Service—NIS control research

Defense Department
  Military Customs Inspection—inspects military transport arriving from foreign areas under Customs and APHIS authority

State Agencies

Governor's Office
  Agricultural Coordinating Committee

Department of Agriculture
  Board of Agriculture
  • Technical Advisory Committee—advises on plant and animal imports, based on input from five technical subcommittees
  Plant Industry Division
  • Plant Quarantine Branch—inspects arriving passengers and cargo to prevent entry of pests; reviews requests to import plants and animals; regulates movement of biological material among islands; provides clearance for export of plant material to meet quarantine standards
  • Plant Pest Control Branch—carries out eradication and control of plant pests through two sections: Biological Control and Chemical/Mechanical Control
  Animal Industry Division
  • Inspection and Quarantine Branch—inspects animals entering Hawaii, manages animal quarantines

Department of Land and Natural Resources
  Division of Forestry and Wildlife—manages State forests, natural area reserves, wildlife sanctuaries; involves watershed protection, natural resources protection, control/eradication of pest species.

SOURCE: Office of Technology Assessment, 1993.

National Park system as a whole, less than 10 percent of the budget is directed to natural resource management, a figure OTA finds to be low (ch. 6).) The budget for the State's Division of Forestry and Wildlife, which oversees State-owned forests, natural areas, public hunting areas, and wildlife sanctuaries (table 8-3), has been substantially increased in recent years. In 1991, it spent $2.8 million for pest control activities (86).

AGRICULTURE

Agriculture is Hawaii's third largest source of revenue—$551 million in 1991 (farmgate value)—behind tourism and military-related spending. Although declining in importance, sugar and pineapple remain Hawaii's two main agricultural products, respectively generating about $200 million and $100 million in recent years. "Diversified" agriculture—macadamia nuts, papayas, flowers, beef, dairy, coffee, and other products—

provides the rest and represents a growth industry for Hawaii.

All these products are derived from imported species, and virtually all the agricultural pests (primarily insects) are non-indigenous as well (8). (By contrast, estimates of non-indigenous agricultural pests on the U.S. mainland range from 40 to 90 percent of all pests.) Many pests arrived in Hawaii on agricultural material that was imported to improve genetic stocks or to introduce new crops. All of today's pineapple pests, for example, were brought in on vegetative material for propagation. The pests not only destroy crops but also limit markets in mainland and foreign areas that have imposed quarantines on produce from Hawaii because of the threat of new pests. This loss of export markets is often cited as the main barrier to the expansion of Hawaii's diversified crops, such as avocados (46).

The Governor's Agriculture Coordinating Committee spent $3.8 million from 1987 to 1990 on research to control or eliminate pest impacts on agricultural commodities (86). The Federal Animal Damage Control (ADC) unit (table 8-3) in Hawaii spent $181,000 (36 percent Federal funds) in 1989 to minimize feral animal damage to agriculture, as well as to natural resources, human health, and property (about half of ADC's work involves controlling bird strike hazards at airports). Agricultural and nonagricultural damage by non-indigenous animal pests confirmed by or reported to ADC in 1989 amounted to $6.9 million (126).

Specific pest-control or -damage costs borne by various types of agriculture follow. Instances where agriculture has contributed to Hawaii's NIS problem are also noted. In general, about half of Hawaii's non-indigenous established plants are thought to have been introduced as crops or ornamentals (133).

Crops—Costs of pest control and damage are best documented for sugar cane, Hawaii's main crop. Throughout its 150-year history, the sugar cane industry has been confronted with a series of damaging insect pests, most of which were eventually controlled biologically. In 1904, the sugar cane leafhopper (*Perkinsiella saccharicida*) from Australia was responsible for the loss of 70,000 tons of sugar, at a cost of $25 million in 1990 prices ($350 per ton), according to the Hawaiian Sugar Planters' Association (91). By 1907, the leafhopper was subdued by several predators imported from Australia.

The sugar cane beetle borer (*Rhabdoscelus obscurus*) from New Guinea was first found in 1865 and remains an important pest of sugar cane. Damage from the insect is exacerbated in areas where rats are a problem, since damaged stalks are favorable for egg laying. A study of losses at two plantations in the 1960s estimated that borers destroyed 2.2 percent of the crop. Industry-wide losses from this pest amount to about 3,000 tons of sugar per year, or about $1 million annually (1990 prices).

Since 1985, at least four new insect pests of sugar cane have become established in the State (90). The lesser cornstalk borer (*Elasmopalpus lignosellus*) has exacted an estimated $9 million in lost yields and other costs since it appeared in 1986 (124). A parasitoid from Bolivia was established in 1991 and is now suppressing the borer in sugar cane fields.

Chemical controls are used on weeds, which are even more costly to the sugar cane industry than are insect pests (91). (Chemical pesticide manufacturers have generally not addressed the needs of Hawaii's agriculture, however, because of its small size and the expense involved in obtaining clearance for new pesticides by the Environmental Protection Agency.) Research costs for all types of pest control in the sugar cane industry in recent years have approached $1 million annually (table 8-4). Development of sugar cane resistance to recently introduced diseases, primarily sugarcane smut and rusts, accounts for another large portion of the industry's research (an estimated $400,000 in 1991 and 1992).

Table 8-4—Research Costs for Sugar Cane Pest Control in Hawaii, 1986-1992

Pest	1986-87	1988-89	1991-92
Weeds	$ 60,000	$214,000	$280,000
Rats	$104,400	$281,000[a]	$232,500[a]
Insects	$101,000	$224,600	$179,000
Diseases	$152,700	$208,000	$172,000
Total	$418,100	$927,600	$863,500

a includes $220,000 from USDA.

SOURCE: Sugar Industry Analyses, 1986, 1988, 1991.

Quarantines imposed on Hawaii's fresh produce because of established pest species have been a substantial cost to growers by limiting markets. The most serious market-limiting pests are the Mediterranean fruit fly (*Ceratitis capitata*), the melon fly (*Dacus cucurbitae*), and the Oriental fruit fly (*Dacus dorsalis*), known as the trifly complex (box 8-A). The financial impact of such quarantines are difficult to gauge; it has been conservatively estimated that Hawaii's export market could increase by 30 percent if quarantines on tropical fruits were lifted (46).

Ranching—Hawaii's pastures and rangelands are vulnerable to invasions by non-indigenous plants, such as the ornamental fountain grass (*Pennisetum setaceum*), which are unpalatable and lower livestock (primarily cattle) productivity. Grasses planted on rangelands themselves are imported and have been plagued by such pests as the armyworm (*Pseudaletia unipuncta*) and grass webworm (*Herpetogramma licarsisalis*). Since its discovery in Kona in 1988, the highly invasive yellow sugarcane aphid (*Sipha flava*) has spread to all the islands and exacted several million dollars in losses annually from State ranchers and $200,000 in biological control research (124).

Seeds, grasses, and animal feed imported by ranchers are believed to have been the avenue for the introduction of some weeds, as in the case of broomsedge (*Andropogon virginicus*) (27), an invasive North American grass that is adapted to fire. Many sugar cane weeds are believed to have arrived in imported rangeland materials (91).

Kikuyu grass (*Pennisetum clandestinum*), a rangeland cover imported from Africa, has itself become a weed in natural areas (109). Finally, browsing cattle have been a destructive force in natural forests and other habitats (27).

Ornamentals—The ornamental plant and floral industry in Hawaii has grown in recent years, although it too has been limited by quarantines on specific fresh products. Based predominantly on NIS, the industry has also been affected by new diseases and pests. A bacterial blight was responsible for a drop in revenues from anthuriums (*Anthurium* spp.), a shiny, brilliantly colored flower from Central America, and a lucrative commodity for the State ($8 million in 1988, the sixth largest crop). A sample of some 50 farms lost $5.5 million in 1987 revenue and $1.6 million in 1989 revenue because of the disease (124).

Two non-indigenous birds, the red-vented bulbul (*Pycnonotus jocosus*) and the red-whiskered bulbul (*P. cafer*), are responsible for significant damage to orchids, a leading product in the cut flower industry, as well as to fruits and other horticultural products. In 1989 the total cost of damaged orchids on Oahu, the only island to be invaded thus far, was $300,000 (46). Indigenous to India and prohibited from entry by State law, bulbuls probably were smuggled into Hawaii as pets, which then escaped or were released in the mid-1960s.

In turn, horticultural activities have been responsible for much of Hawaii's non-indigenous plant problem. Several hundred non-indigenous plants introduced for landscaping or cultivation have escaped and become established (138).

One of Hawaii's worst weeds, the banana poka, a pink-flowered vine, was introduced as an ornamental early in this century and today infests about 100,000 acres of forest. It is notorious for engulfing indigenous trees, killing them or breaking branches and altering the understory. About $1 million in State and Federal funds was spent between 1981 and 1991 on research for the biocontrol of banana poka and Koster's curse

Box 8-A—Costs of Hawaii's Major Fruit Fly Pests and Their Eradication

Three of Hawaii's insect pests—the Mediterranean fruit fly (medfly) (*Ceratitis capitata*), the Oriental fruit fly (*Dacus dorsalis*), and the melon fly (*Dacus cucurbitae*)—were responsible for $300 million in lost markets in 1989, according to the Hawaii Agricultural Alliance. In addition, the so-called trifly complex cost $3.5 million in damaged produce and $1 million in fumigation or other postharvest treatments. The trifly complex has "imposed strong constraints on the development and diversification of agriculture in Hawaii and has provided a large reservoir for the unwanted and increasingly frequent introduction of fruit flies into the mainland United States and other areas of the world via contraband fruit," according to the Agricultural Research Service. Consequently, ARS is conducting a series of technology demonstration tests to help determine the feasibility of statewide eradication of the fruit fly pests.

The three flies became established in Hawaii beginning with the melon fly in 1895, the medfly in 1907, and the oriental fruit fly in 1945. Their establishment was aided by the spread in Hawaii of non-indigenous plants that serve as host plants for the pests. The medfly alone—considered one of the world's worst agricultural pests—infests 250 fruit and vegetable crops. A 1980-1982 effort to eliminate the medfly from seven California counties cost $100 million, according to the U.S. Department of Agriculture (USDA).

California agricultural interests have been strong proponents, if not the strongest, of the proposed eradication project in Hawaii, as well as of the inspection of first-class mail from Hawaii, since the islands are assumed to be a major source of medfly arrivals in California. But preliminary DNA analysis of medflies trapped in California during its 1989 and 1991 infestations indicates the flies very likely did not come from Hawaii; genetically they resemble medflies from Argentina and Guatemala. While the finding does not rule out the possibility that Hawaii may be a source of medfly introductions in the future, it also raises the possibility that Hawaii's role in medfly introductions to the mainland may be overemphasized. Additional genetic studies should help clarify where new infestations are coming from and hence where resources should be targeted.

In the meantime, Hawaii's first demonstration project, slated to end in 1993 at a 3-year cost of $5 million, is attempting to eradicate a large established medfly population on the island of Kauai through the release of sterile insects, although no eradication has been achieved with this technique alone; traps with lures and the insecticide malathion are expected to have to be used against the more abundant oriental fruit fly and melon fly. Demonstration projects for eradication of these fruit fly species are scheduled to run into the next century, at which point the decision is expected to be made on whether to proceed with statewide eradication.

Statewide eradication plans have been controversial because of concerns for public health, as well as for the diverse endemic fruit fly populations in Hawaii, given the likely use of insecticide. Objections have also been raised over the enormous cost of such an undertaking—perhaps $200 million or more for medfly eradication alone—and, if it succeeds, the strong possibility that the pests could become reestablished unless Hawaii's and USDA's inspection and quarantine efforts are substantially improved. The Malaysian fruit fly (*Bactrocera latifrons*), which is also targeted in the eradication plans, was introduced as recently as 1983.

SOURCES: J.R. Carey, "The Mediterranean Fruit Fly in California: Taking Stock," *California Agriculture*, Jan.-Feb. 1992, pp. 12-17; W.S. Sheppard, G.J. Steck, and B.A. McPheron, "Geographic Populations of the Medfly May Be Differentiated by Mitochondrial DNA Variation," *Experientia*, vol. 48, No. 10, October 1992, pp. 1010-1013; U.S. Department of Agriculture, Agricultural Research Service, Tropical Fruit and Vegetable Research Laboratory, "I. ARS Perspective for Fruit Fly Eradication in Hawaii and Pilot Test Requirements for Demonstration of Technology," and "II. Pilot Test to Eliminate Mediterranean Fruit Fly from the Islands of Kauai and Niihau: Detailed Work Plan," drafts (Honolulu, HI: December 1989); R.I. Vargas, research scientist, ARS, personal communications, Dec. 18, 1991, Feb. 10, 1992.

(*Clidemia hirta*), another forest weed (46) (table 8-2); additional sums are spent by public and private groups in pulling weeds or applying herbicide. A 2-year poka eradication effort on Maui was allotted $244,000 by the State (56).

Other ornamentals that have escaped to become problems in natural areas are the fire tree (*Myrica fava*), fountain grass (table 8-2), and other grasses. In some cases, botanic gardens have been the source of the escapees (109). For example, the velvet tree (*Miconia calvescens*), an incipient invader described as the botanical equivalent of rabbits, probably escaped from a private botanic garden.

TOURISM

The large volume of traffic associated with tourism is often cited as a factor behind the influx of harmful NIS to the islands. At the same time, the $9.9 billion visitor industry (in 1991) is the State's biggest source of revenue and largest employer. Consequently, some observers believe there has been resistance in Hawaii to implementing controls that may be perceived as deterring visitors.

The number of visitors in 1990 was 6.9 million, according to the Hawaii Visitors Bureau, an increase of about 50 percent from 1980. Most of the visitors are from the U.S. mainland and Canada, especially the West Coast, with an increasing number from Japan. The remainder come from other countries in Asia and western Europe (78).

According to an opinion survey of State agriculture inspectors, airline passengers are thought to be the most common pathway for insect pests and illegal animals to be introduced, on undeclared plants hidden in carry-on or checked baggage (49) (figure 8-1). For domestic arrivals, this pathway may become less important if a 1992 State law is well enforced. Previously, the State's agricultural declaration process was easily bypassed; the law now requires all passengers to fill out a declaration form, with increased penalties for bringing in prohibited organisms.

Development catering to the large number of visitors may also contribute to the NIS problem by disturbing natural habitats, providing inroads for invasive species. Unauthorized importations of grass materials for golf courses are thought to be the inadvertent avenue for the recent increase in the number of introductions of sugar cane (also a grass) and rangeland pests (91,124). The yellow sugarcane aphid, for example, was first found in 1988 near a new golf course development.

Figure 8-1—Perceived Importance of Pathways in the Introduction of Insect Pests and Illegal Animals in Hawaii

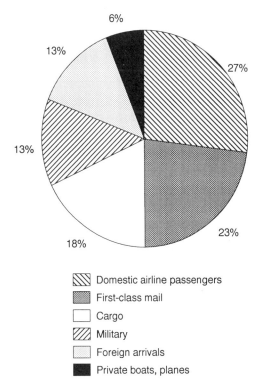

Domestic airline passengers

First-class mail

Cargo

Military

Foreign arrivals

Private boats, planes

SOURCE: Based on an opinion survey of State agriculture inspectors in the Department of Agriculture, State of Hawaii, Honolulu, "Report to the 15th Legislature, 1989 Regular Session."

Many observers point out that Hawaii's tourism depends on the unique natural beauty of the islands and that it would be harmed if the indigenous natural resources are further diminished by harmful NIS (12,78). But there is also said to be little emphasis within the visitor industry on ecotourism or the distinctiveness of Hawaii's indigenous plant and animal life (109,113). Resorts and residences are typically landscaped with tropical plants from around the world: bougainvillea (*Bougainvillea buttiana*) (from Central America); bird-of-paradise flower (*Strelitzia reginae*) (from Africa); palms from other tropical areas. Even the traditional Hawaiian lei is usually made with non-indigenous plants.

MILITARY

Defense spending accounts for about $2 billion, or 10 percent, of State revenues, the second largest share. The military is also believed to be a significant contributor of new introductions to the State and among the islands (figure 8-1) because of the large volume of traffic associated with it. Military personnel traveling from Fiji may have been responsible for the introduction of bulbuls, for example (135).

Military transport in recent years is thought to have been responsible for bringing in from Guam one of the most serious non-indigenous pest threats to Hawaii, the brown tree snake (*Boiga irregularis*). Although the snakes were dead or seized, the possibility of their introduction remains a serious concern (box 8-B), especially with the relocation of military personnel from closed bases in the Philippines to Singapore and Guam. Traffic between Guam and Hawaii is projected to increase accordingly (11).

OTHER SECTORS

Two additional groups are often highlighted for their impact on the NIS problem in Hawaii: sport hunters and pet keepers.

Sport hunting—All of the legally hunted game birds and mammals in Hawaii are introduced, and the maintenance of these populations—including feral ungulates—has often conflicted with conservation of natural areas. Negative impacts on natural areas have been documented for many of the game species (27). The kalij pheasant (*Lophura leucomelana*), for example, feeds on and disperses the seeds of the invasive banana poka, enhancing its spread. Game and other non-indigenous birds are also the source of introduced diseases afflicting indigenous birds (131). On the other hand, sport hunting provides the State with one means of reducing feral ungulates and generates almost $100,000 annually from the sale of licenses (51).

The conflict may have peaked in 1988, when a Federal court found that the State Department of Land and Natural Resources had "demonstrated susceptibility" to hunters by not protecting the habitat of one of Hawaii's endangered birds, the palila (*Loxioides bailleui*), from destruction by feral goats and sheep (120). Under the ruling, the State was required to remove the animals from the palila's habitat (see ch. 7). More recently, the State has begun to address the issue of feral ungulate removal from other especially sensitive natural areas (86).

Pet trade—Animals escaped from their cages or dumped by their owners are a common source of vertebrate introductions today, particularly of birds and reptiles (80). Several species of aquarium fish have also found their way into Hawaii's streams (71). According to the Hawaii Department of Agriculture, about 22,000 birds from U.S. and foreign sources were imported in 1989, primarily for pet stores. They also sell thousands of rabbits (*Oryctolagus cuniculus*) each year.

In October 1989, a resident released six unwanted rabbits at Haleakala National Park. Feral rabbits can severely damage indigenous plants and birds (by attracting predators), and the rabbits' eradication became the park's top priority once the population was discovered. By May 1991, 100 rabbits had been snared, shot, or trapped. The emergency eradication cost $15,000 (National Park money) (66). Although the rabbits were considered eradicated in 1992, future releases of escaped pets are expected to be a recurring problem, with no Federal, State, or island agency mandated to prevent rabbits from establishing (67).

■ Searching for Solutions

Finding:

Hawaii's geographic isolation makes it the state most in need of a comprehensive policy to address NIS—virtually a separate "national" policy with its own programs and resources. The greatest challenge is to coordinate this need with Federal priorities, which can differ. For example, Federal port inspections and

Box 8-B—The Potential Invasion and Impact of the Brown Tree Snake in Hawaii

The brown tree snake has been singled out as one of the more serious—and perhaps imminent—new biological invasions facing Hawaii. It also illustrates how approaches to such threats are often cobbled together, with unclear lines of authority or responsibility among agencies.

Indigenous to the Solomon Islands, Papua New Guinea, and northern Australia, the snake (*Boiga irregularis*) has been accidentally dispersed—usually as a stowaway on planes and ships—to several Pacific Islands, including Hawaii. So far, however, the snake is only known to be established on Guam, where the social cost has been great and the ecological impact disastrous.

As on most Pacific Islands, the indigenous birds of Guam evolved in a snake-free habitat (the island has only one small, blind, wormlike snake species) and consequently lack the protective behaviors of other birds. They were easy prey for the bird- and egg-eating brown tree snake when it arrived sometime around 1950. Of 11 species of indigenous forest birds present in 1945—some of which were unique to the island—9 have gone extinct on Guam. The remaining species have been drastically reduced. Experts attribute the extinctions and declines to the brown tree snake.

Along with birds, the snake also feeds on introduced rats and shrews, whose numbers have also declined. Today the snake is sustained primarily by introduced lizards. The large number of introduced species and other ecological disturbances on Guam have facilitated the snake's invasion of the island. With a diverse and vulnerable prey base and no natural predators, the snake population has soared, reaching densities of 10,000 to 30,000 per square mile.

An able climber, the brown tree snake damages power lines, frequently interrupting service and costing Guam millions of dollars a year. Although it is not considered dangerous to human adults, it is mildly venomous and can poison small children. During a 14-month period in Guam, 27 people were treated for snake bites at one hospital emergency room. The 8-foot-long adult snake commonly enters homes through sewer lines, air conditioning vents, and other openings.

Several characteristics of the brown tree snake make it a likely candidate for invading other islands from Guam. "It is tolerant of disturbed habitats and can maintain dense populations near shipping ports. It is nocturnal [hiding during the day] and readily escapes detection in or around cargo. It is able to live for long periods of time without food, and is thus able to survive for long periods in ships' holds or cargo bays of aircraft. Finally, the broad range of feeding habits ensures that snakes arriving in new environments will adapt to available lizard, bird, and mammal prey species and will therefore be likely to successfully colonize [a new] island" (32). Several reports in 1992 of snake sightings on Saipan in the Marianas, a U.S. Trust Territory, have raised suspicions that the brown tree snake may be colonizing that island.

The increased threat to Hawaii—where the climate is hospitable, habitats have been extensively disturbed, and many indigenous and introduced species exist as a potential prey base—is seen to be the result of the high snake densities on Guam and the frequent number of military and civilian flights from the island. The brown tree snake has turned up in Hawaii at least six times between 1981 and 1991, at Honolulu International Airport, Barbers Point Naval Air Station, and Hickam Air Force Base. Two snakes were found on the same day in September 1991: one crushed on an airport runway, the other live, coiled underneath a military transport that had arrived 12 hours earlier.

Pest problems are best contained by interceptions at the points of departure, and inspection of military flights departing Guam for Hawaii (typically five per week) is said to have been tightened as awareness of the threat has increased. Jurisdictional questions remain, however, about inspection of the 10 to 15 civilian flights per week—whether it is a Federal, Territorial, or State (Hawaii) responsibility. Such questions have resulted in a generally uncoordinated response to the problem.

(continued on next page)

Box 8-B—The Potential Invasion and Impact of the Brown Tree Snake in Hawaii—Continued

The main vehicle for the Federal Government's response has been a line item in the budget for the Office of Territorial and International Affairs in the Department of the Interior. Beginning in 1990, the office has received $500,000 to $600,000 annually for brown tree snake research and control, with $100,000 to $200,000 earmarked for the Hawaii Department of Agriculture, to explore the use of dogs in detecting snakes. The remainder has been disbursed to Guam; a Fish and Wildlife Service research program; and, beginning in fiscal year 1992, the Animal Damage Control unit of the U.S. Department of Agriculture's Animal and Plant Health Inspection Service. Also beginning in 1992, the Department of Defense (DOD) was appropriated $1 million in new money for brown tree snake research and control, in addition to funds available for the brown tree snake through its Legacy program (which provides for natural resources management on DOD lands).

In addition to these appropriations, Congress has addressed the brown tree snake in several pieces of legislation. The Nonindigenous Aquatic Nuisance Prevention and Control Act (NANPACA) of 1990,[1] which focuses on the zebra mussel, directs that a program be developed to control the snake in Guam and other areas. Two other bills direct that the Secretary of Defense[2] and the Secretary of Agriculture[3] take steps to prevent the introduction of the brown tree snake into Hawaii. In Hawaii, in addition to the federally funded airport dog teams for snake detection, State-run SWAT teams have been established on each of the islands to respond in the event of snake sightings.

Despite these actions—as well as a memorandum of agreement intended to coordinate the various State, Territorial, and Federal departments involved—the overall Federal response to the brown tree snake is perceived in Hawaii to have been uneven and sometimes slow. A committee to carry out the NANPACA-directed activities was not in place until 1993, and no agency has taken on the crucial task of inspecting civilian aircraft in Guam before departure.

Ultimately, safeguarding Hawaii and the Pacific basin will depend on establishment of long-term control on Guam. Research by the Fish and Wildlife Service is aimed at an ecological control, along with more immediate controls such as the use of methyl bromide for fumigating cargo and the use of toxicants, baits, and traps. Costs for the various controls that would need development have been estimated to be about $2.5 million annually over several years.

---

[1] P.L. 101-646, sec. 1209.

[2] Department of Defense authorization, P.L. 102.190, sec. 348.

[3] Farm Bill Technical Corrections, P.L. 102.237, sec. 1012.

SOURCES: T.H. Fritts, U.S. Fish and Wildlife Service, *The Brown Tree Snake: A Harmful Pest Species* (Washington, DC: U.S. Government Printing Office, 1988); J. Engbring and T.H. Fritts, "Demise of an Insular Avifauna: The Brown Tree Snake on Guam," *Transactions of the Western Section of the Wildlife Society*, vol. 24, 1988, pp. 31-37; T.H. Fritts, personal communications to the Office of Technology Assessment, Jan. 10, Jan. 30, and December 1992; G.R. Leong and P. McGarey, legislative assistants to Sen. D.K. Akaka, personal communications to Office of Technology Assessment, Jan. 6, 1992, and Dec. 3, 1992, respectively; P. Delongchamps, Office of Territorial and International Affairs, personal communications to Office of Technology Assessment, May 22 and December 1992; L. Nakahara, Plant Quarantine Manager, Hawaii Department of Agriculture, personal communication to Office of Technology Assessment, Apr. 16, 1992 and June 23, 1993.

---

quarantines are directed at protecting mainland agriculture and enforcing international trade agreements, sometimes at the expense of Hawaii's natural resources and agriculture.

FEDERAL INVOLVEMENT

Hawaii's experience with NIS is also distinctive in terms of Federal involvement. Hawaii is the only State where all passengers and cargo enroute to other States (to the U.S. mainland) are subject to "preclearance activity" by Federal agricultural inspectors, a function of Hawaii's geographic isolation and a Federal quarantine imposed before Hawaiian statehood. Agricultural inspection of traffic from the mainland to Hawaii, however, is for the most part left to the State; the nature of mainland pest problems do not meet the existing criteria to warrant Federal inspection of Hawaii-bound passengers and goods.

The domestic quarantine on Hawaii has in turn led to Federal inspection of first-class mail leaving Hawaii and a recent proposal (which failed) to collect inspection fees from passengers departing the State for the mainland. The Federal intent of all these actions, along with the proposed fruit fly eradication program (box 8-A), has been protection of mainland agriculture. An unintended effect, however, has been creation of a double standard, since reciprocal protective measures have not been applied to Hawaii. In 1992, Congress took action to begin to redress this imbalance; any changes in the system have yet to be evaluated.

Details about the Hawaii quarantine, inspection fee, and first-class mail issues follow.

Hawaii quarantine—Passage of the Plant Quarantine Act[4] led to the quarantine of Hawaii to prevent importation of the Mediterranean fruit fly and other agricultural pests.[5] The U.S. Department of Agriculture (USDA) began inspecting goods bound for the U.S. mainland in 1910 and goods arriving in the islands from foreign ports in 1949. Hawaii's own plant and animal quarantines were begun before the turn of the century.

The Federal quarantine regulations stipulate that cargo and passengers from Hawaii to the U.S. mainland are to be inspected by USDA's Animal and Plant Health Inspection Service (APHIS) for prohibited materials (fresh produce, cut flowers, and other plant materials). Certain products are allowed provided they are treated or handled according to prescribed methods to kill any pests.

This preclearance activity, aimed at preventing pests from reaching the mainland, accounts for about 85 percent of APHIS's Plant Pest Quarantine activity in Hawaii (106). Inspection of ships and planes arriving from foreign countries

accounts for 15 percent. The division of resources is said to be roughly proportional to the number of domestic and foreign passengers.

APHIS inspection of foreign arrivals focuses on federally prohibited agricultural pest species, which in turn reflects the temperate climate that predominates in the United States (110). This policy may allow new pests into Hawaii that could otherwise be avoided. For example, State officials tried unsuccessfully to have a mealybug pest (*Pseudococcus elisae*) of bananas declared a federally prohibited species after it repeatedly turned up in the mid-1980s on bananas from Central America that were shipped from the U.S. mainland, where they are inspected by APHIS. The mealybug eventually slipped into Hawaii, became established, and has resulted in lost markets: California rejected shipments of cut flowers from Hawaii because of mealybug infestation (124).

Since the State has no authority over foreign traffic, State agricultural inspectors rely on Federal inspectors (table 8-3) for referrals in order to intercept State-prohibited species. Cooperation among the agencies in this regard is generally said to be good, although neither State nor Federal inspection staffing has kept pace with the growth in traffic through Hawaii in recent years. Between 1971 and 1988, for example, State inspection activities on Oahu increased by a total of 138 to 1000 percent, while staffing increased by 15 percent (49). In the last 5 years, APHIS has received less than its requested budget, and staffing has remained constant, although the 1992 budget allowed for an increase (52).

Over the past decade, Customs has undergone a change in policy, from one of inspection of all foreign arrivals to ''profiling''—inspection of only a fraction of arrivals—in order to facilitate the movement of passengers. In Honolulu, which is said to be one of the stricter ports of entry into the United States, APHIS and Customs each

---

[4] Plant Quarantine Act of 1912, as amended (7 U.S.C.A. 161)

[5] 7 CFR Ch. III part 318 (Jan. 1, 1991).

manage to check about 15 percent of the international baggage passing through the airport. (A goal is to check all of the baggage originating from high-risk areas such as the Philippines.) In contrast, APHIS inspects all of the baggage bound for the mainland by x ray. Many observers maintain that goods and people coming into Hawaii should be as thoroughly inspected as is mainland-bound baggage to minimize the flow of unwanted new species into the State and, in turn, the rest of the country.

Pests found on the U.S. mainland may be as threatening to Hawaii as those brought in from foreign points of origin: seven of the eight new insect pests of grasses that have appeared in Hawaii in the last decade occur in the continental United States, including the economically important yellow sugarcane aphid and the lesser cornstalk borer (124). The transit of goods and people from Florida and the Caribbean through the mainland to Hawaii is thought to be an increasingly common pathway of harmful new pests (7).

Domestic quarantine user fees—In 1991, APHIS proposed to collect user fees from inspected passengers and vessels departing Hawaii for the mainland. The user fee, of $2 per passenger, was intended to cover the cost of agricultural inspections,[6] in order to meet deficit reduction goals. The fee would have been similar to the fees collected by U.S. Customs and Immigration and Naturalization services.

But the fee was interpreted as a ''tourist tax'' that discriminated against Hawaii, being the only State subject to domestic agricultural quarantine and inspection activities. After the rule had been made final,[7] the Hawaii congressional delegation took the unusual step of inserting a provision in the 1992 Federal budget that prohibits such domestic inspection user fees (45). Again, the proposed action was seen as benefiting the

PROTECT HAWAII'S AGRICULTURE AND ENVIRONMENT FROM UNWELCOME VISITORS
A Guide for People Importing Plants & Animals into Hawaii or Exporting Plants from Hawaii

DEPARTMENT OF AGRICULTURE
Plant Quarantine Branch
Plant Industry Division

Inspections of foreign arrivals are intended to intercept harmful non-indigenous species, while educational materials are often aimed at decreasing the number that reach inspection stations.

---

[6] 56 *Federal Register* 8148 (Feb. 27, 1991).

[7] 58 *Federal Register* 18496 (Apr. 23, 1991).

mainland at the expense of Hawaii's tourists and residents.

First-class mail—First-class mail and express mail delivery services have been identified as an important pathway for the introduction of new pests to Hawaii (figure 8-1). Plant material mailed into the State is possibly responsible for the introduction of the large number of whiteflies established in the last 25 years, since these pests can only be transported long distances on living plants (7). Similarly, prohibited seeds, plants, fruits, other insects, and small animals have all made their way into Hawaii through the mail.

Prohibited materials have been intercepted only when suspicious packages were noticed and the State informed, since domestic first-class mail is federally protected from inspection. (Foreign mail may be inspected.) Congress, however, following passage of the Agricultural Quarantine Enforcement Act,[8] which prohibits mailing of quarantined agricultural material, authorized a trial first-class mail inspection program in Hawaii, but only of pieces departing for the mainland. The intent was to determine if fruit flies were arriving on the mainland through domestic first-class mail.

The trial program, originally proposed to run for 60 days at a cost of $30,000 in USDA funds, involved use of an APHIS dog at the main Honolulu post office to sniff parcels for any biological material. Reportable fruit flies, the target of the program, and other insect pests were found on produce seized from 130 parcels (94), most of which were bound for California, Oregon, or Washington. According to another report on the program, fruit flies were found in 29 of the 2 million packages processed between June and October 1990; five contained the Mediterranean fruit fly. The report concluded that first-class domestic mail from Hawaii is a means of trans-

port for the medfly larvae, ''but that the rates are low'' (16).

The trial program has been indefinitely extended, entailing three additional staff positions (107), at an estimated cost of $100,000 annually. The use of Federal funds to conduct the one-way inspection was again perceived as discriminatory in Hawaii, given the importance of first-class mail as a pathway for introduction to the islands (93). Consequently, legislation readdressing the issue for Hawaii was introduced and signed in 1992. The Alien Species Prevention and Enforcement Act[9] is intended to prevent the introduction of new pests to Hawaii through first-class mail by allowing inspection of incoming parcels as well.

With each of these issues, the historical lack of reciprocal protection for Hawaii's agriculture and especially for the large number of federally listed endangered species has created the perception of a Federal bias, with the $17 billion California agriculture industry seen as the primary beneficiary. It is frequently observed as well that the growing national interest in conserving tropical forests in the developing world should be extended to U.S. tropical forests—namely, those in Hawaii (2,85).

A greater Federal role in protecting Hawaii from new damaging introductions may also be warranted because of the large military presence in the State. All military arrivals from foreign ports, as well as military departures for the mainland, are inspected in Hawaii under the authority of Customs and APHIS. Military customs inspectors collaborate with APHIS on foreign arrivals and routinely spray plane cabins with insecticide. Military arrivals from the mainland, however, are a State responsibility, and inspections are said to be limited (49).

On the other hand, the Federal Government—namely the National Park Service—has been considered the most effective manager in terms of

---

[8] Agricultural Quarantine Enforcement Act of 1988, Public Law 100-574.

[9] Alien Species Prevention and Enforcement Act of 1992, Public Law 102-393, Part 3015.

preserving Hawaii's habitats through the control of harmful NIS (112,114).

Finding:

The National Park Service devotes considerable resources to eradicating or controlling harmful NIS in Hawaii within and outside park boundaries. The impact of these efforts are limited, however, because State management on its own lands has been less aggressive. Influx of a significant number of new species annually, despite Hawaii's relatively strict system of regulating introductions, compounds the problem.

STATE ROLE

State laws governing the entry of new plant and animal species specify protection of agriculture, the natural environment, and public health. Natural resources, however, are said to rank behind agriculture and other economic issues, especially tourism, as a priority for the State (61,108). Comparison with other States' spending levels bears out this observation.

Hawaii's Division of Forestry and Wildlife in the Department of Land and Natural Resource, which oversees the State-owned natural areas (table 8-3), ranks 8th out of 50 States in terms of the area it is responsible for (900,000 acres), but 38th in permanent staff and 45th in funding (13). Similarly, Hawaii ranks 44th in terms of its annual expenditures on natural resources and the environment (0.85 percent of the State budget), although this ranking may reflect the State's small size and relative lack of "brown" environmental problems associated with heavily industrialized States. In per capita spending, it ranks 29th ($25.35) (10).

Hawaii spends almost $1.9 million annually on its agricultural quarantine program, 90 percent of which involves inspection of incoming passengers and goods and other preventive measures (50,124). But coverage of incoming traffic to the islands is still incomplete. A 1989 assessment by the Hawaii Department of Agriculture estimated that the additional cost of extra staffing and 16 x-ray units (for 16 baggage claims) to ensure complete inspection of incoming domestic baggage alone would be about $2.25 million (49). In contrast to Federal inspection of mainland-bound baggage, which is all x rayed, State inspectors have relied on agriculture declaration forms to bring to light any incoming produce, plants, or animals.

Opinion differs on the efficacy of the State's importation and quarantine system. In one high-profile example, the importation of Christmas trees each year, the likelihood of harmful new insect introductions has taken a backseat to a traditional societal demand. Because there is no effective fumigant that does not damage the trees, they are only visually inspected. Christmas trees were very likely the vehicle on which yellow jackets arrived in Hawaii, as might gypsy moths (*Lymantria dispar*), according to some observers.

Other prevention efforts are improving. In 1990, State inspectors began to use beagles to sniff baggage and cargo arriving from the mainland. Use of one portable x-ray unit for random inspection of domestic baggage was also instituted. Penalties for smuggling in prohibited species have been substantially increased, and the State list of prohibited plant species is being updated for the first time in 10 years. To emphasize protection of natural areas, the Department of Land and Natural Resources, with the support of environmental groups, is exploring the possibility of creating a separate list of State-prohibited plant species that threaten natural areas.

Many observers point out that the most cost-effective approach to dealing with new pests anywhere is to prevent their introduction (86). Hawaii clearly needs tightened inspection and quarantines to minimize the number of harmful new introductions. Neither State nor Federal efforts have been up to the task.

Harmful new introductions are expected to be reduced once the recently authorized program for inspection of first-class mail from the mainland to Hawaii is in place. New pests could be further reduced by inspection of:

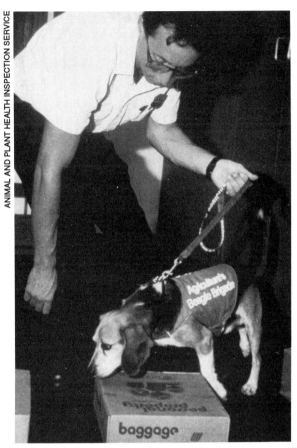

ANIMAL AND PLANT HEALTH INSPECTION SERVICE

In 1990, State inspectors began using beagles to sniff baggage and cargo for prohibited soil, agricultural products, and other biological materials.

- all arriving domestic airline passengers and baggage. Complete inspection by x ray or beagles would require reconfiguration of Honolulu's airport, or that agricultural monitoring be made along with security checks at the main U.S. points of departure for Hawaii. Federal involvement in domestic arrival inspections would require a change in APHIS's mandate; complete inspection by the State would require a redoubling of current efforts and a clarified legal mandate.
- military transport arriving from the mainland, requiring increased State effort and/or military effort or a change in APHIS's mandate.

- all arriving international airline passengers and baggage. Complete inspection by x ray or beagles would require increased APHIS staffing and airport reconfiguration.

A more controversial option, because of objections by the public to pesticides, would involve treating planes arriving from the Pacific region with insecticide, since visual inspection of a plane is not fail-safe. Such treatment was once routine for mosquito (malaria) control.

Shortcomings exist in the State's efforts to control and eradicate NIS. Responsibility is divided, depending on the type of species (insect, plant, or other animal); whether it has an economic impact; and where the infestation is occurring. Response to emergencies is said to be slow for this reason. The jurisdictional difficulties of controlling pest species on private land is a particular problem (86).

Monitoring to detect pests before they become too widespread to eradicate is also incomplete. The Hawaii Department of Agriculture maintains a program using traps, sweepings, and surveys to detect new insect pests, but there is no clear authority for monitoring in cases like feral rabbits.

EDUCATION

Finding:

Public education is considered central to solving problems involving NIS in Hawaii. These efforts are better developed in Hawaii than elsewhere in the United States.

Education is repeatedly cited as the primary tool for enlisting the public's cooperation in containing the problem of harmful NIS. The state of public understanding about the issue in Hawaii is probably no different than anywhere else, but the ecological repercussions of a lack of public understanding are more severe, as in the case of the released rabbits in Haleakala National Park.

The rabbit case also indicates how effective public education can be. Park-generated publicity and media attention resulted in calls from the public about rabbit sightings. The pet owner responsible for the release was unaware of the rabbits' impact and was said to be apologetic. The incident led to a proposal to create a National Park Service public outreach position devoted to such issues. The idea was praised, although it did not receive funding.

Other public and private groups in Hawaii have begun educational campaigns related to NIS, including the Alien Species Alert Program (ASAP) of the Hawaii State office of the National Audubon Society; publicity about prohibitions of mailing fruits and vegetables to the mainland by the USDA and the U.S. Postal Service; informational outreach about indigenous species by the Division of Forestry and Wildlife; and the Bishop Museum's Ohia project (named for a common indigenous tree), a grade school curriculum designed to increase understanding of Hawaii's ecology.

In February 1992, the Hawaii Department of Agriculture publicized a 1-week amnesty program encouraging residents to turn in illegal animals. The campaign netted 53 animals, including snakes, other reptiles and amphibians, harvester ants, hamsters, and birds (82).

The traveling public is singled out as an important target for educators. As one botanist puts it: "Tourists come for the scenery, but unless they've been educated, they won't care if the plants are native or not, just as long as the hills are green." There has been little effort to inform visitors of Hawaii's NIS problem by posters, amnesty buckets, or other means upon arrival, although a State-funded educational video began to be shown on flights of a few domestic carriers in 1992.

The brief video ("It Came From Beyond") takes a decidedly friendly approach to informing visitors about NIS and is expected to reduce the number of "innocent" introductions; some observers believe a stern approach emphasizing the law with its steep fines and penalties is necessary to reduce the potentially more harmful flow of smuggled species, which are probably more commonly brought in by residents with commercial or hobby interests.

Educational efforts in Hawaii also need to be developed and targeted to the State's diverse cultural and ethnic groups. An edible gourd-producing vine (*Coccinia grandis*) that has recently become a weed in Hawaii might have been intentionally brought in as a delicacy from Southeast Asia, for example.

COOPERATIVE EFFORTS
Finding:

In recent years, various groups in Hawaii—from State and Federal agencies, nongovernmental organizations, agriculture, and universities—have taken a strong interest in NIS. Increasingly, they view harmful NIS as a unifying threat.

Awareness of the widespread impact of damaging NIS in Hawaii has prompted a high degree of cooperation across diverse groups. One such effort involves an interagency agreement to research the biological control of forest weeds, an area that no agency was adequately addressing despite the spread of weeds like banana poka. The agreement involves the National Park Service; U.S. Forest Service; Hawaii's Division of Forestry and Wildlife and Department of Agriculture; and the University of Hawaii.

There is growing interest in Hawaii in expanding interagency cooperation to address the larger jurisdictional and informational gaps in the present system. Most of the agencies involved are supporting a plan by the Nature Conservancy of Hawaii and the Natural Resources Defense Council on improving interagency cooperation (86) (box 8-C). A single interagency system may prove more effective for Hawaii's particular needs than applying stop-gap measures to the existing approach.

NON-INDIGENOUS SPECIES IN FLORIDA
Finding:

The problems caused by non-indigenous species (NIS) in Florida are among the most severe in the United States. Certain features of

---

Box 8-C—A View From Hawaii: Recommendations of the Nature Conservancy and Natural Resources Defense Council

In 1992, the Nature Conservancy of Hawaii and the Natural Resources Defense Council released a detailed analysis of the "alien pest species invasion in Hawaii" and offered a plan to create a coordinated multiagency response to the problems, to be led by the Hawaii Department of Agriculture. It does not, however, advocate centralizing all inspection or other activities under one agency. The report stresses public education and involvement in curbing Hawaii's pest problems and identifies the following areas that need initial attention:

- Pre-entry prevention. Visa applications, importation permits, travel and tourist materials, mail order and shipping instructions, and similar materials should be reviewed with an eye to stopping pests at their origin. Similarly, international inspections and trade agreements should be reviewed and improved.
- Port-of-entry sampling and inspection. Methods for sampling and inspection should be developed to meet a standard of pest interceptions.
- Statutes, policy, and rules. Conflicts and gaps in authority should be identified and resolved. A clear system for allowing and prohibiting species should be created.
- Rapid response. Specific plans for dealing with new infestations should be created, including central reporting mechanisms, staffing and equipment concerns, contingency funding, and identification of priority pests.
- Statewide control. Federal, State, and private groups should collaborate in developing strategies to isolate or eradicate selected major pests.

The report further identifies several long-range needs, namely, joint training among agencies for inspection and response activities, coordinated information systems, coordinated research for prevention and control methods, and expanded public awareness campaigns. The pest prevention and control systems of New Zealand and Australia are highlighted as instructive models for Hawaii (see box 1-D).

SOURCE: The Nature Conservancy of Hawaii and Natural Resources Defense Council, "The Alien Pest Species Invasion in Hawaii: Background Study and Recommendations for Interagency Planning," July 1992.

---

the State have contributed to the problems: the subtropical climate; major ports of entry; burgeoning pet, aquarium, and ornamental plant industries; high rates of human immigration; increasing urbanization; and extensive environmental manipulation.

■ The Nature of the Problem

Florida is renowned for its mild climate, abundant waterways, beaches, and other natural attractions. Its freshwater lakes and streams afford recreation, navigation, commercial fishing, and wildlife habitat (57). Its major forest types, various mixtures of oak and pine (22), are crucial for wildlife as well as timber. South Florida contains one of the largest complexes of preserved ecosystems in the eastern United States, totaling about 3,500 square miles: Everglades

National Park, Big Cypress National Preserve, Loxahatchee National Wildlife Refuge, and Fakahatchee Strand Preserve (figure 8-2).

South Florida also contains troublesome infestations of several aggressive non-indigenous plants, most of which were deliberately introduced (30). The State has approximately 925 established non-indigenous plant species (130). Non-indigenous plants and land mammals constitute about 25 percent of all species in the State (table 8-5). Sixty-three percent of the introduced non-indigenous bird species in the continental United States are found in Florida (1), which also has the largest number of established non-indigenous amphibian and reptile species in the United States (136).

Non-indigenous species cause severe ecological, economic, and resource management prob-

Figure 8-2—Protected Areas in Southern Florida

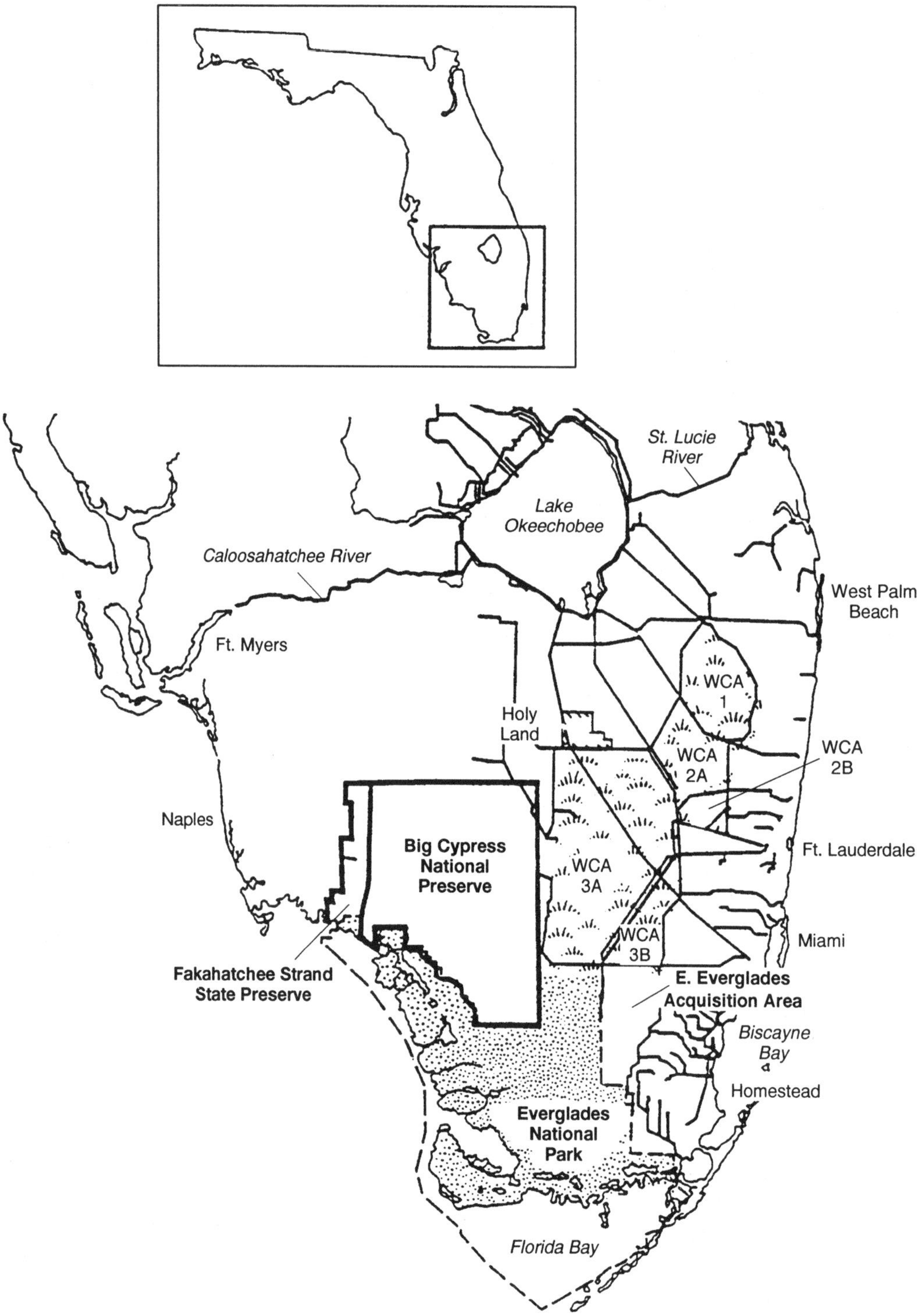

SOURCE: Adapted by OTA from M. Bodle, South Florida Water Management District, West Palm Beach, FL.

Table 8-5—Estimated Numbers of Non-Indigenous Species in Florida

Group	Established NIS	Total species
Plants	≈925	3,450
Insects	271	
Freshwater snails .	6	98
Land snails	40	140
Freshwater fish . . .	19[a]	80
Amphibians	3	55
Reptiles	22	100
Birds	11[b]	607[c]
Land mammals . . .	17	70

[a] Described as "established" and including one transplant; 4 other species are "possibly established," 9 are "formerly reproducing," and 41 are "collected without evidence of reproduction."

[b] Although only 11 are considered established, at least 140 have been classified as "free-flying exotics."

[c] Many birds found in Florida are migratory and do not breed there.

SOURCES: Compiled by the Office of Technology Assessment from: R. Ashton and P. Ashton, *Handbook of Reptiles and Amphibians of Florida*, Parts 1,2,3 (Miami, FL: Windward Publishers Inc., 1981, 1985, 1988); J.H. Frank and E.D. McCoy, "The Immigration of Insects to Florida, With A Tabulation of Records Published Since 1970," *Florida Entomologist*, vol. 75, No. 1, 1992, pp. 1-28; J.N. Layne, Checklist of Recent Florida Mammals, MS, 1987, 10 pp.; W.B. Robertson, Jr. and G.E. Woolfenden, *Florida Bird Species: An Annotated List*, Special Publication No. 6 of the Florida Ornithological Society, Gainesville, Florida, 1992, 260 pp.; P.L. Shafland, "Management of Introduced Freshwater Fishes in Florida," Proceedings of the 1990 Invitational Symposium/Workshop: New Directions in Research, Management and Conservation of Hawaiian Stream Ecosystems, Hawaii Dept. of Natural Resources, Div. of Aquatic Resources, Honolulu, HI, 1991; L.A. Stange, "Snails and Slugs of Florida," *Florida Garden Guide*, January/February 1990, pp. 1-2; D.R. Thompson, APHIS/USDA, personal communication, May 27, 1992; D.B. Ward, "How Many Plant Species Are Native to Florida?" *Palmetto*, winter 89/90, 1989-90; and L D. Wilson, Professor of Biology, Miami Dade Community College, Miami, FL, personal communication to D.W. Johnston.

lems in the State. They have had negative impacts on fishing and water sports and have degraded wildlife habitat, decreased biological diversity, and altered natural ecosystems. Future harmful effects on agriculture and human health can be anticipated from continued immigrations of insects and plant pathogens (39), as well as continued range expansion of established NIS (81).

Disturbed areas—construction sites, abandoned farm land, drained or stressed wetlands, roadsides, and canals and ditches—are often the sites where NIS gain footholds and eventually become established. In such areas NIS often displace indigenous forms, thus altering ecosystem dynamics. Debate persists as to whether NIS become established by actively out-competing and displacing indigenous species even in undisturbed areas or whether they primarily colonize disturbed habitats that are no longer optimum sites for indigenous species. In many south Florida urban and suburban sites, a lizard, the invasive Cuban brown anole (*Anolis sagrei*) has out-competed, and thereby replaced, the indigenous green anole (*Anolis carolinensis*) (136). Undisturbed areas are difficult for many NIS to colonize, but most of Florida's natural areas and waterways have experienced disturbance in some varying degrees, thus making them prone to NIS invasions (35,81).

Other conditions in Florida favor the introduction and establishment of NIS. The State has a subtropical climate and prolonged growing season; abundant freshwater resources; large and growing industries of aquaculture, ornamental and nursery plants, and the pet trade; a thriving tourist industry; and cargo flights originating in Central and South America (102).

* **Subtropical Climate.** Florida's subtropical climate is attractive to people and to certain industries, such as those dealing with ornamental and aquarium plants. The climate is moderated by large bodies of water on three sides. Furthermore, Florida is as close to the equator as is any conterminous State, so that most of it is in the humid subtropical climatic zone; the southern tip, from approximately Lake Okeechobee southward, is tropical savanna, the only such zone in the United States (22). Areas in this last zone are always hot, with alternate dry and wet seasons.

The State has an average annual maximum temperature of 82 degrees F and an average annual minimum temperature of 63 degrees F (137). Winter temperatures (40 degrees F and lower), especially in south-central Florida, probably limit the northward dispersal of many NIS (100,103,136). Florida is one of the wettest States, with an average annual rainfall of 53 inches (60 or more inches in southeastern and panhandle parts). This climate is conducive to the establishment of many NIS of tropical origin. Florida is also subject to tropical weather systems, such as 1992's Hurricane Andrew, which can facilitate the spread of NIS through disturbance (box 8-D).

- **Routes of Entry.** Florida has numerous pathways of entry for NIS. Large numbers of plants (333 million in 1990) and animals pass through Miami International Airport each year, the shipments originating chiefly in Latin America; 85 percent of all plant shipments into the United States pass through the Miami Inspection Station (118). The shipments are destined for a great variety of ornamental, nursery, and landscaping businesses; the aquarium industry; and commercial pet trade. This influx of NIS sets the stage for potential escapes and unintentional and intentional releases.

  Unintentional releases and escapes from animal dealers, aquaculture, subsequent purchasers, public and private collections, and tourist attractions have been documented (92,95). Specific examples of harmful or potentially harmful species are the African giant snail (*Achatina fulica*) (111), cane toad (*Bufo marinus*) (136), and monk parakeet (*Myiopsitta monachus*) (95).

  Deliberate introductions for sport, biological control, food, pharmaceutical material or dyestuffs, ornamental uses, and aesthetics are also well known in Florida (98). In the 1800s and early 1900s, botanist David Fairchild imported large volumes of non-indigenous plants into Florida (96). Since 1900, the most disastrous deliberate introduction has been that of melaleuca (*Melaleuca quinquenervia*), a fast-growing tree brought in to dry out the swamplands of south Florida. Another tree, Brazilian pepper (*Schinus terebinthifolius*), introduced for its showy foliage, is also spreading rapidly in south Florida. At least two introduced aquatic plants continue to cause extensive ecological and economic damage: hydrilla (*Hydrilla verticil-*

lata) and the showy water hyacinth (*Eichhornia crassipes*) (97). Plant pathogens and other stowaways have concomitantly gained entry through importation of foodstuffs and plants on ships or aircraft (28).

In the 19th century and as late as 1941, several insects, such as mole crickets (*Scapteriscus vicinus* and *S. acletus*) and a variety of beetles, probably arrived in ship ballast (96). For most non-indigenous plants and some animals, however, the exact path of entry into the State is unknown.

- **Industries Dealing With NIS.** Several industries have played large direct or indirect roles in the introduction of harmful NIS into Florida. A $1 billion woody ornamental industry continues to import large numbers of plants for landscaping and shade. A few woody ornamentals, such as Australian pine (*Casuarina equisetifolia*) and Brazilian pepper, have become major pest plants in Florida (79). Florida's aquaculture industry is the largest of any state; tropical fish and aquarium plants shipped from Florida are valued at $170 million annually, according to the Florida Tropical Fish Farms Association. Most of Florida's 19 non-indigenous fish species escaped from aquarium fish culture facilities (25). The aquarium plant trade introduced hydrilla into canals near Tampa about 1950, and later into Miami canals and the Crystal River (58). Pet merchants and pet owners have been implicated in the escape of tropical birds, reptiles, and mammals (92,122).

- **Human Population Growth.** Florida continues to be one of the fastest growing States: its 1990 population totaled 12.9 million, an increase of 32.8 percent since 1980 (127). Population growth over the years has increased pressure to develop more land and to make adequate water supplies available. Most of the natural ecosystems of south Florida have been severely altered. The disturbed areas—urban, suburban, and rural—have become prime sites for colonization by non-indigenous plants and animals.

Box 8-D—Non-Indigenous Species and the Effects of Hurricane Andrew

On the morning of August 24, 1992, the small but intense Hurricane Andrew cut a 25-mile swath across south Florida from the Dade County coast westward to Monroe County's west coast. Although total rainfall was relatively light (5 inches or less), maximum sustained winds were 135 to 140 miles per hour and gusts exceeded 164 miles per hour. Estimates of property damage to urban and suburban sites reached $20 billion, thus ranking Hurricane Andrew as among the costliest natural disasters in U.S. history. Natural areas were also affected. The hurricane caused an estimated $51 million in damage at Everglades and Biscayne National Parks and Big Cypress National Preserve.

A large number of non-indigenous animals escaped from captivity when zoos, pet stores, and tropical fish farms were destroyed. Escapees included fish, lizards, nonvenomous snakes, birds, and primates (e.g., some 500 macaque monkeys and 20 baboons).

Based on knowledge of the ecology of non-indigenous trees in south Florida and their invasions enhanced by two previous hurricanes (Donna in 1960 and Betsy in 1965), a significant increase in the spread of some non-indigenous plants can be predicted for the next few years. The hurricane spread melaleuca seeds (*Melaleuca quinquenervia*) and other non-indigenous plants in its path, thus setting back years of efforts to control melaleuca in the East Everglades. Newly disturbed natural communities in south Florida will be more susceptible to invasions. Other potential problems might come from escaped non-indigenous invertebrates and plants that are not already established in south Florida.

As a direct result of the hurricane, Florida's Department of Natural Resources estimates that mechanical and chemical control of non-indigenous plants over the next 5 years will cost $14 million, approximately tripling costs. Because those control measures might not completely eliminate harmful NIS, the Department recommends that biological control agents be introduced as quickly as possible. For species of primary concern in the aftermath of the hurricane—melaleuca, Australian pine (*Casaurina equisetifolia*), Brazilian pepper (*Schinus terebinthifolius*), lather leaf (*Colubrina asiatica*), and air potato (*Dioscorea bulbifera*)—funding for research, quarantine and grow-out facilities are estimated to be $53 million over the next 10 years.

SOURCES: A. DePalma, "Storm Offers Chance to Rethink Everglades," *The New York Times*, Sept. 29, 1992, p. A14; G.E. Davis et al. (eds.), "Assessment of Hurricane Andrew Impacts on Natural and Archaeological Resources of Big Cypress National Preserve," Biscayne National Park, and Everglades National Park, Draft Report, U.S. National Park Service, Atlanta, GA, Sept. 15-24, 1992; *Exotic Pest Plant Council Newsletter*, vol. 2, No. 3, fall 1992; Florida Game and Fresh Water Fish Commission, "Effects of Hurricane Andrew on Fish and Wildlife of South Florida: A Preliminary Assessment," Tallahassee, FL, Sept. 25, 1992; D. Schmitz, personal communication to Office of Technology Assessment, Jan. 21, 1993.

■ Causes and Consequences

Findings:

Natural habitats, especially in south Florida, have been altered or lost by drainage and water storage projects, urban and suburban land development, and land reclamation for agriculture. Harmful NIS often invade and become established in altered ecosystems from which they can invade surrounding areas.

Invasive NIS in the State have disrupted navigation and recreational activities, displaced indigenous wildlife and their habitats, and reduced biological diversity. Severe ecological and economic impacts from several aquatic plants, such as hydrilla and water hyacinth, and trees, such as melaleuca and Brazilian pepper, have been documented.

The most conspicuous non-indigenous plants in Florida are aquatic weeds (e.g., water hyacinth and hydrilla) and trees (melaleuca, Australian pine, and Brazilian pepper). Their success is due to their ecological characteristics as well as the condition of the ecosystem being invaded. In disturbed ecosystems, NIS are sometimes better adapted than indigenous species. Aquatic plants have clogged waterways, hindered navigation, disrupted fishing and water sports, and smothered natural vegetation. In drier habitats, invasive trees have often created monocultures, displacing indigenous species, decreasing biological diversity, and destroying wildlife habitats. Insects, pathogens, and nematodes have caused damage to agricultural crops. Several invading plants and insects have created public health problems.

Invasion and establishment of many non-indigenous plants and animals is closely related to the degree of ecosystem disruption. Alterations to accommodate water management projects, human population growth, and agriculture have been especially important (81,98).

WATER MANAGEMENT IN SOUTH FLORIDA

Water management programs in the southeastern part of the State have greatly contributed to the spread of non-indigenous plants and fishes (83). Waterways and marshes were among the first natural systems in Florida to be affected by increasing numbers of people because of demands for irrigation, urban water supplies, and recreation.

As early as 1907, drainage of south Florida's Everglades was promoted for land reclamation, to reduce flooding, and to supply water to developing southeastern coastal cities (42). Drainage was accelerated in the 1930s, and by 1947, the U.S. Army Corps of Engineers had created the Everglades Agricultural Area and a plan for management of Everglades' waters, thus laying the base for the vast urban areas now found on Florida's southeast coast. Areas along the eastern margin of the Everglades, critical to movement of its waters underground, are now drained and paved.

Today, a complex network of canals, dams, pumping stations, and levees stretches from Lake Okeechobee to southern Dade County, just east of Everglades National Park (119). This network— 80 percent of it federally funded and built by the Corps of Engineers—now controls flooding and diverts large volumes of water for agriculture and coastal urban areas. Half the Everglades—once occupying about 3,600 square miles, perhaps the largest wetland in North America—is now farms, groves, pastures, and cities. The remaining frag-

DON SCHMITZ

Altered hydrology in south Florida has been linked to the spread of non-indigenous fish, aquatic plants, and trees—such as melaleuca (Melaleuca quinquenervia).

ments of natural communities now function so poorly that plant and animal life suffers as water and food supplies are diminished, distorted, and polluted (132).

Altered hydrology in the East Everglades has been linked to the spread of non-indigenous trees such as melaleuca (104). This alteration of the natural water flow has decreased populations of nesting wading birds (92) and accelerated the proliferation and spread of non-indigenous fishes and aquatic plants (24,59,60,102).

Some 700,000 acres of agricultural land just south of Lake Okeechobee—nearly two-thirds of it in sugar cane—not only use much of south Florida's water, but also release run-off contaminated with nitrogen and phosphorus (105). Excessive growth of hydrilla and other plants has been linked to this increased pollution (15).

URBANIZATION

Florida's population in 1990 was concentrated in three principal areas: Miami-Fort Lauderdale (3.19 million), Tampa-St. Petersburg (2.1 million), and Orlando (1.1 million) (127). Natural areas, such as the Atlantic Coastal Ridge and scrub communities, have been developed to supply urban demands for house sites, municipal services, and landscaping. Many urban sites in south Florida have become dominated by NIS, especially ornamental plants, birds, and fishes (23,59,122,136).

Many non-indigenous animal species are today found chiefly or entirely in urban and suburban areas of south Florida. Collectors, hobbyists, and pet owners have deliberately or accidentally released tropical fish, mammals, birds, reptiles, and invertebrates into urban and suburban settings where they find plentiful food, breeding sites, shelter, and a subtropical climate conducive to growth and reproduction (25,31,72,95,136). In cities, non-indigenous birds such as parrots have few predators, diseases, or parasites (122). At ports of entry, such as Miami, stowaway insects and other invertebrates have escaped from their imported hosts (28). The Asian tiger mosquito (*Aedes albopictus*) commonly breeds in water that collects in waste tire dumps and flower pots in cemeteries (89).

THE SPREAD OF MELALEUCA

The last three decades have been marked by an explosive invasion of melaleuca across south Florida (53), where some 450,000 acres are infested (73). In 1983, its estimated rate of spread was 8 acres per day, but less than a decade later the rate is estimated to be 50 acres per day. Thus, melaleuca has the potential to invade all of south Florida's wetlands within the next 50 years (37).

Indigenous to Australia, melaleuca's release from natural competitors, predators, and disease and its characteristics of prolific seed production and adaptation to fire have facilitated its spread. Its monocultures have replaced sawgrass marshes, sloughs, forests, and other natural habitats to the extent that melaleuca is now regarded as the most serious threat to the integrity of all south Florida's natural systems (74).

Because of its proximity to the numerous melaleuca plantings in the urban areas of the Palm Beaches, Loxahatchee National Wildlife Refuge has one of the most severe infestations of melaleuca anywhere in the Everglades. The trees

were rare in the 1960s, but by 1990, 14 percent of the refuge was moderately to heavily infested (36). Moderate to heavy infestations also occur in Big Cypress National Preserve, the eastern half of the East Everglades Acquisition area, in marshes of Okeechobee, in large areas of Broward and Dade counties east of the Everglades, and in an area designated Water Conservation Area 2-B. Equally severe problems exist on the west coast of Florida from Charlotte Harbor to U.S. Highway 41 (74).

ECONOMIC COSTS

The various control programs for melaleuca have been expensive. Since 1986, 2 million melaleuca and Australian pine stems have been treated in the East Everglades at a cost of $287,000 for helicopter services and herbicides (104). Melaleuca management costs in the Big Cypress National Preserve were $60,000 in 1989. Costs for mechanical removal of trees range from $500 to $2,000 per acre. Estimated melaleuca management costs in recent years for Water Conservation Areas 2-A, 2-B, 3 in south Florida, and Lake Okeechobee have been nearly $1 million annually (74).

One estimate in 1991 placed the cost of melaleuca removal in Florida at $1.3 million. For fiscal year 1992 the estimated expenditures for herbicide and mechanical control of melaleuca were $720,000 in the South Florida Water Management District, $150,000 in Loxahatchee National Wildlife Refuge, and $180,000 in Everglades National Park (117). Based on the current rate of expansion, in one water conservation area alone, complete eradication of melaleuca with herbicides and mechanical removal would cost $12.9 million over 5 years (117).

The benefits and costs for removal of melaleuca have been estimated (29). The total annual benefits, especially to tourism, of preventing a complete infestation of melaleuca would be $168.6 million, whereas the resulting losses in honey production and pollination services (the tree provides honey bees with nectar) would cost only $15 million. Thus, eradication of melaleuca would greatly benefit the State's economy, according to this analysis, although some of its assumptions may inflate the benefits (21).

Florida has experienced severe economic impacts from other NIS as well. The economic impact of hydrilla on tourism and recreational fishing can be staggering. For example, a study of Orange Lake in north central Florida indicated that the economic activity on the lake was almost $11 million annually, but in years when hydrilla covers the lake, these benefits are all but lost (63). During the 1980s, statewide costs for controlling hydrilla totaled approximately $50 million (98). Today hydrilla is the most costly aquatic plant to manage, with an annual expenditure of $7 million. Since 1980, management of all non-indigenous aquatic plants by State and Federal agencies has cost $120 million (98).

Consequences to the State's agriculture also have been documented. The value of citrus crops in Florida from 1955 to 1985 totaled $13.5 billion. An estimated 15 percent of the citrus was lost because of the burrowing and citrus nematodes (*Radopholus similis*, *Tylenchulus semipenetrans*), with an average annual estimated cost of $77 million (33). While the nematodes' origins are not certain, experts speculate that one or both are non-indigenous. Fire ants (*Solenopsis invicta*) from South America have extensively damaged eggplants, soybeans, and potatoes. Brazilian pepper growing in proximity to agricultural areas is believed to support large populations of vegetable-damaging insects, especially when vegetable crops are nearing harvest (19). In 1984, the cost of damage and control of mole crickets in Florida, Georgia, Louisiana, and Alabama was about $45 million, with most of the cost to Florida. By 1986, the losses had risen to $77 million for turf grasses alone (38).

From 1957 to 1991, NIS eradication and control programs cost $31 million for citrus canker (*Xanthomas camestris* pv. *citri*), $11 million for fire ants, and $10 million for citrus blackfly (*Aleurocanthus woglumi*). In 1990 and 1991, Mediterranean fruit fly (medfly) eradication programs totaled $0.5 million, according to the Florida Department of Agriculture and Consumer Services.

POTENTIAL OR ACTUAL HEALTH CONSEQUENCES

Many NIS have been linked to human health problems, and an increasing number of incidents are reported annually in the growing urban areas. Very common trees, such as melaleuca and Brazilian pepper, can cause contact dermatitis, allergies, and respiratory problems. A large number of other cultivated and established plants in Florida contain some poisonous compounds (3).

The Asian tiger mosquito, now in virtually all Florida counties, can carry dengue fever and a form of equine encephalitis virus (39) (ch. 10). In addition to their agricultural impacts, non-indigenous fire ants can cause stings, allergic reactions, and secondary infections in people.

EFFECTS ON ENDANGERED SPECIES

Non-indigenous aquatic plants are threatening the integrity of habitats occupied by certain endangered and threatened species in Florida. Both water hyacinth and water lettuce (*Pistia stratiotes*) can cover surface waters, thus hampering efforts of the endangered snail kite (*Rostrhamus sociabilis*) to find its prey (116). Non-indigenous trees are invading habitats of the endangered Cape Sable seaside sparrow (*Ammodramus maritimus mirabilis*). Australian pines have interfered with nesting of endangered and threatened sea turtles (84); on the other hand, they have improved nesting conditions for the American oyster catcher (*Haematopus palliatus*) (121). The endangered beach mouse (*Peromyscus polionotus phasma*) and key deer (*Odocoileus virginianus clavium*) are subject to predation by feral cats or dogs (4). Populations of the endangered Okaloosa darter (*Etheostoma okaloosae*) have been reduced because of competition from the introduced brown darter (*E. edwini*) (14).

CONFLICTING INTERESTS ON NON-INDIGENOUS SPECIES

The introduction of certain NIS into Florida has resulted in conflicts between agencies and user groups. Grass carp (*Ctenopharyngodon idella*) were introduced to control aquatic weeds (115), but the carp shows a preference for important waterfowl food plants, thus apparently causing declines in waterfowl populations (134). Peacock bass (*Cichla* spp.) were introduced to control other non-indigenous fish and as a game fish in southeast Florida canals (101), but the bass is slowly reducing populations of indigenous bass and bream (73). Perhaps the most troublesome of the 19 non-indigenous fish species is the blue tilapia (*Tilapia aurea*), introduced by the Florida Game and Fresh Water Fish Commission as a possible weed-control and sport fish. Blue tilapia competes directly with indigenous fishes and is now established in 18 Florida counties (73).

Hunters value wild hogs (*Sus scrofa*) as game, and management and relocation programs are common in Florida. Yet wild hogs have detrimental effects on terrestrial habitats and are probable public health threats (parasites and diseases) (9).

Certain aquatic plants frequently categorized as pest species may be beneficial for wildlife. Despite extensive, costly efforts to control or eradicate hydrilla, some hunters like the plant because it is an important duck food and its mats provide habitats for wintering waterfowl (44,57). At least in small amounts, it is also believed to improve sport fishing (76).

Aside from those species introduced for biological control or sport, some NIS in Florida benefit people and wildlife. The aesthetic values of colorful tropical birds are intangible, but are important to urban dwellers in an otherwise less colorful environment (92). Avid birdwatchers travel to the Miami area to observe its non-indigenous avifauna (122). The importance of NIS as food for indigenous wildlife is only partly understood, but the endangered Florida panther (*Felis concolor coryi*) feeds on non-indigenous

*Blue tilapia (*Tilapia aurea*) is among the most troublesome of Florida's 19 non-indigenous fish species.*

wild hogs and nine-banded armadillos (*Dasypus novemcinctus*), whose negative environmental impacts have been documented (18,72).

Non-indigenous ornamental shrubs and trees are in great demand for landscaping (because of their showy leaves or flowers), fruit, and shade from the intense sunlight of south Florida (79). Many species of introduced fig trees (*Ficus* spp.) line southeastern Florida's roadsides, and Australian pines offer shade along beach fronts.

POTENTIAL FUTURE IMPACTS OF NON-INDIGENOUS SPECIES

Biologists and ecologists caution that many poorly studied NIS have the potential of becoming agricultural pests, transmitting diseases, or displacing indigenous species. Potentially serious pests include Cogon grass (*Imperata cylindrica*), which is invading pine forests (81); about 20 recent insect immigrants (39); the Asiatic clam (*Corbicula manilensis*) (87); catclaw mimosa (*Mimosa pigra* var. *pigra*), a highly invasive plant of disturbed areas; the disease-carrying Asian tiger mosquito; and African honey bees (*Apis mellifera scutellata*), predicted to be in Florida by 1994.

■ Searching for Solutions

Findings:

Florida's Exotic Pest Plant Council has provided an effective forum for the exchange of ideas and conflict resolution concerning NIS. It has identified the most invasive NIS and involved policymakers in its discussions.

Florida's extensive problems with NIS and its high human immigration rate suggest that public education is vital to the management or eradication of NIS in the State.

SPECIFIC MANAGEMENT PROGRAMS

The Exotic Pest Plant Council (EPPC) was the first multiorganizational effort in Florida to control non-indigenous water weeds because of the growing environmental threats posed by pest plants that were crossing political and jurisdictional boundaries. EPPC is an organization of 40 member agencies, and local and private groups. Through frequent meetings, a newsletter, and other publications, EPPC promotes coordinated efforts in developing management programs. It also assists in writing appropriate legislation; pushes for State and Federal funds to manage invasive plants in wetlands and upland forests; and organizes symposia to bring together scientists, policymakers, and the public to exchange information and formulate plans (30).

EPPC assisted in coordinating efforts by the National Park Service, Dade County Department of Environmental Resource Management, South Florida Water Management District, and the Florida Department of Corrections to establish and maintain a melaleuca-free buffer zone along the eastern boundary of Everglades National Park (the East Everglades).

Because of melaleuca's highly invasive nature, its control and eradication have received top priority in the East Everglades, South Florida Water Management District, Loxahatchee National Wildlife Refuge, and other sites in south Florida. At least three techniques are currently in use: manual removal of seedlings and young trees, mechanical removal of older trees, and herbicides (62).

The future use of biological control agents has been identified as one of the keys to effective, long-lasting management of melaleuca (5). Major efforts are under way to identify natural controls for melaleuca, both in the United States and Australia. Even after biological control agents are identified, several years must pass before their effectiveness can be determined. Meanwhile, herbicidal and mechanical control will be needed to arrest further spread of the tree (74).

Control of Australian pine and Brazilian pepper demands a combination of mechanical removal and herbicides. Hydrilla is currently managed at considerable cost with herbicides and mechanical removal and in some cases with sterile triploid grass carp. At one time, water hyacinth infested more than 120,000 acres of Florida waterways. Herbicidal and mechanical controls have limited the plant to less than 3,000 acres in public waters (98). Three natural enemies, the bagoine weevil (*Bagous affinis*) and two leaf-mining flies (*Hydrellia* spp.), also show some promise in controlling hydrilla (62). Management of these and other species would benefit from increased coordination.

Several other control and eradication projects have been successful in Florida. In the mid-1980s at least 18 million young citrus trees were destroyed to eradicate citrus canker (99). Other species successfully eradicated include the medfly; the giant African snail; and 13 species of insects, viruses, and rusts, according to the Division of Plant Industry in Florida.

LONG-TERM NEEDS

Resource managers in Florida stress that successful management and eradication programs for existing and future problem NIS in Florida will require an educated public along with coordination among agencies, long-range planning, and consistent funding.

Inventories of existing harmful NIS, their distribution, and impacts in the State are needed

E. CHRISTENSEN, NATIONAL PARK SERVICE

*The critically endangered Florida panther (*Felis concolor coryi) *and other indigenous species rely on remnants of undisturbed habitat that are susceptible to damage by non-indigenous species.*

to develop priorities for management. Early detection of damages enhances the probability of success in controlling any pest (20). Because the establishment and spread of any NIS may be due to a lack of natural enemies, the search for biological control agents is an important consideration.

Relatively undisturbed ecosystems in Florida are fast disappearing and are usually represented by small fragments of their original extent. These areas warrant special attention to protect them from injurious NIS. The State needs to enhance strategies for controlling or eradicating injurious non-indigenous animals such as wild hogs (75).

Ample evidence indicates that the existing management of water flow through the Everglades has altered hydroperiods and contributed to the invasion of non-indigenous trees. A new design and management of water flow would be needed to restore a natural water regime, one that would protect the quality and quantity of water feeding the Everglades (34).

Some aspects of water quality management in the Everglades, especially those related to phosphorus, are being addressed now. In 1988, the U.S. Department of Justice sued the Florida

Department of Environmental Regulation and the South Florida Water Management District for not enforcing water quality standards for water entering Everglades National Park. In July 1993, these parties, along with agricultural interests, environmental groups, and Indian tribes, agreed to a mediated framework for a 20-year, $465 million restoration and clean up plan. The impact of these efforts on harmful NIS will not be clear for some time.

COORDINATED EFFORTS FOR MANAGING NIS

Centers or councils to coordinate the work of various agencies and industries could be of help in developing and implementing effective management of harmful NIS. They might also encourage statewide resource protection, public awareness, and consistency in policies, goals, administration, and control methods. The structure and operations of the Exotic Pest Plant Council could be used as a model for coordinating work on pestiferous fish and insects, for example. A planned ''Center for Excellence,'' combining expertise from the University of Florida, Division of Plant Industry, and the U.S. Department of Agriculture, also shows promise in coordinating biological control research and implementation in the State, especially for agricultural crops.

FUNDING FOR RESEARCH, MANAGEMENT, AND BIOLOGICAL CONTROL

Except for a few highly invasive aquatic plants and trees, little biological and ecological information is available for most of Florida's NIS. Equally lacking are data on natural enemies of the species and ecological data for the ecosystem likely to be invaded. Without the necessary research to reveal this information, effective programs of control, management, and eradication cannot be fully developed nor expected to be successful.

For the most part, funding for management and research of NIS in Florida has been piecemeal and often inadequate for programs to achieve maximum success. For example, management programs for noxious weeds and biological control research are said to have been underfunded and short-term. Current quarantine facilities for biological control research are inadequate, thus hampering efforts to control melaleuca and other species. Development and implementation of strategies to arrest further spread of NIS and to decrease their environmental impacts would require consistent, adequate funding.

PUBLIC EDUCATION

Florida's continuing population growth and tourist influx plus the magnitude of the impacts from harmful NIS suggest that public education and awareness programs could be intensified to prevent new introductions. Such programs could be targeted toward unintentional and intentional introductions, including ornamental plants, aquarium fishes, other pets, and insects. Attempts could be made to discourage the planting of invasive ornamental species and to warn of the need to control their spread. The major biological and economic impacts of melaleuca, water hyacinth, and hydrilla could be widely publicized to encourage support for management issues. The importance of protecting remaining natural communities warrants emphasis, especially since undisturbed ecosystems can serve as barriers against the spread of NIS.

CHAPTER REVIEW

Virtually all parts of the country face problems related to harmful NIS, but Hawaii and Florida have been particularly hard hit. Both States have large numbers of established NIS, constituting significant proportions of their flora and fauna, and including numerous high-impact species. Many harm natural areas that are unique or otherwise special reservoirs of the Nation's biological heritage. Both Hawaii and Florida have turned to cooperative, interagency mechanisms and public education to address their particular problems with NIS. Federal action and inaction have sometimes hindered the States' efforts. Lessons learned in these States are likely to serve well elsewhere. The situation in Hawaii and Florida, while unusual in some ways, nevertheless heralds what other States face as numbers of harmful NIS climb and people become more aware of their damage.

Genetically Engineered Organisms as a Special Case | 9

In requesting this assessment, Congress asked OTA to compare non-indigenous species (NIS) and genetically engineered organisms (GEOs; box 9-A)—specifically, whether and how pre-release evaluations can reduce the risks of unwanted introductions (41). The comparison makes sense because the central issues for NIS and GEOs are the same, namely, making decisions regarding intentional introductions, devising strategies to prevent unintentional introductions, and planning eradication and control programs should releases have unexpected harmful effects.

Moreover, according to OTA's definition of non-indigenous, all GEOs are non-indigenous. OTA has defined NIS to include species beyond their natural ranges, domesticated and feral species, and non-naturally occurring hybrids (see ch. 2, box 2-A). Most species used in genetic engineering research today are domesticated species and fall within this definition. When domesticated species long cultivated in the United States are genetically engineered and then released, they become new varieties of these NIS. Just as the products of domestication are non-indigenous, regardless of origin, so too are the products of genetic engineering. Indigenous species that have been altered via genetic engineering and introduced into the environment become non-naturally occurring, and therefore non-indigenous, varieties.

The overlap between GEOs and NIS goes beyond such functional and definitional issues, however. Federal agencies apply many of the same laws to NIS and GEOs, and some of the same legislative gaps and ambiguities hold for both categories. Overlap also occurs in the risk assessment procedures used for

Box 9-A—What Do You Call an Organism With New Genes?

Terms Used by OTA

OTA uses the adjectives **genetically engineered** and **transgenic** to describe plants, animals, and microorganisms modified by the insertion of genes using genetic engineering techniques. GEO is used in this chapter as an abbreviation for "genetically engineered organism."

Genetic engineering refers to recently developed techniques through which genes can be isolated in a laboratory, manipulated, and then inserted stably into another organism. Gene insertion can be accomplished mechanically, chemically, or by using biological vectors such as bacteria or viruses. The bacterium *Agrobacterium tumefaciens* is commonly used to carry genes into plant cells.

A GEO potentially contains genetic material from three types of organisms. Genes from one or more **donor** organisms are isolated for insertion into a **recipient** organism. A biological **vector** may be used to insert the genes. Genetic material in the resulting GEO thus includes all of the recipient's genes, the isolated donor genes, and sometimes genetic material from the vector as well.

Many of the organisms being genetically engineered today are **domesticated** species. Domestication occurs when organisms are selectively bred by humans for desired characteristics. The term "domesticated" often is used in discussions of genetic engineering to indicate how likely an organism is to establish a free-living population. However, this usage can be misleading since domesticated organisms vary greatly in this regard. Some, like corn (*Zea mays*), are incapable of living beyond human cultivation, whereas others, such as goats (*Capra hircus*), readily form free-living populations.

Related Terms

Genetically modified organisms have been deliberately modified by the introduction or manipulation of genetic material in their genomes. They include not only organisms modified by genetic engineering, but also those modified by other techniques such as traditional breeding, chemical mutagenesis, and manipulation of sets of chromosomes.

Biotechnology refers to the techniques, including both genetic engineering and traditional methods, used to make products and extract services from living organisms and their components.

SOURCES: Office of Science and Technology Policy, "Principles for Federal Oversight of Biotechnology: Planned Introduction into the Environment of Organisms with Modified Hereditary Traits," 55 *Federal Register* 31118 (July 31, 1990); U.S. Congress, Office of Technology Assessment, *A New Technological Era for American Agriculture*, OTA-F-474 (Washington, DC: U.S. Government Printing Office, August 1992).

---

the GEOs and NIS, although in the recent past methods have developed more rapidly for GEOs. This chapter takes a closer look at these two areas—regulation and risk assessment—related to Federal review of GEO releases. The analysis draws heavily on the previous assessment of Federal coverage for NIS (ch. 6) and of risks associated with introductions (chs. 2, 3, and 4). The chapter begins, however, with a brief discussion of why comparisons between GEOs and NIS are sometimes controversial.

SOURCES OF CONTROVERSY

Despite the overlap between GEOs and NIS, comparisons between the two can arouse strong objections, especially among those in the executive branch charged with reviewing environmental releases of GEOs (20). Such reactions have origins in the technical and policy issues discussed below. They are complicated by the historical context—the rapid development over the past decade of Federal policies on GEOs (table 9-1) and the continuing dialogue among scientists, policymakers, and the public regarding the potential benefits and risks of GEO releases.

Table 9-1—Federal Policies and Regulations Related to the Environmental Release of GEOs Since 1984

Office of Science and Technology Policy

1992	Exercise of Federal Oversight Within Scope of Statutory Authority: Planned Introductions of Biotechnology Products into the Environment, 57 *Federal Register* (FR) 6753 *(Policy Statement)*
1990	Principles for Federal Oversight of Biotechnology: Planned Introduction into the Environment of Organisms with Modified Hereditary Traits, 55 FR 31118 *(Proposed Policy)*
1986	Coordinated Framework for Regulation of Biotechnology, 51 FR 23302 *(Policy Statement and Request for Public Comment)*
1985	Coordinated Framework for the Regulation of Biotechnology; Establishment of the Biotechnology Science Coordinating Committee, 50 FR 47174
1984	Proposal for a Coordinated Framework for Regulation of Biotechnology, 49 FR 50856 *(Proposed Policy)*

The President's Council on Competitiveness

1991	Report on National Biotechnology Policy *(Policy Statement and Recommendations for Implementation)*

U.S. Department of Agriculture, Animal and Plant Health Inspection Service

1993	Geneticaally Engineered Organisms and Products; Notification Procedures for the Introduction of Certain Regulated Articles; and Petition for Nonregulated Status, 58 FR 17044 *(Final Rule)*
1992	Genetically Engineered Organisms and Products; Notification Procedures for the Introduction of Certain Regulated Articles; and Petition for Nonregulated Status, 57 FR 53036 *(Proposed Rule)*
1987	Introduction of Organisms and Products Altered or Produced Through Genetic Engineering Which Are Plant Pests or Which There Is Reason to Believe Are Plant Pests, 7 CFR 340 *(Final Rule)*
1986	Final Policy Statement for Research and Regulation of Biotechnology Processes and Products. 51 FR 23336 *(Final Policy Statement)*
1986	Plant Pests: Introduction of Organisms and Products Altered or Produced Through Genetic Engineering Which are Plant Pests or Which There is Reason to Believe are Plant Pests, 51 FR 23352 *(Proposed Rule and Notice of Public Hearings)*

U.S. Department of Agriculture, Office of Agricultural Biotechnology

1990	Proposed USDA Guidelines for Research Involving the Planned Introduction into the Environment of Organisms with Deliberately Modified Hereditary Traits, 56 FR 4134 *(Proposed Voluntary Guidelines)*
1986	Advanced Notice of Proposed USDA Guidelines for Biotechnology Research, 51 FR 13367 *(Notice for Public Comment)*

U.S. Environmental Protection Agency

1993	Microbial Pesticides; Experimental Use Permits and Notifications, 58 FR 5878 *(Proposed Rule)*
1989	Biotechnology: Request for Comment on Regulatory Approach, 54 FR 7027 *(Notice)*
1989	Microbial Pesticides; Request for Comment on Regulatory Approach, 54 FR 7026 *(Notice)*
1986	Statement of Policy: Microbial Products Subject to the Federal Insecticide, Fungicide, and Rodenticide Act and the Toxic Substances Control Act (TSCA), 51 FR 23313 *(Policy Statement)*
—	EPA has not yet issued proposed or final rules for the regulation of genetically engineered microbes under TSCA.

SOURCE: Office of Technology Assessment, 1993.

∎ Technical Sources

Considerable controversy surrounded the first releases of GEOs because of concerns over their potential effects and how they should be evaluated before their release. In the absence of experience with GEOs, some scientists argued that experience with ''exotic'' (i.e., non-indigenous) species might help provide guidance (29,32). However, the comparison of GEOs to NIS itself provoked debate.

The approach was criticized because GEOs introduced to the same environment as the parent non-engineered organism differ by only a few genes. Effects of the gene changes in GEOs might be well characterized, allowing better prediction of how they affect the organism's ecology. In contrast, most NIS differ from indigenous organisms by many genes that generally are not well characterized. Further, some comparisons of GEOs to harmful NIS, such as kudzu (*Pueraria lobata*) and the sea lamprey (*Petromyzon marinus*), were alarmist, inappropriately suggesting that all GEOs are potentially like the worst NIS.

These limitations, however, do not address the basic similarity between the *process* of introducing a NIS and the *process* of introducing a GEO. Both involve the release of a living organism potentially capable of reproduction, establishment, and ecological effects beyond the initial release site (36). The specific characteristics of the organism and the receiving environment will determine the consequences of either type of introduction (18,36,37). In this regard, experience with NIS has proven quite useful in defining the types of ecological questions that should be raised before releasing a GEO into the environment (box 9-B) (23,37).

■ Policy Sources

A recurring theme in policy discussions of GEOs has been whether effective regulation can be accomplished under existing Federal statutes or whether new legislation is needed (25,41,42). For the interim, at least, this issue has been tabled by the development of the "Coordinated Framework for the Regulation of Biotechnology" by the White House Office of Science and Technology Policy (OSTP).

OSTP has announced policies related to Federal regulation of biotechnology several times since 1984 (table 9-1). General goals of these policy statements include:

- coordinating and streamlining Federal regulation, in part by clarifying the roles of various agencies;
- giving guidance to Federal agencies in their regulatory approach and scope; and
- ensuring such regulation adequately balances protection of human health and the environment along with the national interest in fostering growth of the biotechnology industry.

An important early conclusion was that existing legislation was generally sufficient to cover planned releases of GEOs to the environment (25). The President's Council on Competitive-

ness strongly reiterated this position in 1991: "The Administration should oppose any efforts to create new or modify existing regulatory structures for biotechnology through legislation" (28). This policy reflected, in part, a desire to support commercial development of biotechnology by reducing the regulatory burden on the industry (28).

Although both proponents and critics of genetic engineering agree that Federal agencies exercise sufficient oversight of most current GEO releases, the adequacy of the Coordinated Framework may be challenged in the future. Certain GEO releases may not be adequately covered by Federal statutes. In some cases, the application of existing statutes to genetic engineering requires application of laws beyond their initial intent. The result has been confusing regulations based on convoluted interpretations of legal definitions.

It is important to note that the Coordinated Framework is an executive branch policy and has no explicit basis in Federal law. This imparts a sometimes counter-productive flexibility. For example, repeated changes since 1984 in how OSTP defined which GEOs should be regulated helped stymie efforts by the Environmental Protection Agency to issue regulations under the Toxic Substances Control Act (39).

The Federal agencies that review environmental releases of GEOs have been faced with the practical reality of regulating an activity where political pressures are strong to allow releases, technical information for decisionmaking is sometimes insufficient, and legislative authority imperfectly matches the problems at hand. The procedures currently in place reflect compromises hard won over the past decade. And for the present, at least, the system generally works. In this light, the reluctance of regulators to revisit debates of the past concerning the risks of GEO releases is understandable. It may, however, leave them unprepared for the future when technical advances, the application of genetic engineering to a wider array of organisms, and the move to

Box 9-B—The Risks of Genetically Engineered Organisms: Lessons from Non-Indigenous Species

Can the Species Become Established Outside of Human Cultivation?

The risks associated with a NIS depend in part on whether it can become free-living. Species requiring human cultivation (e.g., many agricultural crops) are unlikely to become pests or harm natural ecosystems. GEOs formed by the insertion of genes into cultivated species similarly pose little risk, unless the inserted genes affect the organism's reliance on human cultivation or cause it to unintentionally harm other organisms.

Greater risks are associated with introductions of NIS that do not require human cultivation. Some can establish free-living populations and cause environmental or economic harm. Certain significant pests of agriculture and natural areas are escaped crop and horticultural plants (e.g., crabgrass, *Digitaria* spp., and Japanese honeysuckle, *Lonicera japonica*) or livestock (e.g., feral goats (*Capra hircus*)). NIS directly introduced to less managed systems, such as rangelands and forests, can affect other species in these systems. Melaleuca (*Melaleuca quinquenervia*), a major cause of habitat degradation in the Florida Everglades wetland system, was initially introduced for water management. Grass carp (*Ctenopharyngodon idella*), widely introduced for aquatic weed control, also increase water turbidity and destroy habitats of young fish. Thus, GEOs resulting from insertion of genes into potentially free-living species similarly are of greater concern because they might affect natural areas.

Can Genes Spread Through Hybridization?

A potential risk factor common to NIS and GEOs is that of gene spread to other species through hybridization (interbreeding). Genes can move from some cultivated crops that otherwise pose low risk. Notable examples include hybridization between rapeseed (*Brassica napus*) and wild mustards (*B. kaber, B. juncea, B. nigra*); cultivated and free-living squash (*Cucurbita pepo*); and between domesticated tomatoes (*Lycopersicon esculentum*) and wild tomato (*Lycopersicon pimpinnellifolium*) in South America. Hybridization between crop and weed species has sometimes given rise to new weeds like the Bolivian weed potato (*Solanum sucrense*). Moreover, the potential for hybridization between cultivated and wild and weedy relatives varies greatly among species. For example, although there is no evidence that genes move from carrots (*Daucus carota sativa*) to wild relatives like Queen Anne's lace (*Daucus carota*) in North America, gene exchange between alfalfa (*Medicago sativa*) planted for forage and wild relatives appears to be widespread.

The opportunity for hybridization also varies geographically. Most major agricultural crops lack free-living relatives (and therefore the opportunity for hybridization) in the United States because they originated in other areas of the world. Some exceptions are sorghum (*Sorghum bicolor*), sunflower (*Helianthus* spp.), clover (*Trifolium* spp.), and tobacco (*Nicotiana tabacum*). Wild cotton (*Gossypium tomentosum*), which potentially might hybridize with genetically engineered cotton, exists in Hawaii, but not elsewhere in the United States.

The potential for gene spread from GEOs to other species is thus an important consideration in risk assessments. All else being equal, GEOs lacking free-living relatives in the area of release pose fewer risks. The consequences of gene movement from GEOs to other species depend on what traits they confer. Some, like genes affecting fruit color, pose little risk. Greater concerns center on genes that might transfer harmful traits to free-living species. For example, much current research involves insertion of genes for herbicide resistance into crop plants to allow control of weeds without harm to the crop. Should this trait be transferred to weedy relatives, the usefulness of a particular herbicide for weed control could be lost.

SOURCES: N.C. Ellstrand and C.A. Hoffman, "Hybridization as an Avenue of Escape for Engineered Genes," *BioScience*, vol. 40, No. 6, pp. 438-442, June 1990; R.S. Grossman, "Biotechnology Products in the Field: Bringing Regulation Closer to Home," *American Journal of Public Health*, vol. 82, No. 8, August 1992, pp. 1165-1166; K.H. Keeler and C.E. Turner, "Management of Transgenic Plants in the Environment," *Risk Assessment in Genetic Engineering*, M.A. Levin and H.S. Strauss (eds.) (New York, NY: McGraw-Hill, Inc., 1990); E. Small, "Hybridization in the Domesticated-Weed-Wild Complex," *Plant Biosystematics*, W.F. Grant (ed.) (New York, NY: Academic Press, 1984), pp. 195-210; H.D. Wilson, "Gene Flow in Squash Species," *BioScience*, vol. 40, No. 6, June 1990, pp. 449-455.

Table 9-2—Who Regulates Which GEO Releases?

Agency	Regulated category	Authority[a]	Types of approved releases thus far	Number
APHIS	Plant pests	Federal Plant Pest Act Plant Quarantine Act Federal Noxious Weed Act	Transgenic plants	327 contained field tests at 660 sites in 37 States and Puerto Rico[b]
	Veterinary biologics	Virus-Serum Toxin Act	Live animal vaccines (microbes)	25 controlled releases; 7 licenses for commercial distribution[c]
EPA	Pesticides	Federal Insecticide, Fungicide and Rodenticide Act	Pesticidal microbes	42 small-scale field tests[d]
			Pesticidal plants	3 releases of over 10 acres[e]
	Other microbes	Toxic Substances Control Act	Microbes modified for improved detection or enhanced nitrogen fixation	19 small-scale field releases[f]

[a] For full citations of Federal laws see text.

[b] As of October 1992. The "Flavr Savr" tomato was recently exempted from regulation. Permitting requirements were relaxed in 1993 for 5 categories of GEOs, to allow notification of APHIS rather than requirement of a permit before release.

[c] Number permitted during fiscal years 1989 through 1992.

[d] As of July 1993, covering the period 1984 through 1993.

[e] Experimental Use Permits were issued for large-scale tests of *Bacillus thuringiensis* delta endotoxin produced in cotton, corn, and potato. As of July 1993.

[f] As of Feb. 3, 1993.

SOURCES: F. Betz, Environmental Fate and Effects Division, EPA, FAX to E.A. Chornesky, Office of Technology Assessment, Aug. 2, 1993; D.E. Giamporcaro, Section Chief, TSCA Biotechnology Program, letter to P.N. Windle, OTA, Apr. 29, 1993; J.H. Payne, Associate Director Biotechnology, Biologics, and Environmental Protection, APHIS, letter to P.N. Windle, Office of Technology Assessment, Nov. 10, 1992; B. Slutsky, "Pesticidal Transgenic Plants: Risk Issues," *Pesticidal Transgenic Plants: Product Development, Risk Assessment, and Data Needs* (U.S. EPA Conference Proceedings: Nov. 6 and 7, 1990), pp. 127-132.

commercialization of GEOs broaden the scope of regulatory issues.

FEDERAL REGULATION OF GEO RELEASES

Under the Coordinated Framework, two Federal agencies, the U.S. Department of Agriculture's Animal and Plant Health Inspection Service (APHIS) and the U.S. Environmental Protection Agency (EPA) oversee most environmental releases of GEOs (table 9-2).

APHIS regulates releases of GEOs for which the donor, recipient, or vector of new genetic material is a potential or actual plant pest (box 9-C). In the past, anyone wishing to move or release such organisms needed to apply for a permit certifying the action did not pose a significant risk to agriculture or the environment. APHIS then evaluated the ecological risks of release by conducting an in-house environmental assessment for each permit granted. APHIS recently relaxed these permitting requirements for transgenic potatoes (*Solanum tuberosum*), tomatoes (*Lycopersicon esculentum*), cotton (*Gossypium hirsutum*), soybean (*Glycine max*), tobacco (*Nicotiana tabacum*), and corn (*Zea mays*) that fulfilled certain eligibility criteria and released according to specified performance standards.[1] These cases now require only that APHIS be notified in advance of field trials. In practice, APHIS has overseen releases of a wide array of genetically engineered plants because the bacterial vector used to insert genes is itself a plant

[1] 58 *Federal Register* 17044 (March 31, 1993)

Box 9-C—Which Categories of GEOs APHIS Regulates as "Plant Pests"[1]

Definition of a Regulated Article

"Any organism that has been altered or produced through genetic engineering, if the donor organism, recipient organism, or vector or vector agent belongs to any genera or taxa designated in 340.2 of this part and meets the definition of a plant pest, or is an unclassified organism and/or an organism whose classification is unknown, or any product which contains such an organism, or any other organism or product altered or produced through genetic engineering which the Deputy Administrator determines is a plant pest or has reason to believe is a plant pest. Excluded are recipient microorganisms which are not plant pests and which have resulted from the addition of genetic material from a donor organism where the material is well characterized and contains only non-coding regulatory regions."

Taxa Listed in 340.2

Viruses (all members of groups containing plant viruses, and all other plant and insect viruses); **Bacteria** (13 genera; gram-negative phloem-limited bacteria associated with plant diseases; gram-negative xylem-limited bacteria associated with plant diseases; all other bacteria associated with plant or insect diseases);

Other disease-causing organisms (all rickettsial-like organisms associated with insect-diseases; members of the genus *Spiroplasma*; mycoplasma-like organisms associated with plant diseases; mycoplasma-like organisms associated with insect diseases);

Algae (three genera of green algae);

Fungi (3 classes; 16 orders; 33 families; and all other fungi associated with plant or insect diseases);

Plants (parasitic species in 13 families and 27 genera);

Animals (nematodes—20 families; snails—6 superfamilies and 1 subfamily; spiders, mites, and ticks—13 superfamilies; millipedes—1 order; insects—4 orders, 8 superfamilies, 53 families, 5 subfamilies, 3 genera)

Definition of a Plant Pest

"Any living stage (including active and dormant forms) of insects, mites, nematodes, slugs, snails, protozoa, or other invertebrate animals, bacteria, fungi, other parasitic plants or reproductive parts thereof; viruses; or any organisms similar to or allied with any of the foregoing; or any infectious agents or substances, which can directly or indirectly injure or cause disease or damage in or to any plants or parts thereof, or any processed, manufactured, or other products of plants."

---

[1] APHIS has exempted from permitting requirements interstate movement of certain GEOs containing less than the complete genome of a plant pest and field releases of a set of tomatoes having altered softening properties. The agency recently relaxed the permitting requirements for several other categories of GEOs.

SOURCES: 7 CFR 340 (June 16, 1987) as amended, "Introduction of Organisms and Products Altered or Produced Through Genetic Engineering Which Are Plant Pests or Which There is Reason to Believe are Plant Pests;" J.H. Payne, Associate Director, Biotechnology, Biologics, and Environmental Protection, APHIS, letter to P.N. Windle, Office of Technology Assessment, Nov. 10, 1992; U.S. Department of Agriculture, Animal and Plant Health Inspection Service, "Genetically Engineered Organisms and Products; Notification Procedures for the Introduction of Certain Regulated Articles; and Petition for Nonregulated Status," proposed rule, 57 *Federal Register* 53036 (Nov. 6, 1992).

pathogen or because plant pathogen genes have been inserted to promote expression of other inserted genes.

Uncontained uses of live animal vaccines (veterinary biologics) are also regulated by APHIS. A permit is required for experimental use of a vaccine, and vaccines must fulfill standards of product purity, efficacy, and safety (to the environment, human health, and animal health) before licensing and wider distribution. APHIS has not issued specific regulations for GEOs in this category, but has instead relied on existing regulations for live vaccines.

EPA regulates releases of genetically engineered microbes under the Federal Insecticide, Fungicide, and Rodenticide Act (FIFRA)[2] and the Toxic Substances Control Act (TSCA).[3] Final regulations have not yet been promulgated under either Act for small-scale releases; consequently, the agency is operating under interim policy. The GEOs regulated under FIFRA are pesticide-producing microbes. Users must notify EPA before small-scale field tests. Following notification, the agency may require submission of materials for an Experimental Use Permit before release. EPA also intends to regulate under FIFRA the commercial distribution and sale of transgenic plants engineered for pest and disease resistance (i.e., because of the pesticidal substances they produce) (34). This category eventually is likely to include agricultural crops, ornamental plants, aquatic plants, and species for forest and rangeland management (48).

Under TSCA, EPA regulates transgenic microbes not covered by any other statute, for example, nitrogen-fixing bacteria or microbes used for environmental remediation. This regulation rests on extension of TSCA's definition of "chemical substance" to live organisms—an interpretation that has been a source of continuing debate and could be subject to legal challenge in the future. Transgenic microbes constructed by transferring genes between genera or higher taxonomic categories are considered "new chemical substances" under the agency's current policy (unless they are on the TSCA inventory). Notification of EPA is voluntary before experimental releases, but required before full general commercial use (6).

The National Institutes of Health (NIH) historically has had a role in evaluating environmental releases through its Recombinant DNA Advisory Committee. However, this committee has not reviewed any deliberate releases of GEOs since 1987 and voted in May 1991 to terminate over-

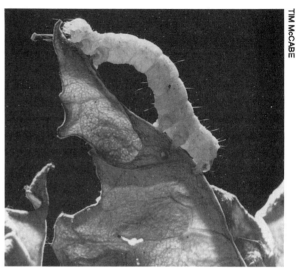

Several corporations hope to genetically engineer insect viruses—such as the celery looper virus that infects cabbage loopers (Trichoplusia ni) and several other insect pests—into more potent insecticides.

view in this area that overlaps with APHIS and EPA. The issue is now under consideration by the director of NIH (43).

■ Holes in the Coordinated Framework
Finding:

Some of the same gaps in current Federal authority and regulation that exist for NIS also apply to GEOs under the Coordinated Framework. In the foreseeable future, commercial development is likely to proceed for several categories of GEOs that lack Federal or State regulation of experimental release or commercial distribution. Similar gaps for NIS continue to allow some ill-advised introductions resulting in economic costs or environmental harm.

Because environmental releases of GEOs currently are regulated under many of the same statutes that cover NIS, several gaps in Federal coverage identified by OTA for NIS also apply to GEOs. Most of the gaps raise few "real-world" concerns at present: environmental releases of GEOs through October 1992 primarily have been

---

[2] Federal Insecticide, Fungicide, and Rodenticide Act (1947), as amended (7 U.S.C.A. 135 *et seq.*).

[3] Toxic Substances Control Act (1976), as amended (15 U.S.C.A. 2601 *et seq.*).

of only a few types of organisms (table 9-3). These generally have presented relatively low risks and are clearly covered by current Federal oversight. However, the gaps may become increasingly important as the range of biological origins and applications of GEOs expands over the next 5 to 10 years. This is especially worrisome given the rapid advances in genetic engineering technologies and the growing numbers of field releases. Between 1987 and 1991 alone, applications to APHIS for field testing of transgenic plants increased more than six-fold (49).

Some observers anticipate that Federal oversight under the Coordinated Framework will evolve to fill these gaps as needs arise (6,43). Experience with NIS has shown, in contrast, that under the constraints of budgetary limitations, Federal agencies sometimes hesitate to expand their regulatory domains, even where clear needs and authority exist (see boxes 3-A, 4-B). Moreover, statutory authority does not exist to fill certain of these gaps. Voluntary compliance by GEO producers—motivated by a desire to quell public concerns—also might help limit future problems resulting from regulatory gaps. One limitation may be that, as the number of releases grows ever larger, public scrutiny of individual releases is likely to decline, potentially decreasing the incentives for producers to seek voluntary approval.

The following sections describe some areas where Federal authority to review GEO releases is lacking or ambiguous. This is not to say that every release of a GEO in these categories necessarily poses a risk. But these are areas where there is no experience on which to evaluate riskiness nor mechanisms yet in place to gain such experience. Moreover, the track record of harmful introductions of NIS in these same categories suggests a need for some level of review before GEO releases (chs. 2 and 4). These potential limits to the Coordinated Framework were addressed by Congress during consideration of the Omnibus Biotechnology Act of 1990 (40). The bill, however, was not enacted.

Table 9-3—Current and Potential Future Releases of GEOs

GEOs Already Released In Field Experiments

Microbes:
  pesticidal microbes
  nitrogen-fixing microbes
  marker microbes for tracking environmental dispersal
  live animal vaccines

Plants:
  agricultural crops (e.g., tomato (*Lycopersicon esculentum*), cotton (*Gossypium hirsutum*), corn (*Zea mays*))
  agricultural crops producing pharmaceuticals or specialty chemicals
  forage crops (e.g., alfalfa (*Medicago sativa*))
  trees (e.g., poplar (*Populus* spp.), walnut (*Juglans* spp.))

Geos Currently Under Research for Future Releases

Microbes:
  microbes that break down chemicals for bioremediation

Plants:
  ornamental plants
  plants for range management
  trees for timber production
  trees for urban plantings
  erosion control plants

Fishes:
  game fish for fisheries management
  fish for aquaculture (rapid growth, disease resistance, cold tolerance)

Invertebrate animals:
  shellfish for aquaculture
  crustaceans for aquaculture
  nematodes (roundworms) for biological control
  insects and arachnids for biological control

SOURCES: M. Fischetti, "A Feast of Gene-Splicing Down on the Fish Farm," *Science*, vol. 253, No. 5019, Aug. 2, 1991, pp. 512-513; P.K. Gupta et al., "Forestry in the 21st Century," *Bio/Technology*, vol. 11, No. 4., pp. 454-463, April 1993; E.M. Hallerman et al., "Gene Transfer in Fish," *Advances in Fisheries Technology and Biotechnology for Increased Profitability*, M.N. Voight and J.R. Bottia (eds.) (Lancaster, PA: Technomic Publishing Co., 1990), pp. 35-49; L.F. Elliot and R.E. Wildung, "What Biotechnology Means for Soil and Water Conservation," *Journal of Soil and Water Conservation*, vol. 47, No. 1, January-February 1992, pp. 17-20.

FISH AND WILDLIFE

No law directly provides for Federal oversight of interstate transport or release of genetically engineered fish (finfish and shellfish) or wildlife. Under the Lacey Act, controls over environmental releases of fish and game are State functions, although the U.S. Fish and Wildlife

Service (FWS) can play a role in limiting the interstate transport of species listed by States as prohibited or injurious (chs. 6, 7). Few States compensate for this lack of a Federal presence with comprehensive laws covering release of GEOs. Moreover, States have been discouraged from developing such laws by those concerned that States might obstruct the testing and development of agricultural GEOs like transgenic crops (9).

Future implementation of the Non-Indigenous Aquatic Nuisance Prevention and Control Act of 1990[4] could narrow this gap slightly by restricting the unintentional importation or transport of harmful aquatic GEOs. However, the Federal interagency task force implementing the Act has not yet addressed GEOs in any context.

Other significant areas remain uncovered by Federal law. No Federal authority exists to directly limit the interstate transport or release of aquaculture species, although this is an active area of genetic engineering research (19). Similarly, should genetic engineering techniques be applied to game species of fish and wildlife, there presently are no Federal requirements for review before release. Moreover, the agencies most likely to be involved, FWS and the National Oceanic and Atmospheric Administration, lack applicable policies on GEOs.

Some experts estimate genetically engineered fish will enter commercial distribution within this decade (11). Two have already been field tested in holding ponds. This category raises particular concerns because many fish can establish free-living populations.

CERTAIN PLANTS

APHIS's current regulations for GEOs do not explicitly include large categories of plants (box 9-C). Listed as regulated are parasitic plants in 13 families and 27 genera that fulfill the definition of plant pest. Not included are numerous taxa containing species that are weeds or can become

U.S. FISH AND WILDLIFE SERVICE

*A genetically engineered variety of striped bass (*Morone saxatilis*) is likely to be among the first transgenic fish released.*

weeds in some habitats. Examples of the latter are Bermuda grass (*Cynodon dactylon*), which is an important turf grass and forage plant but also one of the worst weeds in many parts of the United States (4), as well as many plants used in ornamental horticulture, such as purple loose-strife (*Lythrum salicaria*). Should genetic engineering be used to develop new varieties of species for range management or ornamental horticulture (21), it is unclear whether they would be reviewed before release under the category of organisms "altered or produced through genetic engineering which the Deputy Administrator determines is a plant pest or has reason to believe is a plant pest."

Many genetically engineered plants (including some forage and ornamental plants) presently fall under APHIS's review because the plant pathogen *Agrobacterium tumefaciens* was used as a vector for gene insertion (boxes 9-A, 9-C). New mechanical and chemical techniques for inserting genes into plants do not involve plant pathogens. Consequently, some genetically engineered plants produced by such methods also will not fall squarely under APHIS's authority. Again, it is unclear how the agency will choose to deal with these GEOs.

---

[4] Nonindigenous Aquatic Nuisance Prevention and Control Act of 1990, as amended (16 U.S.C.A. 4705 *et seq.*, 18 U.S.C.A. 42).

Users are required to contact APHIS regarding planned releases of unregulated GEOs only if they have reason to believe the GEO poses a risk of being a plant pest (44). Given the historical complacency regarding introductions of non-indigenous plants, expecting users to rigorously evaluate the risks of transgenic plant introductions may be unrealistic.

CERTAIN INSECTS AND INVERTEBRATES USED FOR BIOLOGICAL CONTROL

In the future, should genetic engineering techniques be applied to insects, nematodes, or other invertebrates, environmental releases of some products might fall outside APHIS's purview. The key criterion defining APHIS's authority is whether an organism is a potential plant pest (box 9-C). Some insects and invertebrates used in biological control clearly fall outside this category since they injure neither plants nor plant products, for example, an insect that eats or parasitizes another insect that is itself a plant pest (40). Given that the agency's present coverage of this category is uneven (ch. 6), and its authority is ambiguous, it is unclear how APHIS would deal with GEOs in this category. The Environmental Protection Agency has exempted such non-microbial biological control agents from regulation under FIFRA.[5] The agency still could step in to assume this role (6), although it has not yet shown any interest in doing so.

RESEARCH

In general, research releases of GEOs are subject to the same restrictions as non-experimental releases. Further, research conducted or funded by Federal agencies is subject to the National Environmental Policy Act.[6] The U.S. Department of Agriculture's Office of Agricultural Biotechnology recently released proposed voluntary research guidelines that apply only to USDA funded research (47). The guidelines rely heavily on input from the Institutional Biosafety Committees that exist at many public and private sector institutions conducting genetic engineering research. The committees originated to ensure that researchers follow guidelines developed by NIH. Their main role has been in the review of contained laboratory research on GEOs. The committees are predominantly composed of members with expertise in genetic engineering (38); an important issue will be whether the committees expand their membership to include ecologists and others with technical backgrounds more appropriate for evaluating the safety of field releases.

Research releases falling within the gaps listed above (fish and wildlife, certain plants, biological control agents) and not funded by Federal dollars may not be covered by the current framework. For example, no Federal agency would review the research release of a genetically engineered fish where the research is privately funded. The Toxic Substances Control Act does not cover non-commercial and strictly academic research releases of non-pesticidal transgenic microbes (30). Concerns over research gaps are not purely hypothetical, as was demonstrated when a researcher at Auburn University moved to conduct experiments involving releases of transgenic carp (*Cyprinus carpio*) in ponds where there was a risk of fish escape (box 9-D).

COMMERCIAL DISTRIBUTION AND SALE

Certain laws, such as the Federal Seed Act;[7] Virus, Serum, and Toxin Act (VSTA);[8] and FIFRA, set standards for accurate labeling and assurance of product purity and efficacy for live organisms in commerce. The Federal Seed Act covers agricultural seed, VSTA covers live mi-

---

[5] 40 CFR 152.20(a) (May 4, 1988).

[6] National Environmental Policy Act of 1969, as amended (42 U.S.C.A. 4321 *et seq.*).

[7] Federal Seed Act (1939), as amended 7 U.S.C.A. 1551 *et seq.*).

[8] Virus, Serum, and Toxin Act (1913) (21 U.S.C.A. 151 *et seq.*).

Box 9-D—Transgenic Fish: Events Surrounding the Auburn Experiments

Considerable controversy erupted in 1989 when a researcher at Auburn University in Alabama moved to conduct experiments with transgenic fish in outdoor holding ponds where there was a risk of escape. After some initial confusion over the appropriate Federal forum for review of the proposal's safety, oversight fell to the Cooperative State Research Service of USDA, which partly funded the experiments. The agency's first Environmental Assessment and its associated finding of no significant environmental impact was met with strong criticism. This prompted the agency to conduct a second assessment with assistance from APHIS. While this assessment also found no significant impact, the finding was contingent on substantial modifications at the site to prevent fish escape. Modifications included construction of new ponds at a higher elevation and filtration of pond effluent, in addition to the existing preventative measures of an 8-foot fence and bird netting above the ponds. No Federal scrutiny necessarily would have occurred had this research been funded by the private sector. In this case, the researcher voluntarily sought Federal oversight even prior to receiving Federal funding.

SOURCES: U.S. Department of Agriculture, Office of the Secretary, "Environmental Assessment of Research on Transgenic Carp in Confined Outdoor Ponds," Nov. 15, 1990; J.L. Fox, "Fish Drift Between Agencies' Guidelines," *Biotechnology*, vol. 7, September 1989, p. 865.

crobes in animal vaccines, and FIFRA regulates microbial pesticides and pesticidal transgenic plants. These laws aim to protect against product misrepresentation and the distribution and sale of contaminants. The lack of equivalent protection for other types of organisms in commerce may become important as the living products of genetic engineering move toward commercialization. Flower seeds, for example, are not covered by the Federal Seed Act. Nor do any Federal laws or regulations currently specify labeling requirements for grown plants or insects and other macroorganisms used in biological control.

An additional role of commercial statutes is to regulate usages of potentially harmful products like pesticides—only allowing certain uses under specified conditions. As agricultural GEOs move toward commercial sale, they will not be subject to such regulation. Under the Federal Plant Pest Act[9] and the Plant Quarantine Act,[10] the mechanism APHIS uses to allow commercial sales of GEOs is to formally exempt them at this stage from regulation.[11] For certain GEOs it may be more appropriate to place constraints on commercial applications; for example, it might be prudent to limit planting of certain transgenic cottons in Hawaii where the potential for hybridization with free-living cotton (*Gossypium tomentosum*) exists.

GAP FILLING BY THE STATES

A perceived lack of adequate Federal regulation has been the driving force behind State efforts to develop laws on GEOs. As of February 1991, nine States had laws specifically dealing with the release of GEOs, and about 30 percent of the States were in the process of developing GEO release and product policies (3). A total of six States introduced, and three enacted, legislation related to the environmental release of GEOs in 1991 (15).

In at least some cases, State laws may cover all releases of GEOs. Under the North Carolina Genetically Engineered Organisms Act, for example, "A genetically engineered organism may not be released into the environment, or sold, offered for sale, or distributed for release into the environment unless a permit for its release has been issued pursuant to this article."[12] Thus,

---

[9] Federal Plant Pest Act (1957), as amended (7 U.S.C.A. 147a *et seq.*).

[10] Plant Quarantine Act (1912), as amended (7 U.S.C.A. 151 *et seq.*).

[11] 7 CFR 340 (June 16, 1987) as amended.

[12] General Stat. of North Carolina, sec. 106-64.

releases of transgenic fish in the State of North Carolina currently would require a State, but no Federal, permit (33). North Carolina, however, is an exception among the States in this regard.

Similar to the patchwork of State fish and wildlife laws (ch. 7), current State laws on GEOs vary widely in scope and rigor (43). Such inconsistency could create burdensome requirements for researchers and industry (13). One representative of the seed industry clearly expressed some of the potential hazards of multiple States' regulation:

> Few engineered crop varieties or hybrids, if any, could bear the cost and time involved in multiple registrations in 50 individual States. Environmentally this approach would also fall short, as environmental problems, should they occur, can hardly be expected to respect State boundaries. Thus, a Federal lead in regulation of engineered crop plants is essential, but can only become a reality if the final system gains the confidence of the public and the States (35).

A SURPRISE CONSEQUENCE OF APPLYING THE SAME LEGAL AUTHORITY TO NIS AND GEOs

Applying the same laws to NIS and GEOs may have some unanticipated results. A case in point is APHIS's recent move to relax permitting requirements for releases of certain transgenic plants. APHIS's authority here derives from the Federal Plant Pest Act and the Plant Quarantine Act, both of which were designed to protect U.S. agriculture from pests. Historically, this is an area where Federal preemption of the States is common; for example, the Federal Government may impose quarantines unsupported by the States or, alternatively, it may allow for more liberal interstate transport of commodities that the States would prefer to curtail (ch. 7). In a recent rule, APHIS asserted its authority to exercise this preemptive power in the area of GEO releases;[13] that is, where the Federal Government has moved

to allow a release, States cannot prevent the release from occurring.

Whether APHIS's position here would withstand a challenge in the courts is open to question (8). The issue may be largely theoretical, however: legal challenge is unlikely since most States lack the technical expertise to evaluate planned releases of GEOs and rely heavily on APHIS's judgment (17,33). Moreover, the new regulations provide for notifying the States before GEO releases. Nevertheless, the example demonstrates an important point. As long as the same sections of the same laws are used as authority for both NIS and GEOs, any amendments to these laws will need to anticipate how they will affect Federal actions regarding both categories of organisms. Moreover, legal precedents established for one category may eventually be applied to the other (7).

ECOLOGICAL RISK ASSESSMENT

Since the first environmental release of a GEO in 1986, Federal agencies have reviewed, authorized, or permitted several hundred additional releases of genetically engineered plants and microbes under final or interim rules. The general approach has been to treat each release as allowed only after case-by-case evaluation (i.e., on a "not sure" list; see ch. 4). Central to the evaluation process is some form of risk assessment. The potential for high profits combined with vocal public concern has driven the rapid development of risk assessment methods for GEOs and a growing scientific literature in this area (table 9-4).

As with NIS, assessments of GEO risk usually center on characteristics of the organism, the environment into which it will be released, and the likelihood the GEO or new genes will spread to other locales. Of particular concern has been characterization of the effects of the genetic modification, specifically its stability and whether

---

[13] "Genetically Engineered Organisms and Products: Notification Procedures for the Introduction of Certain Regulated Articles; and Petition for Nonregulated Status," Final Rule, 58 *Federal Register* 17044 (March 31, 1993).

Table 9-4—Selected Recent Discussions of the Environmental Effects of Releasing GEOs

L.R. Ginzburg (ed.), *Assessing Ecological Risks of Biotechnology* (Butterworth-Heinemann: Boston, 1991).

M. A. Levin and H.S. Strauss (eds.), *Risk Assessment in Genetic Engineering* (New York, NY: McGraw-Hill, Inc., 1990).

D.R. MacKenzie and S.C. Henry (eds.), *International Symposium on the Biosafety Results of Field Tests of Genetically Modified Plants and Microorganisms* (Agricultural Research Institute: Bethesda, MD, 1990).

H.A. Mooney et al. (eds.), *Ecosystem Experiments*, Published on behalf of the Scientific Committee on Problems of the Environment (SCOPE) of the International Council of Scientific Unions (ICSU) (Chichester, England; New York, NY: John Wiley and Sons, 1991).

National Research Council, *Field Testing Genetically Modified Organisms: Framework for Decisions* (Washington, DC: National Academy Press, 1989).

J.M. Tiedje et al., "The Planned Introduction of Genetically Engineered Organisms: Ecological Considerations and Recommendations," *Ecology*, vol. 70, No. 2, 1989, pp. 298-315. (Report from the Ecological Society of America).

U.S. Department of Agriculture, Animal and Plant Health Inspection Service, "Workshop on Safeguards for Planned Introduction of Transgenic Potatoes," Conference Report, 1991.

U.S. Department of Agriculture, Animal and Plant Health Inspection Service, "Workshop on Safeguards for Planned Introduction of Transgenic Corn and Wheat," Conference Report, April 1992.

U.S. Department of Agriculture, Animal and Plant Health Inspection Service, "Workshop on Safeguards for Planned Introduction of Transgenic Oilseed Crucifers," Conference Report, 1990.

U.S. Environmental Protection Agency, *Pesticidal Transgenic Plants: Product Development, Risk Assessment, and Data Needs* (U.S. EPA Conference Proceedings: Nov. 6 and 7, 1990).

SOURCE: Office of Technology Assessment, 1993.

inserted genes might confer unwanted characteristics on the GEO or other species to which they might spread. Factors affecting the GEO or gene spread include how likely the GEO is to establish a free-living population outside of human cultivation and the presence of free-living relatives that might hybridize with GEOs.

A far greater number of authorized releases has occurred for plants than for microbes. Although the same categories of risk apply to both, development of general risk assessment methods has been less tractable for microbes. The biology and ecology of microbes in nature is relatively poorly understood (16), and predicting environmental effect and dispersal potential is difficult (2). Microbes present special problems in evaluating the potential spread of genes since gene exchange in nature can occur not only between different species, but also between different genera (27). In addition, populations of microbes evolve rapidly, complicating predictions of the possible long-term effects of inserted genes.

■ Comparing the Current Level of Review for NIS and GEO Releases

Finding:

While some categories of GEOs actually pose lower risks than similar NIS, pre-release evaluations for certain GEOs have been more rigorous. This inconsistency reflects the chronic underestimation of risk for NIS introductions in the past. Some of the approaches being instituted for evaluating risks of GEOs might usefully be transferred to NIS.

Comparison of the current level of review by the Federal Government for various categories of NIS and GEOs shows that greater scrutiny often is applied to GEOs, even though some may pose lower risks than NIS (table 9-5) (see ch. 4). For example, until 1993, APHIS conducted an environmental assessment for each permitted release of a genetically engineered plant, even for plants highly dependent on human cultivation and lacking free-living relatives in the United States. In contrast, non-indigenous plants are routinely introduced in the United States for applications in

Table 9-5—Federal Pre-Release Requirements for Small-Scale Releases of Certain Non-Indigenous Species (NIS) and GEOs

	NIS	GEOs
Crop and forage plants	No systematic review	**If within APHIS's definition of a "regulated article" (box 9-C):** Most require application to APHIS for a permit; APHIS conducts an environmental assessment; EPA reviews APHIS's assessments for pesticidal plants **For certain regulated articles:** no permit is required, instead requires notification of APHIS at least 30 days before the day of release **If not a regulated article:** same as for NIS
Live animal vaccines	Requires application to APHIS for a permit; APHIS reviews application	Same as for NIS
Pesticidal microbes	Requires notification of EPA; EPA may require additional information or application for an Experimental Use Permit; EPA reviews submitted material **For "plant pests":** APHIS also reviews material before release	Same as for NIS
Non-pest, non-pesticidal microbes (e.g., nitrogen-fixing bacteria)	No systematic review	Voluntary notification of EPA; EPA may request additional information; EPA reviews submitted material

SOURCE: Office of Technology Assessment, 1993.

soil conservation and wildlife forage with no systematic review of the potential environmental consequences of release—although such species may be chosen specifically for the ability to establish free-living populations (ch. 6). Similarly, EPA does a case-by-case review of certain releases of transgenic microbes, such as nitrogen-fixing bacteria, but releases of equivalent non-indigenous microbes are not subject to any Federal oversight. If more rigorous standards are applied to under-evaluated categories of NIS in the future, methods already developed for GEOs could provide a useful model.

■ Impending Scale-Up of Releases for Agricultural GEOs

Finding:
Experience with NIS overwhelmingly has shown that an organism's effects and ecological role can change when it is transferred to new environments. This suggests a need for caution in extrapolating from the results of small-scale field tests of GEOs to larger scale releases. Also GEOs that pose a low risk in the United States sometimes may pose a higher risk in other countries.

Most releases of GEOs in the United States thus far have been small field tests (table 9-2). The geographic area of release will inevitably increase for approved GEOs, particularly as they enter the phase of commercial production, distribution, and sale. This issue looms large especially for agricultural releases: estimates are that commercial distribution for some crops under development could occur as early as 1994 or 1995 (5). The impending scale-up raises several as yet unanswered questions, recently illustrated by the case of transgenic squash (*Cucurbita pepo*) (box 9-E).

**Box 9-E—Controversy Erupts as Upjohn's Transgenic Squash ("ZW-20")
Moves Towards Commercialization**

The case of squash (*Cucurbita pepo*) genetically engineered for disease resistance illustrates several impending issues: the complexity of some of the decisions ahead; needs for better use of field tests to evaluate the risks of large-scale releases; and potential problems in applying domestic decisions internationally.

In September 1992, APHIS announced its intent to rule that a transgenic squash produced by the Upjohn Co.—ZW-20—is not a plant pest and therefore is not subject to further regulation by the agency. This variety contains genes from two plant viruses that confer enhanced disease resistance. APHIS's ruling would be essential to the squash's commercial distribution. Calgene's Flavr Savr™ tomato (*Lycopersicon esculentum*)is the only other transgenic plant that the agency has ruled is not a plant pest.

Response to APHIS's plan, especially from environmental organizations, was strongly negative. Upjohn's petition was criticized for its scientific inaccuracies and failure to cite important research. Further concerns were that APHIS apparently took the scientific content of Upjohn's petition at face value, and, in the absence of outside reaction, might have allowed commercialization of ZW-20 without additional analysis.

Instead, however, APHIS issued a second call for public comment in March 1993. The agency specifically requested further information on the potential for hybrization between ZW-20 and free-living squash and whether transfer of disease resistance genes to free-living populations would affect their weediness. APHIS also contracted with Hugh Wilson, an expert on squash genetics at Texas A&M University, to prepare a report addressing these issues.

Wilson's report clearly identified several important risks. The potential for hybridization with ZW-20 would be great throughout the 12-State range of free-living squash. Moreover, free-living squash are already significant agricultural weeds in some areas and the transfer of new disease resistance genes to these populations could enhance their weediness. Gene transfer might also erode the genetic diversity of the free-living squash populations—a potential gene source for future squash breeding.

(continued)

WHAT IS THE ACCEPTABLE LEVEL OF RISK FOR GEO RELEASES?

Finding:

Proposals approved to date by APHIS for small-scale field releases of GEOs have been low risk. For the most part, APHIS has not yet been challenged to evaluate proposals posing intermediate risk levels. It is unclear how the agency plans to deal with this difficult task of setting acceptable levels of risk, especially as APHIS has not yet standardized its procedures for evaluating the risks associated with NIS.

Permit applications to date primarily have involved low-risk GEOs, such as those lacking free-living relatives in the United States, or involving genes that would pose negligible risk even if transferred to free-living relatives. Decisions concerning which releases to allow will become increasingly complex as the numbers and types of GEOs increase (table 9-3) and GEOs posing more intermediate levels of risk begin to be proposed for release.

APHIS is operating under statutes designed to protect U.S. agriculture from harmful pest species. Neither the Federal Plant Pest Act nor the Plant Quarantine Act contains any specification of what level of "harm" might be acceptable. This is in contrast to commercial statutes like FIFRA and TSCA, which give explicit instructions on how benefits should be weighed against risks. APHIS's current regulations give no indication of how acceptable levels of risk are to be set.

Some perspective on how the agency balances such issues might be gleaned from its experience with NIS. Here APHIS weighs preventing entry of new plant pests against the economic desirability of free trade (see ch. 4). Critics complain the agency often errs in the wrong direction by

Although hybridization between free-living and domesticated squash has probably occurred throughout history, hybridization involving the transgenic squash poses special concerns. According to Wilson, the novel *source* of the disease resistance genes (viruses) "represents, within the biological and historical context . . . an unknown and untested factor. The process of injecting a foreign genetic element . . . that has no precedent within the phylogenetic history of a complex crop-weed system such as *C. pepo*, constitutes a biological risk." Further, the magnitude and impacts of this risk are "difficult—if not impossible—to predict."

APHIS's final ruling on ZW-20 is expected sometime during the fall of 1993. In the interim, Upjohn is conducting additional field tests to address many of the important issues. According to one USDA official, APHIS plans to make its decision regarding ZW-20 according to the same criteria used to judge varieties produced by traditional breeding. However, the consequences of gene transfer from domesticated to free-living plants have not been examined in the past. So, even traditional plant breeding provides little experience on which to base a regulatory decision.

If APHIS rules to allow commercialization of ZW-20, another issue will arise. Free-living squash also occur in Mexico and the export of ZW-20 seed to Mexico could pose additional potential risks.

SOURCES: R. Goldburg, Senior Scientist, Environmental Defense Fund, letter to Chief, Regulatory Analysis and Development, U.S. Department of Agriculture, Animal and Plant Health Inspection Service, Oct. 19, 1992; J. Payne, Senior Microbiologist, Biotechnology, Biologics, and Environmental Protection, Animal and Plant Health Inspection Service, U.S. Department of Agriculture, personal communication to E.A. Chornesky, Office of Technology Assessment, July 13, 1993; J. Rissler et al., "National Wildlife Federation Comments to USDA APHIS on a Proposed Interpretive Ruling Concerning Upjohn's Transgenic Squash," Oct. 19, 1992; U.S. Department of Agriculture, Animal and Plant Health Inspection Service, "Notice of Proposed Interpretive Ruling in Connection With the Upjohn Company Petition for Determination of Regulatory Status of ZW-20 Virus Resistant Squash," 57 *Federal Register* 40632-40633 (Sept. 4, 1992); U.S. Department of Agriculture, Animal and Plant Health Inspection Service, "Proposed Interpretive Ruling in Connection with the Upjohn Company Petition for Determination of Regulatory Status of ZW-20 Virus Resistant Squash," 58 *Federal Register* 15323 (March 22, 1993); H.D. Wilson, "Free-Living *Cucurbita pepo* in the United States: Viral Resistance, Gene Flow, and Risk Assessment," contractor report prepared for the Animal and Plant Health Inspection Service, U.S. Department of Agriculture, May 27, 1993; H.D. Wilson, Professor, Department of Biology, Texas A&M University, personal communication to E.A. Chornesky, Office of Technology Assessment, July 16, 1993.

allowing new species and products to enter the country with few restrictions until risks are clearly demonstrated. Further, APHIS gives far greater attention to effects of its actions on agriculture, often neglecting effects on natural areas. This is of particular concern since upcoming GEO releases may have the potential to invade natural areas, or to affect populations of non-target species through their pesticidal properties.

RESULTS OF CONFINED FIELD TESTS AND POTENTIAL RISKS OF LARGER SCALE RELEASE

In approving the hundreds of test releases of transgenic plants thus far, APHIS has placed considerable emphasis on confinement— requiring that special precautions be incorporated into experimental protocols to prevent gene spread. Such precautions include destroying the plants before they flower or removing the flowers. Sometimes non-engineered plants are planted

around the perimeter of an experiment to "trap" pollen from the transgenic plants. Test fields also may be isolated a certain distance from other fields to minimize the chance of pollen transfer.

General agreement exists that confinement will become infeasible for many GEOs when they are released on a large-scale or go into commercial sale. The range of different environments into which a GEO is released will also increase. If changes in environment influence such risk factors as likelihood of establishment or dispersal, the relative risk of a release may increase with scale-up. Evidence from experiments with transgenic crucifers (plants in the mustard family) in England already has demonstrated variation among sites in the plants' reproduction and other features that affect the potential for establishment (10).

Confined experimental releases conducted thus far demonstrate the characteristics, stability, and performance of GEOs—attributes important to evaluate during product development. They do not, however, necessarily provide any additional information on the ecological risks posed by a GEO under unconfined conditions or whether these risks will change as the scale of release increases (49). An analysis by the National Wildlife Federation showed that, for the 115 field releases permitted by APHIS from 1987 through 1990, the required final report was filed for only half (24). And most lack data on potential environmental effects that could be used for scale-up decisions. Nevertheless, proponents of genetic engineering have used the approval of, and low risk attributed to, small-scale experimental releases as evidence that permitting requirements for field tests are far too stringent (1).

In new regulations issued in 1993,[14] APHIS used the same reasoning to justify why certain releases of GEOs should require only agency notification rather than receipt of a permit. This probably poses few problems for the bulk of low-risk GEOs that will fall under the new regulations. It does, however, establish a poor precedent for higher risk GEOs. Especially for these, small field trials will need to better incorporate research and monitoring designed to evaluate the ecological risks of larger scale releases.

In the absence of such research, it is unclear what information will be used to make scale-up decisions. APHIS assumes that petitions to exempt an organism from regulation (i.e., allow commercial distribution) will include the necessary information to judge whether a GEO will cause significant environmental impacts when grown under unconfined conditions (26). However, the existing data applicable to such decisions are patchy at best.

Some groups in the private sector also have conducted or funded experiments to determine whether genes are likely to spread from transgenic crops by hybridization with wild and weedy

DAVID NANCE

The cotton boll at left (Gossypium hirsutum) *was protected from pests by a gene from Bt* (Bacillus thuringiensis). *Domesticated cotton has wild relatives* (G. tomentosum) *in Hawaii and elsewhere in the world that potentially could hybridize with the genetically engineered form.*

relatives (22,49). But, Federal investment in basic research in this area has not occurred in the United States until quite recently. The 1990 Farm Bill[15] required USDA to allocate 1 percent of its research budget to "biotechnology risk assessment research." The Cooperative State Research Service administers the program, which is expected to provide about $1 million annually in research grants (14).

HOW TO DEAL WITH INTERNATIONAL TRADE IN GEOS?

An even greater level of scale-up will occur when GEOs enter international commerce. Current Federal regulations do not address export of GEOs (44), although the risks associated with releases in other countries sometimes may be substantially greater than in the United States (box 9-B) (18, 23). Further, recipient countries for exports may themselves lack laws or regulations requiring oversight of GEO releases (12).

---

[14] 58 *Federal Register* 17044 (March 31, 1993).

[15] The Food, Agriculture, Conservation, and Trade Act of 1990, Public Law 101-624.

Most important crops lack wild and weedy relatives in the United States because they originated elsewhere. However, in countries closer to these crops' centers of origin, wild and weedy relatives generally are common. Close relatives of corn, tomatoes, and potatoes are common in Central and South America. In these areas the risk would be far higher that engineered genes might spread through hybridization (45,46). Moreover, the small fields surrounded by vegetation typical of farming in developing countries provide greater opportunity for contact and hybridization with wild and weedy relatives (4).

■ A Question of Values: The Hazards of Our Successes

Objections to the first releases of GEOs commonly addressed the intrinsic merit of altering the natural world. This issue has been less prominent recently probably because it is less germane for agricultural releases to environments already highly modified by human manipulation. It may, however, reemerge as GEOs begin to be released into natural areas.

In many cases, NIS are valued by natural resource managers because of their ability to live in stressed, polluted, or otherwise degraded habitats where comparable indigenous species cannot dwell. Concerns have been voiced that genetic engineering may pose a similar opportunity to deal with environmental degradation not by fixing the problem but by changing the managed species.

In the past, we tried to control pollution to accommodate plants and animals. Now, new [genetic engineering] techniques give us the power to control plants and animals to accommodate pollution. . . . In the past, petrochemical companies engineered pesticides to make them compatible with crops. Now they can engineer crops to make them compatible with pesticides (31).

The potentially vast opportunities genetic engineering brings also will pose certain implicit questions about the biological future of the country. As with NIS, managers of natural areas may need to decide between indigenous species and GEOs, or between improving habitats and stocking degraded habitats with GEOs that are more stress tolerant. As with NIS, explicit articulation of such choices and the development of clear policies is needed at a national level.

CHAPTER REVIEW

This chapter examined how the Federal Government oversees the environmental release of GEOs. Many low risk GEOs have been subject to a level of review never applied to potentially harmful NIS. However, other important issues—such as the need for better research on higher risk GEO releases and post-release monitoring—have received scant attention. The current Federal framework for regulating release of GEOs employs laws that were not designed for this purpose. As for NIS, a patched-together approach has resulted—one that leaves significant areas unaddressed and creates confusion for industry, academia, and government.

The kinds of GEOS discussed here seemed futuristic only a few years ago. In the next chapter, OTA takes a closer look at the future and the kinds of global changes that may further shape the impacts of harmful NIS.

The Context
of the Future:
International
Law and
Global Change | 10

M uch of the debate about non-indigenous species (NIS) concerns the future—what trends related to the movement of species are inevitable and desirable. This debate takes place in the context of increasing "globalization" of national economies and environmental problems. In the face of these changes, many consider unilateral regulation of the movement of NIS inadequate, especially because international trade is among the most important pathways for harmful introductions. This chapter broadens our point of view by examining a few global socioeconomic and technological trends related to harmful NIS and evaluating pertinent international law. Then, the chapter highlights specific predictions regarding the future status of NIS, including scenarios related to species movement and global climate change.

INCREASING GLOBAL TRADE AND OTHER SOCIOECONOMIC TRENDS

Finding:

As international trade relationships change, new pathways for species exchange will open. Similarly, the increasing volume of international commerce in biological commodities— in part because of liberalized trade—is likely to increase the number of new species entering the United States.

Global social and economic trends have long affected the kinds, numbers, and pathways of NIS that move around the world (ch. 3, table 3-5). Global population growth and economic expansion contribute to ever-greater demands on natural ecosystems, on agriculture, and on governmental institutions. Greater U.S. demand for particular kinds of foreign imports generates

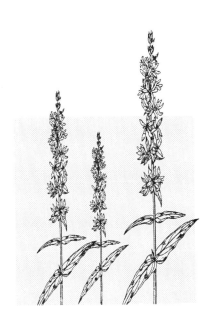

Box 10-A—U.S. Exports of Non-Indigenous Aquatic Species

The United States, as a trading partner and home base to many travelers, exports as well as imports harmful NIS. OTA has not systematically examined the United States' role as an exporter. However, some scientists and officials express concern that Federal and State authorities are not accountable for damaging species intentionally sent outside the United States.

A number of harmful or accidental U.S. exports have occurred. The slipper limpet (*Crepidula fornicata*) was inadvertently exported to Europe with a shipment of American oysters in the 1880s; also Canadian scientists know or suspect U.S. origins for coho salmon (*Oncorhychus kisutch*) in Nova Scotia, an oyster disease in Prince Edward Island, and a trout disease from certified Idaho trout. *Bonamia ostreae*, a parasite of European oysters, probably originated in oysters shipped from California in the 1970s. R.L. Welcomme, of the Food and Agriculture Organization of the United Nations, lists 64 fish and other aquatic species that were introduced to other countries from the United States for ornamental, sport fisheries, aquaculture, or other purposes. Not all established reproducing populations; nor have all been harmful. According to his records, the United States accounted for 240 of the 996 separate international introductions with known countries of origin.

Other kinds of species have also been exported. A North American moth is defoliating trees in large parts of central China. A pine wood nematode (*Bursaphelenchus lignicolus*), probably from the Southeastern United States, is killing black pines (*Pinus nigra*) in Japan. And ragweed (*Ambrosia* spp.) is spreading on the Russian steppes.

SOURCES: R.A. Elston, "Effective Applications of Aquaculture Disease-Control Regulations: Recommendations From an Industry Viewpoint," *Dispersal of Living Organisms into Aquatic Ecosystems*, A. Rosenfield and R. Mann (eds.) (College Park, MD: Maryland Sea Grant, 1992), pp. 353-359; K. Langdon, Great Smoky Mountains National Park, U.S. Department of the Interior, Gatlinburg, TN, personal communication to K.E. Bannon, Office of Technology Assessment, Aug. 17, 1993; D.J. Scarratt and R.E. Drinnan, "Canadian Strategies for Risk Reductions in Introductions and Transfers of Marine and Anadromous Species," *Dispersal of Living Organisms into Aquatic Ecosystems*, A. Rosenfield and R. Mann (eds.) (College Park, MD: Maryland Sea Grant, 1992), pp. 377-385; R.L. Welcomme, *International Introductions of Inland Aquatic Species*, FAO Fishieries Technical Paper No. 294 (Rome: Food and Agriculture Organization of the United Nations, 1988).

---

new and more heavily used pathways for accidental introductions. Foreign demand stimulates U.S. exports of species (box 10-A). Socioeconomic trends also drive the processes by which ecosystems become vulnerable to invasion. For instance, clearing land often eliminates indigenous vegetation and creates pathways for invaders; more recreational visitors to natural areas increases the likelihood that harmful NIS will invade them (105).

From the standpoint of harmful NIS, the continuing increase in global trade is among the most significant trends of the 1990s. Harmful NIS move via intentional commercial imports of live animals, live plants, seeds, and plant products, together with unintended ''hitchhikers'' on these products or in the ships, planes, and trucks that transport them (ch. 3). The United States is a major market for these biologically based prod-

ucts, and imports of many are increasing. The opening of trade relationships through free trade agreements with Canada and Mexico and the General Agreement on Tariffs and Trade (GATT) will mean increased volumes of trade, as well as new trade routes. Climatic and ecological similarities between regions of the United States, Russia, China, and Chile, for example, suggest great potential for species exchange as trade increases.

■ The General Agreement on Tariffs and Trade

The United States recognizes the General Agreement on Tariffs and Trade (GATT)—the post-World War II agreement that liberalized global trade. GATT set rules to eliminate national practices that distort free global markets and provided mechanisms for dispute settlement. The parties to this Agreement have been renegotiating since 1986 (the "Uruguay round"), with no final resolution yet.

GATT declares trade restraints invalid if they do not protect legitimate domestic interests. Article XX(b) acknowledges the need for parties to protect themselves from harmful NIS in that it legitimizes trade restraints, such as quarantine regulations, that are "necessary to protect human, animal, or plant life or health." However, some quarantines are alleged to be protectionist barriers designed to spare domestic products from foreign competition.

Pacific Northwest apple growers contend that Japan's quarantine of their apples is an example. They claim to be shut out of the lucrative Japanese markets by a quarantine against the lesser apple worm (*Enorminia prunivora*) (1), a pest that is indigenous to the eastern United States. The insect exists in very low numbers in Northwest orchards; no outbreaks of quarantine significance have occurred since the 1950s. According to a Washington State University agricultural economist, the Japanese quarantine is scientifically "indefensible" (71). Meanwhile, high-quality apples sell in Japan's markets for the equivalent of $7 or $8 each.

Allegations have been raised by other countries about protectionist U.S. pest regulations as well. These include:

- restrictions on imports of cut flowers and potted plants from the Netherlands (2);
- a ban on seed potatoes from some Canadian provinces (4); and
- a ban on imports of Mexican avocados (81).

GATT has rarely been invoked to resolve these sorts of allegations.

Also, GATT authorities have only resolved a few disputes about whether environmental measures violate its norms of liberal trade (98). Under GATT, trade restraints are not to be imposed by

Increased trade is likely to distribute more harmful non-indigenous species among nations but these changes have received scant attention in free trade agreements—like that proposed with Canada and Mexico.

one party to compel another to change its environmental practices. In 1992, a GATT dispute settlement panel decided that provisions of the U.S. Marine Mammal Protection Act[1] amounted to an unfair trade restraint (98). These provisions banned imports of Mexican tuna caught using methods that kill dolphins (34). Under GATT, the United States may impose bans on such imports only if their very presence is harmful, that is, if the imports could introduce pests. However, GATT does not allow quarantines if they discriminate against foreign imports without scientific justification.

Little systematic analysis of the environmental impacts of different trade patterns or policies has been done (98). Some groups have proposed that U.S. acceptance of future changes to GATT or other trade agreements be subject to formal environmental review. The applicability of the National Environmental Policy Act (NEPA)[2]— the law that requires environmental impact as-

---

[1] Marine Mammal Protection Act of 1972, as amended (16 U.S.C.A. 1361 *et seq.*)

[2] National Environmental Policy Act of 1969 (42 U.S.C.A. 4321 *et seq.*)

sessments for Federal actions—to trade agreements is not resolved legally.[3]

GATT's solution to unfairly restrictive quarantine standards is to encourage parties to "harmonize" their standard-setting criteria. All parties need not regulate the same pests. However, they should recognize common principles, adopt equivalent definitions of key terms like "economic pest," and use comparable criteria for deciding whether to quarantine imports (69). This would make quarantine decisions more amenable to objective scrutiny.

Harmonization does not in and of itself lead to more liberal importation. It could, however, reduce the cases of protectionism disguised as quarantine standards. Reaching agreements on acceptable levels of pest risk presents great difficulty in practice. The proposed harmonized risk analysis prepared for the Food and Agriculture Organization of the United Nations (FAO) concedes this: "it is not possible to define a level of risk that is acceptable for all situations" (69). Currently, determining acceptable levels of risk is a sovereign decision made by individual governments (11). In addition, pest risk analysis often entails high uncertainty (ch. 4). Given these obstacles to achieving international consensus, complete harmonization of pest risk standards is probably not achievable, although agreeing on analytical processes may be.

Greater international harmonization raises two main concerns. First, many developing countries lack the resources or expertise for the sophisticated risk analyses that are feasible for developed countries (63). Second, an overriding GATT approach could preempt national, State, and local NIS laws (84,107).

The concern is whether the United States would be obligated to strike down or preempt a State law that requires a more rigorous pest risk analysis for imports than the international "harmonized" approach under GATT. GATT's current draft language would support the State's case, as long as its laws use "science-based" risk analysis (108). A State might, however, ban a class of imports on the grounds that uncertainty prevented determining which should be allowed and which prohibited. At the same time, State officials might be unwilling or unable to undertake the research necessary to remove those uncertainties. Then the foreign exporter could argue that the State's ban was not based on scientific evidence and therefore violated GATT.[4] GATT's current emphasis on harmonization generally—including pesticide and food safety standards—has been criticized by some legislators and environmental groups as sacrificing national, State, and local environmental controls for the ideal of global free trade (78,98).

■ Free Trade Agreements With Canada and Mexico

Canada and Mexico are the top two suppliers of U.S. agricultural imports (100). Considerable effort has been expended to coordinate pest prevention approaches with both. The pest-related provisions of the existing Canada-U.S. Trade Agreement (signed in 1988) constitutes a continuation of these efforts (101). The proposed North American Free Trade Agreement (NAFTA), which would create a Canada-U.S.-Mexico free trade bloc, includes language on harmonization of pest risk approaches similar to that in the current GATT draft (108).

---

[3] On Sept. 15, 1992, a lawsuit was filed in the U.S. District Court for the District of Columbia challenging the Government's lack of environmental analysis under NEPA for the proposed North American Free Trade Agreement. *Public Citizen, et al., v. Office of the United States Trade Representative, et al.*, Cause No. 92-2102. On June 30, 1993, the court ruled that NEPA applied. However, the United States filed an appeal on July 2, 1993, in the Court of Appeals for the District of Columbia, *Public Citizen et al., v. Espy et al.*, Cause No. 93-5212. The appeal has yet to be decided.

[4] The issue of preemption of U.S. national, State, and local NIS laws under GATT and the North American Free Trade Agreement is analogous to constitutional preemption of State and local laws by Federal laws (see ch. 7) and their potential unconstitutionality under the Interstate Commerce Clause (see box 7-A on the key U.S. Supreme Court decision, *Maine v. Taylor*).

NAFTA will increase the prospects of importing new non-indigenous pests by increasing the volume of agricultural and horticultural imports from Mexico (52). Programs to prevent pest exports traditionally have been weaker in Mexico, although the country recently strengthened its approach and capabilities (3,11). By one estimate, Mexican agricultural exports to the United States would increase by only a few percent (41). By another estimate, commercial truck traffic across the U.S.-Mexico border could expand more than four-fold (to 8 million crossings) from 1990 and the year 2000 (104).

Extensive controversy and information have been generated regarding the environmental impacts of NAFTA. Little of this information relates specifically to the consequences of harmful NIS, however.

■ Other Socioeconomic Trends

Additional socioeconomic trends are likely to shift the movements and impacts of harmful NIS (table 10-1). International travel is also expected to increase and play a key role in the emergence of new threats to human health (54), some of which are carried by insects or other vectors that are not indigenous to the United States.

Both the biological control and aquaculture industries are poised to expand (9,19,25,51). Rates of introduction linked to both of these industries are likely to increase in the future.

Consumer demand exceeds the capacities of catch-fisheries. The proportion of aquatic organisms raised by aquaculture is expected to climb from 11 percent to 25 percent of the global harvest by the year 2000 (72). Likewise, sport fishing is projected to double by the year 2030 (72).

As the aquaculture industry expands—and as researchers, commodity distributors, and the general public also transport fish and shellfish—some fisheries experts expect that species movements are likely to diversify, with the increased risk of spreading pathogens (31). On the other hand, some observers envision that new introduc-

tions will come to be judged by more consistent standards and that aquaculture and non-indigenous fish will be managed "in a manner that preserves the biological integrity of native and desired naturalized fish communities" (42).

Growing interest in environmentally sound methods of pest control is spurring development of commercial biological control. Interest also is growing in applying biological control to new environments, for example, the use of blue crabs (*Callinectes sapidus*) to control zebra mussels (*Dreissena polymorpha*) in lakes and rivers (76). Biological control brings the risk of new species introductions and unexpected effects. Biological control agents, like other introductions, also can carry associated pests unintentionally, although quarantines are in place to prevent this.

Gardening is already the most popular leisure activity in the United States—involving 1 in 3 adults—and most surveys predict that gardening will grow. Nursery stock, seeds, equipment and so on amount to a $9 billion industry (109). Gardeners, in their search for plants that are novel, that reflect particular cultures, or that reflect fashion trends, are spurring changes in the seed and plant industry (40). For example, demand for wildflower seed is so keen that supplies do not meet demand, and some seeds are imported from Europe (40). Drought- and heat-tolerant species are especially popular. Gardening trends could have a variety of implications for NIS. Wildflower seeds are a largely unregulated potential source for the unintentional import and interstate movement of harmful NIS (ch. 1). Gardeners' demands could spur removal of technical and marketing bottlenecks to the use of indigenous species and thus decrease demand for potentially risky NIS imports.

Predicting changes in species use is an uncertain proposition. Even the more exhaustive studies tend not to evaluate species use at this level of detail. For example, agricultural economist Pierre Crosson's (22) future scenarios for U.S. agriculture focuses broadly on production of wheat, major grains, and soybeans. Other recent analyses

Table 10-1—How Social and Economic Factors and Technological Innovations Could Change the Status of Non-Indigenous Species in the Future

Social and Economic Factors	
Factors	Potential effects
Seed exchanges between previously isolated regions, e.g., Russia and the United States	Could increase international spread of pests and pathogens
Increased cross-border movement of material and refugees due to regional wars	Could break down national inspection and quarantine systems and increase the spread of NIS regionally
Doubling of U.S. air passengers by the year 2000	Could increase interstate spread of harmful species
Broadened interest in ornamental uses of indigenous plants	Could decrease incentives for foreign plant exploration and importation; could spread non-indigenous plants of U.S. origin throughout the country
Increased interest in smaller pets for urban areas	Could increase demand for non-indigenous fish and birds
Increased interest in planting forage for wildlife	Could increase introduction of non-indigenous plants to natural areas
Increased concerns regarding risks of chemical pesticide use	Could result in loss of some effective techniques to exclude, manage, or eradicate NIS
Increased interest in protecting endangered species	Could lead to relocations of species and additional introductions

Possible Technological Innovations	
Innovations	Potential effects
Further development of biological control for NIS	Could increase imports of control agents
Improvements in pest eradication mentods	Could cut needs for widespread pesticide spraying in urban areas
Improvements in detection equipment at ports of entry, e.g., molecular probes and biomarkers	Could increase interception of contaminated seed lots, microbes, and other small NIS
Upgraded ballast water exchange systems	Could reduce likelihood of unintentional introductions of aquatic NIS
Progress in genetic engineering	Could blur distinctions between indigenous and NIS as traits are traded
Domestication of "microlivestock" such as the black iguana (*Ctenosaura* spp.) and giant rat (*Cricetmys gambianus, C. emini*).	Could create new pathways for introductions and could spread vertebrate diseases
Development of new plant species to replace shrinking traditional supplies of wood	Could cut imports of raw timber and associated pests
Use of "constructed wetlands" for wastewater treatment	Could increase direct planting of otherwise harmful NIS, such as the water hyacinth (*Eichhornia crassipes*)
Environmental remediation using bacteria, algae, and fungi	Could increase release of non-indigenous microbes

SOURCES: Anonymous, "Wildlife Market On the Rise," *Seed World*, November 1991, p. 26; M.J. Bean, "The Role of the U.S. Department of the Interior in Nonindigenous Species Issues," contractor report prepared for the Office of Technology Assessment, November 1991; G. Bria, "Newsletter Seeks Seed Swaps with Russians," *The Washington Post*, Dec. 28, 1991, p. D3; A. Gibbons, "Small Is Beautiful," *Science*, vol. 253, No. 5018, July 26, 1991, p. 378; L.A. Hart, Director, Human-Animal Program, University of California-Davis, cited in "Smaller Pets," *The Futurist*, vol. 24, No. 2, March/April 1990, p. 5; R. Keeler, "Bioremediation: Healing the Environment Naturally," *R & D Magazine*, July 1991, pp. 34-40; D. Morris, "We Should Make Paper From Crops, Not Trees," *The Seattle Times*, May 5, 1991, p. A12; National Research Council, *Microlivestock: Little-Known Small Animals With a Promising Economic Future*, Board on Science and Technology for International Development, N.D. Vietmeyer (ed.) (Washington, DC: National Academy Press, 1991); Partnership for Improved Air Travel, Washington, DC, cited in "Ailing Aviation Intrastructure Threatens U.S. Economy," *The Futurist*, vol. 23, No. 6, November/December 1989, p. 7; S. Reed, "Constructed Wetlands for Wastewater Treatment," *Biocycle*, vol. 32, January 1991, pp. 44-49.

of the nursery, greenhouse, and turf grass industries; floriculture; and forestry do not distinguish between indigenous and non-indigenous species (10,46,92).

TECHNOLOGICAL CHANGES

Finding:

Technological changes and other means will continue to add non-indigenous organisms to the United States, sometimes by new pathways. At the same time, certain technological innovations, e.g., improved predictive models and more biologically sophisticated pesticides, are likely to provide more effective ways to detect, eradicate, and manage NIS.

Technology, like social, economic, and political changes, will continue to alter the movement, survival, and impacts of NIS (table 10-1). Indeed, experts predict that technical innovation will proceed at increasing rates (18,86) and provide new approaches to preventing and solving environmental problems (16,20). Based on past experience, breakthroughs in transportation, pest control, and information management are most likely to affect NIS directly.

More complex pest control methods seem virtually certain as biotechnology expands (chs. 5, 9). Phytotoxins—plant-damaging compounds produced biologically by microbes—may form the next generation of herbicides; combinations of other biological control methods, the use of modified cultivation practices, and lowered chemical herbicide use may also be increasingly common (91). A host of new methods might ultimately be available to manage NIS more effectively. One biologist predicts: ''[i]t probably will be possible to eliminate most exotic species in less than a decade after the initiation of a program'' with methods such as the release of sterile males; genetically engineered, host-restricted pathogens; repression of pests' immune systems; manipulation of reproduction; and the use of sexual attractants (86). Not all of these are near-term possibilities, however. And insect pests

ANIMAL AND PLANT HEALTH INSPECTION SERVICE

Flourishing air travel is likely to bring more harmful non-indigenous species to the United States and spur technological innovations in detection and baggage handling that will have additional impacts.

have proved to be difficult to eradicate, even with sophisticated technology (30), despite repeated predictions that better methods were on the way.

Biotechnology will also shape the way indigenous and non-indigenous germ plasm is used and combined (ch. 9). Conventional breeders and specialists in biotechnology are increasingly turning their attention to fish. Fish with new adaptations to specific environments can be expected, along with larger fish and the use of more complex reproductive technologies to isolate new strains from indigenous species (72). Technology now allows more fish and shellfish species to be manipulated to limit their post-release reproduction—technology unavailable even 10 years ago (31). Likewise, plant breeders expect novel additions of genes through biotechnology (29,37). Management of some non-indigenous weeds will change, for example, when genes for herbicide resistance are introduced into crops.

Improved methods to assess risk and make decisions are underway and likely to develop further (ch. 4) (14). Other improved means of gathering and managing information remain tantalizing, but remote, possibilities. For example, computerized systems might enable worldwide tracking of pests and other species. The National Aeronautic and Space Administration uses extremely sensitive biomarkers to detect and identify microbes that might contaminate space missions (68). These techniques might eventually be adaptable to detecting NIS at ports of entry, although they require complex and expensive laboratory methods now. Medical technology might have new applications, e.g., nuclear magnetic resonance imaging (MRI) might be used to identify and classify previously unknown species (86) but cost prohibits such uses currently.

High-speed trains are already in service in some parts of the world and high-speed magnetic levitation systems are under development—other examples of technological innovation. In the past, higher speed transportation has increased the survival of intentionally and unintentionally transported NIS (ch. 3). High-speed ground transportation could accentuate this trend. Ultimately, experts envision that high-speed ground transportation would interconnect with highways and air travel (93). Difficulties in restricting NIS of foreign origin could increase if international airports become hubs for multiple high-speed ground transportation systems that automatically transfer baggage.

The caliber of international restrictions on the movement of harmful or potentially damaging NIS is significant, given the increasingly global nature of the socioeconomic and technological trends cited here. Many damaging NIS already in the United States, such as zebra mussels, arrived circuitously, sometimes crossing several international borders. The United States has vast agricultural and other natural resources that are vulnerable to damaging NIS. Thus, this country would be a major beneficiary of an international system that is as effective as possible. In the next section, OTA examines how well international treaty obligations protect the United States and others from damaging NIS.

TREATIES AND THE MOVEMENT OF HARMFUL NON-INDIGENOUS SPECIES

Finding:

Generally, the international regulation of NIS is weak. Except for plant protection, no multilateral treaty to which the United States is a party directly addresses the risks of NIS imports, although the new Convention on Biological Diversity includes a weak provision on NIS.

International environmental laws have multiplied in the last 20 years but they remain weak compared with national prerogatives, as the laws tend to lack enforceability (96). International legal obligations can be important, however, and they are becoming more comprehensive.[5] A number of treaties address harmful NIS directly, with specific provisions. Others deal with related environmental issues and indirectly affect NIS (box 10-B). Only the former are discussed in detail here.

Some experts have called for more effective international laws regarding NIS, particularly to regulate aquatic releases (13,111). Of the three directly relevant multilateral treaties, one has only vague provisions on NIS (the Convention on Biological Diversity) and another has not been ratified by enough countries to take effect (the Convention on the Law of the Sea).

---

[5] Additional international mechanisms also relate to NIS. For example, the United States is a member of the International Council for the Exploration of the Seas (ICES) and a signatory to its Code of Practice. The Code is not an international law or regulation but a protocol and, thus, is discussed in ch. 4.

Box 10-B—Main International Treaties with Provisions Related to Non-Indigenous Species

Multilateral Treaties Directly Affecting NIS

- International Plant Protection Convention, signed by the United States in 1972
- Convention on Biological Diversity, signed by the United States in 1993
- Convention on the Law of the Sea, United States has not signed

Bilateral Treaties Directly Affecting NIS

- Convention on Prevention of Diseases in Livestock (U.S.-Mexico), signed in 1928
- Boundary Waters Treaty of 1909 (U.S.-Canada), in particular the Great Lakes Water Quality Agreement of 1978, as amended in 1987
- Convention on Great Lakes Fisheries (U.S.-Canada), signed in 1954
- Convention Concerning the Conservation of Migratory Birds and their Environment (U.S.-U.S.S.R), signed in 1976

Multilateral and Bilateral Treaties With Indirect Effects on NIS.

These generally protect habitats or groups of indigenous species deemed to have major conservation significance.

- Convention Concerning the Protection of the World Cultural and Natural Heritage, signed in 1973
- Convention on International Trade in Endangered Species of Wild Fauna and Flora (CITES), signed in 1975; (see box 10-C).
- Convention on Wetlands of International Importance Especially as Waterfowl Habitat, signed in 1985
- Convention on Nature Protection and Wildlife Preservation in the Western Hemisphere, signed in 1942
- Convention for Protection and Development of Marine Resources of the Wider Caribbean Region, signed in 1983
- Convention for the Protection of Migratory Birds (U.S.-Canada), signed in 1916
- Convention for the Protection of Migratory Birds and Game Mammals (U.S.-Mexico), signed in 1936

The bilateral migratory bird treaties focus on harvest restrictions and include general provisions to preserve important habitats. The United States would be obligated to protect such habitats if they were threatened by NIS. However, these older treaties tend to be less comprehensive and to lack adequate legal mechanisms to enforce obligations.

NOTE: Dates given are for U.S. signature. Agreements were established and opened for signature either in the same year or up to several years earlier. The Convention on Biological Diversity has not yet been ratified by the Senate.

SOURCES: S. Lyster, *International Wildlife Law* (Cambridge, England: Grotius Publications, 1985); U.S. Congress, Office of Technology Assessment, *Technologies to Maintain Biological Diversity*, OTA-F-330 (Washington, DC: U.S. Government Printing Office, March 1987).

■ The International Plant Protection Convention (IPPC)

IPPC covers agricultural pests. Created under United Nations auspices, this major multilateral treaty has been signed by 94 countries, including the United States in 1972. It establishes a framework for cooperation in agricultural pest regulation; lays out general and specific quarantine principles; standardizes terminology and permits; and provides a process for resolving disputes (47). It aims to:

- strengthen international efforts to prevent the introduction and spread of pests of plants and plant products,
- secure international cooperation to control pests and to promote measures for pest control, and
- ensure adoption by each country of the legislative, technical, and administrative measures to carry out the Convention's provisions (15).

IPPC requires each signatory to establish a plant protection organization to undertake certification, inspection, control, and research; to conduct surveys; and to share information. This does not guarantee uniform performance by all parties. Training, equipment, and facilities differ among the parties and are lacking altogether in some (15).

From the U.S. perspective, this unevenness means that agricultural agencies in many exporting countries cannot be relied on to keep potentially harmful pests out of shipments. In some cases, it has been advantageous to assist developing countries in improving their pest prevention infrastructures, as with economically important Mexico.

The Food and Agriculture Organization of the United Nations (FAO) administers IPPC with input from regional plant protection organizations, such as the North American Plant Protection Organization, to which the United States belongs. Proposals for changes to IPPC include: the need for its own secretariat, separate from and stronger than the current FAO administration (48); and expanded coverage beyond commercial plants, that is, to explicitly protect indigenous plants in non-agricultural areas (12).

No convention comparable to IPPC exists for animals or their pests, but livestock disease prevention terminology and information is coordinated by the International Office on Epizootics. Based in France, it is the international standard setting organization for animal health.

■ The Convention on Biological Diversity

Plans for a global multilateral convention on international protection of biological diversity have been discussed since 1982 (53). At the request of the U.N. Environment Programme, the International Union for the Conservation of Nature and Natural Resources's (IUCN) Environmental Law Centre prepared the initial draft (44). The goal was to present a convention at the U.N.

Conference on Environment and Development (UNCED) in Rio de Janeiro, Brazil, in June 1992.

Initially, a detailed "alien species" article would have obligated the parties to: prevent introductions harmful to biological diversity; attempt eradication of existing harmful introductions; and be attentive to the determinations of a new international expert body (to be created by the Convention) as to harmful species, risk management, and eradication. Several preparatory meetings for UNCED considered the alien species article and weakened the IUCN draft, reducing the specificity of the obligations and eliminating the proposed expert body. In the version of the "Convention on Biological Diversity" presented at Rio de Janeiro, and signed by almost all countries except the United States, the alien species provision reads:

> Each Contracting Party shall, as far as possible and appropriate: . . . (h) Prevent the introduction of, control or eradicate those alien species which threaten ecosystems, habitats or species (95).

The initial obstacles to U.S. signature were financial and legal concerns concerning biotechnology; language regarding property rights; and inadequate provisions for financial oversight by donors (103). The alien species provision did not contribute to the U.S. refusal. The United States later signed the Convention in June 1993 but the Senate has not yet acted on ratification.

The Convention on Biological Diversity does not hold much promise for significantly reducing unwanted international exchanges. The alien species provision is vague and probably unenforceable. This approach contrasts significantly with the detailed requirements of the Convention on International Trade in Endangered Species of Wild Fauna and Flora (CITES), an important and relatively successful treaty (57). Paul Munton, who chairs the Introductions Specialist Group for IUCN's Species Survival Commission, suggested that CITES could serve as a model for international regulation of harmful non-agricultural NIS, i.e., those not covered by IPPC (67). However,

CITES has both strengths and weaknesses as a model (box 10-C). U.N. officials, other international experts, and the U.S. International Trade Commission have suggested recently that monitoring compliance with CITES and other international agreements needs more attention (96). Suggested improvements include monitoring efforts like those used by GATT, the International Labor Organization, or other groups.[6]

■ The Convention on the Law of the Sea

The United States has not signed the sole multilateral convention with provisions specific to marine introductions, the Convention on the Law of the Sea. Indeed, the Convention has not taken effect because fewer than the required number of countries have ratified it. The United States refused to sign the Convention primarily because of concerns over distribution of revenues from deep sea-bed mining (53). However, the Reagan administration did express its intent to voluntarily comply with the non-mining provisions (102).

The Convention proposes an international approach to marine introductions:

> States shall take all measures necessary to prevent, reduce, and control pollution of the marine environment resulting from the use of technologies under their jurisdiction or control, or the intentional or accidental introduction of species, alien or new, to a particular part of the marine environment, which may cause significant and harmful changes thereto (Article 196) (94).

Articles 197 and 200 call for formulation of standards on a cooperative global or regional basis to prevent harmful introductions and to conduct coordinated research on "pathways, risks, and remedies."

■ Bilateral Treaties

The United States has adopted several bilateral agreements on agricultural quarantines and animal health with Canada and Mexico. These are agreements between corresponding agency departments, without treaty status. The United States and Mexico did sign a convention to protect livestock in 1928 that has facilitated mutually advantageous veterinary programs, such as U.S. participation in the control of foot and mouth disease in Mexico in the 1940s and 1950s (66).

The Boundary Waters Treaty of 1909 covers the Great Lakes. The International Joint Commission co-manages the treaty and has overseen agreements on NIS such as the zebra mussel (39). The invasion of the sea lamprey (*Petromyzon marinus*) in the early 1900s, which devastated indigenous fish populations, precipitated the establishment of another treaty in 1955—the Convention on Great Lakes Fisheries (33). The Great Lakes Fishery Commission administers this treaty and coordinates sea lamprey control. Also, the Commission coordinates fish stocking with such NIS as Pacific salmon (*Oncorhynchus* spp.). Disputes among the parties (States, Provinces, and Federal Governments) regarding fish restoration were anticipated by the Joint Strategic Plan adopted in 1980 (38). The Plan calls for consensus before unilateral actions, and the Commission can arbitrate if a dispute cannot be resolved otherwise.

Outside the Great Lakes, disputes have occurred between individual States and the Canadian and Provincial governments over fish releases. North Dakota's experimental release of the European zander (*Stizostedion lucioperca*) raised concerns not only because of uncertainty over impacts from the fish itself, but also from two potentially associated non-indigenous fish diseases (5). No direct legal mechanism like the

---

[6] In January 1991, Senator Daniel Moynihan introduced Senate bill S.59, the General Agreement on Tariffs and Trade for the Environment Act of 1991. This bill proposed using GATT to monitor and enforce international environmental agreements (96).

Box 10-C—CITES as a Model for International Regulation of Non-Indigenous Species

The Convention on International Trade in Endangered Species of Wild Fauna and Flora (CITES) is credited with saving many species from extinction and has been called the most successful international treaty concerned with wildlife conservation. It has had its share of difficulties, too—many involving political disagreements. CITES regulates and monitors international wildlife trade, business that grosses between $5 billion and $17 billion annually.

CITES detailed approach is quite different from that of the Convention on Biological Diversity and thus represents an alternate model for regulating those harmful NIS not already covered by the International Plant Protection Convention or other agreements. However, CITES is intended to prevent harm in the *exporting* country. The major threat from trade in NIS is harm in the *importing* country. (Trade in some species, though, may cause harm for both parties. For example, exporting rare parrots can diminish South American fauna and threaten indigenous U.S. birds with disease if they escape here. Tree ferns are rare and protected in Australia, but they are invasive (e.g., *Cyathea cooperi*) when imported in Hawaii.) Also, CITES regulates only intentional movements; unintentional movements are important pathways for harmful NIS.

Positive features of CITES that are potentially applicable to trade in NIS are:

- regularly updated lists of hundreds of species for which trade is prohibited and over 27,000 species for which trade is monitored by a permit system;
- an independent secretariat, with staff and budget;
- a trust fund to finance the secretariat and biennial meetings of the parties;
- a network of national Management Authorities to address the mechanics of trade, and Scientific Authorities to address biological aspects, in most signatory countries; these communicate directly with each other and the secretariat;
- international forums for governments and non-governmental groups;
- technical advice from various expert organizations, including the IUCN's Wildlife Trade Specialist Group; and
- facilitation of enforcement against CITES violators (including non-parties) via trade sanctions adopted by the parties.

(continued)

Great Lakes agreement, existed for Canada to challenge the action by North Dakota.

In sum, international agreements that control the movement of harmful NIS are quite limited. What kind of future can be predicted, given the continuing, and probably increasing, numbers and kinds of NIS in international transit?

FROM TRENDS TO PREDICTIONS

Finding:

Many experts anticipate increasingly negative impacts from unintentional introductions of NIS in the long term. OTA concurs that there is considerable cause for concern. At the same time, future problems associated with intentional importations and releases could be reduced if appropriate screening and regulatory programs are adopted and implemented.

■ A Pessimistic View of the Future

Many researchers are strikingly pessimistic about slowing and managing harmful introductions. Some anticipate:

a future "... with invasion sure to play an increasingly important role in the ecology of the biosphere ..." (106)

"continued mixing of the regions' biotas ..." (36)

an "... inexorable invasion of all biotas by alien species from other regions, biomes and continents." (87)

Negative features of CITES that detract from it as a model for international regulation of NIS are:
- a narrow focus on trade, which excludes non-commercial pathways;
- the tendency, by restricting all trade in a given species, to penalize the countries that manage species carefully along with those that manage carelessly;
- creation of a harmful underground trade (approximately one-fourth to one-third of threatened and endangered species trade is estimated to be illegal);
- lack of scientific knowledge and/or political will in many countries to make appropriate listings and to enforce permits;
- the opportunity for countries that disagree with CITES on particular listings to exempt themselves by entering "reservations;"
- limited compliance with reporting requirements and lack of enforcement measures specified in the treaty itself; and
- lack of uniform documentation for importing and exporting countries, making misrepresentation and forgery easier;

For the United States, in particular, CITES' weaknesses include: insufficient importation inspection capability, lack of information on enforcement, excessive allowance of imports through non-designated ports, and inadequate assessment and collection of penalties.

SOURCES: C. Beasley, Jr., "Live and Let Die," *Buzzworm*, vol. 4, July-August 1991, p. 28-36; F. Campbell, Natural Resources Defense Council, Washington, DC, personal communication to the Office of Technology Assessment, Dec. 24, 1992; G. Hemley, "International Wildlife Trade," *Audubon Wildlife Report 1988/1989* (San Diego, CA: Academic Press, 1988); S. Lyster, *International Wildlife Law* (Cambridge, England: Grotius Publications, 1985); J.A. McNeely et al., *Conserving the World's Biological Diversity* (Gland, Switzerland: IUCN et al., 1990); P. Munton, "Problems Associated With Introduced Species," paper presented at the Workshop on Feral Animals at the Third International Theriological Conference, Helsinki, Finland, August 1982; U.S. Congress, General Accounting Office, *International Environment: International Agreements are Not Well Monitored*, GAO/RCED-92-32 (Gaithersburg, MD: U.S. General Accounting Office, January, 1992); Worldwatch Institute, *State of the World -1992* (New York, NY: W.W. Norton Co., 1992).

"In the face of ongoing habitat alteration and fragmentation, this implies a biota increasingly enriched in wide-spread, weedy species—rats, ragweed and cockroaches . . . ." (45)

". . . that the circumstances conducive to the invasion of introduced species will become more widespread in the future, not less widespread." (59)

"Because of increasing contact and exchange throughout the world, introductions of exotic pests will take place with increasing frequency . . ." (23).

". . . as species are introduced or move in response to environmental changes, some of today's desirable species may become pests in their new environmental context, while some pests may become more pernicious." (56)

"If we look far enough ahead, the eventual state of the biological world will become not more complex but simpler—and poorer. Instead of six continental realms of life with all their minor components of mountain tops, islands and fresh waters, separated by barriers to dispersal, there will be only one world, with the remaining wild species dispersed up to the limits set by their genetic characteristics, not to the narrower limits set by mechanical barriers as well." (32)

When people speculate about the future, they tend to be predominantly pessimists or optimists; the work of futurists has even been categorized on that basis (62). Whether these experts are unduly pessimistic or not, they picture a serious problem that is getting worse. One prominent conservation biologist sees the spread of NIS as the only high impact threat to biological diversity that affects both richer and poorer countries at every level of biological organization—from single genes to whole landscapes (88). In order to supplement these views, OTA asked its Advisory Panel to envision the world's future regarding NIS also.

Box 10-D—OTA's Advisory Panel Envisions the Future

OTA's Advisory Panelists (p. iv) have been dealing with NIS for much of their professional lives and are more expert than most in assessing what the future might hold. Following are some of the fears and hopes they identified when asked to ponder the best and worst that might be ahead.

Life Out of Bounds . . .

"The future will bring more reaction to zebra mussels (*Dreissena polymorpha*) and inaction to the massive alteration of natural habitats and natural flora and fauna . . . By the mid-21st Century, biological invasions become one of the most prominent ecological issues on Earth . . . A few small isolated ecosystems have escaped the hand of [humans] and in turn NIS . . . One place looks like the next and no one cares . . . The homogeneity may not be aesthetically or practically displeasing, but inherently it diminishes the capacity of the biotic world to respond to changing environments such as those imposed by global warming . . . The Australian melaleuca tree (*Melaleuca quinquenervia*) continues its invasive spread and increases from occupying half a million acres in the late 1980s to more than 90 percent of the Everglades conservation areas."

. . . Or Life In Balance

"An appropriate respect for preserving indigenous species becomes a national goal by consensus . . . All unwanted invasions are treated with species-specific chemicals or by vast releases of 100 percent sterile triploids (created quickly) that depress the exotic populations. Invasions slow to a trickle and fade away like smallpox . . . Jobs for invasion biologists fade away . . . [There is] an effective communication network, an accessible knowledge base, a planned system of review of introductions, and an interactive, informed public . . . Native [species] are still there in protected reserves . . . The contribution of well-mannered NIS—for abuse-tolerant urban landscaping, for ornamentals in gardens, for biological control of pests, for added interest, for increased biodiversity, for new food and medicine—is appreciated. The overarching criterion for judging the value of a species is its contribution to the health of its host ecosystem."

SOURCE: Advisory Panel Meeting, Office of Technology Assessment, July 29-30, 1992, Washington, DC.

Their worst case scenarios are similar to those excerpted above (box 10-D).

Such scenarios would have substantial financial, as well as environmental, costs. The worst case scenarios of future U.S. economic losses from 15 harmful NIS could total $134 billion[7] (table 10-2). These figures are based on available economic projections, ranging from 1 to 50 years. However, far more than these 15 harmful NIS are likely to create losses in the future (ch. 2, 3). For example, if leafy spurge (*Euphorbia esula*) is allowed to spread unrestricted throughout Montana, South Dakota, and Wyoming, annual impacts could reach $46 million by 1995 (6). Similar cost estimates are not available for most harmful NIS, however. Nor do these projections account

for analogs to the zebra mussel—those surprise species that radically and rapidly alter economic outlooks.

Island species, as well as those inhabiting long-isolated bodies of freshwater, will remain at high environmental risk from non-indigenous predators, diseases, and competitors from continental regions (ch. 8) (86). Generally, however, island-like continental areas (such as isolated mountains) have not experienced the same degree of evolutionary isolation and thus are less likely to be as vulnerable to NIS-caused extinctions as oceanic islands are (59).

In the future, larger proportions of harmful introductions will be unplanned as controls are likely to tighten on intentional ones. Most ani-

---

[7] Past economic losses are provided in ch. 3.

Table 10-2—Worst Case Scenarios: Potential Economic Losses From 15 Selected Non-Indigenous Species[a]

Group	Species studied	Cumulative loss estimates (in millions, $1991)[b]
Plants	melaleuca, purple loosestrife, witchweed	4,588
Insects	African honey bee, Asian gypsy moth, boll weevil, Mediterranean fruit fly, nun moth, spruce bark beetles	73,739
Aquatic invertebrates..........	zebra mussel	3,372
Plant pathogens	annosus root disease, larch canker, soybean rust fungus	26,924
Other	foot and mouth disease, pine wood nematodes	25,617
Total	15 species	134,240

[a] See index for scientific names.

[b] Estimates are net present values of economic loss projections obtained from various studies and reports on selected potentially harmful NIS. Many of the economic projections are not weighted by the probability that the invasions would actually occur. Thus, the figures represent worst case scenarios. The periods of the projections range from 1 to 50 years.

SOURCE: M. Cochran, "Non-Indigenous Species in the United States: Economic Consequences," contractor report prepared for the Office of Technology Assessment, March 1992.

mals are intentionally imported and released and heightened awareness should limit further releases of harmful entries (59) (box 3-b).

Some observers expect that weed problems are likely to become greater and more complex (35,59). The significance of woody weeds is likely to increase in Mediterranean-like regions of the world, including California, while better management may cause other types to decline (36). Many U.S. weeds have close relatives overseas. As more of these weeds reach the United States, hybridization could sufficiently alter the non-indigenous weeds to make identification of their natural enemies difficult; as a result, biological control would progress more slowly (30).

Likewise, some forests may experience more severe insect and disease outbreaks as new pests add to the cumulative damage of current ones (7). One prominent conservation biologist expects control of various NIS to be a growth industry and anticipates calls for massive spraying of pesticides (87).

■ **The Next Pests**

OTA identified 205 foreign species that were introduced or detected in the United States between 1980 and 1993; 59 are known to be harmful (table 3-1). For those that become established, impacts are likely to increase as their ranges expand. This kind of data could alert managers to new and potentially damaging NIS. However, such information is scattered and of highly variable quality.

USDA's APHIS tracks certain overseas pests and diseases. This is a daunting task. Thousands of organisms are potential pests, but a smaller number are most likely to reach the United States and become established. The USDA Veterinary Service identified 61 diseases of livestock and poultry and 4 fish diseases of particular concern; veterinarians receive these reports and are asked to look out for new diseases (52). The most recent, comprehensive assessment of future plant pests for the United States identified 1,200 species still restricted to other countries (75). Plants worth watching include:

- 23 species plus 8 multispecies genera of aquatic, parasitic, or terrestrial plants prohibited from entry by the Federal Noxious Weed Act (FNWA); and
- 29 species, 10 genera, plus 6 families of weedy plants that are not listed on the FNWA (60).

Preventing damage by the first group depends on the effectiveness of the system back stopping the Federal Noxious Weed Act. The second group of plants also poses significant economic and ecological hazards, but the FNWA provides no protection from them. Many species in this group have close relatives in the United States that are already troublesome weeds (60). Another 29 species and one genus of non-indigenous weedy plants are present in the United States but not yet widespread. Examples of the most significant of these 3 categories of potential U.S. weeds appear in table 10-3.

Neither the Federal Government nor most State Governments make systematic efforts to evaluate such imminent problems. Within USDA, support has been sporadic for the databases that would provide early warnings (ch. 5). On the other hand, Minnesota recently completed an assessment of threats from NIS; 11 plant, 8 insect, 5 fish, 2 invertebrate, and 7 vertebrate species were identified as potential pests not known to occur in Minnesota as of January 1991 (65). Also, the Federal interagency Aquatic Nuisance Species Task Force is developing an information system on non-indigenous aquatic species and their effects. This part of the Task Force's proposed Aquatic Nuisance Species Program is intended to provide timely notification of the detection and dispersal of these organisms.

■ Climate Change: the Wild Card in Predictions

Finding:

Projected ecological disruption from climate change would increase the probability of invasions by NIS. Also, it would inject great complexity into defining what is and is not indigenous and cause even more policy-making difficulty than currently exists. In particular, new policies would be needed to address whether movements by populations in response to climate change should be treated passively—as if they were natural—or actively.

Scientists are confident that human activity is dramatically changing the chemical makeup of the Earth's atmosphere (97). Atmospheric concentrations of the greenhouse gases that trap heat in the atmosphere—carbon dioxide, methane, nitrous oxide, and chlorofluorocarbons—have increased rapidly over the last 100 years, generally as a result of human activity. The Intergovernmental Panel on Climate Change concluded that the global mean temperature could increase from today's level by roughly 2.2 °F (1.0 °C) by 2030 and 6.6 °F (3.0 °C) by 2100 if present emission levels continue (43). Because greenhouse gases may persist in the atmosphere for up to a century, some amount of global warming appears unavoidable even if countries take stringent measures to limit further emissions today (43).

Temperature changes of this magnitude would have significant effects on the distribution of indigenous and non-indigenous species. Any predictions about the future status of harmful NIS need to account for the possibility of global climate change.

Many uncertainties surround predictions of climate change. However, 100-year increases of 1.5 to 5.5 °C fall in the middle range of most models' predictions. If realized, these levels would make the Earth warmer than at any time in the past 200,000 years (110), with temperatures rising at a rate perhaps 15 to 40 times as fast as past natural changes (80).

Living organisms are quite sensitive to temperature and temperature-related parameters such as precipitation, humidity, and soil moisture. To find the same temperature, a 3 °C increase requires a northward shift of 250 kilometers or an upward elevational movement of 500 meters (58). Different species will shift ranges at different rates. Estimates for individual species project larger range shifts: 350 km northward for loblolly pine (*Pinus taeda*) (64) and 500 to 1,000 km for 4 common North American trees[8] (24). Such species relocation may be possible for highly mobile organisms or for those that readily colonize new

---

[8] Beech (*Fagus grandifolia*), yellow birch (*Betula alleghaniensis*), eastern hemlock (*Tsuga canadensis*), and sugar maple (*Acer saccharum*).

Table 10-3—Examples of Weedy Plants With Potential for Significant Economic or Ecological Harm

Common (when available) and scientific names	Comments
Weedy plants not established in the United States and prohibited from entry by the Federal Noxious Weed Act (FNWA)	
Monochoria vaginalis	aquatic weed of rice fields throughout Asia
African payal (*Salvinia auriculata*)	South American aquatic weed now troublesome in Africa; closely related to one of world's worst weeds (*S. molesta*)
Dodders (*Cuscuta* spp.)	parasitic plant of many crops; worldwide problem (some species in warmer parts of United States)
Broomrapes (*Orobanche* spp.)	parasitic plant of many crops; worldwide problem
Witchweeds (*Striga* spp.)	parasite mostly of grasses; widespread in India, Africa
Couchgrasses (*Digitaria* spp.)	terrestrial weed of fields, disturbed areas in Africa
African boxthorn (*Lycinum ferocissimum*)	terrestrial hedge plant--escaping; serious weed of natural areas and fields in South Africa, Australia, New Zealand
Serrated tussock (*Nassella trichotoma*)	terrestrial weed of fields, waste places; one of worst weeds in New Zealand
Weedy plants not established in the United States and not listed by the FNWA	
Tutsan (*Hypericum androsaemum*)	close relative to already troublesome U.S. weed; found in roadsides, damp places in Europe, North Africa, Australia
(*Oxylobium parviflorum*)	one of worst poisonous plants; found in western Australia
Bromegrasses (*Bromus* spp.)	close relatives to already troublesome U.S. weed; weeds of arid sites in central Asia, Russia, Mediterranean region
(*Avena strigosa*)	close relative to already troublesome U.S. weed; weed in corn and oats fields in central Europe and Mediterranean region
Panic grasses (*Panicum* spp.)	climbs over vegetation; problem in tropical Africa and Asia
Sedges (Cyperaceae)	weeds of waste places, cultivated fields, and wet areas near ponds, streams, rivers; worldwide; multiple genera
Weedy plants in the United States but not yet widespread	
Common crupina (*Crupina vulgaris*)	listed under FNWA; problem along roadsides and in waste places in Idaho; eradicated in California
Catclaw mimosa (*Mimosa pigra* var. *pigra*)	listed under FNWA; wide-spread in tropical Africa; quarantined in Australia; present in Florida
Persian darnel (*Lolium persicum*)	not listed under FNWA; weed of fields and waste places in North Dakota
Cudweed (*Filago arvensis*)	not listed under FNWA; weed of fields, waste places, overgrazed rangeland in Washington and Oregon
Tall fescue (*Festuca arundinacea*)	not listed under FNWA; poisonous to livestock; found in roadsides, gullies, canals in Florida and the Pacific Northwest
(*Thladiantha dubia*)	not listed under FNWA; vine that climbs over vegetation; found in New Hampshire and Minnesota
Raoul grass (*Rottboellia exaltata*)	listed under FNWA; invading sugarcane fields in Florida and Louisiana
(*Medicago polymorpha*)	not listed under FNWA; weed of cultivated areas and waste places; found in Hawaii and almost worldwide

SOURCES: R.N. Mack, "Additional Information on Non-Indigenous Plants in the United States," contractor report prepared for the Office of Technology Assessment, October 1992. Compiled from: R.D. Blackburn, L.W. Weldon, R.R. Yeo, and T.M. Taylor, "Identification and Distribution of Certain Similar-Appearing Submersed Aquatic Weeds in Florida," *Hyacinth Control Journal*, vol. 8, 1969, pp. 17-21; R.A. Creager, "Seed Germination, Physical and Chemical Control of Catclaw Mimosa (*Mimosa pigra* var. *pigra*), *Weed Technology*, vol. 6, 1992, pp. 884-891; T. Miller and D. 20Thill, "Today's Weed: Common Crupine (*Crupina vulgaris*), *Weeds Today*, vol. 14, 1983, pp. 10-11; C.F. Reed and R.O. Hughes, *Economically Important Foreign Weeds*, Agriculture Handbook Number 498 (Washington, DC: U.S. Department of Agriculture, 1977); R.G. Westbrooks, "Introduction of Foreign Noxious Plants into the United States," *Weeds Today*, vol. 12, 1981, pp. 16-17.

areas. Some evidence indicates that northward range shifts are already happening (79).

Species that cannot relocate fast enough may be able to adapt genetically to climate changes. However, most species' physiological adaptations to climate are highly conservative. They are unlikely to evolve fast enough to fit such rapidly changing conditions and extinctions of populations and species can be expected (74).

Also, biological, geographic, or human-caused factors such as habitat destruction may prevent many species from adjusting their ranges or otherwise responding successfully. Even those species capable of spreading rapidly to cooler sites may not flourish given new soil conditions, changes in day length, or different food sources and they may also be extirpated (74). Indeed, the most successful species are likely to be those adept at invading new habitats, including many current pests and pathogens (26).

New species may arrive from overseas or spread north from Mexico, the Caribbean, or from the southern United States. For example, models predict that at least a few agricultural pests and pathogens, such as the potato leafhopper (*Emposasca fabae*), which feeds on soybeans and other crops, are likely to experience expanded areas in which they can survive winter temperatures (90). The species compositions of aquatic communities will change with increasing water temperatures. Many water bodies, such as the Chesapeake Bay, will probably become poorer in terms of diversity and size of harvest (50). Other water bodies could become more productive, e.g., populations of warm-, cool-, and cold-water fish in the Great Lakes are expected to increase because of longer growing seasons, although biological diversity overall could decline (61). Forest pests and pathogens may spread (74). Tropical livestock diseases, such as Rift Valley fever and African swine fever, will be more likely to spread (73).

Increases in the incidences of several human diseases and parasites could result from the northward movement of their vector species in the United States. These include:

1. ascariasis, caused by the nematode *Ascaris lumbricoides*;
2. Chagas' disease, caused by a protozoan parasite (*Trypanosoma cruzi*) transmitted by temperature-sensitive insects (*Triatoma sp.*) *(28);*
3. dengue, caused by a virus carried by the temperature-sensitive mosquitoes (*Aedes aegypti*, *A. albopictus*, and *A. triseriatus*);
4. malaria, caused by *Plasmodium* spp., with mosquito vectors (*Anopheles* spp.); and
5. arthropod-borne encephalitis, a group of viral diseases carried by a variety of mosquitoes (55).

Rapid ecological changes set the stage for speeding up the process by which new diseases emerge or by bringing humans in contact with new agents (83).

Today's biological communities will break apart as some species relocate or are lost and others are added (74). These newly re-sorted biological groups could be more vulnerable to further invasions by NIS (49). Some observers predict that climate change could become the dominant driving force for new biological invasions in the next century (26). Assuming climate change occurs, and significantly affects North America, the changing biological communities will greatly complicate issues relating to NIS, compelling increasingly difficult decisions (21).

Understanding the complex forces that drive large-scale movements of animal and plant populations will be critical to unraveling particular invasions (56). But, in one scientist's view, ''it is hopelessly optimistic to expect that the scientific understanding that can be obtained over the next 100 years will enable us to predict the kind and extent of changes in the distribution and abundance of dominant plant species'' (21). Others are less pessimistic regarding biologists' predictive capabilities (49). Several studies outline at least

Understanding the rapid spread of non-indigenous species such as dyer's woad (Isatis tinctoria) *might help predict and manage the biological shifts that would accompany global climate change.*

the general effects of unprecedented warming on future species ranges and different ecosystems (99,113).

Global climate change would also scramble policies related to NIS. Under the commonly used historical definition of indigenous or its equivalents, a individual becomes non-indigenous when it leaves its species' range at some particular point in time. That time would need to be steadily reassessed for this definition to remain meaningful during climate changes. Otherwise, an increasing proportion of species would be considered non-indigenous, "exotic," or "alien" and subject to the statutes, regulations, and policies that use these terms.

Under OTA's definition (ch. 2), an individual remains indigenous as long as it is within its species' natural range or natural zone of potential dispersal—areas determined in the absence of "significant human influence." Natural ranges and zones can and do change over ecological and evolutionary time. Climate change would alter the specific location of species' ranges and dispersal zones but species would retain their designation as indigenous if their movements were treated as "natural." For this definition to remain meaningful, there must be some way to

distinguish between phenomena that involve lesser and greater human intervention. This is increasingly difficult.

In time, global climate change could render both definitions obsolete, along with policies based on them. Therefore, management and policy flexibility is likely to be increasingly important. A number of options regarding species movement have been suggested. Each presents difficult, and often expensive, choices. Many lessons learned from managing harmful NIS could inform such choices. These include decisions to:

1. **Block Species' Movements**—Managers might want to block movements of particularly harmful species. However, USDA's $6 million attempt to slow the African honey bee's (*Apis mellifera scutellata*) advance in southern Mexico has proven impractical (77). It is not clear whether other such efforts would be more successful.

2. **Conduct Intensive Habitat Creation or Restoration**—Managers might try to create artificial habitats for some species unable to adjust on their own (8). This could also entail controlling invaders from the south or lower elevations (70). However, the science of ecological restoration is in its infancy (ch. 5), managers would face great difficulty in anticipating changes and implementing plans (21), and some sites may change so much that habitat restoration or creation is impossible.

3. **Provide Movement Corridors and New Protected Areas**—Farmland, highways, cities, forest clear cuts, and other human-made areas can interrupt the movement of populations adjusting to climate change. Managers might acquire and maintain either movement corridors through these areas or new protected areas (74). Movement corridors might provide new pathways for harmful species as well (85), however. The practicality of corridors is not known because few have been intentionally created and studied. Data on pathways for harmful NIS might suggest plausible approaches.

4. Translocate Species—Impassable barriers to population movements may compel managers to physically move individual organisms, or their germ plasm (70). Perhaps only a few commercial, recreational, or otherwise popular species would receive the political and financial support for such expensive efforts. Unanticipated ecological and economic consequences could result from releases into new environments, as have other releases discussed in this report. Large-scale species translocations to prevent extinction remain largely theoretical.

5. Emphasize Ecosystem Functions—Managers might aim to preserve desirable ecosystem services—such as erosion control or providing forage, timber, or other commodities—rather than preserving particular species or communities (89). In some cases, NIS may be the only species capable of providing these services during climate change. However, little is known about the functional substitutability of species, in part because many critical species are decomposer microbes and soil invertebrates (112).

Expanding international trade and other 20th Century changes have increased the numbers of species being moved worldwide. Climate change would be likely to accelerate these trends further. Many more species would be shifting ranges and people could have additional reasons to import and release species into new areas. Climate change is the wild card in predicting the future status of NIS.

WRAP-UP: THE CHOICES BEFORE US

Certainly parts of the future pictured in this chapter will come to pass—the trends toward loss of indigenous species and greater global movement of non-indigenous ones are firmly in place.

These trends may be slowed but they would be very difficult and costly to reverse. As a result, some observers find that a profound transition is under way. The metaphors that guide natural resource management are shifting—from the self-sustaining wilderness to the managed garden (27. The world is being defined more in terms of the "unnatural" rather than the "natural" (82). This change is just one part of a general trend toward a more managed globe, whether such management relates to trade, pollution, telecommunications, or international conflict (17). To some, this shift represents a grave loss. To others, it represents greater willingness to undertake responsible action. Issues regarding indigenous and non-indigenous species underscore these different points of view.

In thinking about the future, the distinction between forecasts and visions is significant. Forecasts are concerned with the probable and possible. Much of this chapter, and this report, resides in that analytical realm. Visions, however, appraise the desirable, the imagined, the intended, and compelling (114). In their best-case scenarios, OTA's Advisory Panelists envisioned a future in which beneficial NIS contributed a great deal to human well-being, indigenous species were preserved, and harmful NIS were brought under control (box 10-D). Much of this report is designed to provide the background and means for Congress to achieve such a vision. But deciding the vision's worthiness—and choosing whether to pursue it—are not choices that science can make. Nor does nature provide answers. Which species to import and release, which to exclude, and which to control are ultimately cultural and political choices—choices about the kind of world in which we want to live.

Appendix A: List of Boxes, Figures, and Tables

Appendix B: Authors, Workshop Participants, Reviewers, and Survey Respondents

■ Part 1. The Assessment's Contractors' Reports and Their Reviewers

Pathways and Consequences of the Introduction of Non-Indigenous Species in the United States

Report: *Pathways and Consequences of the Introduction of Non-Indigenous Freshwater, Terrestrial and Estuarine Mollusks in the United States—October 1991*

Author: Joseph C. Britton, Department of Biology, Texas Christian University, Fort Worth, TX

Advisory panel reviewers: William B. Kovalak and Rudolph A. Rosen

Outside reviewers:

Robert Hershler, Department of Invertebrate Zoology, Smithsonian Institute, Washington, DC

Robert F. McMahon, Department of Biology, University of Texas, Arlington, TX

Barry Roth, Consultant, San Francisco, CA

Report: *Pathways and Consequences of the Introduction of Non-Indigenous Plants in the United States—September 1991*

Author: Richard N. Mack, Department of Botany, Washington State University, Pullman, WA

Advisory panel reviewers: Faith T. Campbell, John D. Lattin, Don C. Schmitz, Howard M. Singletary, Jr., and Clifford W. Smith

Outside reviewers:

Richard R. Old, Department of Plant, Soil and Entomological Sciences, University of Idaho, Moscow, ID

Marcel Rejmanek, Department of Botany, University of California, Davis, CA

Kevin J. Rice, Department of Agronomy and Range Science, University of California, Davis, CA

Report: *Pathways and Consequences of the Introduction of Non-Indigenous Fishes in the United States—August 1991*

Author: Walter R. Courtenay, Jr., Department of Biological Sciences, Florida Atlantic University, Boca Raton, FL

Advisory panel reviewers: William B. Kovalak and Rudolph A. Rosen

Outside reviewers:

James E. Deacon, Department of Biological Sciences, University of Nevada, Las Vegas, NV

Christopher C. Kohler, Cooperative Fisheries Research Laboratory, Southern Illinois University, Carbondale, IL

James A. McCann, National Fisheries Research Center, Fish and Wildlife Service, U.S. Department of the Interior, Gainesville, FL

Peter B. Moyle, Department of Wildlife and Fisheries Biology, University of California, Davis, CA

Richard T. Noble, Departments of Zoology and Forestry, North Carolina State University, Raleigh, NC

Bruce R. Schmidt, Utah Division of Wildlife Resources, Department of Natural Resources, Salt Lake City, UT

Report: *Pathways and Consequences of the Introduction of Non-Indigenous Plant Pathogens in the United States—December 1991*

Author: Calvin L. Schoulties, Regulatory and Public Service Program, Clemson University, Clemson, SC

Advisory panel reviewers: Faith T. Campbell, Robert P. Kahn, and Clifford W. Smith

Outside reviewers:

Stella Melugin Coakley, Department of Botany and Plant Pathology, Oregon State University, Corvallis, OR

Conrad J. Krass, Division of Plant Industry, California Department of Food and Agriculture, Sacramento, CA

Kurt J. Leonard, Cereal Rust Laboratory, University of Minnesota, Agricultural Research Service, U.S. Department of Agriculture, St. Paul, MN

George A. Zentmyer, Department of Plant Pathology, University of California, Riverside, CA

Report: *Pathways and Consequences of the Introduction of Non-Indigenous Insects and Arachnids in the United States—December 1991*

Authors: Ke Chung Kim and A.G. Wheeler, Jr., Frost Entomological Museum, Department of Entomology, Pennsylvania State University, University Park, PA

Advisory panel reviewers: Faith T. Campbell, Lester E. Ehler, Robert P. Kahn, John D. Lattin, Jerry D. Scribner, Clifford W. Smith, William S. Wallace, and Reggie Wyckoff

Outside reviewers:

James R. Carey, Department of Entomology, University of California, Davis, CA

Ernest S. Delfosse, National Biological Control Institute, Animal and Plant Health Inspection Service, U.S. Department of Agriculture, Hyattsville, MD

Mark A. Deyrup, Archbold Biological Station, Lake Placid, FL

Ronald L. Johnson, Plant Protection and Quarantine, Animal and Plant Health Inspection Service, U.S. Department of Agriculture, Morestown, NJ

Douglass R. Miller, Agricultural Research Service, U.S. Department of Agriculture, Beltsville, MD

Report: *Pathways and Consequences of the Introduction of Non-Indigenous Vertebrates in the United States—October 1991*

Authors: Stanley A. Temple and Dianne M. Carroll, Department of Wildlife Ecology, University of Wisconsin, Madison, WI

Advisory panel reviewers: J. Baird Callicott and Faith T. Campbell

Outside reviewers:

Thomas H. Fritts, National Museum of Natural History, Fish and Wildlife Service, Washington, DC

Michael P. Moulton, Department of Biology, Georgia Southern University, Statesboro, GA

Gordon H. Orians, Department of Zoology, University of Washington, Seattle, WA

Charles van Riper, III, Cooperative National Park Resources Studies Unit, National Park Service, U.S. Department of the Interior, Flagstaff, AZ

Decisionmaking Models

Report: *Risk Analysis As a Tool for Making Decisions About the Introduction of Non-Indigenous Species Into the United States—July 1991*

Authors: Peter Kareiva, Martha Groom, Ingrid Parker, and Jennifer Ruesink, University of Washington, Seattle, WA

Advisory panel reviewers: Faith T. Campbell, Lester E. Ehler, Robert P. Kahn, John D. Lattin, and Clifford W. Smith

Outside reviewers:

Ronald D. Hiebert, National Park Service-Midwest, U.S. Department of Interior, Omaha, NE

Gary H. Johnston, Wildlife and Vegetation Division, U.S. Department of the Interior, Washington, DC

Matthew H. Royer, Plant Protection and Quarantine, Animal and Plant Health Inspection Service, U.S. Department of Agriculture, Hyattsville, MD

Daniel Simberloff, Department of Biological Sciences, Florida State University, Tallahassee, FL

Report: *The Role and Limits of Economics in Decisionmaking Regarding Non-Indigenous Species—August 1991*

Authors: Alan Randall and Michael H. Thomas, Department of Agricultural Economics and Rural Sociology, Ohio State University, Columbus, OH

Advisory panel reviewers: J. Baird Callicott and Katherine H. Reichelderfer

Outside reviewers:

Richard C. Bishop, Department of Agricultural Economics, University of Wisconsin, Madison, WI

Gardner Brown, Economics Department, University of Washington, Seattle, WA

Richard B. Norgaard, Energy and Resources Group, University of California, Berkeley, CA

Richard C. Ready, Department of Agricultural Economics, University of Kentucky, Lexington, KY

Mark Sagoff, Institute for Philosophy and Public Policy, University of Maryland, College Park, MD

Reports:

A. *A Framework for Special Accounting of the Effects of Introductions of Non-Indigenous Species—March 1992*

B. *The Application of Benefit Cost Analysis to Social Accounting of Non-Indigenous Species Introductions—March 1992*

C. *Economic Impact Tables—March 1992*

D. *Two Case Studies Applying the General Framework for Benefit Costs Analysis on Non-Indigenous Species:*

   1: *Melaleuca—March 1992*

   2: *Sea Lamprey—March 1992*

Author: Mark J. Cochran, Department of Agricultural Economics, University of Arkansas, Fayetteville, AR

Advisory panel reviewer: Katherine H. Reichelderfer

Outside reviewers:

William G. Boggess, Food and Resource Economics Department, University of Florida, Gainesville, FL

Craig Osteen, Economic Research Service, U.S. Department of Agriculture, Washington, DC

Alan Randall, Department of Agricultural Economics and Rural Sociology, Ohio State University, Columbus, OH

Micheal E. Wetzstein, Department of Agricultural Applied Economics, University of Georgia, Athens, GA

Federal Policy Regarding Non-Indigenous Species

Report: *Federal Policy on Non-Indigenous Species: The Role of the United States Department of Agriculture's Animal and Plant Health Inspection Service—December 1991*

Author: Donald H. Kludy, Consultant, Richmond, VA

Advisory panel reviewers: Faith T. Campbell, William Flemer, III, Robert P. Kahn, Jerry D. Scribner, and William S. Wallace

Outside reviewer:

Craig J. Regelbrugge, Regulatory Affairs and Grower Services, American Association of Nurseryman, Washington, DC

Report: *The Role of the United States Department of the Interior in Non-Indigenous Species Issues—November 1991*

Author: Michael J. Bean, Environmental Defense Fund, Washington, DC

Advisory panel reviewers: Faith T. Campbell, Lynn Greenwalt, Gary H. Johnston, and Clifford W. Smith

Outside reviewers:

R. Joseph Abrell, Great Smoky Mountain National Park Headquarters, National Park Service, U.S. Department of the Interior, Gatlinburg, TN

Donald J. Barry, Land and Wildlife, World Wildlife Fund, Washington, DC

David Cottingham, Ecology and Conservation Office, National Oceanic and Atmospheric Administration, U.S. Department of Commerce, Washington, DC

Gary B. Edwards, Fisheries Division, Fish and Wildlife Service, U.S. Department of the Interior, Washington, DC

Amos S. Eno, Conservation Programs, National Fish and Wildlife Foundation, Washington, DC

Dennis Lassuy, Fisheries Division, Fish and Wildlife Service, U.S. Department of the Interior, Arlington, VA

Lewis H. Waters, Bureau of Land Management, U.S. Department of the Interior, Washington, DC

Report: *Federal Policy on Non-Indigenous Species: An Overview of the U.S. Department of Agriculture—December 1991*

Author: John Schnittker, Schnittker Associates, Santa Ynez, CA

Advisory panel reviewers: Robert P. Kahn, Katherine H. Reichelderfer, William S. Wallace, and Melvyn J. Weiss

Outside reviewers:

Ray Brush, Ray Brush Horticultural Consulting Services, Madison, VA

Dan Laster, U.S. Meat Animal Research Center, Agricultural Research Service, U.S. Department of Agriculture, Clay Center, NE

Special Topics

Report: *Bioengineered Organisms and Non-Indigenous Species—November 1991*

Authors: Sheldon Krimsky, Department of Urban and Environmental Policy, Tufts University, Medford, MA

Richard Wetzler, Center for Environmental Management, Tufts University, Medford, MA

Advisory panel reviewers: Robert P. Kahn, Philip J. Regal, and William S. Wallace

Outside reviewers:

Norman C. Ellstrand, Department of Botany and Plant Sciences, University of California, Riverside, CA

Rebecca Goldburg, Environmental Defense Fund, New York, NY

Eric M. Hallerman, Department of Fisheries and Wildlife Sciences, Virginia Polytechnic Institute and State University, Blacksburg, VA

Greg Simon, United States Senate, Office of the Honorable Albert Gore, Washington, DC

Report: *Ecological Restoration and Non-Indigenous Species—August 1991*

Authors: John J. Berger, Restoring the Earth, Berkeley, CA and Lawrence A. Riggs, GENREC, Oakland, CA

Advisory panel reviewers: William Flemer, III, Don C. Schmitz, Clifford W. Smith

Outside reviewers:

Robert F. Doren, Everglades National Park and Fort Jefferson National Monument, National Park Service, U.S. Department of the Interior, Homestead, FL

John M. Randall, Nature Conservancy, Consumnes River Preserve, Galt, CA

Report: *Non-Indigenous Species in Hawaii—January 1992*

Author: Christine Mlot, Milwaukee, WI

Advisory panel reviewer: Clifford W. Smith

Outside reviewers:

Michael G. Buck, Department of Land and Natural Resources, State of Hawaii, Honolulu, HI

Thomas Fritts, Fish and Wildlife Service, U.S. Department of the Interior, Washington, DC

Alan Holt, Nature Conservancy of Hawaii, Honolulu, HI

Charles H. Lamoureux, Harold L. Lyon Arboretum, University of Hawaii at Manoa, Honolulu, HI

Lloyd L. Loope, Haleakala National Park, National Park Service, U.S. Department of the Interior, Makawao, HI

Ahser K. Ota, Hawaiian Sugar Planters' Association, Aiea, HI

Ilima Piianaia, Hawaii Department of Agriculture, Honolulu, HI

Charles P. Stone, Hawaii Volcanoes National Park, National Park Service, U.S. Department of Interior, Hawaii National Park, HI

Roger I. Vargas, Agricultural Research Service, U.S. Department of Agriculture, Honolulu, HI

Report: *Non-Indigenous Species in Florida: A Survey of Pathways, Consequences, and Economic and Environmental Impacts—December 1991*

Author: David W. Johnston, Fairfax, VA

Advisory panel reviewer: Don C. Schmitz

Outside reviewers:

Mark A. Deyrup, Archbold Biological Station, Lake Placid, FL

Robert F. Doren, Everglades National Park and Fort Jefferson National Monument, National Park Service, U.S. Department of the Interior, Homestead, FL

Brian A. Millsap, Nongame Wildlife Section, Florida Game and Fresh Water Fish Commission, Tallahassee, FL

Report: *Selected Analysis of Introduced Animal Species Survey Data—December 1991*

Author: Peter T. Schuyler, Division of Forestry and Wildlife, Honolulu, HI

Outside reviewers:

Gary M. Fellers, Point Reyes National Seashore, National Park Service, U.S. Department of the Interior, Point Reyes, CA

Report: *Public Educational Efforts in the United States Regarding Prevention and Management of Non-Indigenous Species—December 1991*

Authors: Andrea Shotkin and Edward J. McCrea, North American Association for Environmental Education, Washington, DC

Advisory panel reviewers: John D. Lattin and Clifford W. Smith

Outside reviewers:

John F. Disinger, School of Natural Resources, Ohio State University, Columbus, OH

Diana L. King, Hawaii Nature Center, Honolulu, HI

R. Ben Peyton, Department of Fisheries and Wildlife, Michigan State University, East Lansing, MI

Report: *Additional Information on Non-Indigenous Plants in the United States—October 1992*

Author: Richard N. Mack, Department of Botany, Washington State University, Pullman, WA

Report: *Addendum to Report: Bioengineered Organisms and Non-Indigenous Species—March 1992*

Authors: Sheldon Krimsky, Department of Urban and Environmental Policy, Tufts University, Medford, MA

Richard Wetzler, Center for Environmental Management, Tufts University, Medford, MA

Report: *Additional Information on Recently Detected Non-Indigenous Insects—June 1993*

Author: E. Richard Hoebeke, Department of Entomology, Cornell University, Ithaca, NY

Report: *Review of OTA's Script for Scenarios Exercise—July 1992*

Author: Ann Z. Kulp, Annandale, VA

The following individuals provided additional reviews on certain papers:

Advisory Panel Reviewers:

Marion Cox, William Flemer, III, Joseph P. McCraren, and Howard M. Singletary, Jr.

Outside Reviewers:

John O. Browder, Department of Urban Affairs and Planning, Virginia Polytechnic Institute and State University, Blacksburg, VA

Ted Center, Aquatic Plant Management Laboratory, Agricultural Research Service, U.S. Department of Agriculture, Ft. Lauderdale, FL

Michael J. Crawley, Department of Biology, Imperial College at Silkwood Park, U.K.

Luther Val Giddings, Biotechnology Biologics and Environmental Protection, Animal and Plant Health Inspection Service, U.S. Department of Agriculture, Hyattsville, MD

Francis G. Howarth, Department of Entomology, Bishop Museum, Honolulu, HI

■ Part 2: Workshop Participants: Workshop on Management and Policy Decisionmaking

R. Joseph Abrell, Great Smoky Mountain National Park Headquarters, National Park Service, U.S. Department of the Interior, Gatlinburg, TN

Faith Thompson Campbell, Natural Resources Defense Council, Washington, DC

Mark J. Cochran, Department of Agricultural Economics, University of Arkansas, Fayetteville, AR

M. Lynne Corn, Environment and Natural Resources, Congressional Research Service, Library of Congress, Washington, DC

Peter Kareiva, Department of Zoology, University of Washington, Seattle, WA

Stephen R. Kellert, School of Forestry and Environmental Studies, Yale University, New Haven, CT

Lynn A. Maguire, School of Forestry and Environmental Studies, Duke University, Durham, NC

Marshall Meyers, Pet Industry Joint Advisory Council, Washington, DC

Alan Randall, Department of Agricultural Economics and Rural Sociology, Ohio State University, Columbus, OH

Craig J. Regelbrugge, Regulatory Affairs and Grower Services, American Association of Nurseryman, Washington, DC

Matthew H. Royer, Plant Protection and Quarantine, Animal and Plant Health Inspection Service, U.S. Department of Agriculture, Hyattsville, MD

Mark Sagoff, Institute for Philosophy and Public Policy, University of Maryland, College Park, MD

Michael Shannon, Plant Protection Quarantine, Animal and Plant Health Inspection Service, U.S. Department of Agriculture, Hyattsville, MD

Terry R. Steinwand, North Dakota Department of State Game and Fish, Bismarck, ND

■ Part 3: The Assessment's Reviewers

This assessment was reviewed either totally or in part by the study's Advisory Panel (listed on p. iv); certain authors of its contractor reports; and another group of external experts.

These authors took part
(see Part 1 for affiliations):

Michael J. Bean
Joseph C. Britton
Mark J. Cochran
Walter R. Courtenay, Jr.
Peter Kareiva
Ke Chung Kim
Donald H. Kludy
Sheldon Krimsky
Richard N. Mack
Alan Randall
John Schnittker
Calvin L. Schoulties
Stanley A. Temple
Richard Wetzler

Also, the following people contributed reviews:

Fred S. Betz, Environmental Fate and Effects Division, U.S. Environmental Protection Agency, Washington, DC

Michael E. Buck, Division of Forestry and Wildlife, Department of Land and Natural Resources, State of Hawaii, Honolulu, HI

D. Scot Campbell, International Services, Animal and Plant Health Inspection Service, U.S. Department of Agriculture, Hyattsville, MD

Joseph F. Coates, Coates and Jarratt, Inc., Washington, DC

David Cottingham, Ecology and Conservation Office, National Oceanic and Atmospheric Administration, U.S. Department of Commerce, Washington, DC

Michael J. Crawley, Department of Biology, Imperial College at Silwood Park, Ascot, Berksire, U.K.

Albert Cutress, Ministry of Forestry, Wellington, New Zealand

Gary B. Edwards, Fisheries Division, Fish and Wildlife Service, U.S. Department of the Interior, Washington, DC

Robert E. Eplee, Whiteville Plant Methods Center, Animal and Plant Health Inspection Service, U.S. Department of Agriculture, Whiteville, NC

David E. Giamporcaro, TSCA Biotechnology Program, Chemical Control Division, Office of Pollution Prevention and Toxics, U.S. Environmental Protection Agency, Washington, DC

Rebecca Goldburg, Environmental Defense Fund, New York, NY

Eric M. Hallerman, Department of Fisheries and Wildlife Sciences, Virginia Polytechnic Institute and State University, Blacksburg, VA

Charles A. Havens, Plant Protection and Quarantine, Animal and Plant Health Inspection Service, U.S. Department of Agriculture, Hyattsville, MD

Ginette Hemley, World Wildlife Fund, Washington, DC

Francis G. Howarth, Department of Entomology, Bishop Museum, Honolulu, HI

Will Kissinger, Plant Industry Division, Montana Department of Agriculture, Helena, MT

Dennis Lassuy, Fisheries Division, Fish and Wildlife Service, U.S. Department of the Interior, Arlington, VA

Lloyd L. Loope, Haleakala National Park, National Park Service, U.S. Department of the Interior, Makawao, HI

Lynn A. Maguire, School of the Environment, Duke University, Durham, NC

James A. McCann, National Fisheries Research Center, Fish and Wildlife Service, U.S. Department of the Interior, Gainesville, FL

Elizabeth Milewski, Office of Prevention, Pesticides and Toxic Substances, U.S. Environmental Protection Agency, Washington, DC

Edward L. Mills, Biological Field Station, Department of Natural Resources, Cornell University, Bridgeport, NY

Abdul Moeed, Ministry for the Environment, Wellington, New Zealand

Barbra Mullin, Agricultural and Biological Sciences Division, Montana Department of Agriculture, Helena, MT

John G. Nickum, Division of Fish Hatcheries, Fish and Wildlife Service, U.S. Department of Interior, Washington, DC

John M. Randall, Nature Conservancy Consumnes River Preserve, Galt, CA

Mark J. Reeff, International Association of Fish and Wildlife Agencies, Washington, DC

Craig J. Regelbrugge, Regulatory Affairs and Grower Services, American Association of Nurseryman, Washington, DC

Jay Rendall, Minnesota Interagency Exotic Species Task Force, Division of Fish and Wildlife, Department of Natural Resources, St. Paul, MN

Matthew H. Royer, Plant Protection and Quarantine, Animal and Plant Health Inspection Service, U.S. Department of Agriculture, Hyattsville, MD

Mark Sagoff, Institute for Philosophy and Public Policy, University of Maryland, College Park, MD

William Schneider, Office of Pesticide Programs, U.S. Environmental Protection Agency, Washington, DC

W. Curtis Sharp, Soil Conservation Service, U.S. Department of Agriculture, Washington, DC

Stanwyn G. Shetler, National Museum of Natural History, Smithsonian Institution, Washington, DC

Calvin R. Sperling, Agricultural Research Service, U.S. Department of Agriculture, Beltsville, MD

Timothy A. Sullivan, World Conservation Union, Species Survival Commission, Brookfield, IL

Lewis H. Waters, Bureau of Land Management, U.S. Department of Interior, Washington, DC

Randy G. Westbrooks, Whiteville Plant Methods Center, Animal and Plant Health Inspection Service, U.S. Department of Agriculture, Whiteville, NC

∎ Part 4: State Officials Who Reviewed State Laws and Regulations and Who Responded to the OTA Survey (alphabetically by State)

Kathleen Meddleton, Wildlife Biologist-Permit Section, Division of Wildlife Conservation, Department of Fish and Game, Juneau, AK

Charles D. Kelley, Director, Division of Game and Fish, Department of Conservation and Natural Resources, Montgomery, AL

Steve N. Wilson, Director, Arkansas Game and Fish Commission, Little Rock, AR

Duane L. Shroufe, Director, Game and Fish Department, Phoenix, AZ

Richard Kahn, Senior Biologist, Division of Wildlife, Department of Natural Resources, Denver, CO (now retired)

Robert M. Brantly, Executive Director, Game and Fresh Water Fish Commission, Department of Natural Resources, Tallahassee, FL

Bill Fletcher, Regional Game Management Supervisor, Game and Fish Division, Department of Natural Resources, Gainesville, GA

Calvin Lum, Animal Industry Administrator, Hawaii Department of Agriculture, Honolulu, HI

Larry M. Nakahara, Manager, Plant Quarantine Branch, Division of Plant Industry, Hawaii Department of Agriculture, Honolulu, HI

Rick McGeough, Bureau Chief, Department of Natural Resources, Des Moines IA

Lloyd Oldenburg, Wildlife and Research Manager, Bureau of Wildlife, Fish and Game Department, Boise, ID

John Tranquilli, Director, Office of Resource Management, Department of Conservation, Springfield, IL

Gregg McCollam, Chief of Administrative Services, Division of Fish and Wildlife, Department of Natural Resources, Indianapolis, IN

Bill D. Hlavachick, Chief, Wildlife Management Section, Fish and Wildlife Division, Department of Wildlife and Parks, Pratt, KS

Hugh A. Bateman, Administrator, Wildlife Division, Department of Wildlife and Fisheries, Baton Rouge, LA

Thomas W. French, Assistant Director, Division of Fisheries, Natural Heritage and Endangered Species Program, Boston, MA

Mary Jo Scanlan, Wildlife Permits Coordinator, Wildlife Division, Department of Natural Resources, Annapolis, MD

R. Steven Early, Assistant to the Director, Fisheries Division, Department of Natural Resources, Annapolis, MD

Henry Hilton, Staff Wildlife Biologist, Department of Inland Fisheries and Wildlife, Augusta, ME

Jay Rendall, Exotic Species Coordinator, Minnesota Interagency Exotic Species Task Force, Division of Fish and Wildlife, Department of Natural Resources, St. Paul, MN

Ollie Torgerson, Chief, Wildlife Division, Department of Conservation, Jefferson City, MO

Bill Thomason, Chief of Game, Department of Wildlife, Fisheries and Parks, Jackson, MS

Heidi Youmans, Special Projects Coordinator, Department of Fish, Wildlife, and Parks, Helena, MT

Gary Burke, Special Investigations Unit, Department of Fish, Wildlife, and Parks, Helena, MT

R.B. Hamilton, Assistant Director, Wildlife Resources Commission, Raleigh, NC

Rex Sohn, Disease Research Supervisor, Department of Game and Fish, Bismarck, ND

Ken Johnson, Chief, Wildlife Division, Game and Parks Commission, Lincoln, NE

Nancy L. Girard, Legal Coordinator, Fish and Game Department, Concord, NH

Robert McDowell, Director, Fish, Game and Wildlife Division, Environmental Protection Department, Trenton, NJ

Patrick P. Martin, Senior Wildlife Biologist, New York State Department of Environmental Conservation, Albany, NY

Richard T. Scott, Executive Administrator, Law Enforcement, Division of Wildlife, Department of Natural Resources, Columbus, OH

Larry Taylor, Chief, Law Enforcement, Department of Wildlife Conservation, Oklahoma City, OK

J.R. Fagan, Director, Law Enforcement, Pennsylvania Game Commission, Harrisburg, PA

Michael Lapisky, Deputy Chief, Wildlife Division, Department of Environmental Management, Fish, and Wildlife, Wakefield, RI

Ron Fowler, Game Staff Specialist, Game, Fish and Parks Department, Pierre, SD

Walter Cook, Wildlife Officer II, Tennessee Wildlife Resources Agency, Nashville, TN

Randy Radant, Chief, Nongame Management, Division of Wildlife Resources, Salt Lake City, UT

Timothy VanZandt, Commissioner, Department of Fish and Wildlife, Agency of Natural Resources, Waterbury, VT

S. Shapiro Hurley, Wildlife Health Specialist, Department of Natural Resources, Madison, WI

James Ruckel, Assistant Chief, Division of Natural Resources, Charleston, WV

Francis Petera, Director, Department of Game and Fish, Cheyenne, WY

Appendix C: References

CHAPTER 1: SUMMARY, ISSUES, AND OPTIONS

1. Alfieri, S.A., Jr., Assistant Director, Division of Plant Industry, Florida Department of Agriculture and Consumer Services, Tallahassee FL, "Regulatory Pest Control Philosophy: Views and Assessment," paper presented at the meeting of the National Plant Board, Kalispell, MT, August 1991.

2. Arner, D.H., Jones, J., and Bucciantini, C., "Carolina Clover: High Prospect for the Conservation Reserve Program," *Journal of Soil and Water Conservation*, vol. 47, No. 4, July-August 1992, pp. 292-293.

3. Asbury, C.H., "The Orphan Drug Act," *Journal of the American Medical Association*, vol. 265, No. 7, Feb. 20, 1991, pp. 893-897.

4. Backiel, A. et al., "The Major Federal Land Management Agencies: Management of Our Nation's Lands and Resources," CRS Report for Congress, 93-197 ENR, Congressional Research Service, Library of Congress, Washington, DC, Feb. 8, 1993.

5. Barnard, J., "Forest Slams Shut Wolf-trapping Plans," *Seattle Times*, Jan. 19, 1990.

6. Bean, M.J., "The Role of the U.S. Department of the Interior in Non-Indigenous Species Issues," contractor report prepared for the Office of Technology Assessment, November 1991.

7. Bergsman, J., "Changes in Pet Regulations Raise Howls," *Seattle Times*, May 12, 1992, p. C-1.

8. Britton, J.C., "Pathways and Consequences of the Introduction of Non-Indigenous Freshwater, Terrestrial, and Estuarine Mollusks in the United States," contractor report prepared for the Office of Technology Assessment, October 1991.

9. Brown, L.L., "A Legislative History of Outdoor Recreation User Fees," CRS Report for Congress, 92-645 ENR, Congressional Research Service, The Library of Congress, Washington, DC, Aug. 14, 1992.

10. Browning, H.R. and Charudattan, R., Executive Summary of the National Workshop on Regulatory Issues," *Regulations and Guidelines: Critical Issues in Biological Control—Proceedings of a USDA/CSRS National Workshop*, R. Charudattan and H.W. Browning (eds.) (Gainesville, FL: Institute of Food and Agricultural Sciences, University of Florida, 1992), pp. 199-201.

11. California Department of Food and Agriculture, Division of Plant Industry, "Annual Report," Sacramento, CA, 1990, pp. 16-17.

12. Carr, D.A., "Prosecutors Out of Control," *ECO*, vol. 1, No. 1, June 1993, pp. 56, 58.

13. Carter, P.C.S., "Risk Assessment and Pest Detection Surveys for Exotic Pests and Diseases Which Threaten Commercial Forestry in New Zealand," *New Zealand Journal of Forestry Science*, vol. 19, Nos. 2/3, 1989, pp. 353-74.

14. Chadwell, J., "Marshall Meyers: The Environment and the Pet Industry," *Pet Product News* (Mission Viejo, CA), vol. 47, No. 5, May 1993, pp. 6-8.

15. Comp, T.A. (ed.), *Blueprint for the Environment—A Plan for Federal Action* (Salt Lake City, UT: Howe Brothers, 1989).

16. Cook, A., "Grub Attack Refund," *The Washington Post*, Home Section, June 4, 1992, p. 4.

17. Cottingham, D., Director, Ecology and Conservation Office, National Oceanic and Atmospheric Administration, U.S. Department of Commerce, Washington, DC, letter to P.N. Windle, Office of Technology Assessment, July 31, 1992.

18. Cottingham, D., Director, Ecology and Conservation Office, National Oceanic and Atmospheric Administration, U.S. Department of Commerce, Washington, DC, memorandum for agency distribution, Jan. 6, 1993.

19. Coulson, J.R. and Soper, R.S., "Protocols for the Introduction of Biological Control Agents in the United States," *Plant Protection and Quarantine. Volume III. Special Topics*, R.P. Kahn (ed.) (Boca Raton, FL: CRC Press, Inc, 1989), pp. 1-35.

20. Council for Agricultural Science and Technology, "Ecological Impacts of Federal Conservation and Cropland Reduction Programs," Task Force Report No. 117, Ames, IA, September 1990.

21. Counts, C.L. III, "The Zoogeography and History of the Invasion of the United States by *Corbicula Fluminea* (Bivalvia: Corbiculidae), *American Malacological Bulletin*, Special Edition No. 2, 1986, pp. 7-39.

22. Courtenay, W.R., Jr., "Fish Introductions and Translocations, and Their Impacts in Australia," *Introduced and Translocated Fishes and Their Ecological Effects*, Proceedings No. 8, Australian Society for Fish Biology Workshop, Magnetic Island, Aug. 24-25, 1989, D.A. Pollard (ed.) (Canberra: Australian Government Publishing Service, 1990), pp. 171-179.

23. Courtenay, W.R., Jr., "Pathways and Consequences of the Introduction of Non-Indigenous Fishes in the United States," contractor report prepared for the Office of Technology Assessment, September 1991.

24. Courtenay, W.R., Jr., Professor of Zoology, Department of Biological Sciences, Florida Atlantic University, letter to P.N. Windle, Office of Technology Assessment, July 27, 1992.

25. Craighead, F.C., Jr. and Dasmann, R.F., "Exotic Big Game on Public Lands" (Washington, DC:

U.S. Department of the Interior, Bureau of Land Management, 1966).

26. Doren, R.F. and Whiteaker, L.D., "The Exotic Pest Plant Council: An Example of Interagency Cooperation To Solve Resource Related Problems," *Proceedings of the Symposium on Exotic Pest Plants*, T.D. Center et al. (eds.), Technical Report NPS/NREVER/NRTR-91/-6 (Washington, DC: National Park Service, U.S. Department of the Interior, September 1991), pp. 111-114.

27. Dottavio, F.D. et al. (eds.), *Protecting Biological Diversity in the National Parks: Workshop Recommendations*, Transactions and Proceedings Series No. 9 (Washington, DC: U.S. Department of the Interior, National Park Service, 1990).

28. Doyle, P., *States as Water Quality Financiers: Legislative Options for the 1990s* (Denver, CO: National Conference of State Legislatures, May 1991).

29. Drost, C.A. and Fellers, G.M., National Park Service, U.S. Department of the Interior, "Draft Handbook for the Removal of Non-Native Animals," Point Reyes, CA, 1992.

30. Edwards, G.E., Assistant Director, Fisheries, U.S. Fish and Wildlife Service, personal communication to Office of Technology Assessment, letter to P.N. Windle, Office of Technology Assessment, Aug. 7, 1992.

31. Elston, R.A., "Effective Applications of Aquaculture Disease-Control Regulations: Recommendations From an Industry Viewpoint," *Dispersal of Living Organisms into Aquatic Ecosystems*, A. Rosenfield and R. Mann (eds.) (College Park, MD: Maryland Sea Grant, 1992), pp. 353-359.

32. Environmental Law Review Committee, "Environmental Legislation: The Increasing Costs of Regulatory Compliance to the City of Columbus," report to the Mayor and City Council of the City of Columbus, OH, May 13, 1991.

33. Fleming, P., Rock Creek Park, National Park Service, U.S. Department of the Interior, Washington, DC, personal communication to Office of Technology Assessment, Oct. 9, 1991.

34. Gallagher, J.E. and Haller, W.T., "History and Development of Aquatic Weed Control in the United States," *Review of Weed Science*, vol. 5, December 1990, pp. 115-192.

35. Geist, V., "Endangered Species and the Law," *Nature*, vol. 357, No. 6376, May 28, 1992, pp. 274-276.

36. Glosser, J.W., Administrator, USDA Animal and Plant Health Inspection Service, U.S. Department of Agriculture, testimony at hearings before the Senate Subcommittee on Agricultural Research and General Legislation, Committee on Agriculture, Nutrition, and Forestry, "Preparation for the 1990 Farm Bill: Noxious Weeds," Mar. 28, 1990, pp. 350-354, 376-379.

37. Glosser, J.W., Director, Animal and Plant Health Inspection Service, U.S. Department of Agriculture, testimony at hearings before the U.S. House Committee on Appropriations, Subcommittee on Agriculture, Rural Development, and Related Agencies, *Hearings on Agriculture, Rural Development and Related Agencies Appropriations for 1992: Part 4*, Serial No. 43-171 O (Washington, DC: Government Printing Office, May 1991), pp. 271-467.

38. Hallerman, E.R., Assistant Professor, Department of Fisheries and Wildlife Sciences, Virginia Polytechnic Institute and State University, Blacksburg, Virginia, personal communication to the Office of Technology Assessment, Mar. 7, 1991.

39. Harty, F.M., "How Illinois Kicked the Exotic Habit," conference on Biological Pollution: the Control and Impact of Invasive Exotic Species, Indiana Academy of Sciences, Indianapolis, IN, Oct. 26, 1991.

40. Havens, C., Chief Operations Officer, Animal and Plant Health Inspection Service, U.S. Department of Agriculture, Hyattsville, MD, personal communication to S.M. Fondriest, Office of Technology Assessment, July 23, 1992.

41. Hester, F.E., "The U.S. National Park Service Experience With Exotic Species," *Natural Areas Journal*, vol. 11, No. 3, July 1991, pp. 127-128.

42. Howarth, F.G. and Medeiros, A.C., "Non-Native Invertebrates," *Conservation Biology in Hawai'i*, C.P. Stone and D.B. Stone (eds.) (Honolulu, HI: University of Hawaii Cooperative National Park Resources Studies Unit, 1989), pp. 82-87.

43. Hubbell, S., "Ladybugs," *The New Yorker*, Oct. 7, 1991, pp. 103-111.

44. International Union for the Conservation of Nature and Natural Resources, Species Survival Commission in collaboration with the Commission on Ecology and the Commission on Environmental Policy, Law and Administration, "The IUCN Position Statement on Translocation of Living Organisms Introductions, Re-Introductions and Re-Stocking" (Gland, Switzerland: IUCN, 1987).

45. Johnson, G., National Park Service, Washington, DC, letter to E.A. Chornesky, Office of Technology Assessment, Sept. 27, 1991.

46. Kareiva, P. et al., "Risk Analysis as a Tool for Making Decisions About the Introduction of Non-Indigenous Species Into the United States," contractor report prepared for the Office of Technology Assessment, July 1991.

47. Keystone Center, *Final Consensus Report of the Keystone Policy Dialogue on Biological Diversity on Federal Lands* (Keystone, CO: The Keystone Center, April 1991).

48. Kim, K.C. and Wheeler, A.G., Jr., "Pathways and Consequences of the Introduction of Non-Indigenous Insects and Arachnids in the United States," contractor report prepared for the Office of Technology Assessment, December 1991.

49. Kludy, D.H., "Federal Policy on Non-Indigenous Species: The Role of the United States Department of Agriculture's Animal and Plant Health Inspection Service," contractor report prepared for the Office of Technology Assessment, Washington, DC, December 1991.

50. Kludy, D.H., Consultant, Richmond, VA, letter to P.N. Windle, July 30, 1992.

51. Kohler, C.C. and Courtenay, W.R., Jr., "American Fisheries Society Position on Introductions of Aquatic Species," *Fisheries*, vol. 11, No. 2, March/April 1986, pp. 39-42.

52. Koller, G., "Native Dictates," *Arnoldia*, vol. 52, No. 4, winter 1992, pp. 23-32.

53. Koller, G., Senior Horticulturist, Arnold Arboretum, personal communication to K.E. Bannon, Office of Technology Assessment, June 2, 1993.

54. Kucewicz, W.P., "Grime and Punishment," *ECO*, vol. 1, No. 1, June 1993, pp. 50-55.

55. Kurdila, J., "The Introduction of Exotic Species into the United States: There Goes the Neighbor-

hood!'' *Boston College Environmental Affairs Law Review*, vol. 16, No. 1, 1988, pp. 95-118.

56. Langston, A., Animal and Plant Health Inspection Service, Hyattsville, MD, personal communication to E.A. Chornesky, Office of Technology Assessment, May 6, 1991.

57. Lattin, J., Professor of Entomology, Oregon State University, Corvallis, OR, memorandum to Bill Wright, Administrator, Plant Division, Oregon Department of Agriculture, Salem, OR, Nov. 26, 1991.

58. Lederberg, J., Shope, R.E., and Oaks, S.C., Jr. (eds.), *Emerging Infections—Microbial Threats to Health in the United States* (Washington, DC: National Academy Press, 1992).

59. Macdonald, I.A.W. et al., ''Wildlife Conservation and the Invasion of Nature Reserves by Introduced Species: Global Perspective,'' *Biological Invasions: A Global Perspective*, J.A. Drake et al. (eds) (New York, NY: John Wiley and Sons, 1989), pp. 215-255.

60. Mack, R.N., ''Pathways and Consequences of the Introduction of Non-Indigenous Plants in the United States,'' contractor report prepared for the Office of Technology Assessment, September 1991.

61. Mack, R.N., Professor and Chair, Department of Botany, Washington State University, Pullman, WA, letter to P.N. Windle, Office of Technology Assessment, Aug. 4, 1992.

62. Mack, R.N., ''Additional Information on Non-Indigenous Plants in the United States,'' contractor report prepared for the Office of Technology Assessment, October 1992.

63. McCann, J., Center Director, National Fisheries Research Center, U.S. Fish and Wildlife Service, Gainesville, FL, letter to P.N. Windle, Office of Technology Assessment, Aug. 3, 1992.

64. McGregor, R.C., *The Emigrant Pests*, a report to F.J. Mulhern, Administrator, Animal and Plant Health Inspection Service, U.S. Department of Agriculture, May 1973.

65. Mech, L.D., ''Wolf/Dog Hybirds,'' *Species—Newsletter of the Species Survival Commission* (International Union for the Conservation of Nature), No. 15, December 1990, pp. 66-67.

66. Metterhouse, W.M., ''Biological Control: State Regulatory View Point,'' *Regulations and Guidelines: Critical Issues in Biological Control—Proceedings of a USDA/CSRS National Workshop*, R. Charudattan and H.W. Browning (eds.) (Gainesville, FL: Institute of Food and Agricultural Sciences, 1992).

67. Meyers, M., General Counsel, Pet Industry Joint Advisory Council, Washington, DC, letter to P.T. Jenkins, Office of Technology Assessment, July 29, 1992.

68. Miller, D., Research Leader, Systematic Entomology Laboratory, U.S. Department of Agriculture, Agricultural Research Service, Hyattsville, MD, personal communication to E.A. Chornensky, Office of Technology Assessment, Apr. 16, 1991.

69. Miller, M. and Aplet, G., ''Biological Control: A Little Knowledge Is a Dangerous Thing,'' *Rutgers Law Review*, vol. 45, No. 2, winter 1993, pp. 285-334.

70. Minnesota Exotic Species Interagency Task Force, *Report and Recommendations* (St. Paul, MN: State of Minnesota, Department of Natural Resources, July 1991).

71. Moody, M.E. and Mack, R.N., ''Controlling the Spread of Plant Invasions: the Importance of Nascent Foci,'' *The Journal of Applied Ecology*, vol. 25, 1988, pp. 1009-1021.

72. Nagel, S.S., ''Projecting Trends in Public Policy,'' *Policy Theory and Policy Evaluation*, S.S. Nagel (ed.) (Westport, CT: Greenwood Press, 1990), pp. 161-204.

73. National Association of State Departments of Agriculture, ''Honey Bee Pests—A Threat to the Vitality of U.S. Agriculture,'' Washington, DC, February 1991.

74. National Fish and Wildlife Foundation, *U.S. Fish and Wildlife Service. Fiscal Year 1992 Wildlife and Fisheries Assessment*, Washington, DC, April 1991.

75. National Parks and Conservation Association, *A Race Against Time: Five Threats Endangering America's National Parks and the Solutions to Avert Them* (Washington, DC: National Parks and Conservation Association, Aug. 20, 1991).

76. National Research Council, *Science and the National Parks* (Washington, DC: National Academy Press, 1992).

77. Nature Conservancy of Hawaii and the Natural Resources Defense Council, ''The Alien Pest Species Invasion in Hawaii: Background Study and Recommendations for Interagency Planning,'' July 1992.

78. Neesen, J., U.S. Department of Agriculture, Animals and Plant Health Inspection Service, Budget and Accounting Division, personal communication to Office of Technology Assessment, Nov. 9, 1992.

79. North American Native Fishes Association, Introductions Committee, memorandum to D. Lassuy, Intentional Introductions Review Committee, U.S. Aquatic Nuisance Species Task Force, Jan. 25, 1992.

80. Noxious Weed Technical Advisory Group, ''The Federal Noxious Weed Program Evaluation and Recommendations,'' report to Deputy Administrator, Animal and Plant Health Inspection Service, U.S. Department of Agriculture, Nov. 16, 1983.

81. Oberheu, J., Staff Specialist, U.S. Fish and Wildlife Service, testimony at hearing on Proposed Injurious Wildlife Regulations before the House Subcommittee on Fisheries and Wildlife, Dec. 12, 1974, p. 108.

82. O'Brien, S.J., ''Bureaucratic Mischief: Recognizing Endangered Species and Subspecies,'' *Science*, vol. 251, No. 4998, Mar. 8, 1991, pp. 1187-1188

83. Peoples, R.A., Jr., McCann, J.A., and Starnes, L.B., ''Introduced Organisms: Policies and Activities of the U.S. Fish and Wildlife Service,'' *Dispersal of Living Organisms Into Aquatic Ecosystems*, A. Rosenfield and R. Mann (eds.) (College Park, MD: Maryland Sea Grant, 1992), pp. 325-352.

84. Reichelderfer, K.H., ''The Expanding Role of Environmental Interests in Agricultural Policy,'' *Resources*, No. 102, winter 1991, pp. 4-7.

85. Repetto, R. et al., *Green Fees: How a Tax Shift Can Work for the Environment and the Economy* (Washington, DC: World Resources Institute, November 1992).

86. Schmitz, D.C., Florida Department of Natural Resources, Tallahassee, FL, statement submitted at hearings before the Senate Subcommittee on Agricultural Research and General Legislation, Committee on Agriculture, Nutrition, and Forestry, ''Preparation for the 1990 Farm Bill: Noxious Weeds,'' Mar. 28, 1990, pp. 357-364.

87. Schmitz, D.C., Department of Natural Resources, State of Florida, Tallahassee, FL, personal communication to P.N. Windle, Office of Technology Assessment, June 14, 1993.

88. Schneider, K., ''Where We Went Wrong on Regulation,'' *Eco*, June 1993, pp. 16-22.

89. Schoulties, C.L., ''Pathways and Consequences of the Introduction of Non-Indigenous Plant Pathogens in the United States,'' contractor report prepared for the Office of Technology Assessment, Washington, DC, December 1991.

90. Schoulties, C.L., Director, Division of Regulatory and Public Service Programs, Clemson University, Clemson, SC, personal communication to Office of Technology Assessment, July 31, 1992.

91. Schoulties, C.L., Director, Division of Regulatory and Public Service Programs, Clemson University, Clemson, SC, personal communication to P.N. Windle, Office of Technology Assessment, June 14, 1993.

92. Schuyler, P.T., *Introduced Animal Species Issues: Current Status and Problems Including a Feasibility Study for the Establishment of a Multi-Disciplinary Introduced Animal Species Center*, TRI Working Papers No. 56, 56A (New Haven, CT: Yale University School of Forestry and Environmental Studies, 1991).

93. Schuyler, P.T., ''Non-Indigenous Species in the United States: Selected Analysis of Introduced Animal Species Survey Data,'' contractor report prepared for the Office of Technology Assessment, Washington, DC, December 1991.

94. Sellars, R., ''The Roots of National Park Management,'' *Journal of Forestry*, vol. 90, January 1992, pp. 16-19.

95. Shields, E., *Funding Environmental Programs: An Examination of Alternatives* (Washington, DC: National Governors' Association, 1989).

96. Shotkin, A. and McCrea, E.J., ''Public Educational Efforts in the United States Regarding Prevention and Management of Non-Indigenous Species,'' contractor report prepared for the Office of Technology Assessment, December 1991.

97. Siddiqui, I.A., Assistant Director, California Department of Food and Agriculture, testimony at hearings before the Senate Subcommittee on Federal Services, Post Office, and Civil Service, Committee on Governmental Affairs, ''Oversight of the Agricultural Quarantine Enforcement Act (Washington, DC: U.S. Government Printing Office, June 5, 1991), pp. 36-39, 107-111.

98. Singletary, H.M., Director, Plant Industry Division, North Carolina Department of Agriculture, Statement submitted with testimony at hearings before the Senate Subcommittee on Agricultural Research and General Legislation, Committee on Agriculture, Nutrition, and Forestry, Mar. 28, 1990, pp. 354-356, 380-382.

99. Smart, B. (ed.), ''XII. A Look Into the Future,'' *Beyond Compliance—A New Industry View of the Environment* (Washington, DC: World Resources Institute, 1992), pp. 235-245.

100. Smith, C., Unit Leader, Cooperative National Parks Resources Studies Unit, University of Hawaii at Manoa, Honolulu, HI, letter to Office of Technology Assessment, July 1992.

101. Sprengelmeyer, E., ''Killing to Preserve,'' *Santa Fe Reporter*, Feb. 10-16, 1993, pp. 17-19.

102. Steering Committee for the 75th Symposium, *Report of National Parks for the 21st Century—The Vail Agenda*, Report and Recommendations to the Director of the National Park Service, National Park Service Document No. D-726, 1992.

103. Temple, S.A., ''The Nasty Necessity: Eradicating Exotics,'' *Conservation Biology*, vol. 4, No. 2, pp. 113-115, June 1990.

104. Temple, S.A. and Carroll, D.M., ''Pathways and Consequences of Introduction of Non-Indigenous Vertebrates in the United States,'' contractor report prepared for the Office of Technology Assessment, Washington, DC, October 1991.

105. Thomas, C., ''Reorganizing Public Organizations: Alternatives, Objectives, and Evidence,'' unpublished report prepared for the Secretary of Energy Advisory Board, U.S. Department of Energy, Washington, DC, August 1992.

106. U.S. Congress, House Committee on Appropriations, Subcommittee on Agriculture, Rural Development and Related Agencies, *Hearings on Agriculture, Rural Development and Related Agencies Appropriations for 1991: Part 4*, Serial No. 43-171 O, May 1, 1991.

107. U.S. Congress, House Committee on Appropriations, Subcommittee on the Department of the Interior and Related Agencies, *Hearings on Department of the Interior and Related Agencies Appropriations for 1992, Part 1, Justification of the Budget Estimates* (Washington, DC: U.S. Goverment Printing Office, 1991).

108. U.S. Congress, General Accounting Office, *Parks and Recreation: Recreation Fee Authorizations, Prohibitions, and Limitations*, GAO/RCED-86-149 (Washington, DC: GAO, 1986).

109. U.S. Congress, General Accounting Office, *Wildlife Protection: Enforcement of Federal Laws Could Be Strengthened*, GAO/RCED-91-44 (Washington, DC: GAO, April 1991).

110. U.S. Congress, General Accounting Office, *Pesticides: USDA's Research To Support Registration of Pesticides for Minor Crops*, GAO/RCED-92-190BR (Washington, DC: GAO, June 1992).

111. U.S. Congress, Joint Committee on Taxation, *Present Law and Background Relating to Federal Environmental Tax Policy*, JCS-6-90 (Washington, DC: U.S. Government Printing Office, Mar. 1, 1990).

112. U.S. Congress, Senate Committee on Agriculture, Nutrition, and Forestry, Subcommittee on Agricultural Research and General Legislation, *Preparation for the 1990 Farm Bill* (Washington, DC: Mar. 28, 1990), pp. 349-436.

113. U.S. Department of Agriculture, Forest Service, *The Report: A Forum for Cooperation*, Noxious Weed Workshop, Billings, MT, Feb. 14-17, 1989 (Washington, DC: U.S. Department of Agriculture, May 1989).

114. U.S. Department of the Interior, Bureau of Land Management, *Vegetation Diversity Project: A Research and Demonstration Program Plan. Restoration and Maintenance of Native Plant Diversity on Deteriorated Rangelands. Great Basin and Columbia Plateau* (Portland, OR: U.S. Department of the Interior, Bureau of Land Management, Oregon State Office, May 1990).

115. U.S. Department of the Interior, Bureau of Land Management, draft report, ''Evaluation: Bu-

reauwide Noxious Weed Program," Washington, DC, December 1991.

116. U.S. Department of the Interior, Fish and Wildlife Service, internal memorandum from Director to Regional Directors 1-8, Washington, DC, Oct. 20, 1987.

117. U.S. Department of the Interior, National Park Service, *State of the Parks 1980: A Report to Congress* (Washington, DC: Office of Science and Technology, May 1980).

118. U.S. Department of the Interior, National Park Service, and U.S. Department of Agriculture, Soil Conservation Service, *Native Plants for Parks*, D-425 (Washington, DC: U.S. Government Printing Office, 1989).

119. U.S. Department of the Interior, National Park Service, *Federal Recreation Fee Report to Congress 1991*, Washington, DC, 1992.

120. U.S. Department of the Interior, Office of Inspector General, *Semiannual Report*, Washington, DC, April 1992.

121. U.S. Fish and Wildlife Service Law Enforcement Advisory Commission, *Report of Findings and Recommendations*, Washington, DC, June 15, 1990.

122. U.S. Interagency Aquatic Nuisance Species Task Force, Washington, DC, *Proposed Aquatic Nuisance Species Program*, Sept. 28, 1992.

123. Van't Woudt, B.D., "Roaming, Stray, and Feral Domestic Cats and Dogs as Wildlife Problems," *Proceedings of the 14th Vertebrate Pest Conference*, L.R. Davis and R.E. Marsh (eds.) (Davis, CA: University of California, 1990), pp. 291-295.

124. Washington State Department of Agriculture, Plant Services Division, *Plant Quarantine Manual*, Seattle, WA, 1992.

125. Waterworth, H.E., "Our Plants' Ancestors Immigrated, Too," *BioScience*, vol. 31, No. 9, October 1981, p. 698.

126. West, A.J., Deputy Chief, State and Private Forestry, USDA Forest Service, Washington, DC, personal communication to Office of Technology Assessment, Aug. 24, 1992.

127. Westbrooks, R.G., "Interstate Sale of Aquatic Federal Noxious Weeds as Ornamentals in the United Sates," *Aquatics*, June 1990, pp. 16-24.

128. Westbrooks, R.G., Station Leader, Whiteville Noxious Weed Station, USDA Animal and Plant Health Inspection Service, Whiteville, NC, letter to P.W. Windle, Office of Technology Assessment, Aug. 11, 1992.

129. Westman, W.E., "Managing for Biodiversity," *BioScience*, vol. 40, No. 1, January 1990, pp. 26-33.

130. White, R.J., "We're Going Wild: A 30-year Transition from Hatcheries to Habitat," *Trout*, 30 year anniversary series, summer 1989, pp. 15-49.

131. Working Group on Non-*Apis* Bees, Memorandum to State Departments of Agriculture, W.P. Stephen and R.W. Thorp, Secretaries Pro-Tem., Department of Entomology, Oregon State University, Corvallis, OR, June 2, 1993.

132. Yount, J.D. (ed.), *Ecology and Management of the Zebra Mussel and Other Introduced Aquatic Nuisance Species*, report based on an EPA Workshop, Sept. 26-28, 1990, Saginaw, MI, EPA/600/3-91-003 (Washington, DC: U.S. Environmental Protection Agency, Office of Research and Development, February 1991).

CHAPTER 2: THE CONSEQUENCES OF HARMFUL NON-INDIGENOUS SPECIES

1. Anonymous, "Count it Quick, Before it's Gone," *The Economist*, Sept. 14, 1991, vol. 319, No. 7724, pp. 99-100.

2. Bangsund, D.A. and Leistritz, F.L., Department of Agricultural Economics, North Dakota State University, "Economic Impact of Leafy Spurge on Grazing Lands in the Northern Great Plains," Agricultural Economics Report No. 275-5, November 1991.

3. Bean, M.J., "The Role of the U.S. Department of the Interior in Non-Indigenous Species Issues," contractor report prepared for the Office of Technology Assessment, Washington, DC, November 1991.

4. Bedunah, D.J., "The Complex Ecology of Weeds, Grazing and Wildlife," *Western Wildlands*, vol. 18, No. 2, summer 1992, pp. 6-11.

5. Bergsman, J., "Changes in Pet Regulations Raise Howls," *Seattle Times*, May 12, 1992, p. C-1.

6. Brenchley, G.A. and Carlton, J.T., "Competitive Displacement of Native Mud Snails by Introduced Periwinkles in the New England Intertidal Zone," *Biological Bulletin*, vol. 165, December 1983, pp. 543-558.

7. Britton, J.C., "Pathways and Consequences of the Introduction of Freshwater, Terrestrial, and Estuarine Mollusks in the United States," contractor report prepared for the Office of Technology Assessment, October 1991.

8. Cochran, M.J., "Non-Indigenous Species in the United States: Economic Consequences," contractor report prepared for the Office of Technology Assessment, March 1992.

9. Conant, S., "Saving Endangered Species by Translocation—Are we Tinkering with Evolution?" *BioScience*, vol. 38, No. 4, 1988, pp. 254-57.

10. Courtenay, W., "Pathways and Consequences of the Introduction of Non-Indigenous Fishes in the United States," contractor report prepared for the Office of Technology Assessment, September 1991.

11. D'Antonio, C.M. and Vitousek, P.M., "Biological Invasions by Exotic Grasses, the Grass/Fire Cyclel, and Global Change," *Annual Review of Ecology and Systematics*, vol. 23, 1992, pp. 63-87.

12. Dickerson, W.A., "Gypsy Moth in the Southeastern United States," North Carolina Department of Agriculture report, Raleigh, NC, Apr. 16, 1991.

13. Dowling, T.E. and Childs, M.R., "Impact of Hybridization on a Threatened Trout of the Southwestern United States," *Conservation Biology*, vol. 6, No. 3, September 1992, pp. 355-364.

14. Edwards, G.B. and Cottingham, D., Co-Chairs of the Aquatic Nuisance Species Task Force, letter to E.A. Chornesky, Office of Technology Assessment, Nov. 25, 1992.

15. Engbring, J. and Fritts, T.H., "Demise of an Insular Avifauna: The Brown Tree Snake on Guam," *Transactions of the Western Section of the Wildlife Society*, vol. 24, 1988, pp. 31-37.

16. Ervin, K., "The Brink of Extinction," *The Seattle Times/Seattle Post-Intelligencer*, Nov. 25, 1990, pp. A1, A16.

17. Evangelou, P., Agricultural Economist, U.S. Department of Agriculture, Animal and Plant Health Inspection Service, letters to Office of Technology Assessment, Dec. 9, 1991 and Jan. 7, 1992, with attachment summarizing information from the APHIS Fiscal Year 1990 Budget Witness Book.

18. Fergus, C., "The Florida Panther Verges on Extinction," *Science*, vol. 251, Mar. 8, 1991, pp. 1178-1180.

19. Foy, C.L. et al., "History of Weed Introductions," *Exotic Plant Pests and North American Agriculture*, C.L. Wilson and C.L. Graham (eds.) (New York, NY: Academic Press, 1983), pp. 65-92.

20. Gallagher, J.E. and Haller, W.T., "History and Development of Aquatic Weed Control in the United States," *Review of Weed Science*, vol. 5, 1990, pp. 115-192.

21. Gangstad, E.O. and Cardarelli, N.F., "The Relation Between Aquatic Weeds and Public Health," *Aquatic Weeds: The Ecology and Management of Nuisance Aquatic Vegetation*, A.H. Pieterse and K.J. Murphy (eds.) (New York, NY: Oxford University Press, 1990), pp. 85-90.

22. Ganzhorn J. et al., "Dissemination of Microbial Pathogens Through Introductions and Transfers of Fishfish," *Dispersal of Living Organisms Into Aquatic Ecosystems*, A. Rosenfield and R. Mann (eds.) (College Park, MD: Maryland Sea Grant, 1992), pp. 175-192.

23. Goodman, B., "Keeping Anglers Happy Has a Price," *Bioscience*, vol. 41, No. 5, May 1991, pp. 294-299.

24. Great Lakes Fishery Commission, "Ruffe in the Great Lakes: A Threat to North American Fisheries," Report of the Ruffe Task Force, Ann Arbor, MI, March 1992.

25. Hadfield, M.G. and Miller, S.E., "Alien Predators and Decimation of Endemic Hawaiian Tree Snails," *Pacific Science*, vol. 46, No. 3, 1992, p. 395.

26. Haughton, C.S., *Green Immigrants: The Plants that Transformed America* (New York, NY: Harcourt Brace Jovanovich, 1978).

27. Hester, F.E., "The U.S. National Park Service Experience with Exotic Species," *Natural Areas*

Journal, vol. 11, No. 3, July 1991, pp. 127-128.

28. Horbold, B. and Moyle, P.B., "Introduced Species and Vacant Niches," *The American Naturalist*, vol. 128, No. 5, November 1986, pp. 751-760.

29. Howarth, F.G. and Ramsay, G.W., "The Conservation of Island Insects and Their Habitats," *The Conservation of Insects and Their Habitats*, N.M. Collins and J.A. Thomas (eds.) (New York, NY: Academic Press, 1991), pp. 71-107.

30. Johnston, G.H., Chief of Wildlife and Vegetation Division, Natural Resources Program Branch, National Park Service, U.S. Department of the Interior, personal communications to Office of Technology Assessment, July 10, 1991 and Mar. 13, 1992.

31. Kass, H., "Once a Savior, Moth is Now a Scourge," *Plant Conservation*, vol. 5, No. 2, summer 1990, p. 3.

32. Kay, S.H., "Hydrilla, a Rapidly Spreading Aquatic Weed Problem in North Carolina," Agriculture Bulletin—AG-449, North Carolina Cooperative Extensive Service (Raleigh, NC: North Carolina State University, May 1992).

33. Kim, K.C. and Wheeler, A.G., Jr., "Pathways and Consequences of the Introduction of Non-Indigenous Insects and Arachnids in the United States," contractor report prepared for the Office of Technology Assessment, December 1991.

34. Klassen, W., "Eradication of Introduced Arthropod Pests: Theory and Historical Practice," *Entomological Society of America, Miscellaneous Publications*, vol. 73, 1989, pp. 1-29.

35. Langdon, K.R. and Johnson, K.D., "Alien Forest Insects and Diseases in Eastern USNPS Units: Impacts and Interventions," *The George Wright Forum*, vol. 9, No. 1, 1992, pp. 2-14.

36. Lanka, B., "Analysis and Recommendations on the Applications by Mr. John T. Dorrence III to Import and Possess Native and Exotic Species," Wyoming Game and Fish Department, Cheyenne, Wyoming, Mar. 1, 1990.

37. Leonard, K., Director, Cereal Rust Lab, University of Minnesota, U.S. Department of Agriculture, St. Paul, MN, letter to P.N. Windle, Office of Technology Assessment, Jan. 3, 1992.

38. Lodge, D.M., "Species Invasions and Deletions: Community Effects and Responses to Climate and Habitat Change," *Biotic Interactions and Global Change* (Sunderland, MA: Sinauer and Assoc., 1993).

39. Loope, L.L. and Sanchez, P.G., "Biological Invasions of Arid Land Nature Reserve," *Biological Conservation*, vol. 44, 1988, pp. 95-118.

40. Luoma, J.R., "Boon to Anglers Turns Into a Disaster for Lakes and Streams," *New York Times*, Nov. 17, 1992.

41. Macdonald, I.A.W. et al., "Wildlife Conservation and the Invasion of Nature Reserves by Introduced Species: Global Perspective," *Biological Invasions: A Global Perspective*, J.A. Drake et al. (eds.) (New York, NY: John Wiley and Sons, Ltd., 1989), pp. 215-255.

42. Mace, G.M. and Lande, R., "Assessing Extinction Threats: Toward a Reevaluation of IUCN Threatened Species Categories," *Conservation Biology*, vol. 5, No. 2, June 1991, pp. 148-157.

43. Mack, R.N., "Pathways and Consequences of the Introduction of Non-Indigenous Plants in the United States," contractor report prepared for the Office of Technology Assessment, September 1991.

44. Mack, R.N., Chairman, Department of Botany, Washington State University, Pullman, WA, personal communication to P.N. Windle, Office of Technology Assessment, Aug. 4, 1992.

45. Marsh, P.C. and Langhorst, D.R., "Feeding and Fate of Wild Larval Razorback Sucker," *Environmental Biology of Fishes*, vol. 21, No. 1, 1988, pp. 59-67.

46. Master, L., "Aquatic Animals: Endangerment Alert," *Nature Conservancy*, March/April 1991, pp. 26-27.

47. Mayer, J.J. and Brisbin, I.L., Jr., Savannah River Ecology Laboratory, Aiken, South Carolina, "Wild Pigs in the United States: Their History, Comparative Morphology, and Current Status" (Athens, GA: University of Georgia Press, 1991).

48. McCann, J.A., "Involvement of the American Fisheries Society with Exotic Species, 1969-1982," *Distribution, Biology, and Management of Exotic Fishes*, W.R. Courtenay, Jr. and J.R. Stauffer, Jr. (eds.) (Baltimore, MD: Johns Hopkins University Press, 1984), pp. 1-7.

49. McCrea, E.J. and Shotkin, A., "Public Educational Efforts in the United States Regarding

Prevention and Management of Non-Indigenous Species,'' contractor report prepared for the Office of Technology Assessment, December 1991.

50. McKey, D.B. and Kaufmann, S.C., "Naturalization of Exotic *Ficus* Species (Moraceae) in South Florida," Department of the Interior, National Park Service, Technical Report NPS/NREVER/NRTR-91/06, September 1991.

51. McNab, W.H., "Oriental Bittersweet: Another Kudzu?" *Proceedings of the 16th Annual Hardwood Symposium* (Cashiers, NC: Hardwood Research Council, 1988), pp. 190-191.

52. Mech, L.D., "Wolf/Dog Hybrids," *Species* (Newsletter of the Species Survival Commission of IUCN—World Conservation Union), Brookfield, IL, No. 15, December 1990, pp. 66-67.

53. Minnesota Interagency Exotic Species Task Force, "Report and Recommendations," submitted to the Natural Resources Committees of the Minnesota House and Senate, St. Paul, MN, July 1991.

54. Missouri Department of Natural Resources, "Challenge of the '90s: Our Threatened State Parks," Jefferson City, MO, October 1991.

55. Mitchell, C.J. et al., "Isolation of Eastern Equine Encephalitis Virus From *Aedes albopictus* in Florida," *Science*, vol. 257, July 24, 1992, pp. 526-527.

56. Mooney, H.A. and Drake, J.A., "Biological Invasions: a SCOPE Program Overview," *Biological Invasions: A Global Perspective*, J.A. Drake et al. (eds.) (New York, NY: John Wiley and Sons, 1989), pp. 491-506.

57. Moyle, P.B., Li, H.W., and Barton, B.A. "The Frankenstein Effect: Impact of Introduced Fishes on Native Fishes in North America," *Fish Culture in Fisheries Management*, R.H. Stroud (ed.) (Bethesda, MD: American Fisheries Society, 1986), pp. 415-426.

58. Moyle, P.B. and Williams, J.E., "Biodiversity Loss in the Temperate Zone: Decline of the Native Fish Fauna of California," *Conservation Biology* vol. 4, No. 3, September 1990, pp. 275-284.

59. National Association of State Departments of Agriculture, "Honey Bee Pests—A Threat to the Vitality of U.S. Agriculture," Washington, DC, February 1991.

60. Nature Conservancy of Hawaii and the Natural Resources Defense Council, "The Alien Pest Species Invasion in Hawaii: Background Study and Recommendations for Interagency Planning," July 1992.

61. Neill, W.M., "The Tamarisk Invasion of Desert Riparian Areas," *CalEPPC News*, vol. 1, No. 1, winter 1993, pp. 6-7.

62. O'Brien, S.J., "Bureaucratic Mischief: Recognizing Endangered Species and Subspecies," *Science*, vol. 251, Mar. 8, 1991, pp. 1187-88.

63. Paul, J., Communications Director, National Pest Control Association, personal communication, Oct. 30, 1992.

64. Peine, J.D. and Farmer, J.A., "Wild Hog Management Program at Great Smoky Mountains National Park," *Proceedings of the 14th Vertebrate Pest Conference*, L.R. Davis and R.E. Marsh (eds.) (Davis, CA: University of California, 1990), pp. 221-227.

65. Perdue, R.E., Jr. and Christenson, G.M., "Plant Exploration," *Plant Breeding Reviews: Volume 7 The National Plant Germplasm System of the United States*, J. Janick (ed.) (Portland, OR: Timber Press, 1989), pp. 67-94.

66. Pimentel, D.H. et al., "Environmental and Economic Effects of Reducing Pesticide Use," *Handbook on Pest Management in Agriculture*, D.H. Pimentel (ed.) (Boca Raton, FL: CRC Press, 1991).

67. Sailer, R.I., "History of Insect Introductions," *Exotic Plant Pests and North American Agriculture*, C.L. Wilson and C.L. Graham (eds.) (New York, NY: Academic Press, 1983), pp. 15-38.

68. Savidge, J.A., "Extinction of an Island Forest Avifauna by an Introduced Snake," *Ecology*, vol. 68, No. 3, 1987, pp. 660-668.

69. midt, B.R., Chief of Fisheries, Division of Wildlife Resources, Utah Department of Natural Resources, Salt Lake City, UT, letter to P.N. Windle, Office of Technology Assessment, Dec. 16, 1991.

70. Schmitz, D.C. et al., "The Ecological Impact of Three Invasive Alien Plant Species in Florida: A Review," unpublished manuscript.

71. Schoener, A.A., "The Role of Competition in the Replacement of Native Fishes by Introduced Species," *Fishes in North American Deserts*, R.J. Naiman and D.L. Soltz (eds.) (New York, NY: John Wiley and Sons, 1981), pp. 173-203.

72. Schoulties, C.L., "Pathways and Consequences of the Introduction of Non-Indigenous Plant Pathogens in the United States," contractor report prepared for the Office of Technology Assessment, Washington, DC, December 1991.

73. Showler, A.T. and Reagan, T.E., "Ecological Interactions of the Red Imported Fire Ant in the Southeastern United States," *J. Entomol. Sci. Suppl.*, vol. 1, 1987, pp. 52-64.

74. Soulé, M.E., "Conservation: Tactics for a Constant Crisis," *Science*, vol. 253, Aug. 16, 1991, pp. 744-750.

75. Spaulding, W.M., Jr. and McPhee, R.J.—Bi-National Evaluation Team (submitted by), "The Report of the Evaluation of the Great Lakes Fishery Commission by the Bi-National Evaluation Team: Volume 2, An Analysis of the Economic Contribution of the Great Lakes Sea Lamprey Program" (Twin Cities, MN: U.S. Fish and Wildlife Service, Nov. 28, 1989).

76. Spencer, C.N. et al., "Shrimp Stocking, Salmon Collapse, and Eagle Displacement," *Bioscience*, vol. 41, No. 1, January 1991, pp. 14-21.

77. Struzik, E., "The Game Ranch," *International Wildlife*, vol. 22, July/August 1992, pp. 18-20, 22-24.

78. Taylor, O.R., Jr., "African Bees: Potential Impact in the United States," *Bulletin of the Entomological Society of America*, winter 1985, pp. 14-24.

79. Temple, S.A. and Carroll, D.M., "Pathways and Consequences of Introduction of Non-Indigenous Vertebrates in the United States," contractor report prepared for the Office of Technology Assessment, October 1991.

80. Terborgh, J., "Why American Songbirds are Vanishing," *Scientific American*, May 1992, pp. 98-104.

81. Texas Department of Agriculture, Agriculture Development Program, "Exotic Game in Texas: An Overview of Commercial Potential," Austin, TX, February 1989.

82. Thomas, L.K., Jr., "The Impact of Three Exotic Plant Species on Potomac Island," U.S. Department of the Interior, National Park Service Scientific Monograph Series, No. 13, Washington, DC, 1980.

83. Thompson, D.Q. et al., "Spread, Impact, and Control of Purple Loosestrife (*Lythrum salicaria*) in North American Wetlands," United States Fish and Wildlife Service, Washington, DC, 1987.

84. Toland, B., "Use of Forested Spoil Islands by Nesting American Oystercatchers in Southeast Florida," *Journal of Field Ornithology*, vol. 63, No. 2, spring, 1992, pp. 155-158.

85. Turner, C.E., "Conflicting Interests and Biological Control of Weeds," *Proc. IV Int. Symp. Biol. Contr. Weeds*, August 1984, pp. 203-225.

86. U.S. Congress, General Accounting Office, "Great Lakes Fishery Commission—Actions Needed to Support an Expanded Program," NSIAD-92-108 (Gaithersburg, MD: U.S. General Accounting Office, March 1992).

87. U.S. Congress, Office of Technology Assessment, *Technologies To Maintain Biological Diversity*, OTA-F-330 (Washington, DC: U.S. Government Printing Office, March 1987).

88. U.S. Congress, Office of Technology Assessment, *Bioremediation for Marine Oil Spills—Background Paper*, OTA-BP-O-70 (Washington, DC: U.S. Government Printing Office, May 1991).

89. U.S. Department of Agriculture, "Kudzu for Erosion Control in the Southeast," Farmer's Bulletin No. 1840, 1944.

90. U.S. Department of Agriculture, Animal and Plant Health Inspection Service, "Imported Fire Ant: A Guide for Nursery Operators," program aid no. 1420, 1989.

91. U.S. Department of Agriculture, Forest Service, "Pest Risk Assessment of the Importation of Larch from Siberia and the Soviet Far East," Miscellaneous Publication No. 1495, September 1991.

92. U.S. Department of Agriculture, Forest Service, "Forest Service Manual 2600—Wildlife, Fish, and Sensitive Plant Habitat Management WO Amendment 2600-91-6," Sept. 24, 1991.

93. U.S. Department of Agriculture, Soil Conservation Service, "Improved Conservation Plant Materials Released by the SCS and Cooperators Through December 1990," 1990.

94. U.S. Department of the Interior, Bureau of Reclamation, Lower Colorado Region, "Vegetation Management Study—Lower Colorado River," Phase I, September 1992.

95. U.S. Department of the Interior, Fish and Wildlife Service, *Restoring America's Wildlife*, (Washington, DC: U.S. Government Printing Office, 1987).

96. von Broembsen, S.L., "Invasions of Natural Ecosystems by Plant Pathogens," *Biological Invasions: a Global Perspective* (New York, NY: John Wiley and Sons, 1989), pp. 77-83.

97. Webb, D.A., "What Are the Criteria for Presuming Native Status?," *Watsonia*, vol. 15, 1985, pp. 231-236.

98. Westbrooks, R.G., *Poisonous Plants of Eastern North America* (Columbia, SC: University of South Carolina Press, 1986).

99. Westman, W.E., "Managing for Biodiversity—Unresolved Science and Policy Questions," *BioScience*, vol. 40, No. 1, 1990, pp. 26-33.

100. Whitson, T.D. et al., *Weeds of the West* (Jackson, WY: Pioneer of Jackson Hole, January 1991).

101. Williams, M.C., "Purposefully Introduced Plants that Have Become Noxious or Poisonous Weeds," *Weed Science*, vol. 28, No. 3, May 1980, pp. 300-305.

102. Young, J.A., "Tumbleweed," *Scientific American*, March 1991, pp. 82-87.

CHAPTER 3: THE CHANGING NUMBERS, CAUSES, AND RATES OF INTRODUCTIONS

1. American Fisheries Society, Southern Division, "Newsletter," Virginia Polytechnic Institute and State University, Blacksburg, VA, February 1991.

2. Anonymous, "Hazards of Travel," *Washington Post*, Dec. 31, 1991, p. C2.

3. Anonymous, 'Around the States: Florida," *Ecology USA*, vol. 21, No. 7, Apr. 6, 1992, p. 67.

4. Anonymous, 'Search On for Snail That Eats Up Lawns," *Washington Times*, May 15, 1992, p. B3.

5. Bandel, D.M., Chief, Natural and Cultural Resources Division Directorate of Public Works, Department of the Army, Fort Belvoir, VA, memorandum to the Executive Director, Armed Forces Pest Management Board, Forest Glen Section, Feb. 4, 1992.

6. Bangsund, D.A. and Leistritz, F.L., "Economic Impacts of Leafy Spurge on Grazing Lands in the Northern Great Plains," Department of Agricultural Economics, Agricultural Experiment Station, North Dakota State University, Fargo, ND, Agricultural Economics Report No. 275-S, November 1991.

7. Barton, B.J., *Gardening by Mail: a Source Book* (Boston, MA: Houghton Mifflin Co., 1990).

8. Beauchamp, R.M., President, Pacific Southwest Biological Services, Inc., letter to P.T. Jenkins, Office of Technology Assessment, Dec. 27, 1991.

9. Bedunah, D.J., "The Complex Ecology of Weeds, Grazing, and Wildlife," *Western Wildlands*, vol. 18, No. 2, summer 1992, pp. 6-11.

10. Bossard, C., "The Establishment and Dispersal of *Cytisus scoparius* (Scotch Broom) in California Sierra Nevada Foothill and Northern Coastal Habitats," *Proceedings of the Symposium on Exotic Pest Plants*, T.D. Center et al. (eds.) (Denver, CO: National Park Service, Natural Resources Publication Office, 1991), pp. 35-55.

11. Boyce, J.S., *Forest Pathology* (New York, NY: McGraw-Hill Book Co., Inc., 1961).

12. Britton, J.C., "Pathways and Consequences of the Introduction of Non-Indigenous Freshwater, Terrestrial, and Estuarine Mollusks in the United States," contractor report prepared for the Office of Technology Assessment, October 1991.

13. Burkhead, N.M. and Williams, J.D., "An Intergenic Hybrid of a Native Minnow, the Golden Shiner, and an Exotic Minnow, the Rudd," *Transactions of the American Fishery Society*, vol. 120, 1991, pp. 781-795, as cited in Mills, E.L. et al., 1993 (#71).

14. California Exotic Plant Pest Council, *CalEPPC News*, vol. 1, No. 1, winter 1993, p. 5.

15. California Forest Pest Council, "Forest Pest Conditions in California—1991," Sacramento, CA, Feb. 28, 1992.

16. Cardoza, J.E. et al., "A Compilation of the History and Status of Exotic Vertebrates of Massachusetts," *Fauna of Massachusetts Series*, No. 6, Massachusetts Division of Fisheries and Wildlife, Westborough, MA, September 1992.

17. Carlton, J.T., "Introduced Invertebrates of San Francisco Bay," *San Francisco Bay: The Urbanized Estuary* (San Francisco, CA: Pacific Division of the American Association for the Advancement of Science, 1979), pp. 427-444.

18. Carlton, J.T., Director, Maritime Studies Program, Williams College, Mystic, CT, personal communication to E.A. Chornesky, Office of Technology Assessment, December 1990.

19. Carlton, J.T., "Dispersal of Living Organisms into Aquatic Ecosystems as Mediated by Aquaculture and Fisheries Activities," *Dispersal of Living Organisms Into Aquatic Ecosystems*, A. Rosenfield and R. Mann (eds.) (College Park, MD: Maryland Sea Grant, 1992), pp. 13-46.

20. Carlton, J.T., "Marine Species Introductions by Ship's Ballast Water: An Overview," *Introductions and Transfers of Marine Species*, M.R. DeVoe (ed.) (Charleston, SC: South Carolina Sea Grant, 1992), pp. 23-29.

21. Carlton, J.T. et al., "Remarkable Invasion of San Francisco Bay (California, USA) by the Asian Clam *Potamocorbula amurensis*. I. Introduction and Dispersal," *Marine Ecology Progress Series*, vol. 66, 1990, pp. 81-94.

22. Carlton, J.T. and Geller, J.B., "Ecological Roulette: The Global Transport of Nonindigenous Marine Organisms," *Science*, vol. 261, July 2, 1993, pp. 78-82.

23. Center for Conservation Biology, *Center for Conservation Biology Update*, Stanford University, Stanford, CA, vol. 6, No. 2, winter 1993, p. 12.

24. Cochran, M.J., "Non-Indigenous Species In the United States: Economic Impact Tables," contractor report prepared for the Office of Technology Assessment, March 1992.

25. Counts, C.L. III, "The Zoogeography and History of the Invasion of the United States by *Corbicula Fluminea* (Bivalvia: Corbiculidae)," *American Malacological Bulletin*, Special Edition No. 2, 1986, pp. 7-39.

26. Courtenay, W.R., "Pathways and Consequences of the Introduction of Non-Indigenous Fishes in the United States," contractor report prepared for the Office of Technology Assessment, September 1991.

27. Courtenay, W.R., Professor of Zoology, Florida Atlantic University, Boca Raton, FL, FAX to E.A. Chornesky, Office of Technology Assessment, Apr. 13, 1993.

28. Courtenay, W.R. and Robins, C.R., "Fish Introductions: Good Management, Mismanagement, or No Management," *Critical Reviews in Aquatic Sciences*, vol. 1, No. 1, 1989, pp. 159-172.

29. Courtenay, W.R. and Williams, J.D., "Dispersal of Exotic Species From Aquaculture Sources, With Emphasis on Freshwater Fishes," *Dispersal of Living Organisms into Aquatic Ecosystems*, A. Rosenfield and R. Mann (eds.) (College Park, MD: Maryland Sea Grant, 1992), pp. 49-81.

30. Craven, R.B. et al., "Importation of *Aedes albopictus* and Other Exotic Mosquito Species into the United States in Used Tires From Asia," *Journal of the American Mosquito Control Association*, vol. 4, No. 2, 1988, pp. 138-142.

31. Deitrich, B., "State Girds To Battle Alien Invader: A Moth," *Seattle Times*, Nov. 19, 1991, p. A-1.

32. Dickerson, W.A., "Gypsy Moth in the Southeastern United States," North Carolina Department of Agriculture, Raleigh, NC, Apr. 16, 1991.

33. Dold, C., "Exotic Birds, at Risk in Wild, May be Banned as Imports to U.S.," *The New York Times*, Oct. 20, 1992.

34. Dowell, R.V. and Gill, R., "Exotic Invertebrates and Their Effects on California," *Pan-Pacific Entomologist*, vol. 65, No. 2, 1989, pp. 132-145.

35. Dowell, R.V., State Entomologist, California Department of Food and Agriculture, Sacramento, CA, FAX to E.A. Chornesky, Office of Technology Assessment, Apr. 12, 1993.

36. Ebinger, J.E. and McClain, W.E., "Naturalized Amur Maple (*Acer ginnala* Maxim.) in Illinois," *Natural Areas Journal*, vol. 11, No. 3, 1991, pp. 170-171.

37. Ezzell, C., "Strangers in Paradise," *Science News*, vol. 142, No. 45, Nov. 7, 1992, pp. 314-319.

38. Fisher, T.W., personal communication to J.C. Britton, as cited in Britton, J.C., 1991 (#12).

39. Fleming, P. and Kanal, R., "Newly Documented Species of Vascular Plants in the District of Columbia," *Castanea*, vol. 57, No. 2, June 1992, pp. 132-146.

40. Frankie, G.W. et al. "Some Considerations for the Eradication and Management of Introduced Insect Pests in Urban Environments," S.L. Battenfield (ed.) *Proceedings of a Symposium on the Imported Fire Ant*, Atlanta, GA, June 7-10, 1982 (Washington, DC: U.S. Department of Agriculture, Animal and Plant Health Inspection Service/U.S. Environmental Protection Agency, 1982) as cited in K.C. Kim and A.G. Wheeler, Jr., 1991 (#53).

41. Fritts, T.H., "The Brown Tree Snake, *Boiga irregularis*, A Threat to Pacific Islands," *U.S. Fish and Wildlife Service Biological Report*, vol. 88, No. 31, September 1988.

42. Ganzhorn, J. et al., "Dissemination of Microbial Pathogens Through Introductions and Transfers of Finfish," *Dispersal of Living Organisms into Aquatic Ecosystems*, A. Rosenfield and R. Mann (eds.) (College Park, MD: Maryland Sea Grant, 1992), pp. 175-192.

43. Great Lakes Fishery Commission, Ruffe Task Force, "Ruffe in the Great Lakes: A Threat to North American Fisheries," Ann Arbor, MI, March 1992.

44. Greathead, D.J. and Greathead, A.H., "Biological Control of Insect Pests by Insect Parasitoids and Predators: the BIOCAT Database," *Biocontrol News and Information 1992*, vol. 13, No. 4, 1992, pp. 61N-68N.

45. Gregg, M., Administrator, Metro Parks Natural Areas Management, State of Florida, memo to K. Hale, Director, Office of Emergency Management, Oct. 11, 1992.

46. Hanna, G.D., "Introduced Mollusks of Western North America," *Occasional Papers of the California Academy of Sciences*, No. 48, 1966, as cited in Britton, J.C., 1991 (#12).

47. Hobbs, R.J. and Huenneke, L.F., "Disturbance, Diversity, and Invasion: Implications for Conservation," *Conservation Biology*, vol. 6, No. 3, September 1992, pp. 324-337.

48. Hoffman, G.L., "The Asian Tapeworm, *Bothriocephalus gowkongensis*, in the United States, and Research Needs in Fish Parasitology," *Proceedings of the 1976 Fish Farming Conference and Annual Convention of Catfish Farmers of Texas*, Texas A&M University, College Station, TX, July 20-21, 1976, pp. 84-90.

49. Hofstetter, R.H., "The Current Status of *Melaleuca quinquenervia* in Southern Florida," *Proceedings of the Symposium on Exotic Pest Plants*, T.D. Center et al. (eds.) (Denver, CO: National Park Service, Natural Resources Publication Office, 1991), pp. 159-176.

50. Howard, L.O., "Gipsy Moth in America: A Summary Account of the Introduction and Spread of *Porthetria dispar* in Massachusetts and of the Efforts Made by the State to Repress and Exterminate It," U.S. Department of Agriculture, Division of Entomology, Bulletin No. 11—New Series, 1897, pp. 1-39.

51. Hubbell, S., "A Reporter at Large: Ladybugs," *The New Yorker*, Oct. 7, 1991, pp. 103-111.

52. Kaplan, J.K., "Africanized Honey Bees," *Agricultural Research*, vol. 38, December 1990, pp. 4-11.

53. Kim, K.C. and Wheeler, A.G., "Pathways and Consequences of the Introduction of Non-Indigenous Insects and Arachnids in the United States," contractor report prepared for the Office of Technology Assessment, December 1991.

54. Kindler, S.D. and Springer, T.L., "Alternate Hosts of Russian Wheat Aphid (Homoptera: Aphididae)," *Journal of Economic Entomology*, vol. 82, No. 5, 1989, pp. 1358-1362.

55. Kohler, C.C. and Courtenay, W., Jr., "American Fisheries Society Position on Introductions of Aquatic Species," *Fisheries*, vol. 11, No. 2, March-April 1986, pp. 39-42.

56. Langdon, K.R. and Johnson, K.D., "Alien Forest Insects and Diseases in Eastern USNPS Units: Impacts and Interventions," *The George Wright FORUM*, vol. 9, No. 1, 1992, pp. 2-14.

57. Lanka, B. et al., "Analysis and Recommendations on the Application by Mr. John T. Dorrance III to Import and Possess Native and Exotic Species," Exotic Species Committee, Game Division, Wyoming Game and Fish Department, Cheyenne, WY, Mar. 1, 1990.

58. Lewis, V.R. et al., "Imported Fire Ants: Potential Risk to California," *California Agriculture*, vol. 46, No. 1, January-February 1992, pp. 29-31.

59. Lieberman, S.S., "Awareness Grows on the Problems of Trading Live Birds and Other Live Wildlife," *Endangered Species Technical Bulletin*, vol. 15, No. 11, 1990, pp. 13-14.

60. Lightner, D.V. et al., "Geographic Dispersion of the Viruses IHHN, MBV, and HPV as a Consequence of Transfers and Introductions of Penaeid Shrimp to New Regions for Aquaculture Purposes," *Dispersal of Living Organisms into Aquatic Ecosystems*, A. Rosenfield and R. Mann (eds.) (College Park, MD: Maryland Sea Grant, 1992), pp. 155-173.

61. Lombardi, A.K., Director, Office of Cargo Enforcement and Facilitation, Department of Customs, U.S. Department of the Treasury, Washington, DC, personal communication to Office of Technology Assessment, Jan. 24, 1992.

62. Mack, R.N., "Catalog of Woes," *Natural History*, March 1990, pp. 44-53.

63. Mack, R.N., "Pathways and Consequences of the Introduction of Non-Indigenous Plants in the United States," contractor report prepared for the Office of Technology Assessment, September 1991.

64. Mack, R.N., "The Commercial Seed Trade: An Early Disperser of Weeds in the United States," *Economic Botany*, vol. 45, No. 2, 1991, pp. 257-273.

65. Mack, R.N., "Additional Information on Non-Indigenous Plants in the United States," contractor report prepared for the Office of Technology Assessment, October 1992.

66. McCann, J.A., "Involvement of the American Fisheries Society With Exotic Species, 1969-1982," *Distribution, Biology, and Management of Exotic Fishes*, W.R. Courtenay and J.R. Stauffer, Jr. (eds.) (Baltimore, MD: The Johns Hopkins University Press, 1984), pp. 1-7.

67. McCann, J.A., Director, National Fisheries Research Center, Fish and Wildlife Service, U.S. Department of the Interior, Gainesville, FL, personal communication to Office of Technology Assessment, Apr. 17, 1991.

68. McClintock, E., "Escaped Exotic Weeds in California," *Fremontia*, vol. 12, No. 4, January 1985, pp. 3-6.

69. McCoid, M. and Fritts, T.H., "Growth and Fatbody Cycles in Feral Populations of the African Clawed Frog, *Xenopus laevis* (Pipidae), in California With Comments on Reproduction," *The Southwestern Naturalist*, vol. 34, No. 4, December 1989, pp. 499-505.

70. McMullen, M.M., Inspector—Program Officer, Office of Cargo Enforcement and Facilitation, Department of Customs, U.S. Department of the Treasury, Washington, DC, personal communication to Office of Technology Assessment, Jan. 24, 1992.

71. Mills, E.L. et al., "Exotic Species in the Great Lakes: A History of Biotic Crises and Anthropogenic Introductions," *Journal of Great Lakes Research*, vol. 19, No. 1, 1993, pp. 1-54.

72. Morrison, J., "Cockroaches on the Move," *Agricultural Research*, vol. 35, No. 2, February 1987, pp. 6-9.

73. Moyle, P.B., "Fish Introductions in California: History and Impact on Native Fishes," *Biological Conservation*, vol. 9, No. 2, 1976, pp. 101-118.

74. National Association of State Departments of Agriculture, Washington, DC, "Honey Bee Pests —A Threat to the Vitality of U.S. Agriculture: A National Strategy," February 1991.

75. Nichols, F.H. et al., "Remarkable Invasion of San Francisco Bay (California, USA) by the Asian Clam *Potamocorbula amurensis*: II. Displacement of a Former Community," *Marine Ecology Progress Series*, vol. 66, 1990, pp. 95-101.

76. Nuzzo, V.A., "Current and Historic Distribution of Garlic Mustard (*Alliaria petiolata*) in Illinois," report prepared for the Illinois Department of Conservation, Springfield, IL, July 1991.

77. Randall, J., Invasive Weed Specialist, Consumnes River Preserve, The Nature Conservancy, Galt, CA, letter to P.N. Windle, Office of Technology Assessment, July 23, 1992.

78. Regelbrugge, C.J., Director of Regulatory Affairs and Grower Services, American Association of Nurserymen, Washington DC, letter to

P.N. Windle, Office of Technology Assessment, Aug. 10, 1992.

79. Rosenberry, B., "Shrimp Farming in the United States," *Aquaculture Digest*, February 1988.

80. Sailer, R.I., "History of Insect Introductions," *Exotic Plant Pests and North American Agriculture*, C. Wilson and C.L. Graham (eds.) (New York, NY: Academic Press, 1983), pp. 15-38.

81. Schafland, P.L., "A Proposal for Introducing Peacock Bass (*Cichla* spp.) in Southeast Florida Canals," Contribution No. 40, Non-Native Fish Research Laboratory, Florida Game and Fresh Water Fish Commission, Boca Raton, FL, 1984.

82. Schoulties, C.L., "Pathways and Consequences of the Introduction of Non-Indigenous Plant Pathogens in the United States," contractor report prepared for the Office of Technology Assessment, December 1991.

83. Schmitz, D.C., Biological Scientist III, Permitting Section, Florida Department of Natural Resources, Tallahassee, FL, memorandum to Rob Kipker, June 28, 1990.

84. Schwegman, J.E., "Newsletter of the Illinois Native Plant Conservation Program Division of Natural Heritage, Illinois Department of Conservation," *Illinoisensis*, vol. 7, No. 2, July 1991.

85. Schwegman, J.E., Plant Conservation Manager, Illinois Department of Conservation, Springfield, IL, letter to P.T. Jenkins, Office of Technology Assessment, Apr. 6, 1992.

86. Seed World, "Wildlife Market on the Rise," *Seed World*, November 1991, p. 26.

87. Sharp Bros. Seed Co., "Buffalo Brand: Select Native Grasses" (seed catalog), Healy, KS, 1989.

88. Shelton, A.M. and Wyman, J.A., "Insecticide Resistance of Diamondback Moth in North America," *Journal of Economic Entomology*, vol. 79, No. 1, 1986, pp. 11-19.

89. Shelton, W.L. and Smitherman, R.O., "Exotic Fishes in Warmwater Aquaculture," *Distribution, Biology, and Management of Exotic Fishes*, W.R. Courtenay and J.R. Stauffer (eds.) (Baltimore, MD: The Johns Hopkins University Press, 1984), pp. 262-301.

90. Shotts, E.B., Jr. and Gratzek, J.B., "Bacteria, Parasites, and Viruses of Aquarium Fish and Their Shipping Waters," *Distribution, Biology,* *and Management of Exotic Fishes*, W.R. Courtenay, Jr. and J.R. Stauffer, Jr. (eds.) (Baltimore, MD: Johns Hopkins University Press, 1984), pp. 215-232.

91. Siddiqui, I.A., Assistant Director, California Department of Food and Agriculture, Sacramento, CA, testimony at hearings before the Senate Committee on Governmental Affairs, Subcomitteee on Federal Services, Post Offices, and Civil Services, *Postal Implementation of the Agricultural Quarantine Enforcement Act*, June 5, 1991.

92. Sugg, I.C., "To Save an Endangered Species, Own One," *The Wall Street Journal*, Aug. 31, 1992, p. A10.

93. Taylor, D.W., "An Eastern American Freshwater Mussel, *Anodonta*, Introduced into Arizona," *The Veliger*, vol. 8, No. 3, 1966, pp. 197-198.

94. Teer, J.G., "Non-Native Large Ungulates in North America," *Wildlife Production: Conservation and Sustainable Development*, L.A. Renecker and R.J. Hudson (eds.) (Fairbanks, AK: Agricultural and Forestry Experiment Station, 1991), pp. 55-66.

95. Temple, S.A. and Carroll, D.M., "Pathways and Consequences of the Introduction of Non-Indigenous Vertebrates in the United States," contractor report prepared for the Office of Technology Assessment, October 1991.

96. Torchio, P.F. et al., "Introduction of the European Bee, *Osmia cornuta*, into California Almond Orchards (Hymenoptera: Megachilidae)," *Environmental Entomology*, vol. 16, No. 3, June 1987, pp. 664-667.

97. Turbak, G., "A Pleasant Bird is the Pheasant," *National Wildlife*, vol. 30, October/November, 1992, pp. 14-20.

98. Tyser, R.W. and Worley, C.A., "Alien Flora in Grasslands Adjacent to Road and Trail Corridors in Glacier National Park, Montana (U.S.A.)," *Conservation Biology*, vol. 6, No. 2, June 1992, pp. 253-262.

99. U.S. Aquatic Nuisance Species Task Force, "Proposed Aquatic Nuisance Species Program," draft report to the U.S. Congress, Sept. 28, 1992.

100. U.S. Department of Agriculture, Forest Service, Northeastern Forest Experiment Station, Radnor, PA, "Gypsy Moth Research and Develop-

ment Program'' (Washington, DC: U.S. Government Printing Office, October 1990).

101. U.S. Department of Agriculture, Forest Service, ''Pest Risk Assessment of the Importation of Larch from Siberia and the Soviet Far East,'' Miscellaneous Publication No. 1495, September 1991.

102. U.S. Department of the Interior, Fish and Wildlife Service, *Restoring America's Wildlife* (Washington, DC: U.S. Government Printing Office, 1987).

103. U.S. District Court, District of Columbia, ''Memorandum Opinion in *Pennington Enterprises, Inc.* v. *United States*,'' Civil Action No. 90-1067, filed Mar. 31, 1992.

104. U.S. Federal Coordinating Council on Science, Engineering, and Technology, Joint Subcommittee on Aquaculture, ''National Aquaculture Development Plan: Volume II,'' Washington, DC, September 1983.

105. Westbrooks, R.G., ''Interstate Sale of Aquatic Federal Noxious Weeds as Ornamentals in the United States,'' *Aquatics*, vol. 2, No. 3, June 1990, pp. 16-24.

106. Westbrooks, R.G., U.S. Department of Agriculture, Animal and Plant Health Inspection Service, ''Federal Noxious Weed Inspection Guide: Noxious Weed Inspection System,'' September 1991, pp. 3.5.1—4.1.5.

107. Whitson, T.D. et al., *Weeds of the West* (Jackson, WY: Pioneer of Jackson Hole, January 1991).

108. Wildlife Nurseries, Inc., ''What Brings Them In'' (catalogue of animals and plants for wildlife habitat) (Oshkosh, WI: 1989).

109. Wildseed Farms, ''Wildflowers: 1992 Seed Catalogue'' (Eagle Lake, TX: 1992).

110. Williams, J., Project Leader, National Fisheries Research Center, Fish and Wildlife Service, U.S. Department of the Interior, personal communication to E.A. Chornesky, Office of Technology Assessment, Apr. 16, 1993.

111. Williams, M.C., ''Purposefully Introduced Plants that Have Become Noxious or Poisonous Weeds,'' *Weed Science*, vol. 28, No. 3, May 1980, pp. 300-305.

112. Yee, R.S.N., personal communication to R.N. Mack, as cited in Mack, R.N., 1991 (#63).

113. Zamora, D.L. et al., ''An Eradication Plan for Plant Invasions,'' *Weed Technology*, vol. 3, No. 1, January/March 1989, pp. 2-12.

CHAPTER 4: THE APPLICATION OF DECISIONMAKING METHODS

1. Associated Press, ''Forest Bugaboo—Alarm Over Discovery of Asian Gypsy Moths,'' *Seattle Times/Post Intelligencer*, Nov. 24, 1991, p. B-8.

2. Blockstein, D., Executive Director, Committee for the National Institute for the Environment, Washington, DC, ''Sample Guidelines for Consideration of Biological Diversity in Environmental Impact Statements,'' draft unpublished memorandum, 1989.

3. Bullis, C.A. and Kennedy, J.J., ''Values Conflicts and Policy Interpretation,'' *Policies Studies Journal*, vol. 19, No. 3-4, 1991, pp. 542-552.

4. Bullock, G., ''A Love Affair with Exotic Plants,'' *Bulletin of the Pacific Tropical Botanical Garden*, vol. 18, No. 2-3, April/July, 1988, pp. 58-9.

5. Carlton, J.T., ''Man's Role in Changing the Face of the Ocean,'' *Conservation Biology*, vol. 3, No. 3, 1989, pp. 270-272.

6. Carter, P.C.S., ''Risk Assessment and Pest Detection Surveys for Exotic Pests and Diseases which Threaten Commercial Forestry in New Zealand,'' *New Zealand Journal of Forestry Science*, vol. 19, Nos. 2/3, 1989, pp. 353-374.

7. Clark, A., Program Supervisor, Pest Exclusion Branch, California Department of Agriculture, Sacramento, CA, personal communication to P.T. Jenkins, Office of Technology Assessment, Feb. 14, 1991.

8. Coates, J.F., ''Factors Shaping and Shaped by the Environment: 1990-2010,'' *Futures Research Quarterly*, vol. 7, No. 3, 1991, pp. 5-55.

9. Cochran, M., ''Non-Indigenous Species in the United States: Economic Consequences,'' contractor report prepared for the Office of Technology Assessment, March 1992.

10. Coulson, J.R. and Soper, R.S., ''Protocols for the Introduction of Biological Control Agents in the U.S.,'' *Plant Protection and Quarantine—Vol. III*, R.P. Kahn (ed.) (Boca Raton, FL: CRC Press, 1989), pp. 1-35.

11. Courtenay, W.R., "Pathways and Consequences of Introductions of Non-Indigenous Fishes in the United States," contractor report prepared for the Office of Technology Assessment, August 1991.

12. Crawley, M.J., "What Makes A Community Invasible?" *Colonization, Succession and Stability*, A.J. Gray et al. (eds.) (Oxford, England: Blackwell Scientific Publ., 1987), pp. 429-453.

13. Cropper, M.L. and Oates, W.E., 'Environmental Economics: A Survey," *Journal of Economic Literature*, vol. 30, June 1992, pp. 675-740.

14. DeFazio, P., Member of U.S. House of Representatives et al., letter to Clayton K. Yeutter, Secretary, U.S. Department of Agriculture, Dec. 5, 1990.

15. Drost, C.A. and Fellers, G.M., National Park Service, U.S. Department of the Interior, Washington, DC, "Draft Handbook for the Removal of Non-Native Animals," 1992.

16. Edelson, D., Staff Attorney, Natural Resources Defense Council, San Francisco, CA, personal communication to P.T. Jenkins, Office of Technology Assessment, Feb. 5, 1991.

17. Edwards, G.B., Assistant Director, Fisheries Division, Fish and Wildlife Service, U.S. Department of the Interior, Washington, DC, testimony at hearings before the House Committee on Merchant Marine and Fisheries, Subcommittee on Oversight and Investigations, July 12, 1991, Serial No. 102-28, pp. 32-33.

18. Environmental Law Institute, *Law of Environmental Protection—Vol. I* (New York, NY: Clark Boardman Co., 1990), p. 6-24.

19. Gleick, J., *Chaos. Making a New Science* (New York, NY: Viking Press, 1987).

20. Gregory, R., "A Decision Framework for Managing the Risks of Deliberate Releases of Genetic Materials," *Dispersal of Living Organisms into Aquatic Ecosystems*, A. Rosenfeld and R. Mann (eds.) (College Park, MD: Maryland Sea Grant, 1992), pp. 421-434.

21. Gregory, R. et al., "Adapting the Environmental Impact Statement Process to Inform Decision Makers," *Journal of Policy Analysis and Management*, vol. 11, No. 1, 1992, pp. 58-75.

22. Harding, M., Arizona Department of Game and Fish, Phoenix, AZ, "Public Interest in Conserva-

tion: Perceptions of Native Fishes in Arizona—A Report to the Nongame Branch of the Arizona Department of Game and Fish," June 1991.

23. Hiebert, R.D., Chief Scientist, National Park Service—Midwest Region, U.S. Department of the Interior, Omaha, NE, personal communication to P.T. Jenkins, Office of Technology Assessment, Sept. 27, 1991.

24. Kahn, R.P., "Quarantine Significance: Biological Considerations," *Plant Protection and Quarantine—Vol. I*, R.P. Kahn (ed.) (Boca Raton, FL: CRC Press, 1989), pp. 201-216.

25. Kahn, R.P., Consultant, Rockville, MD, letter to P.N. Windle, Office of Technology Assessment, Dec. 2, 1991.

26. Kareiva, P., Groom, M., Parker, I., and Ruesink, J., "Risk Analysis as Tool for Making Decisions about the Introduction of Non-Indigenous Species into the United States," contractor report prepared for the Office of Technology Assessment, July 1991.

27. Kareiva, P., Professor, Department of Zoology, University of Washington, contract paper prepared for OTA Workshop on Management and Policy Decisionmaking for Non-Indigenous Species in the United States, Washington, DC, Nov. 13-14, 1991.

28. Kellert, S.R., *Knowledge, Affection and Basic Attitudes toward Animals in American Society* (Washington, DC: U.S. Government Printing Office, 1980).

29. Kellert, S.R. and Clark, T.W., "The Theory and Application of a Wildlife Policy Framework," *Public Policy Issues in Wildlife Management*, W.R. Mangun (ed.) (New York, NY: Greenwood Press, 1991), pp. 17-38.

30. Klingman, D.L. and Coulson, J.R., "Guidelines for Introducing Foreign Organisms into the United States for the Biological Control of Weeds," *Bulletin of the Entomological Society of America*, vol. 19, No. 3, 1983, pp. 55-61.

31. Kohler, C.C., "Toward a Reasoned Approach to Introduced Aquatic Organisms," *Dispersal of Living Organisms into Aquatic Ecosystems*, A. Rosenfeld and R. Mann (eds.) (College Park, MD: Maryland Sea Grant, 1992), pp. 393-404.

32. Kuchler, F. and Duffy, M., U.S. Department of Agriculture, Economic Research Service, "Con-

trol of Exotic Pests—Forecasting Economic Impacts," *Agricultural Economic Report No. 518,* 1984.

33. Kurdila, J., "The Introduction of Exotic Species into the United States: There Goes the Neighborhood!" *Boston College Environmental Affairs Law Review,* vol. 16, No. 1, 1988, pp. 95-118.

34. Lassuy, D., Fish and Wildlife Biologist, Fisheries Division, Fish and Wildlife Service, U.S. Department of the Interior, Arlington, VA, personal communication to E.A. Chornesky, Office of Technology Assessment, Apr. 7, 1993.

35. Lattin, J.D., Professor of Entomology, Oregon State University, memorandum to Bill Wright, Administrator, Plant Division, Oregon Department of Agriculture, Salem, OR, Nov. 1, 1990.

36. Lattin, J.D., Professor of Entomology, Oregon State University, Corvallis, OR, personal communication to P.T. Jenkins, Office of Technology Assessment, Jan. 31, 1991.

37. Lattin, J.D., Professor of Entomology, Oregon State University, Corvallis, OR, letter to Kathleen Johnson, Oregon Department of Agriculture, Feb. 20, 1991.

38. Lawton, J.H., "Annual Report: 1 April 1990—31 March 1991," Center for Population Biology, Imperial College at Silwood Park, Berkshire, England, June 14, 1991.

39. Levin, S.A., "Analysis of Risk for Invasions and Control Programs," *Biological Invasions: a Global Perspective,* J.A. Drake et al. (eds.) (New York, NY: Wiley & Sons, 1989), pp. 425-435.

40. Mack, R.N., "Pathways and Consequences of Introductions of Non-Indigenous Plants in the United States," contractor report prepared for the Office of Technology Assessment, September 1991.

41. Mack, R.N., "Pathways and Consequences of Introductions of Non-Indigenous Plants in the United States—Additional Information," supplemental contractor report prepared for the Office of Technology Assessment, October 1992.

42. Maguire, L.A., Professor, School of Forestry and Environmental Studies, Duke University, presentation at OTA Workshop on Management and Policy Decisionmaking for Non-Indigenous Species in the United States, Washington, DC, Nov. 13-14, 1991.

43. Maguire, L.A. and Servheen, C., "Integrating Biological and Sociological Concerns in Endangered Species Management: Augmentation of Grizzly Bear Populations," *Conservation Biology,* vol. 6, No. 3, September 1992, pp. 426-432.

44. Mann, R., "Exotic Species in Aquaculture: An Overview of When, Why and How," *Proceedings of a Symposium on Exotic Species in Mariculture,* Woods Hole, MA, Sept. 18-20, 1978, pp. 331-53.

45. McCrea, E.J. and Shotkin, A., "Public Educational Efforts in the United States Regarding Prevention and Management of Non-Indigenous Species," contractor report prepared for the Office of Technology Assessment, December 1991.

46. Minnesota Interagency Exotic Species Task Force, St. Paul, MN, "Report and Recommendations," submitted to the Natural Resources Committees of the Minnesota House and Senate, July 1991.

47. Moody, M.E. and Mack, R.N., "Controlling the Spread of Plant Invasions: the Importance of Nascent Foci," *Journal of Applied Ecology,* vol. 25, 1988, pp. 1009-1021.

48. Mooney, H.A. and Drake, J.A., "The Release of Genetically Designed Organisms in the Environment: Lessons from the Study of Ecology of Invasions," *Introduction of Genetically Modified Organisms into the Environment,* H.A. Mooney and G. Bernardi (eds.) (New York, NY: Wiley and Sons, 1990), pp. 117-129.

49. Morris, R., Division Resources Manager, Louisiana-Pacific Corp., Samoa, CA, internal memorandum to Bill Phillips, Dec. 19, 1990.

50. Nature Conservancy of Hawaii, Honolulu, HI, and the Natural Resources Defense Council, Honolulu, HI, "The Alien Pest Species Invasions in Hawaii: Background Study and Recommendations for Interagency Planning," July 1992.

51. Noble, R., Professor, Fisheries and Wildlife Sciences, North Carolina State University, Raleigh, NC, personal communication to P.N. Windle, Office of Technology Assessment, May 31, 1991.

52. Norgaard, R.B. and Howarth, R.B., "The Rights of Future Generations and the Role of Benefit-

Cost Analysis in the Policy Process,'' paper presented at the Annual Meeting of the Association for Public Policy Analysis and Management, San Francisco, CA, October 1990.

53. Orr, R.L., Entomologist, and Cohen, S.D., Plant Pathologist, Animal and Plant Health Inspection Service, U.S. Department of Agriculture, ''Generic Pest Risk Assessment Process—For Estimating the Pest Risk Associated with Importation of Foreign Plants and Plant Products (Draft),'' Nov. 20, 1991.

54. Peoples, R.A. et al., ''Introduced Organisms: Policies and Activities of the U.S. Fish and Wildlife Service,'' *Dispersal of Living Organisms into Aquatic Ecosystems*, A. Rosenfeld and R. Mann (eds.) (College Park, MD: Maryland Sea Grant, 1992), pp. 325-352.

55. Randall, A., Professor of Economics, Ohio State University, Columbus, OH, letter to P.T. Jenkins, Office of Technology Assessment, Sept. 28, 1992.

56. Randall, A. and Thomas, M., ''The Role and Limits of Economics in Decisionmaking Regarding Non-Indigenous Species,'' contractor report prepared for the Office of Technology Assessment, August 1991.

57. Royer, M.H., ''Global Pest Information Systems—Can We Make Them Work?'' *Plant Protection and Quarantine—Vol. III*, R.P. Kahn (ed.) (Boca Raton, FL: CRC Press, 1989), pp. 37-57.

58. Sagoff, M., ''Some Problems with Environmental Economics,'' *Environmental Ethics*, vol. 10, spring, 1988, pp. 55-74.

59. Sagoff, M., ''On Making Nature Safe for Biotechnology,'' *Assessing Ecological Risks of Biotechnology*, L. Ginsburg (ed.) (Boston, MA: Butterworth, 1991), pp. 341-365.

60. Schaffer, W.M., ''Order and Chaos in Ecological Systems,'' *Ecology*, vol. 66, No. 1, 1985, pp. 93-106.

61. Schuyler, P.T., ''Selected Analysis of Introduced Animal Species Survey Data,'' contractor report prepared for the Office of Technology Assessment, December 1991.

62. Scribner, J.D., Attorney, Sacramento, CA, letter to P.N. Windle, Office of Technology Assessment, Feb. 3, 1992.

63. Scribner, J.D., Attorney, Sacramento, CA, personal communication to Office of Technology Assessment, July 30, 1992.

64. Shannon, M., Chief Operating Officer for Planning and Design, Animal and Plant Health Inspection Service, U.S. Department of Agriculture, Hyattsville, MD, personal communication to P.T. Jenkins, Office of Technology Assessment, Feb. 5, 1991.

65. Shannon, M., Assistant Director for Planning and Design, Animal and Plant Health Inspection Service, U.S. Department of Agriculture, Hyattsville, MD, personal communication to P.T. Jenkins, Office of Technology Assessment, Nov. 13, 1991.

66. Shannon, M., Assistant Director for Planning and Design, Animal and Plant Health Inspection Service, U.S. Department of Agriculture, Hyattsville, MD, personal communication to P.T. Jenkins, Office of Technology Assessment, Mar. 2, 1992.

67. Shelton, W.L., ''Strategies for Reducing Risks from Introductions of Aquatic Organisms: An Aquaculture Perspective,'' *Fisheries*, vol. 11, No. 2, 1986, pp. 16-18.

68. Starfield, A.M. and Herr, A.M., ''A Response to Maguire,'' letter to the editors, *Conservation Biology*, vol. 5, No. 4, 1991, p. 435.

69. Stone, C.P., ''Humane Natural Area Management in Hawai'i,'' *The George Wright Forum*, vol. 9, No. 1, 1992, pp. 32-35.

70. Suter, G., ''Exotic Organisms,'' *Ecological Risk Assessment*, G.W. Suter II (ed.) (Boca Raton, FL: Lewis Publishers, 1993), pp. 391-401.

71. Suter, G. and Barnthouse, L., ''Assessment Concepts,'' *Ecological Risk Assessment*, G.W. Suter II (ed.) (Boca Raton, FL: Lewis Publishers, 1993), pp. 21-47.

72. Tiedje, J.M. et al., ''The Planned Introduction of Genetically Engineered Organisms: Ecological Considerations and Recommendations,'' *Ecology*, vol. 70, No. 2, 1989, pp. 298-315.

73. Towns, D.R. et al., ''Protocols for Translocation of Organisms to Islands,'' *Ecological Restoration of New Zealand Islands*, D.R. Towns et al. (eds.) (Wellington, New Zealand: Department of Conservation, 1990).

74. U.S. Aquatic Nuisance Species Task Force, Intentional Introductions Policy Committee, "Options Paper—Intentional Introductions Policy Review," May 1992.

75. U.S. Aquatic Nuisance Species Task Force, "Proposed Aquatic Nuisance Species Program," Sept. 28, 1992.

76. U.S. Congress, House Committee on Merchant Marine and Fisheries, Subcommittee on Fisheries and Wildlife Conservation and the Environment, *Hearing on Proposed Injurious Wildlife Regulations*, Dec. 12, 1974.

77. U.S. Congress, House Committee on Merchant Marine and Fisheries, Subcommittee on Oversight and Investigations, *Hearings on the Introduction of Non-Indigenous Species into Existing Ecosystems*, Serial No. 102-28, July 12, 1991.

78. U.S. Department of Agriculture, Animal and Plant Health Inspection Service, Washington, DC, press release entitled "USDA Places Temporary Prohibition on Entry of Siberian Logs Because of Pests," Dec. 20, 1990.

79. U.S. Department of Agriculture, Animal and Plant Health Inspection Service, Washington, DC, "An Efficacy Review of Control Measures for Potential Pests of Imported Soviet Timber," Miscellaneous Publication No. 1496, September 1991.

80. U.S. Department of Agriculture, Animal and Plant Health Inspection Service, "Review of OTA's Draft Document: Non-Indigenous Species in the United States," July 30, 1992.

81. U.S. Department of Agriculture, Forest Service, "Pest Risk Assessment of the Importation of Larch from Siberia and the Soviet Far East," Miscellaneous Publication No. 1495, September 1991.

82. U.S. Department of Interior, Bureau of Sports Fisheries and Wildlife, "Draft Environmental Statement on Proposed Importation Regulations—Injurious Wildlife," DES 74-64, 1974.

83. U.S. Department of the Interior, Fish and Wildlife Service, Washington, DC, "Policies for Reducing Risks Associated with Introductions of Aquatic Organisms," executive summary of discussion paper, Oct. 14, 1987.

84. Welcomme, R.L., "International Measures for the Control of Introduction of Aquatic Organisms," *Fisheries*, vol. II, No. 2, April-May 1986, pp. 4-9.

85. Whiteaker, L.D. and Doren, R.F., Southeast Regional Office, National Park Service, U.S. Department of the Interior, "Exotic Plant Species Management Strategies and List of Exotic Species in Prioritized Categories for Everglades National Park," Research/Resources Management Report SER-89/04, 1989.

86. Wood, D.L., Professor of Entomology, and Cobb, F.W., Jr., Professor of Plant Pathology, Univ. of California, Berkeley, letter to Dean Cromwell, California State Board of Forestry et al., Sacramento, CA, Dec. 11, 1990.

87. Yang, X.B. et al., "Assessing the Risk and Potential Impact of an Exotic Plant Disease," *Plant Disease*, vol. 75, No. 10, 1991, pp. 976-82.

CHAPTER 5: TECHNOLOGIES FOR PREVENTING AND MANAGING PROBLEMS

1. Anonymous, "Insect Aside: Beware of the Fire Down Below. Stinging Ants From Farther South Have Begun to Make Inroads in Virginia, Maryland," *Washington Post*, June 2, 1992, pp. B3.

2. Anonymous, "Entomology: Destructive Fly May Be New Species," *Washington Post*, Jan. 4, 1993, pp. A2.

3. Anonymous, "Science Update," *Agricultural Research*, February 1993.

4. Aspelin, A.L., Grube, A.H., and Kibler, V., Environmental Protection Agency, Office of Pesticide Programs, "Pesticide Industry Sales and Usage: 1989 Market Estimates," Washington, DC, July 1991.

5. Berger, J.L. and Riggs, L.A., "Ecological Restoration and Non-Indigenous Species," contractor report prepared for the Office of Technology Assessment, August 1991.

6. Butler, L., "Non-target Impact of Dimilin and *Bacillus thuringiensis*," *Gypsy Moth News*, February 1992, pp. 4-5.

7. California Department of Food and Agriculture, Division of Plant Industry, "1990 Annual Report," Sacramento, CA, 1990.

8. California Department of Food and Agriculture, Divison of Plant Industry, Pest Exclusion, "Cali-

fornia Plant Pest Distribution Report," vol. 9, 1990, pp. 63-66.

9. Carey, J.R., "The Mediterranean Fruit Fly in California: Taking Stock," *California Agriculture*, Jan.-Feb. 1992, pp. 12-17.

10. Charudattan, R. et al., "Special Problems Associated with Aquatic Weed Control," *New Directions in Biological Control: Alternatives for Suppressing Agricultural Pests and Diseases*, R.R. Baker and P.E. Dunn (eds.) (New York, NY: Alan R. Liss, Inc., 1990).

11. Council for Agricultural Science and Technology (CAST), *Pests of Plant and Animals: Their Introduction and Spread*, report No. 112, Ames, IA, March 1987.

12. Courtenay, W.R., "Pathways and Consequences of the Introduction of Non-Indigenous Fishes in the United States," contractor report prepared for the Office of Technology Assessment, August 1991.

13. Cowley, J.M., "A New System of Fruit Fly Surveillance Trapping in New Zealand," *New Zealand Entomologist*, vol. 13, 1990, pp. 81-84.

14. Crooks, E. et al., "Stopping Pest Introductions," *Exotic Plant Pests and North American Agriculture*, C.I. Wilson and C.L. Graham (eds.) (London, U.K.: Academic Press, Inc., 1983), pp. 239-259.

15. Crozier, Y.C., Koulianos, S., and Crozier, R.H., "An Improved Test for Africanized Honeybee Mitochondrial DNA," *Experientia*, vol. 47, 1991, pp. 968-969.

16. DeBach, P. and Rosen, D., *Biological Control by Natural Enemies* (Cambridge, U.K.: Cambridge University Press, 1991).

17. Delfosse, E.S., Director, National Biological Control Institute, Animal and Plant Health Inspection Service, U.S. Department of Agriculture, letter to P.N. Windle, Office of Technology Assessment, Aug. 31, 1992.

18. DeLoach, C.J., "Past Successes and Current Prospects in Biological Control of Weeds in the United States and Canada," *Natural Areas Journal*, vol. 11, No. 3, 1991, pp. 129-142.

19. Drost, C.A. and Fellers, G.M., "Handbook for the Removal of Non-native Animals," draft, 1992.

20. Ehler, L.E., "Revitalizing Biological Control," *Issues in Science and Technology*, vol. 7, No. 1, fall 1990, pp. 91-96.

21. Ehler, L.E. and Endicott, P.C., "Effect of Malathion-Bait Sprays on Biological Control of Insect Pests of Olive, Citrus, and Walnut," *Hilgardia*, vol. 52, No. 5, 1984, pp. 1-47.

22. Ehler, L.E. et al., "Medfly Eradication in California: Impact of Malathion-Bait Sprays on an Endemic Gall Midge and Its Parasitoids," *Entomological Experiments and Applications*, vol. 36, 1984, pp. 201-208.

23. Ehrenfeld, D., "Conservation and the Rights of Animals," *Conservation Biology*, vol. 5, No. 1, March 1991, pp. 1-3.

24. Fagerstone, K.A., Bullard, R.W., and Ramey, C.A., "Politics and Economics of Maintaining Pesticide Registrations," *Proceedings of the 14th. Vertebrate Pest Conference*, L.R. Davis and R.E. Marsh (eds.) (Davis, CA: University of California at Davis, 1990).

25. Fritts, T.H., Scott, N.J., Jr., and Smith, B.E., "Trapping *Boiga irregularis* on Guam Using Bird Odors," *Journal of Herpetology*, vol. 23, No. 2, 1989, pp. 189-192.

26. Gallagher, J.E. and Haller, W.T., "History and Development of Aquatic Weed Control in the United States," *Review of Weed Science*, vol. 5, 1990, pp. 115-192.

27. Geist, V., "Endangered Species and the Law," *Nature*, vol. 357, May 28, 1992, pp. 274-276.

28. Greathead, D.J. and Greathead, A.H., "Biological Control of Insect Pests by Parasitoids and Predators: the BIOCAT Database," *Biocontrol News and Information*, vol. 13, No. 4, 1992, pp. 61N-68N.

29. Hansen, J.D. et al., "Efficacy of Hydrogen Cyanide Fumigation as a Treatment for Pests of Hawaiian Cut Flowers and Foliage after Harvest," *Journal of Economic Entomology*, vol. 84, No. 2, 1991, pp. 532-536.

30. Harris, P., "Environmental Impacts of Weed-Control Insects. Weed Stands have been Replaced by Diverse Plant Communities," *BioScience*, vol. 38, No. 8, September 1988, pp. 542-548.

31. Hester, F.E., "The U.S. National Park Service Experience with Exotic Species," *National Areas*

Journal, vol. 11, No. 3, 1991, pp. 127-128.

32. Holt, M. and Radzanowski, D., Library of Congress, Congressional Research Service, "Memorandum on Potential Environmental Characteristics and Monitoring Technologies Currently Used by Federal Agencies," Dec. 5, 1991.

33. Hopper B.E. and Campbell, W.P., "Crop Protection Information Needs: A NAPPO Perspective," *Crop Protection Information: An International Perspective, Proceedings of the International Crop Protection Information Workshop, 1989*, K.M Harris and P.R. Scott (eds.) (Wallingford, Oxon, U.K.: CAB International, 1989), pp. 225-233.

34. Howarth, F.G., "Environmental Impacts of Classical Biological Control," *Annual Review of Entomology,* vol. 36, 1991, pp. 485-509.

35. Jennings, P. and McCann, J.A., U.S. Fish and Wildlife Service, National Fisheries Research Center, "Research Protocol for Handling Non-Indigenous Aquatic Species," Gainesville, FL, April 1991.

36. Joslin, P., "The Genetic Test that Can Recognize Wolves, Dogs, and Hybrids from Each Other," *Wolftracks* (newsletter by Wolf Haven International, Tenino, WA), vol. 8, No. 4, 1991, p. 6.

37. Jutsum, A.R., "Commercial Application of Biological Control: Status and Prospects," Phil. Trans. R. Soc. Lond. B, vol. 318, *Biological Control of Pests, Pathogens and Weeds: Development and Prospects*, R.K.S. Wood, F.R.S., and M.J. Way (eds.) 1988, pp. 357-376.

38. Kahn, R.P., "Exclusion as a Plant Disease Control Strategy," *Annual Review of Phytopathology*, vol. 29, 1991, pp. 219-246.

39. Kareiva, P. et al., "Risk Analysis as a Tool for Making Decisions about the Introduction of Non-Indigenous Species into the United States," contractor report prepared for the Office of Technology Assessment, July 1991.

40. Kim, K.C. and Wheeler, A.G., Jr., "Pathways and Consequences of the Introduction of Non-Indigenous Insects and Arachnids in the United States," contractor report prepared for the Office of Technology Assessment, December 1991.

41. Kirkpatrick, J.F., "Wildlife Contraception. A New Way of Looking at Wildlife Management,"

Humane Society of the United States News, fall, 1991, pp. 22-25.

42. Klassen, W., "Eradication of Introduced Arthropod Pests: Theory and Historical Practice," Entomological Society of America, Miscellaneous Publications, No. 73, November 1989.

43. Kludy, D.H., "Federal Policy on Non-Indigenous Species: The Role of USDA's Animal and Plant Health Inspection Service," contractor report prepared for the Office of Technology Assessment, December 1991.

44. LaMadeleine, L.A., U.S. Department of Agriculture, Forest Service, "Non-Indigenous Species in the United States—Gypsy Moth Eradication—Utah," Ogden, UT, May 13, 1991.

45. Lambert, B. and Peferoen, M., "Insecticidal Promise of *Bacillus thuringiensis*," *BioScience*, vol. 42, No. 2, February 1992, pp. 112-122.

46. Lanier, G.N., "Principles of Attraction-Annihilation: Mass Trapping and Other Means," *Behavior-modifying Chemicals for Insect Management: Applications of Pheromones and Other Attractants*, R.L. Ridgway, R.M. Silverstein, R.M. Inscoe (eds.) (New York, NY: Marcel Dekker, Inc., 1990), pp. 25-45.

47. Lanka, B. et al., Game Division, Wyoming Game and Fish Department, Cheyenne, WY, "Analysis and Recommendations on the Applications by Mr. John T. Dorrance III to Import and Possess Native and Exotic Species," Mar. 1, 1990.

48. Lindsey G. and Novak, K., "Developments in Information Dissemination Techniques: Directions for International Agricultural Research Centers," *Crop Protection Information: and International Perspective: Proceedings of the International Crop Protection Information Workshop, 1989*, K.M. Harris and P.R. Scott (eds.) (Wallingford, U.K.: CAB International, 1989), pp. 27-48.

49. Mack, R.N., "Pathways and Consequences of Non-Indigenous Plants in the United States," contractor report prepared for the Office of Technology Assessment, September 1991.

50. Mallet, J. and Porter, P., "Preventing Insect Adaptation to Insect-Resistant Crops: Are Seed Mixtures or Refugia the Best Strategy?" *Proceed-*

ings of the Royal Society of London, Series B, vol. 250B, 1992, pp. 165-169.

51. Mastro, V.C., Schwalbe, C.P., and O'Dell, T.M., "Sterile-Male Technique," The Gypsy Moth: Research Toward Integrated Pest Management, C.C. Doane and M.L. McManus (eds.) (Washington, DC: U.S. Department of Agriculture, 1981), pp. 669-679.

52. Mathys, G. and Baker, E.A., "An Appraisal of the Effectiveness of Quarantine," Annual Review of Phytopathology, vol. 18, 1980, pp. 85-101.

53. McGaughey, W.H. and Whalon, M.E., "Managing Insect Resistance to Bacillus thuringiensis Toxins," Science, vol. 258, Nov. 27, 1992, pp. 1451-1455.

54. McGregor, R.C., "The Emigrant Pests," a report to Dr. F.J. Mulhern, Administrator, Animal and Plant Health Inspection Service, U.S. Department of Agriculture, Washington, DC, May 1974.

55. McMahon, R.F. and Tsou, J.L., "Impact of European Zebra Mussel Infestation to the Electric Power Industry," paper presented at the Annual Meeting of the American Power Conference, Chicago, IL, Apr. 23-25, 1990, pp. 1-9.

56. Meehan, B.W., Carlton, J.T., and Wenne, R., "Genetic Affinities of the Bivalve Macoma balthica from the Pacific Coast of North America: Evidence for Recent Introduction and Historical Distribution," Marine Biology, vol. 102, 1989, pp. 235-241.

57. Meeusen, R.L. and Warren, G., "Insect Control with Genetically Engineered Crops," Annual Review of Entomology, vol. 34, 1989, pp. 373-381.

58. Melching, J.S., Bromfield, K.R., and Kingsolver, C.H., "The Plant Pathogen Containment Facility at Frederick, Maryland," Plant Disease, vol. 67, No. 7, July 1983, pp. 717-722.

59. Meyers, J.H., Higgins, C., and Kovacs, E., "How Many Insect Species are Necessary for the Biological Control of Insects?" Environmental Entomology, vol. 18, No. 4, 1989, pp. 541-547.

60. Micales, J.A., "The Use of Isozyme Analysis in Fungal Taxonomy and Genetics," Mycotaxon, vol. 27, October-December 1986, pp. 405-449.

61. Mills, N.J., "Biological Control, A Century of Pest Management," Bulletin of Entomological Research, vol. 80, 1990, pp. 359-362.

62. Minnesota Department of Natural Resources, Exotic Species Programs and Bureau of Information and Education, Exotic Species Handbook (St. Paul, MN: Minnesota Department of Natural Resources, 1992).

63. Morrissey, W.A., Library of Congress, Congressional Research Service, "Methyl Bromide and Stratospheric Ozone Depletion: Scientific Basis for Regulation?" Report for Congress, Aug. 14, 1992.

64. Murphy, K.J. and Pieterse, A.H., "Present Status and Prospects of Integrated Control of Aquatic Weeds," Aquatic Weeds: The Ecology and Management of Nuisance Aquatic Vegetation, A.H. Pieterse and K.J. Murphy (eds.) (Oxford, UK: Oxford University Press, 1990), pp. 223-227.

65. National Research Council, Managing Global Genetic Resources: The U.S. National Plant Germplasm System (Washington, DC: National Academy Press, 1991).

66. National Research Council, Restoration of Aquatic Ecosystems: Science, Technology, and Public Policy (Washington, DC: National Academy Press, 1992).

67. Peacock, J. et al., "Laboratory and Field Studies on the Effects of Bacillus thuringiensis on Non-target Lepidoptera," Gypsy Moth News, February 1992, pp. 7-8.

68. Rosenberg, D.Q., "The Interaction of State and Federal Quarantines in the U.S.," Plant Protection and Quarantine, Volume III Special Topics, R.P. Kahn (ed.) (Boca Raton, FL: CRC Press Inc., 1989), pp. 59-74.

69. Roth, H., "Concepts and Recent Developments in Regulatory Treatments," Plant Protection and Quarantine, Volume III Special Topics, R.P. Kahn (ed.) (Boca Raton, FL: CRC Press, Inc., 1989), pp. 117-144.

70. Sailer, R.I., "Extent of Biological and Cultural Control of Insect Pests of Crops," CRC Handbook of Pest Management in Agriculture, 1st. ed., D. Pimentel (ed.) (Boca Raton, FL: CRC Press, 1981), pp. 57-67.

71. Sand, P.F. and Manley, J.D., "The Witchweed Eradication Program Survey, Regulatory and Control," *Witchweed Research and Control in the United States*, P.F. Sand, R.E. Eplee, and G.W. Westbrooks (eds.) (Champaign, IL: Weed Science Society of America, 1990), pp. 141-150.

72. Schoulties, C.L., "Pathways and Consequences of the Introduction of Non-Indigenous Plant Pathogens in the United States," contractor report prepared for the Office of Technology Assessment, December 1991.

73. Schuyler, P., "Feral Animal Control: Santa Cruz Island, California," *Natural Areas Journal*, vol. 9, No. 4, 1989, p. 265.

74. Sheppard, W.S., Steck, G.J., and McPheron, B.A., "Geographic Populations of the Medfly May Be Differentiated by Mitochondrial DNA Variation," *Experimentia*, vol. 48, No. 10, October 1992, pp. 1010-1013.

75. Shoemaker, A., Chairman of the Mammal Standards Committee, Riverbank Zoological Park, Columbia, SC, personal communication to S. Fondriest, Office of Technology Assessment, Apr. 7, 1992.

76. Shotkin, A. and McCrea, E.J., "Public Educational Efforts in the United States Regarding Prevention and Management of Non-Indigenous Species," contractor report prepared for the Office of Technology Assessment, December 1991.

77. Siddiqui, I.A., Assistant Director, California Department of Food and Agriculture, Sacramento, CA, testimony at hearings before the Senate Committee on Governmental Affairs, Subcommittee on Federal Services, Post Offices, and Civil Services, *Postal Implementation of the Agricultural Quarantine Enforcement Act*, June 5, 1991.

78. Silverstein, R.M., "Practical Use of Pheromones and Other Behavior-modifying Compounds: Overview," *Behavior-Modifying Chemicals for Insect Management: Applications of Pheromones and Other Attractants*, R.L. Ridgway, R.M. Silverstein, and R.M. Inscoe (eds.) (New York, NY: Marcel Dekker, 1990), pp. 1-8.

79. Stace-Smith, R. and Martin, R.R., "Plant Quarantine Diagnostic Problems: Viruses," *Plant Protection and Quarantine Volume II, Selected Pests and Pathogens of Quarantine Significance*, R.P. Kahn (ed.) (Boca Raton, FL: CRC Press, Inc., 1989), pp. 183-204.

80. Stephens, C.H., "Minimizing Pesticide Use in a Vegetative Management System," *HortScience*, vol. 25, No. 2, February 1990, pp. 164-167.

81. Stevens, W.K., "Biologists Plan Survey of Ecosystems," *The Sunday Herald*, Monterey Bay, CA, Mar. 14, 1993, p. 8A.

82. Temple, S.A. and Carroll, D.M., "Pathways and Consequences of Introductions of Non-Indigenous Vertebrates in the United States," contractor report prepared for the Office of Technology Assessment, October 1991.

83. Tisdell, C.A., "Economic Impacts of Biological Control of Weeds and Insects," *Critical Issues in Biological Control*, M. Mackauer, L.E. Ehler, and J. Roland (eds.) (Andover, Hants, UK: Intercept Ltd., 1990), pp. 301-316.

84. U.S. Aquatic Nuisance Species Task Force, *Proposed Aquatic Nuisance Species Program*, Sept. 28, 1992.

85. U.S. Congress, General Accounting Office, *Geographic Information Systems, Status of Selected Agencies*, GAO/IMTEC-90-74FS, August 1990.

86. U.S. Congress, General Accounting Office, *Great Lakes Fishery Commission: Actions Needed to Support an Expanded Program*, March 1992.

87. U.S. Congress, General Accounting Office, *USDA's Research to Support Reregistration of Pesticides for Minor Crops*, June 1992.

88. U.S. Congress, House Committee on Appropriations, *Hearings on the Department of the Interior and Related Agencies Appropriations for 1993*, Serial No. 56-090 (Washington, DC: U.S. Government Printing Office, Apr. 28, 1992), pp. 231-246.

89. U.S. Congress, House of Representatives, Committee on Merchant Marine and Fisheries, *Great Lakes Sea Lamprey Control Program Hearings, Status of Efforts to Control Sea Lamprey Populations in the Great Lakes* (Washington, DC: U.S. Government Printing Office, Sept. 17, 1991).

90. U.S. Congress, Office of Technology Assessment, *Pest Management Strategies, Vol. I— Summary*, October 1979.

91. U.S. Department of Agriculture, Animal and Plant Health Inspection Service, Plant Protection

and Quarantine, "Protecting United States Agriculture from Foreign Pests and Diseases," August 1985.

92. U.S. Department of Agriculture, Animal and Plant Health Inspection Service, Management and Budget Administrative Services Division, Hyattsville, MD, "Master Plan for the USDA National Plant Germplasm Quarantine Center," Sept. 11, 1989.

93. U.S. Department of Agriculture, Animal and Plant Health Inpection Service, "Administrative Instructions Governing the Entry of Apples and Pears from Certain Countries in Europe," 7 CFR Ch. III, 319.56-2r,(1-1-91 Edition), 1991.

94. U.S. Department of Agriculture, Animal and Plant Health Inspection Service, "Papayas From Hawaii," 7 CFR Part 318. 56 (156), Tuesday, Aug. 13, 1991.

95. U.S. Department of Agriculture, Cooperative State Research Service, "Role of IR-4 in the Registration of Biological Control Agents," *Proceedings of the USDA/CSRS National Workshop on Critical Issues in Biological Control*, held at the National Wildlife Federation, Vienna, VA, 1991.

96. U.S. Department of Agriculture, Office of the Secretary, Office of Agricultural Biotechnology, "Environmental Assessment of Research on Transgenic Carp in Confined Outdoor Ponds," Nov. 15, 1990.

97. U.S. Department of Transportation, Coast Guard 33 CFR Part 151, "Ballast Water Management for Vessels Entering the Great Lakes," vol. 57, No. 192, Friday, Oct. 2, 1992.

98. Veitch, C.R. and Bell, B.D., "Eradication of Introduced Animals from the Islands of New Zealand," *Ecological Restoration of New Zealand Islands*, D.R. Towns, C.H. Daugherty, and I. Atkinson (eds.), *Conservation Sciences Publication No. 2*. Department of Conservation, Wellington, 1990, pp. 137-146.

99. Vinson, S.B., "Potential Impact of Microbial Insecticides on Beneficial Arthopods on the Terrestrial Environment," *Safety of Microbial Insecticides*, M. Laird, L.A. Lacey and E.W. Davidson (eds.) (Boca Raton, FL: CRC Press, 1990), pp. 43-64.

100. Wall, C. "Principles of Monitoring," *Behavior-modifying Chemicals for Insect Management: Applications of Pheromones and Other Attractants*, R.L. Ridgway, R.M. Silverstein, and R.M. Inscoe (eds.) (New York, NY: Marcel Dekker, Inc., 1990), pp. 9-23.

101. Wapshere, A.J., Delfosse, E.S., and Cullens, J.M., "Recent Developments in Biological Control of Weeds," *Crop Protection*, vol. 8., August 1989, pp. 227-250.

102. Wilson, E.O., "The Coming Pluralization of Biology and the Stewardship of Systematics," *BioScience*, vol. 39, No. 4, April 1989, pp. 242-245.

103. Witt, I., Policy and Program Development, Animal and Plant Health Inspection Service, U.S. Department of Agriculture, personal communication to E. Chornesky, Office of Technology Assessment, Nov. 9, 1992.

104. Womach, J., Library of Congress, Congressional Research Service, "Regulating U.S. Pesticide Exports: Policy Issues and Proposed Legislation," Report to Congress, July 1991.

105. Zadig, D., "Plant Quarantines: Domestic Strategies Yield to International Policies," *California Agriculture*, January-February 1992, pp. 9-10.

CHAPTER 6: A PRIMER ON FEDERAL POLICY

1. Anonymous, "Preventing Tomorrow's Catastrophes," *Farm Forum*, summer 1990.

2. Backiel, A. et al., Library of Congress, Congressional Research Service, "Major Federal Land Management Agencies: Management of Our Nation's Land and Resources," May 7, 1990.

3. Bandel, D.M., Chief, Natural and Cultural Resources Division, Directorate of Public Works, Department of the Army, Fort Belvoir, VA, memorandum to the Executive Director, Armed Forces Pest Management Board, Feb. 4, 1992.

4. Bean, M.J., "The Role of the U.S. Department of the Interior in Non-Indigenous Species Issues," contractor report prepared for the Office of Technology Assessment, November 1991.

5. Beard, J.D., "Bug Detectives Crack the Tough Cases," *Science*, vol. 25, Dec. 13, 1991, pp. 1580-1581.

6. Betz, F.S., Acting Chief, Science Evaluation and Coordination Staff, Environmental Protection

Agency, Washington, DC, personal communication to E.A. Chornesky, Office of Technology Assessment, January 1992.

7. Betz, F.S., Acting Chief, Science Evaluation and Coordination Staff, U.S. Environmental Protection Agency, Washington, DC, letter to E.A. Chornesky, Office of Technology Assessment, Apr. 10, 1992.

8. Betz, F.S., Forsyth, S.F., and Stewart, W.E., "Registration Requirements and Safety Considerations for Microbial Pest Control Agents in North America," *Safety of Microbial Peticides*, M. Laird et al. (eds.) (Boca Raton, FL: CRC Press, Inc., 1990).

9. Bolsenga, S.J., Great Lakes Environmental Research Laboratory, National Oceanic and Atmospheric Administration, U.S. Department of Commerce, letter to E.A. Chornesky, Office of Technology Assessment, Apr. 1, 1992.

10. Brosten, D., "Tackling Tumbleweeds on CRP Acreage," *Agrichemical Age*, vol. 32, No. 5, May 1988, pp. 12-18.

11. Bushman, J., Office of Environmental Policy and Civil Works, Army Corp of Engineers, Washington, DC, personal communication to S.M. Fondriest, Office of Technology Assessment, Feb. 11, 1992.

12. Campbell, D.S., Director, Operational Support International Services, Animal Plant Health Inspection Service, letter to P.N. Windle, Office of Technology Assessment, Aug. 11, 1992.

13. Carruthers, R.I. and Coulson, J.R., "Ecological Benefits and Risks Associated with the Introduction of Exotic Species for Biological Control of Agricultural Pests," paper presented at workshop of National Academy of Science, Committee on Risk Assessment Methodology, Airlie Foundation, Warrington, VA, Feb. 26—Mar. 1, 1991.

14. Cauchon, S.P., Coastal Information Specialist, National Ocean Service, National Oceanic and Atmospheric Administration, U.S. Department of Commerce, Washington, DC, letter to E.A. Chornesky, Office of Technology Assessment, June 15, 1992.

15. Clay, W., Director, Animal Damage Control, Animal and Plant Health Inspection Service, U.S. Department of Agriculture, Hyattsville, MD, personal communication to the Office of Technology Assessment, Mar. 13, 1991.

16. Clay, W., Director, Animal Damage Control, Animal and Plant Health Inspection Service, FAX to E.A. Chornesky, Office of Technology Assessment, Oct. 23, 1992.

17. Conlin, M., Chief, Fisheries Division, Illinois Department of Conservation, Springfield, IL, letter to R.A. Peoples, Fish and Wildlife Service, U.S. Department of the Interior, Jan. 4, 1993.

18. Coulson, J.R. and Soper, R.S., "Protocols for the Introduction of Biological Control Agents in the U.S.," *Plant Protection and Quarantine: Volume III*, Robert P. Kahn (ed.) (Boca Raton, FL: CRC Press, Inc., 1989), pp. 1-35.

19. Council for Agricultural Science and Technology, "Ecological Impacts of Federal Conservation and Cropland Reduction Programs," Task Force Report No. 117, September 1990.

20. Courtenay, W.R., Jr., "Pathways and Consequences of the Introduction of Non-Indigenous Fishes in the United States," contractor report prepared for the Office of Technology Assessment, August 1991.

21. Craig, G., Professor of Biology, University of Notra Dame, letter to P.T. Jenkins, Office of Technology Assessment, Mar. 14, 1992.

22. Craven, R.B. et al., "Importation of *Aedes albopictus* and Other Exotic Mosquito Species into the United States in Used Tires from Asia," *Journal of the American Mosquito Control Association*, vol. 4, No. 2, 1988, pp. 138-42.

23. Crosby, M., Director, Division of Research, Sanctuaries and Reserves, National Ocean Service, National Oceanic and Atmospheric Administration, U.S. Department of Commerce, Washington, DC, personal communication to E.A. Chornesky, Office of Technology Assessment, Jan. 22, 1992.

24. Cunningham, I.S., "The U.S. National Arboretum: Leader in Ornamental Plant Germplasm Collection," *Diversity*, No. 12, 1988.

25. DeFazio, P. et al., Members of Congress, letter to C.K. Yeutter, Secretary, U.S. Department of Agriculture, Dec. 5, 1990.

26. Deitrich, B., "State Girds to Battle Alien Invader: A Moth," *Seattle Times*, Nov. 19, 1991, p. A-1.

27. Dudley, T.R., Research Botanist and Project Leader, National Arboretum, Agricultural Research Service, U.S. Department of Agriculture, Washington, DC, personal communication to Office of Technology Assessment, Oct. 4, 1991.

28. Duffy, L., Assistant Secretary for Environmental Restoration and Water Management, U.S. Department of Energy, Washington, DC, personal communication to S.M. Fondriest, Office of Technology Assessment, Jan. 2, 1992.

29. Earl, R., Real Property Office, Department of Energy, personal communication to S.M. Fondriest, Office of Technology Assessment, Dec. 13, 1992.

30. Edwards, G.B., Assistant Director, Fisheries Division, Fish and Wildlife Service, U.S. Department of the Interior, Washington, DC, letter to P.N. Windle, Office of Technology Assessment, Mar. 16, 1992.

31. Edwards, G.B. and Cottingham, D., Co-Chairs, U.S. Aquatic Nuisance Species Task Force, letter to E.A. Chornesky, Office of Technology Assessment, Nov. 25, 1992.

32. Elder, E., Chief Operations Officer, Domestic and Emergency Operations, Animal and Plant Health Inspection Service, U.S. Department of Agriculture, personal communication to Office of Technology Assessment, Mar. 13, 1991.

33. Fons, J., Senior Operations Officer, Inspection and Compliance, Animal and Plant Health Inspection Service, personal communication to Office of Technology Assessment, Mar. 13, 1991.

34. Fornasari, L. et al., U.S. Department of Agriculture, Agricultural Research Service, Biological Control of Weeds Laboratory—Rome, Italy, "Overview of the U.S. Department of Agriculture Biocontrol of Weeds Laboratory—Europe," Doc. No. 0018P, 1991.

35. Fritts, T.H., "The Brown Tree Snake, *Boiga irregularis*, A Threat to Pacific Islands," *U.S. Fish and Wildlife Service, Biological Report*, vol. 88, No. 31, September 1988.

36. Gardner, D.E., U.S. Department of the Interior, National Park Service, "Role of Biological Control as a Management Tool in National Parks and Other Natural Areas," Technical Report NPS/NRUH/NRTR-90/01, September 1990.

37. Geiger, J.G., Acting Assistant Director, Fish and Wildlife Service, U.S. Department of the Interior, letter to J.A. Schmid, Mar. 29, 1991.

38. Giamporcaro, D., TSCA Biotechnology Program, Chemical Control Division, Office of Pollution Prevention, U.S. Environmental Protection Agency, Washington, DC, personal communication to E.A. Chornesky, Office of Technology Assessment, Jan. 14, 1992.

39. Gillis, A.M., "Should Cows Chew Cheatgrass on Commonlands?" *BioScience*, vol. 41, No. 10, November 1991, pp. 668-675.

40. Glosser, J.W., Director, Animal and Plant Health Inspection Service, U.S. Department of Agriculture, testimony at hearings before the U.S. Senate Committee on Agriculture, Nutrition, and Forestry, Subcommittee on Agricultural Research and General Legislation, *Preparation for the 1990 Farm Bill: Noxious Weeds*, Mar. 28, 1990, pp. 350-354.

41. Glosser, J.W., Director, Animal and Plant Health Inspection Service, U.S. Department of Agriculture, letter to The Honorable Tom Daschle, Apr. 17, 1990.

42. Glosser, J.W., Director, Animal and Plant Health Inspection Service, U.S. Department of Agriculture, testimony at hearings before the U.S. House Committee on Appropriations, Subcommittee on Agriculture, Rural Development, and Related Agencies, *Hearings on Agriculture, Rural Development and Related Agencies Appropriations for 1992: Part 4*, Serial No. 43-171 O (Washington, DC: Government Printing Office, May 1991), pp. 271-467.

43. Grable, K., Plant Pathologist, U.S. Environmental Protection Agency, personal communication to E.A. Chornesky, Office of Technology Assessment, Mar. 31, 1992.

44. Greater Yellowstone Coordinating Committee, Billings, MT, "Guidelines for Coordinated Management of Noxious Weeds in the Greater Yellowstone Area," undated.

45. Griswold, B., Program Director, Marine Advisory Services, National Sea Grant College Program, National Oceanic and Atmospheric Administration, U.S. Department of Commerce, Silver Spring, MD, FAX to E.A. Chornesky, Office of Technology Assessment, Mar. 31, 1992.

46. Heck, J.A., Library of Congress, Congressional Research Service, "National Wildlife Refuges: Places to Hunt?" July 28, 1992.

47. Hester, F.E., "The U.S. National Park Service Experience with Exotic Species," *Natural Areas Journal*, vol. 11, No. 3, July 1991, pp. 127-128.

48. Hetke, S.F., Associate Director for Research Operations, Environmental Research Laboratory, U.S. Environmental Protection Agency, Duluth, MN, letter to E.A. Chornesky, Office of Technology Assessment, Jan. 15, 1992.

49. Holmes, R., Director, Division of Fish and Wildlife, Minnesota Department of Natural Resources, letter to R.A. Peoples, Fish and Wildlife Service, Department of the Interior, Dec. 28, 1992.

50. Hunsaker, C.T. and Carpenter, D.E., *Ecological Indicators for the Environmental Monitoring and Assessment Program* (Research Triangle Park, NC: U.S. Environmental Protection Agency, Office of Research and Development, 1990).

51. Johnston, G.H., Chief of Wildlife and Vegetation Division, Natural Resources Program Branch, National Park Service, U.S. Deaprtment of the Interior, personal communication, July 10, 1991, Mar. 13, 1992.

52. Johnston, G.H. and Dennis, J.G., Wildlife and Vegetation Division, Natural Resources Program Branch, National Park Service, U.S. Department of the Interior, letter to P.N. Windle, Office of Technology Assessment, Jan. 13, 1992.

53. Kaplan, J.K., "Bring 'Em Back Alive and Growing," *Agricultural Research*, vol. 39, No. 7, July 1991, pp. 4-13.

54. Kim, K.C. and Wheeler, A.G., Jr., "Pathways and Consequences of the Introduction of Non-Indigenous Insects and Arachnids in the United States," contractor report prepared for the Office of Technology Assessment, Washington, DC, December 1991.

55. Kludy, D.H., "Federal Policy on Non-Indigenous Species: The Role of the United States Department of Agriculture's Animal and Plant Health Inspection Service," contractor report prepared for the Office of Technology Assessment, Washington, DC, December 1991.

56. Krugman, S., U.S. Department of Agriculture, Forest Service, personal communication to J.A. Schnittker, 1991.

57. Kutz, R., Deputy Director EMAP, U.S. Environmental Protection Agency, personal communication to E.A. Chornesky, Office of Technology Assessment, Jan. 13, 1992.

58. Lima, P., Entomologist, BATS, Animal and Plant Health Inspection Service, U.S. Department of Agriculture, personal communication to S.M. Fondriest, Office of Technology Assessment, Nov. 12, 1992.

59. Loope, L., Research Biologist, U.S. Department of the Interior, National Park Service, letter to P.N. Windle, Office of Technology Assessment, July 28, 1992.

60. Mack, R.N., "Pathways and Consequences of the Introduction of Non-Indigenous Plants in the United States," contractor report prepared for the Office of Technology Assessment, September 1991.

61. Mack, R.N., "Additional Information on Non-Indigenous Plants in the United States," Contractor Report Prepared for the Office of Technology Assessment, October 1992.

62. Marroquin, R., Entomologist, U.S. Army Engineering and Housing Support Center, Department of the Army, personal communication to S.M. Fondriest, Office of Technology Assessment, Feb. 7, 1992.

63. McCann, J.A., Director, National Fisheries Research Center, Fish and Wildlife Service, U.S Department of the Interior, Gainesville, FL, personal communication to E.A. Chornesky, Office of Technology Assessment, Mar. 23, 1993.

64. Melland, R.B., Director, Animal and Plant Health Inspection Service, U.S, Department of Agriculture, testimony at hearings before the U.S. House Committee on Appropriations, Subcommittee on Agriculture, Rural Development, and Related Agencies, *Hearings on Agriculture, Rural Development, Food and Drug Administration, and Related Agenciew Appropriations for 1993, Part 3*, Serial No. 54-888 O (Washington, DC: Government Printing Office, March 1992), pp. 211-314.

65. Miller, S., Chief, Division of Land and Water Resources, Bureau of Indian Affairs, U.S. Department of the Interior, personal communication to E.A. Chornesky, Office of Technology Assessment, Oct. 26, 1992.

66. National Fish and Wildlife Foundation, Washington, DC, "Fiscal Year 1991 Budget Recommendations for United States Department of the Interior, National Park Service, Natural Resource Management Program," 1990.

67. National Fish and Wildlife Foundation, Washington, DC, "Fiscal Year 1992 Wildlife and Fisheries Assessment," April 1991.

68. National Fish and Wildlife Foundation, Washington, DC, Internal Progress Report on the "Bring Back the Natives Projects," Mar. 1, 1992.

69. Newman, J.B., Director Ecological Sciences, Soil Conservation Service, U.S. Department of Agriculture, personal communication to E.A. Chornesky, Office of Technology Assessment, July 11, 1991.

70. Nickum, J., National Aquaculture Coordinator, Fish and Wildlife Service, U.S. Department of Interior, Washington, DC, personal communication to E.A. Chornesky, Office of Technology Assessment, Aug. 26, 1991.

71. Noldan, H.K., Acting Assistant Director, Land and Renewable Resources, Bureau of Land Management, Washington, DC, internal information bulletin No. 91-266 to all State Directors, Feb. 6, 1991.

72. Peoples, R.A., Resource Analyst, Fish and Wildlife Service, U.S. Department of Interior, personal communication to E.A. Chornesky, Office of Technology Assessment, Aug. 19, 1991, Mar. 4, 1992.

73. Perez, A., Quarantine Division, Center for Disease Control, Public Health Service, Atlanta, GA, personal communication to S.M. Fondriest, Office of Technology Assessment, Jan. 17, 1992.

74. Robertson, F.D., "Rise to the Future: The Forest Service Fisheries Program," Fisheries, vol. 13, No. 3, May-June 1988, pp. 22-23.

75. Schnittker, J., "Federal Policy on Non-Indigenous Species: An Overview of the U.S. Department of Agriculture," contractor report prepared for the Office of Technology Assessment, December 1991.

76. Schwegman, J.E., "Newsletter of the Illinois Native Plant Conservation Program Division of Natural Heritage, Illinois Department of Conservation," Illinoensis, vol. 7, No. 2, July 1991.

77. Shands, H.L., Fitzgerald, P.J., and Eberhart, S.A., "Program for Plant Germplasm Preservation in the United States: the U.S. National Plant Germplasm System," Biotic Diversity and Germplasm Preservation, Global Imperatives, L. Knutson and A.K. Stoner (eds.) (Netherlands: Kluwer Academic Publishers, 1989), pp. 97-115.

78. Shannon, M., Assistant Director for Planning and Design, Animal and Plant Health Inspection Service, U.S. Department of Agriculture, personal communication to P.T. Jenkins, Office of Technology Assessment, May 19, 1992.

79. Sharp, C., "The Soil Conservation Service Perspective," paper presented at the 16th Annual Natural Areas Conference, Knoxville, TN, Oct. 18, 1989.

80. Sharp, C., National Plant Materials Specialist, Soil Conservation Service, U.S. Department of Agriculture, personal communication to E.A. Chornesky, Office of Technology Assessment, Aug. 15, 1991.

81. Sharp, C., National Plant Materials Specialist, Soil Conservation Service, U.S. Department of Agriculture, letter to P.N. Windle, Office of Technology Assessment, July 24, 1992.

82. Siehl, G.H., "Natural Resource Issues in National Defense Programs," Congressional Research Service Report to Congress, 91-781 ENR, Oct. 31, 1991.

83. Singletary, H.M., Director, Plant Industry Division, North Carolina Department of Agriculture, testimony at hearings before the U.S. Senate Committee on Agriculture, Nutrition, and Forestry, Subcommittee on Agricultural Research and General Legislation, Preparation for the 1990 Farm Bill: Noxious Weeds, Mar. 28, 1990, pp. 354-357.

84. Smith, C., Professor, Department of Botany, University of Hawaii at Manoa, personal communication to E.A. Chornesky, Office of Technology Assessment, May 6, 1993.

85. Stanley, J.G., Peoples, R.A., Jr., and McCann, J.A., "Legislation and Responsibilities Related to Importation of Exotic Fishes and Other

Aquatic Organisms,'' *Canadian Journal of Fisheries and Aquatic Organisms*, vol. 48, supl. 1, 1991, pp. 162-166.

86. Steering Committee for the 75th Symposium, *National Parks for the 21st Century: The Vail Agenda Report and Recommendations to the Director of the National Park Service*, National Park Service Document No. D-726, Mar. 25, 1992.

87. Thill, D., Department of Plant Soil and Environmental Sciences, University of Idaho, Moscow, ID, personal communication to R.N. Mack, Pullman, WA, Oct. 21, 1992.

88. Torchio, P.F. et al., ''Introduction of the European Bee, *Osmia cornuta*, into California Almond Orchards (Hymenoptera: Megachilidae), *Environmental Entomology*, vol. 16, No. 3, 1987, pp. 664-667.

89. Troust, J., Senior Environmental Specialist, Bureau of Reclamation, U.S. Department of the Interior, FAX to E.A. Chornesky, Office of Technology Assessment, Nov. 13, 1992.

90. Turner, J.F., Director, Fish and Wildlife Service, U.S. Department of Interior, Washington, DC, internal agency memorandum, Mar. 22, 1990.

91. U.S. Aquatic Nuisance Species Task Force, ''Proposed Aquatic Nuisance Species Program,'' Sept. 28, 1992.

92. U.S. Congress, General Accounting Office, *Forest Service: Difficult Choices Face the Future of the Recreation Program*, RCED-91-115 (Gaithersburg, MD: U.S. General Accounting Office, Apr. 15, 1991).

93. U.S. Congress, General Accounting Office, *Rangeland Management: Forest Service Not Performing Needed Monitoring of Grazing*, RCED-91-148 (Gaithersburg, MD: U.S. General Accounting Office, May 16, 1991).

94. U.S. Congress, General Accounting Office, *National Forests: Funding Fish and Wildlife Projects*, RCED-91-113 (Gaithersburg, MD: U.S. General Accounting Office, June 12, 1991).

95. U.S. Congress, House Committee on Appropriations, Subcommittee on Agriculture, Rural Development, and Related Agencies, *Hearings on Agriculture, Rural Development and Related Agencies Appropriations for 1992: Part 2*, Serial

No. 43-149 O (Washington, DC: Government Printing Office, April 1991).

96. U.S. Congress, House Committee on Appropriations, Subcommittee on Agriculture, Rural Development, and Related Agencies, *Hearings on Agriculture, Rural Development and Related Agencies Appropriations for 1992: Part 5*, Serial No. 43-194 (Washington, DC: Government Printing Office, April 1991).

97. U.S. Congress, House Committee on Appropriations, Subcommittee on Agriculture, Rural Development, and Related Agencies, *Hearings on Agriculture, Rural Development and Related Agencies Appropriations for 1992: Part 4*, Serial No. 43-171 O (Washington, DC: Government Printing Office, May 1991).

98. U.S. Congress, House Committee on Appropriations, Subcommittee on Agriculture, Rural Development, and Related Agencies, *Hearings on Agriculture, Rural Development, Food and Drug Administration, and Related Agenciew Appropriations for 1993, Part 3*, Serial No. 54-888 O (Washington, DC: Government Printing Office, March 1992).

99. U.S. Congress, House Committee on Appropriations, Subcommittee on Agriculture, Rural Development, and Related Agencies, *Hearings on Agriculture, Rural Development, Food and Drug Administration, and Related Agencies Appropriations for 1993: Part 4*, Serial No. 054-898 O (Washington, DC: Government Printing Office, Mar. 25, 1992).

100. U.S. Congress, House Committee on Appropriations, Subcommittee on the Department of the Interior and Related Agencies, *Hearings on Justifications of the Budget Estimates for the Bureau of Land Management, Fish and Wildlife Service, and National Park Service*, Serial No. 40-442 (Washington, DC: Government Printing Office, 1991).

101. U.S. Congress, House Committee on Appropriations, Subcommittee on the Department of the Interior and Related Agencies, *Justification of the Budget Estimates: Part 2*, Serial No. 52-547 O (Washington, DC: Government Printing Office, 1992).

102. U.S. Congress, House Committee on Merchant Marine and Fisheries, Subcommittee on Ocean-

ography, Great Lakes and the Outer Continental Shelf, *Hearings on Present Activities and Future Needs of Federal Programs Directed at Improving the Water Quality of the Great Lakes*, Serial No. 102-15, Apr. 30, 1991.

103. U.S. Department of Agriculture, "Policy of Noxious Weed Management," departmental regulation, 1990.

104. U.S. Department of Agriculture, Agricultural Marketing Service, "This is AMS," March 1991.

105. U.S. Department of Agriculture, Agricultural Research Service, "Seeds for Our Future: The U.S. National Plant Germplasm System," December 1990.

106. U.S. Department of Agriculture, Animal and Plant Health Inspection Service, "National Biological Control Institute," Information Sheet, undated.

107. U.S. Department of Agriculture, Economic Research Service, "The Conservation Reserve Program: Enrollment Statistics for Signup Periods 1-9 and Fiscal Year 1989," Statistical Bulletin No. 811, 1990.

108. U.S. Department of Agriculture, Forest Service, "Forest Service Manual: Chapter 2640—Stocking and Harvesting," June 1, 1990.

109. U.S. Department of Agriculture, Forest Service, "Forest Service Manual: Chapter 2610—Cooperative Relations," June 25, 1990.

110. U.S. Department of Agriculture, Forest Service, "Blue Mountains Forest Health Report: New Perspectives in Forest Health," April 1991.

111. U.S. Department of Agriculture, Forest Service, "Forest Service Manual 2600—Wildlife, Fish, and Sensitive Plant Habitat Management WO Amendment 2600-91-6," Sept. 24, 1991.

112. U.S. Department of Agriculture, Forest Service, "Forest Service Manual: Chapter 2080—Noxious Weed Management," Interim Directive, Aug. 3, 1992.

113. U.S. Department of Agriculture, Soil Conservation Service, "Improved Conservation Plant Materials Released by the SCS and Cooperators Through December 1990," 1990.

114. U.S. Department of Agriculture, Soil Conservation Service, "National Plant Materials Manual (NPMM), 190-V, Second Edition," Sept. 9, 1990.

115. U.S. Department of Agriculture, Soil Conservation Service, "SCS National Plant Materials Program: Accomplishments and Opportunities," September 1990.

116. U.S. Department of Agriculture, Soil Conservation Service, "Plant Materials Centers: Finding Vegetative Solutions to Conservation Problems," June 1991.

117. U.S. Department of Agriculture, Soil Conservation Service, "Draft Strategic Plan for SCS Plant Materials Center Program," Aug. 15, 1991.

118. U.S. Department of Commerce, National Oceanic and Atmospheric Administration, "Great Lakes Environmental Research Laboratory: Annual Report FY 91," undated.

119. U.S. Department of Commerce, National Oceanic and Atmospheric Administration, "The National Sea Grant College Program Annual Report FY 89," 1989.

120. U.S. Department of Commerce, National Oceanic and Atmospheric Administration, "National Estuarine Research Reserve System Site Catalogue," 1990.

121. U.S. Department of Commerce, National Oceanic and Atmospheric Administration, National Sea Grant College Program, "Second Annual Call for Research Proposals Related to Nonindigenous Species With Special Emphasis on the Zebra Mussel," 1992.

122. U.S. Department of Defense, "Department of Defense Directive: Natural Resources Management Program," Directive No. 4700.4, Jan. 24, 1989.

123. U.S. Department of Defense, "Legacy Resource Management Program," Report to Congress, September 1991.

124. U.S. Department of Defense, Department of the Air Force, "AF Regulation 400-21: Retrograde Material Preclearance Program," June 15, 1972.

125. U.S. Department of Defense, Department of the Air Force, "AF Regulation 126-1: Conservation and Management of Natural Resources," Oct. 21, 1988.

126. U.S. Department of Defense, Department of the Army, "Army Regulation 420-74: Natural Resource Management," Feb. 25, 1986.

127. U.S. Department of Defense, Department of the Army, "Army Regulation 420-76: Facilities Engineering Pest Management," June 3, 1986.

128. U.S. Department of Energy, "Real Property Statistics," Sept. 30, 1991.

129. U.S. Department of the Interior, Bureau of Indian Affairs, "1992 Noxious Weed Plan," executive summary, Washington, DC, 1992.

130. U.S. Department of the Interior, Bureau of Land Management, "Fish and Wildlife 2000: A Plan for the Future," undated.

131. U.S. Department of the Interior, Bureau of Land Management, "BLM Manual: 6820—Wildlife Introduction and Transplants," Oct. 20, 1976.

132. U.S. Department of the Interior, Bureau of Land Management, "Northwest Area Noxious Weed Control Program: Final Environmental Impact Statement," December 1985.

133. U.S. Department of the Interior, Bureau of Land Management, "BLM Manual: 1745—Introduction, Transplant, Augmentation, and Reestablishment of Fish, Wildlife and Plants," revision of section 6820, draft, Aug. 15, 1986.

134. U.S. Department of the Interior, Bureau of Land Management, "Draft Environmental Impact Statement: Vegetation Treatment on BLM Lands in Thirteen Western States," 1989.

135. U.S. Department of the Interior, Bureau of Land Management, "BLM Manual: 9014—Use of Biological Control Agents of Pests on Public Lands," Oct. 30, 1990.

136. U.S. Department of the Interior, Bureau of Land Management, "Evaluation: Bureauwide Noxious Weed Program," internal agency review, draft, December 1991.

137. U.S. Department of the Interior, Bureau of Reclamation, Lower Colorado Region, "Vegetation Management Study: Lower Colorado River—Phase I," Boulder City, NV, September 1992.

138. U.S. Department of the Interior, Fish and Wildlife Service, "NEPA in Federal Aid Proposals: Guidance to the States," September 1980.

139. U.S. Department of the Interior, Fish and Wildlife Service, "Federal Aid Manual," 1982.

140. U.S. Department of the Interior, Fish and Wildlife Service, Internal memorandum from the Director to the Regional Directors concerning "potential policy initiative—reducing risks of aquatic organism introductions," Oct. 20, 1987.

141. U.S. Department of the Interior, Fish and Wildlife Service, *Restoring America's Wildlife* (Washington, DC: U.S. Government Printing Office, 1987).

142. U.S. Department of the Interior, Fish and Wildlife Service, "Federal Aid in Fish and Wildlife Restoration," 1988.

143. U.S. Department of the Interior, Fish and Wildlife Service, "Current Federal Aid Research Report: Fish 1990," report, 1990.

144. U.S. Department of the Interior, Fish and Wildlife Service, "Current Federal Aid Research Report: Wildlife 1990," report, 1990.

145. U.S. Department of the Interior, Fish and Wildlife Service, Law Enforcement Advisory Commission, "Report of Findings and Recommendations," June 15, 1990.

146. U.S. Department of the Interior, Fish and Wildlife Service, "Triploid Grass Carp Report," 1990.

147. U.S. Department of the Interior, Fish and Wildlife Service, "Vision for the Future: 1991 Total Quality Management Plan," May 1991.

148. U.S. Department of the Interior, National Park Service, "Management Policies," 1988.

149. U.S. Department of the Interior, National Park Service, and U.S. Department of Agriculture, Soil Conservation Service, *Native Plants for Parks*, D-425 (Washington, DC: U.S. Government Printing Office, 1989).

150. U.S. Department of the Interior, National Park Service, "1989 Inventory of Research Activities in the National Parks," Science Report NPS/NRWV/NRSR-91/03, January 1991.

151. U.S. Department of the Interior, National Park Service, "1990 Inventory of Research Activities in the National Parks," Science report, 1992.

152. U.S. Department of the Interior, Office of Inspector General, "The Endangered Species Program, U.S. Fish and Wildlife Service," Audit Report No. 90-98, September 1990.

153. U.S. Department of the Treasury, U.S. Customs Service, "Master Plan for Passenger Processing at United States Airports for the 1990's," undated.

154. U.S. Department of the Treasury, U.S. Customs Service, "Importing Into the United States," September 1991.

155. U.S. Environmental Protection Agency, "Pesticide Assessment Guidelines: Microbial Pest Control Agents," March 1989.

156. U.S. Environmental Protection Agency, "EMAP Monitor," EPA-600/M-90/022, January 1991.

157. U.S. Public Health Service, "Medical Entomology-Ecology Branch (MEEB)," information brochure, draft, undated.

158. U.S. Public Health Service, "Public Health Screenings at U.S. Ports of Entry: A Guide for U.S. Immigration & Naturalization Service and U.S. Customs Service Inspectors," October 1991.

159. Versar, Inc., "Introduction of Pacific Salmonids into the Delaware River Watershed," draft environmental impact statement prepared for the U.S. Fish and Wildlife Service and the New Jersey Division of Fish, Game and Wildlife, July 25, 1991.

160. Wahle, C.M., Evaluation Team Leader, Office of Coastal Resources Management, National Oceanic and Atmospheric Administration, U.S. Department of Commerce, Washington, DC, personal communication to E.A. Chornesky, Office of Technology Assessment, Mar. 31, 1992.

161. Waters, L.H., Pest Management Specialist, Bureau of Land Management, U.S. Department of the Interior, Washington, DC, personal communication to E.A. Chornesky, Office of Technology Assessment, Sept. 5, 1991.

162. Weiss, M.J., Assistant Director Forest Pest Management, Forest Service, U.S. Department of Agriculture, Washington, DC, letter to P.N. Windle, Office of Technology Assessment, Feb. 5, 1992.

163. West, A.J., Deputy Chief of State and Private Forestry, Forest Service, U.S. Department of Agriculture, Washington, DC, letter to P.N. Windle, Office of Technology Assessment, Aug. 24, 1992.

164. Westbrooks, R.G., Station Leader, Whiteville Noxious Weed Station, Animal and Plant Health Inspection Service, U.S. Department of Agriculture, Whiteville, NC, letter to P.N. Windle,

Office of Technology Assessment, Aug. 11, 1992.

165. White, G.A., "The U.S. Department of Agriculture's Plant Introduction Program—Safeguards Against Introducing Pests," 1991.

166. White, G.A., Shands, H.L., and Lovell, G.R., "History and Operation of the National Plant Germplasm System," *Plant Breeding Reviews: The National Plant Germplasm System of the United States, Vol. 7*, J. Janick (ed.) (Portland, OR: Timber Press, 1989), pp. 5-56.

167. Wilkinson, D., Marine Resource Management Specialist, Office of Protected Resources, National Marine Fisheries Service, National Oceanic and Atmospheric Administration, U.S. Department of Commerce, Silver Spring, MD, personal communications to E.A. Chornesky, Office of Technology Assessment, Jan. 22, 1992 and Mar. 31, 1992.

168. Williamson, W., Director of Range Management, Forest Service, U.S. Department of Agriculture, Washington, DC, personal communication to E.A. Chornesky, Office of Technology Assessment, July 19, 1991.

169. Witfield, P., Deputy Assistant Secretary, Environmental Restoration, U.S. Department of Energy, Washington, DC, personal communication to S.M. Fondriest, Office of Technology Assessment, Jan. 9, 1992.

170. Witt, I., Policy and Program Development, Animal Plant Health Inspection Service, U.S. Department of Agriculture, Washington, DC, FAXes to E.A. Chornesky, Office of Technology Assessment, Nov. 9 and Nov. 10, 1992.

CHAPTER 7: STATE AND LOCAL APPROACHES FROM A NATIONAL PERSPECTIVE

1. Alexander, D., Pest Exclusion Specialist, Washington Department of Agriculture, Olympia, WA, personal communication to P.T. Jenkins, Office of Technology Assessment, Feb. 15, 1991.

2. Alfieri, S.A., Jr., Assistant Director, Division of Plant Industry, Florida Department of Agriculture and Consumer Services, Tallahassee FL, "Regulatory Pest Control Philosophy: Views and Assessment," paper presented at the meeting of the National Plant Board, Kalispell, MT, August 1991.

3. Anonymous, "Guidelines for Coordinated Management of Noxious Weeds in the Greater Yellowstone Area," adopted by parties to Memorandum of Understanding between the Governor of the State of Wyoming et al., dated Sept. 30, 1990.

4. Anonymous, "Wyoming Noxious Weed-Free Forage," pamphlet issued by Wyoming Department of Agriculture et al., Cheyenne, WY, 1991.

5. Anonymous, "Non-Natives Bill Triggers Concern," *American Horticulturist*, May 1992, p. 12.

6. Bean, M.J., "The Role of the United States Department of Interior in NIS Issues," contractor report prepared for the Office of Technology Assessment, November 1991.

7. Britton, J.C., "Pathways and Consequences of Introductions of Non-Indigenous Freshwater, Terrestrial, and Estuarine Mollusks in the United States," contractor report prepared for the Office of Technology Assessment, October 1991.

8. California Department of Food and Agriculture, Division of Plant Industry, "Annual Report," Sacramento, CA, 1990.

9. Clark, J.A., Environmental Defense Fund, written statement submitted to the Intentional Introductions Policy Committee of the Aquatic Nuisance Species Task Force, Feb. 26, 1992.

10. Cohn, D., "Plan to Revive Bay's Oysters Creates a Regional Dispute," *Washington Post*, Nov. 18, 1991, pp. A1, A12.

11. Collins, R.A., "California's Approach to Risk Reduction in the Introduction of Exotic Species," *Dispersal of Living Organisms into Aquatic Ecosystems*, A. Rosenfield and R. Mann (eds.) (College Park, MD: Maryland Sea Grant, 1992), pp. 361-364.

12. Commissioner, Maine Department of Marine Resources, Code of Maine Rules, Ch. 24, Basis Statement, Aug. 15, 1984.

13. Courtenay, W.R., Jr., "Pathways and Consequences of Introductions of Non-Indigenous Fishes in the United States," contractor report prepared for the Office of Technology Assessment, August 1991.

14. Doren, R.F., Assistant Research Director, Everglades National Park, Homestead, FL, personal communication to P.T. Jenkins, Office of Technology Assessment, Jan. 21, 1992.

15. Dowell, R.V. and Krass, C.J., "Exotic Pests Pose Growing Problem for California," *California Agriculture*, vol. 46, No. 1, January-February 1992, pp. 6-12.

16. Edwards, G.B., Assistant Director, Fisheries Division, Fish and Wildlife Service, U.S. Department of the Interior, Washington, DC, testimony at hearings before the House Subcommittee on Oversight and Investigations of the Committee on Merchant Marine and Fisheries, July 12, 1991, Serial No. 102-28, pp. 32-3.

17. Elder, E.W., Chief Operations Officer, Domestic and Emergency Operations, Animal and Plant Health Inspection Service, U.S. Department of Agriculture, Hyattsville, MD, personal communication to P.T. Jenkins, Office of Technology Assessment, Mar. 13, 1991.

18. Environmental Law Institute, *Law of Environmental Protection, Vol. I* (New York, NY: Clark Boardman Co., 1990), p. 6-24.

19. Federal Noxious Weed Committee, Weed Science Society of America, "Points of Discussion on Six Proposed Amendments to the Federal Noxious Weed Act," 1989.

20. Fleck, A., Enforcement Officer, Utah Division of Wildlife Resources, Salt Lake City, UT, personal communication to P.T. Jenkins, Office of Technology Assessment, Sept. 9, 1992.

21. Gilbert, B., "Coyotes Adapted to Us, Now We Have to Adapt to Them," *Smithsonian*, April 1991, pp. 69-78.

22. Girard, N., Counsel, New Hampshire Fish and Game Department, Concord, NH, personal communication to P.T. Jenkins, Office of Technology Assessment, May 6, 1992.

23. Gomez, R., Chairman, U.S. Department of Agriculture Interagency Technical Working Group on the Africanized Honey Bee, Washington, DC, personal communication to P.T. Jenkins, Office of Technology Assessment, Sept. 9, 1992.

24. Greenwalt, L.A., Vice-President, International Affairs, National Wildlife Federation, Washington, DC, personal communication to P.T. Jenkins, Office of Technology Assessment, Dec. 11, 1991.

25. Hardy, T.N., Administrator Coordinator, Quarantine Programs, Louisiana Department of Agriculture and Forestry, Baton Rouge, LA, letter to P.T. Jenkins, Office of Technology Assessment, Dec. 21, 1992.

26. Hocutt, C.H., ''Toward the Development of an Environmental Ethic for Exotic Fishes,'' *Distribution, Biology and Management of Exotic Fishes*, W.R. Courtenay and J.R. Stauffer (eds.) (Baltimore, MD: Johns Hopkins Univ. Press, 1984), pp. 374-386.

27. Kahn, R.P., ''Technologies To Maintain Biological Diversity: Assessment of Plant Quarantine Practices,'' contractor report prepared for the Office of Technology Assessment, November 1985.

28. Kern, F.G. and Rosenfield, A., ''Shellfish Health and Protection,'' *Dispersal of Living Organisms into Aquatic Ecosystems*, A. Rosenfield and R. Mann (eds.) (College Park, MD: Maryland Sea Grant, 1992), pp. 313-323.

29. King, S.T. and Schrock, J.R., *Controlled Wildlife—Vol. III: State Wildlife Regulations* (Lawrence, KS: Assoc. of Systematics Collections, 1985).

30. Krantz, G.E., ''Present Management Position on *Crassostrea virginica* in Maryland with Comments on the Possible Introduction of an Exotic Oyster, *Crassostrea gigas*,'' *Introductions and Transfers of Marine Species, Achieving a Balance Between Economic Development and Resource Protection*, Proceedings of Conference/Workshop at Hilton Head, SC (Charleston, SC: SC Sea Grant Consortium, 1992), pp. 121-126.

31. Kurdila, J., ''The Introduction of Exotic Species Into the United States: There Goes the Neighborhood!'' *Boston College Environmental Affairs Law Review*, vol. 16, No. 1, 1988, pp. 95-118.

32. Lanka, B. et al., Wyoming Game and Fish Department, Game Division, Exotic Species Committee, Cheyenne, WY, ''Analysis and Recommendations on the Applications by Mr. John T. Dorrance III to Import and Possess Native and Exotic Species,'' Mar. 1, 1990.

33. Mack, R.N., ''Pathways and Consequences of Introductions of Non-Indigenous Plants in the United States—Additional Information,'' supplemental contractor report prepared for the Office of Technology Assessment, October 1992.

34. Mann, R., Burreson, E.M., and Baker, P.K., ''The Decline of the Virginia Oyster Fishery in Chesapeake Bay: Considerations for Introduction of a Non-endemic Species, *Crassostrea gigas* (Thunberg),'' *Introductions and Transfers of Marine Species, Achieving a Balance Between Economic Development and Resource Protection*, Proceedings of a Conference/Workshop at Hilton Head, SC (Charleston, SC: South Carolina Sea Grant Consortium, 1992), pp. 107-120.

35. Martin, G., ''Hog Wild,'' *Discover*, vol. 11, No. 6, June 1990, pp. 43-46.

36. Mayer, J.J. and Brisbin, I.L., *Wild Pigs in the United States: Their History, Comparative Morphology, and Current Status* (Athens, GA: University of Georgia Press, 1991).

37. McCann, J.A., Director, National Fisheries Research Center, Fish and Wildlife Service, U.S. Department of the Interior, Gainesville, FL, letter to E.A. Chornesky, Office of Technology Assessment, Oct. 4, 1991.

38. McDowell, R., Director, Division of Fish, Game and Wildlife, New Jersey Department of Environmental Protection, Trenton, NJ, testimony at hearings before the House Committee on Merchant Marine and Fisheries, Subcommittee on Oversight and Investigations, Serial No. 102-28, July 12, 1991, p. 19.

39. Metterhouse, W.M., ''Biological Control: State Regulatory View Point,'' *Regulations and Guidelines: Critical Issues in Biological Control—Proceedings of a USDA/CSRS National Workshop*, R. Charudattan and H.W. Browning (eds.) (Gainesville, FL: Institute of Food and Agricultural Sciences, 1992).

40. Michigan Department of Natural Resources, Lansing, MI, Draft Briefing Statement on Zebra Mussels, August 1992.

41. Minnesota Interagency Exotic Species Task Force, St. Paul, MN, ''Report and Recommendations,'' submitted to the Natural Resources Committees of the Minnesota House and Senate, July 1991.

42. Montana Department of Fish, Wildlife and Parks, Wildlife Division, ''Draft Summary of Background Information Pertaining to Regulation of the Game Farm Industry,'' 1992.

43. Moyle, P.B., Department of Wildlife and Fisheries Biology, University of California, Davis, CA, letter to P.N. Windle, Office of Technology Assessment, Dec. 16, 1991.

44. Mullin, B., Weed Coordinator, Montana Department of Agriculture, Helena, MT, letter to P.T. Jenkins, Office of Technology Assessment, Feb. 18, 1992.

45. Musgrave, R., Program Director, Center for Wildlife Law, University of New Mexico Law School, Albuquerque, NM, personal communication to P.T. Jenkins, Office of Technology Assessment, Jan. 16, 1992.

46. Nettles, V., Director, Southeast Cooperative Wildlife Disease Study Center, Athens, GA, personal communication P.T. Jenkins, Office of Technology Assessment, Feb. 10, 1992.

47. Nickum, J., National Aquaculture Coordinator, Fish and Wildlife Service, Department of the Interior, Washington, DC, personal communication to E.A. Chornesky, Office of Technology Assessment, Aug. 26, 1991.

48. Old, R.R., Weed Diagnostician, University of Idaho, Moscow, ID, letter to P.N. Windle, Office of Technology Assessment, Oct. 28, 1991.

49. Regelbrugge, C., Director of Regulatory Affairs and Grower Services, American Association of Nurserymen, Washington, DC, letter to P.N. Windle, Office of Technology Assessment, Aug. 10, 1992.

50. Rosenberg, D.Y., "The Interaction of State and Federal Quarantines in the U.S.," *Plant Protection and Quarantine—Vol. III*, R.P. Kahn (ed.) (Boca Raton, FL: CRC Press, 1989), pp. 59-74.

51. Scarratt, D.J. and Drinnan, R.E., "Canadian Strategies for Risk Reductions in Introductions and Transfers of Marine and Anadromous Species," *Dispersal of Living Organisms into Aquatic Ecosystems*, A. Rosenfield and R. Mann (eds.) (College Park, MD: Maryland Sea Grant, 1992), pp. 377-385.

52. Schmidt, B.R., Chief of Fisheries, Division of Wildlife Resources, Utah Department of Natural Resources, Salt Lake City, UT, letter to P.N. Windle, Office of Technology Assessment, Dec. 16, 1991.

53. Schmitz, D.C., Bureau of Aquatic Plant Management, Florida Department of Natural Resources, Tallahassee, FL, memorandum to Jeremy Craft, Florida Division of Resource Management, Aug. 9, 1988.

54. Schmitz, D.C., Bureau of Aquatic Plant Management, Florida Department of Natural Resources, Tallahassee, FL, testimony at public hearing before the Intentional Introductions Policy Committee of the U.S. Aquatic Nuisance Species Task Force, Feb. 26, 1992.

55. Schwegman, J., Botany Program Manager, Illinois Department of Conservation, Springfield, IL, "A State Program to Combat Exotic Weeds in Natural Communities," Apr. 1, 1989.

56. Scribner, J.D., Attorney, Sacramento, CA, letter to P.N. Windle, Office of Technology Assessment, Feb. 3, 1992.

57. *Seed World*, Seed Trade Buyers Guide, vol. 129, No. 4, April 1991.

58. Siddiqui, I.A., Assistant Director, California Department of Food and Agriculture, Sacramento, CA, testimony at hearings before the Senate Committee on Governmental Affairs, Subcommittee on Federal Services, Post Office, and Civil Service, *Postal Implementation of the Agricultural Quarantine Enforcement Act*, June 5, 1991.

59. South Florida Exotic Pest Plant Council, Tallahassee, FL, "Draft Model Exotic Species Ordinance for Municipalities and Counties in South Florida," Nov. 20, 1985.

60. Southeast Cooperative Wildlife Disease Study Center, College of Veterinary Medicine, University of Georgia, Athens, GA, "Model for State Regulations Pertaining to Captive Wild and Exotic Animals," Oct. 12, 1988.

61. Taylor, O., Professor of Entomology, University of Kansas, Lawrence, KS, personal communication to P.T. Jenkins, Office of Technology Assessment, Sept. 9, 1992.

62. Temple, S.A. and Carroll, D.M., "Pathways and Consequences of Introductions of Non-Indigenous Vertebrates in the United States," contractor report prepared for the Office of Technology Assessment, October 1991.

63. Thompson, D.Q. et al., U.S. Department of Interior, Fish and Wildlife Service, "Spread, Impact and Control of Purple Loosestrife (*Lythrum Salicaria*) in North American Wetlands," 1987.

64. U.S. Aquatic Nuisance Species Task Force, "Proposed Aquatic Nuisance Species Program," Sept. 28, 1992.

65. U.S. Congress, House Committee on Appropriations, Subcommittee on the Department of the Interior and Related Agencies, *Hearings on Department of the Interior and Related Agencies Appropriations for 1992: Part 3*, Serial No. 40 444 O, 1991.

66. U.S. Department of Interior, Fish and Wildlife Service, "Policies for Reducing Risks Associated with Introductions of Aquatic Organisms," Executive Summary and Discussion Document, Washington, DC, Oct. 14, 1987.

67. Waples, R.S. et al., Department of Commerce, National Oceanic and Atmospheric Administration, National Marine Fisheries Service, Seattle, WA, "NOAA Technical Memorandum NMFS F/NWC-201—Status Review for Snake River Fall Chinook Salmon," June 1991.

68. Waterworth, H.E., "Control of Plant Diseases by Exclusion: Quarantines and Disease-free Stocks" *Handbook of Pest Management in Agriculture, Vol. I*, D. Pimental et al. (eds.) (Boca Raton, FL: CRC Press, Inc., 1981), pp. 269-295.

69. Westman, W.E., "Park Management of Exotic Plant Species: Problems and Issues," *Conservation Biology*, vol. 4, No. 3, 1990, pp. 251-58.

70. Wingate, P.J., Minnesota Department of Natural Resources, Fisheries Section, St. Paul, MN, "U.S. States' Views and Regulations on Fish Introductions," 1990.

71. Wingate, P.J., Fisheries Research Manager, Fisheries Section, Minnesota Department of Natural Resources, St. Paul, MN, letter to Jack Baughman, Environmental Defense Fund, July 16, 1991.

72. Wingate, P.J., Fisheries Research Manager, Fisheries Section, Minnesota Department of Natural Resources, St. Paul, MN, personal communication, P.T. Jenkins, Office of Technology Assessment, Jan. 23, 1992.

CHAPTER 8: TWO CASE STUDIES: NON-INDIGENOUS SPECIES IN HAWAII AND FLORIDA

1. American Ornithologists' Union, *Check-list of North American Birds* (Washington, DC: American Ornithologists' Union, 1983).

2. Anonymous, "Preaching to Brazil From Hawaii," *New York Times*, July 24, 1990.

3. Austin, D.F., "Exotic Plants and Their Effects in Southeastern Florida," *Environmental Conservation*, vol. 5, No. 1, 1978, pp. 25-34.

4. Austin, D.F., "Vegetation on the Florida Atlantic University Ecological Site," *Florida Scientist (Biol. Sci.)*, vol. 53, No. 1, 1990, pp. 11-27.

5. Balciunas, J.K. and Center, T.D., "Biological Control of *Melaleuca quinquenervia*: Prospects and Conflicts," *Proceedings of the Symposium on Exotic Pest Plants*, T.D. Center et al. (eds.) (Washington, DC: U.S. Department of the Interior, National Park Service, 1991).

6. Beardsley, J.W., Jr., "New Immigrant Insects in Hawaii: 1962 through 1976," *Proceedings of the Hawaiian Entomological Society*, vol. 13, No. 1, April 1979, pp. 35-44.

7. Beardsley, J.W., Jr., University of Hawaii at Manoa, Honolulu, HI, "Introduction of Arthropod Pests into the Hawaiian Islands," paper presented at Workshop on Exotic Pests in the Pacific, University of Guam, May 31, 1990.

8. Beardsley, J.W., Jr., Chairman, Entomology Department, University of Hawaii at Manoa, Honolulu, HI, personal communication to K. Desmond, Oct. 19, 1990.

9. Belden, R.C., *Wild Hog Stocking: A Discussion of Issues* (Tallahassee, FL: Florida Game and Fresh Water Fish Commission, 1990).

10. Brown, R.S. et al., "Resource Guide to State Environmental Management," Council of State Governments, 1990, as cited in World Resources Institute, *The 1992 Information Please Environmental Almanac* (Boston, MA: Houghton Mifflin Co., 1992), p. 196.

11. Brown, T., Lieutenant Colonel, Military Customs Inspection, U.S. CINCPAC, Camp Smith, HI, personal communication to C. Mlot, Apr. 13, 1992.

12. Buck, M.E., Administrator, Hawaii Department of Land and Natural Resources, Honolulu, HI,

personal communication to K. Desmond, Oct. 11, 1990.

13. Buck, M.E., Administrator, Hawaii Department of Land and Natural Resources, Honolulu, HI, personal communication to C. Mlot, Feb. 10, 1992.

14. Burkhead, N.M. and Williams, J.D., "Research on the Okaloosa Darter, Focuses on Competition and Habitat Use," *Endangered Species Technical Bulletin*, vol. 15, No. 11, 1990, pp. 5-6.

15. Canfield, D.E., Jr., Maceina, M.J., and Shireman, J.V., "Effects of Hydrilla and Grass Carp on Water Quality in a Florida Lake," *Water Resources Bulletin*, vol. 19, No. 5, September/October, 1983, pp. 773-778.

16. Carey, J.R., "Establishment of the Mediterranean Fruit Fly in California," *Science*, vol. 253, No. 5026, Sept. 20, 1991, p. 1371.

17. Carlquist, S., "Worst-case Scenarios for Island Conservation: The Endemic Biota of Hawaii," *Ecological Restoration of New Zealand Islands, Conservation Sciences Publication No. 2*, D.R. Towns et al. (eds.) (Wellington, New Zealand: Department of Conservation, 1990), pp. 207-212.

18. Carr, A., "Armadillo Dilemma," *Animal Kingdom*, vol. 85, No. 5, September/October 1983, pp. 40-43.

19. Cassani, J.R., "Arthropods on Brazilian Peppertree, (*Schinus terebinthifolius*) (Anacardiaceae), in South Florida," *Florida Entomologist*, vol. 69, No. 1, 1986, pp. 184-196.

20. Clark, R., Steck, G.J., and Weems, H.V., Jr., "Detection, Quarantine, and Eradication of Fruit Flies Invading Florida," Division of Plant Industry, Florida Department of Agriculture and Consumer Services, Gainesville, FL, undated.

21. Cochran, M.J., "Applying the General Framework for Benefit Costs Analysis on Non-Indigenous Species to the Case of Melaleuca," contractor report prepared for the Office of Technology Assessment, March 1992.

22. Council on Environmental Quality, *Environmental Trends* (Washington, DC: Council on Environmental Quality, 1989).

23. Courtenay, W.R., Jr. and Robins, C.R., "Exotic Aquatic Organisms in Florida With Emphasis on Fishes: A Review and Recommendations," *Trans-actions of the American Fisheries Society*, vol. 101, No. 1, January/February 1973, pp. 1-12.

24. Courtenay, W.R., Jr. et al., "Exotic Fishes in Fresh and Brackish Waters of Florida," *Biological Conservation*, vol. 6, No. 4, 1974, pp. 292-302.

25. Courtenay, W.R., Jr. and Stauffer, J.R., Jr., "The Introduced Fish Problem and The Aquarium Fish Industry," *Journal of World Aquaculture Society*, vol. 21, No. 3, September 1991, pp. 145-159.

26. Cuddihy, L.W., Stone, C.P., and Tunison, J.T., "Alien Plants and their Management in Hawaii Volcanoes National Park," *Transactions of the Western Section of the Wildlife Society*, vol. 24, February 1988, pp. 42-46.

27. Cuddihy, L.W. and Stone, C.P., *Alteration of Native Hawaiian Vegetation: Effects of Humans, Their Activities and Introductions* (Honolulu, HI: University of Hawaii Press, 1990).

28. Denmark, H.A. and Porter, J.E., "Regulation of Importation of Arthropods Into and of Their Movement Within Florida," *Florida Entomologist*, vol. 56, No. 4, 1973, pp. 347-358.

29. Diamond, C., Davis, D., and Schmitz, D.C., "Economic Impact Statement: The Addition of *Melaleuca quinquenervia* to the Florida Prohibited Aquatic Plant List," *Proceedings of the Symposium on Exotic Pest Plants*, T.D. Center et al. (eds.), Technical Report NPS/NREVER/NRTR-91/06 (Washington, DC: U.S. Department of the Interior/National Park Service, 1991).

30. Doren, R.F. and Whiteaker, L.D., "The Exotic Pest Plant Council: An Example of Interagency Cooperation to Solve Resource Related Problems," *Proceedings of the Symposium on Exotic Pest Plants*, T.D. Center et al. (eds.), Technical Report NPS/NREVER/NRTR-91-06 (Washington, DC: U.S. Department of the Interior/National Park Service, 1991).

31. Duda, M.D., "Blacktailed Jackrabbit Importation: Potential Impacts on Florida's Natural Environment," Florida Game and Fresh Water Fish Commission, Tallahassee, FL, 1986.

32. Engbring, J. and Fritts, T.H., "Demise of an Insular Avifauna: The Brown Tree Snake on Guam," *Transactions of the Western Section of the Wildlife Society*, vol. 24, 1988, pp. 31-37.

33. Esser, R.P., O'Bannon, J.H., and Riherd, C.C., "The Citrus Nursery Site Approval Program for Burrowing Nematode and Its Beneficial Effect on the Citrus Industry of Florida," *Bulletin OEPP/EPPO(European Plant Protection Organization)*, No. 18, 1988, pp. 579-586.

34. Everglades Coalition, "Everglades in the 21st Century, The Water Management Future," 1992.

35. Ewel, J.J., "Invasibility: Lessons from South Florida," *Ecology of Biological Invasions of North America and Hawaii*, H.A. Mooney and J.A. Drake (eds.) (New York, NY: Springer-Verlag, 1986).

36. Exotic Pest Plant Council, *Newsletter*, vol. 1, No. 3, summer, 1991.

37. Exotic Pest Plant Council, *Newsletter*, vol. 2, No. 2, spring, 1992.

38. Frank, J.H., "Mole Crickets and Arthropod Pests of Turf and Pastures," *Classical Biological Control in the Southern United States*, D.H. Habeck et al. (eds.), Southern Cooperative Series Bull. No. 355 (Gainesville, FL: Institute of Food and Agricultural Sciences, University of Florida, 1990).

39. Frank, J.H. and McCoy, E.D., "The Immigration of Insects to Florida, with a Tabulation of Records Published since 1970," *Florida Entomologist*, vol. 75, No. 1, 1992, pp. 1-28.

40. Funasaki, G.Y. et al., "A Review of Biological Control Introductions in Hawaii: 1890 to 1985," *Proceedings of the Hawaii Entomological Society*, vol. 28, May 31, 1988, pp. 105-160.

41. Gagné, W.C., "Conservation Priorities in Hawaiian Natural Systems," *BioScience*, vol. 38, No. 4, April 1988, pp. 264-271.

42. Gleason, P.J. (ed.), *Environments of South Florida, Present and Past II* (Miami, FL: Miami Geological Society, 1984).

43. Hadfield, M., Zoologist, University of Hawaii, Honolulu, HI, personal communication to C. Mlot, Jan. 6, 1992.

44. Hardin, S. et al., "Food Items of Grass Carp, American Coots, and Ring-Necked Ducks from a Central Florida Lake," *Proceedings of the Annual Conference of Southeastern Associations of Fish and Wildlife Agencies*, vol. 38, 1984, pp. 313-318.

45. Havens, C.A., Chief Operations Officer, Port Operations, Animal and Plant Health Inspection Service, U.S. Department of the Interior, personal communication to C. Mlot, Jan. 3, 1992.

46. Hawaii Agricultural Alliance, Aiea, HI, "Introduced Species: An Overview of Damages Caused by the Introduction of Some Alien Species to Hawaii," draft, 1991.

47. Hawaii Department of Agriculture, Honolulu, HI, "Annual Report," 1985.

48. Hawaii Department of Agriculture, Honolulu, HI, "Annual Report," 1987.

49. Hawaii Department of Agriculture, Honolulu, HI, "Report to the 15th Legislature, 1989 Regular Session," 1989.

50. Hawaii Department of Agriculture, Honolulu, HI, "Annual Report," 1990.

51. Hawaii Department of Land and Natural Resources, Honolulu, HI, "Report to the Governor, vol. 1," 1987-1989.

52. Hinsdale, G., Assistant Western Regional Director, Animal and Plant Health Inspection Service, U.S. Department of Agriculture, personal communication to C. Mlot, Dec. 20, 1991.

53. Hofstetter, R.H., "The Current Status of *Melaleuca quinquenervia* in Southern Florida," *Proceedings of the Symposium on Exotic Pest Plants*, T.D. Center et al. (eds.), Tech. Rept. NPS, NREVER, NRTR-91/06 (Washington, DC: U.S. Department of the Interior, National Park Service, 1991).

54. Howarth, F.G., Sohmer, S.H., and Duckworth, W.D., "Hawaiian Natural History and Conservation Efforts," *BioScience*, vol. 38, No. 4, April 1988, pp. 232-237.

55. Howarth, F.G. and Ramsay, G.W., "The Conservation of Island Insects and their Habitats," *The Conservation of Insects and their Habitats*, N.M. Collins and J.A. Thomas (eds.) (London, England: Harcourt Brace Jovanovich/Academic Press, 1991), pp. 71-107.

56. Hurley, T., "Inmate Labor Worries Kula Residence," *Maui News*, Oct. 10, 1990, p. A3.

57. Johnson, F.A. and Montalbano, F., III, "Selection of Plant Communities by Wintering Waterfowl on Lake Okeechobee, Florida," *Journal of Wildlife Management*, vol. 48, 1984, pp. 174-178.

58. Joyce, J.C., "History of Aquatic Plant Management in Florida," *Aquatic Pest Control Application Training Manual*, K.A. Langeland (ed.) (Gainesville, FL: Institute of Food and Agricultural Sciences, University of Florida, 1990).

59. Kushlan, J.A., "Exotic Fishes in the Everglades, A Reconsideration of Proven Impact," *Environmental Conservation*, vol. 13, 1986, pp. 67-69.

60. Kushlan, J.A., "External Threats and Internal Management: The Hydrologic Regulation of the Everglades, Florida, USA," *Environmental Management*, vol. 11, No. 1, 1987, pp. 109-119.

61. Lamoureux, C., Botanist, University of Hawaii, Honolulu, HI, personal communication to K. Desmond, Oct. 17, 1990.

62. Langeland, K.A., "Exotic Woody Plant Control," Circular No. 868 (Gainesville, FL: Florida Cooperative Extension Service, Institute of Food and Agricultural Sciences, University of Florida, 1990).

63. Langeland, K.A., "Hydrilla, A Continuing Problem in Florida Waters," Circular No. 884, (Gainesville, FL: Center for Aquatic Plants, Institute of Food and Agricultural Sciences, University of Florida, 1990).

64. Long, J.L., *Introduced Birds of the World* (Sydney: A.H. and A.W. Reed, 1981) cited in L.L. Loope and D. Mueller-Dombois, "Characteristics of Invaded Islands, With Special Reference to Hawaii," *Biological Invasions: A Global Perspective*, J.A. Drake et al. (eds.) (Chichester, England: John Wiley and Sons, 1989), pp. 257-280.

65. Loope, L.L., "Haleakala National Park and the 'Island Syndrome': Active Management is Needed for Preservation of Island Biota," *Proceedings of Conference on Science in National Parks, (vol. 5): Management of Exotic Species in Natural Communities*, L.K. Thomas (ed.) (Hancock, MI: George Wright Society/U.S. National Park Service, 1986).

66. Loope, L.L., Research Scientist, Haleakala National Park, Makawao, HI, personal communication to P.T. Jenkins, Office of Technology Assessment, Aug. 21-22, 1991.

67. Loope, L.L., Research Scientist, Haleakala National Park, Makawao, HI, personal communication to P.N. Windle, Office of Technology Assessment, Feb. 10, 1992.

68. Loope, L.L. and Gon, S.M., III, "Biological Diversity and Its Loss," *Conservation Biology in Hawaii*, C.P. Stone and D.B. Stone (eds.) (Honolulu, HI: University of Hawaii Press, 1989).

69. Loope, L.L., Hamann, O., and Stone, C.P., "Comparative Conservation Biology of Oceanic Archipelagoes," *BioScience*, vol. 38, No. 4, April 1988, pp. 272-282.

70. Loope, L.L. and Mueller-Dombois, D., "Characteristics of Invaded Islands, with Special Reference to Hawaii," *Biological Invasions: A Global Perspective*, J.A. Drake et al. (eds.) (Chichester, England: John Wiley and Sons, 1989), pp. 257-280.

71. Maciolek, J.A., "Exotic Fishes in Hawaii and Other Islands of Oceania," *Distribution, Biology, and Management of Exotic Fishes*, W.R. Courtenay, Jr. and J.R. Stauffer, Jr. (eds.) (Baltimore, MD: Johns Hopkins University Press, 1984).

72. Maehr, D.S. and Holler, N.R., "Exotic Terrestrial Wildlife in Florida," unpublished manuscript, 1984.

73. McClelland, M., "Invasion of the Bio-Snatchers," *Florida Environments*, vol. 6, No. 2, February 1992, pp. 5, 13.

74. Melaleuca Task Force, "Melaleuca Management Plan for South Florida, Recommendations from the Melaleuca Task Force," West Palm Beach, FL, April, 1990.

75. Millsap, B., Chief, Florida Bureau of Nongame Wildlife, Tallahassee, FL, personal communication to D.W. Johnston, Jan. 7, 1993.

76. Milon, J.W. and Welsh, R., "An Economic Analysis of Sport Fishing and the Effects of Hydrilla Management in Lake County, Florida," Economics Report No. 118, Institute of Food and Agricultural Sciences, University of Florida, Gainesville, FL, 1989.

77. Mooney, H.A. and Drake, J.A., "The Ecology of Biological Invasions," *Environment*, vol. 29, No. 5, June 1987, p. 12.

78. Morgan, J., "Tourism," *Conservation Biology in Hawaii*, C.P. Stone and D.B. Stone (eds.)

(Honolulu, HI: University of Hawaii Press, 1989), pp. 146-153.

79. Morton, J.F., "Pestiferous Spread of Many Ornamental and Fruit Species in South Florida," *Proceedings of the Florida State Horticultural Society*, vol. 89, 1976, pp. 348-353.

80. Moulton, M.P. and Pimm, S.L., "Species Introductions to Hawaii," *Ecology of Biological Invasions of North America and Hawaii*, H.A. Mooney and J.A. Drake (eds.) (New York, NY: Springer-Verlag, 1986), pp. 231-249.

81. Myers, R.L. and Ewel, J.J. (eds.), *Ecosystems of Florida* (Orlando, FL: University of Central Florida Press, 1990).

82. Nakahara, L., Plant Quarantine Manager, Hawaii Department of Agriculture, Honolulu, HI, personal communication to C. Mlot, Apr. 16, 1992.

83. National Audubon Society, *Report of the Advisory Panel on the Everglades and Endangered Species*, Audubon Conservation Report No. 8. (New York, NY: National Audubon Society, 1992).

84. National Research Council, *Decline of the Sea Turtles* (Washington, DC: National Academy Press, 1990).

85. Natural Resources Defense Council, "Extinction in Paradise: Protecting our Hawaiian Species," Honolulu, HI, April 1989.

86. Nature Conservancy of Hawaii and the Natural Resources Defense Council, "The Alien Pest Species Invasion in Hawaii: Background Study and Recommendations for Interagency Planning," July 1992.

87. Nordlie, F.G., "Rivers and Springs," *Ecosystems of Florida*, R.L. Myers and J.J. Ewel (eds.) (Orlando, FL: University of Central Florida Press, 1990).

88. Olson, S.L. and James, H.F., "Fossil Birds From the Hawaiian Islands: Evidence for Wholesale Extinction by Man Before Western Contact," *Science*, vol. 217, No. 4560, Aug. 13, 1982, pp. 633-635.

89. O'Meara, G.F. et al., "Invasion of Cemeteries in Florida by *Aedes albopictus*," *Journal of the American Mosquito Control Association*, vol. 8, No. 1, March 1992, pp. 1-10.

90. Ota, A.K., Entomologist, Hawaiian Sugar Planters' Association, Aiea, HI, letter to W. Metcalf, Hawaii State Capitol, Apr. 5, 1990.

91. Ota, A.K., Entomologist, Hawaiian Sugar Planters' Association, Aiea, HI, personal communication to K. Desmond, Oct. 16, 1990.

92. Owre, O.T., "A Consideration of the Exotic Avifauna of Southeastern Florida," *Wilson Bulletin*, vol. 85, 1973, pp. 491-500.

93. Peterson, S.D., Chairperson, Hawaii Board of Agriculture, Honolulu, HI, testimony at hearings before the U.S. Senate Committee on Governmental Affairs, Subcommittees on Postal Operation and Services and Postal Personnel and Modernization, Nov. 4, 1987.

94. Reynolds, J.R., Regional Director, Animal and Plant Health Inspection Service, U.S. Department of Agriculture, Sacramento, CA, memo to B.G. Lee on the results of the first 90 days of the Hawaii First Class Mail Pilot Project, Sept. 19, 1990.

95. Robertson, W.B., Jr. and Woolfenden, G.E., *Florida Bird Species: An Annotated List*, Special Publication No. 6 (Gainesville, FL: Florida Ornithological Society, 1992).

96. Sailer, R.I., "History of Insect Introductions," *Exotic Plant Pests and North American Agriculture*, C.L. Wilson and C.L. Graham (eds.) (New York, NY: Academic Press, Inc., 1983).

97. Schardt, J.D. and Schmitz, D.C., "1990 Florida Aquatic Plant Survey," Technical Report #91-CGA (Tallahassee, FL: Florida Department of Natural Resources, Bureau of Aquatic Plants, 1990).

98. Schmitz, D., Schardt, J.D., and Craft, J.J., "Seeds of Destruction," *Florida Environments*, vol. 4, No. 12, December 1990, p. 9.

99. Schoulties, C.L. et al., "A New Outbreak of Citrus Canker in Florida," *Plant Disease*, vol. 69, No. 4, April 1985, p. 361.

100. Schoulties, C.L. et al., "Citrus Canker in Florida," *Plant Disease*, vol. 71, No. 5, May 1987, pp. 388-395.

101. Shafland, P.L., "Study XIX. Peacock Bass Investigations," unpublished report, Florida Game and Fresh Water Fish Commission, Tallahassee, FL, 1990.

102. Shafland, P.L., Management of Introduced Freshwater Fishes in Florida," *Proceedings of the 1990 Invitational Symposium/Workshop: New Directions in Research, Management and Conservation of Hawaiian Stream Ecosystems* (Honolulu, HI: Hawaii Department of Natural Resources, Division of Aquatic Resources, in press).

103. Shafland, P.L. and Pestrak, J.M., "Lower Lethal Temperatures for Fourteen Non-native Fishes in Florida," *Environmental Biology of Fish*, vol. 7, No. 2, February 1982, pp. 149-156.

104. Sharp, D.U., Doren, R.F., and Anderson, J.N., "East Everglades Exotic Plant Control, Annual Report," South Everglades Research Center, Homestead, FL, 1991.

105. Shelton, M.L., "Surface-Water Flow to Everglades National Park," *Geographical Review*, vol. 80, No. 4, October 1990, pp. 356-369.

106. Shiroma, E., Officer in Charge, Honolulu International Airport, Animal and Plant Health Inspection Service, U.S. Department of Agriculture, personal communication to K. Desmond, Oct. 15, 1990.

107. Shiroma, E., Officer in Charge, Honolulu International Airport, Animal and Plant Health Inspection Service, U.S. Department of Agriculture, personal communication to C. Mlot, Jan. 6, 1992.

108. Smith, C.W., "Weed Management in Hawaii's National Parks," *Monogr. Syst. Bot. Missouri Botanical Gardens*, vol. 32, 1990, pp. 223-234.

109. Smith, C.W., "The Alien Plant Problem in Hawaii," *Proceedings of the Symposium on Exotic Pest Plants*, Technical Report NPS/NREVER/NRTR-91/06 (Washington, DC: U.S. Department of the Interior/National Park Service, 1991), pp. 327-337.

110. Smith, C.W., Botany Department, University of Hawaii at Manoa, Honolulu, HI, personal communication to C. Mlot, Oct. 21, 1991.

111. Stange, L.A., Taxonomic Entomologist, Division of Plant Industry, Florida Department of Agriculture and Consumer Services, Gainesville, FL, personal communication to D.W. Johnston, June 1992.

112. Stone, C.P., "Alien Animals in Hawaii's Native Ecosystems: Toward Controlling the Adverse Effects of Introduced Vertebrates," *Hawaii's Terrestrial Ecosystems: Preservation and Management* (Honolulu, HI: Cooperative National Park Resources Study Unit, University of Hawaii at Manoa, 1985), pp. 251-288.

113. Stone, C.P., Research Scientist, Hawaii Volcanoes National Park, "Environmental Education in Hawaii," paper presented at workshop on Critical Resource Issues in the Pacific, Guam, April 1990.

114. Stone, C.P. and Holt, R.A., "Managing the Invasions of Alien Ungulates and Plants in Hawaii's Natural Areas," *Monogr. Syst. Bot. Missouri Botanical Gardens*, vol. 32, 1990, pp. 211-221.

115. Sutton, D.L. and Vandiver, V.V., Jr., "Grass Carp, a Fish for Biological Management of Hydrilla and Other Aquatic Weeds in Florida," Florida Agricultural Experiment Stations, Bulletin No. 867 (Gainesville, FL: Institute of Food and Agricultural Sciences, University of Florida, 1986).

116. Sykes, P.W., "The Feeding Habits of the Snail Kite in Florida, USA," *Colonial Waterbirds*, vol. 10, 1987, pp. 84-92.

117. Thayer, D., Director, Aquatic Plant Management Division, South Florida Water Management District, West Palm Beach, FL, personal communication to D.W. Johnston, May 27, 1992.

118. Thompson, D.R., Operations Officer, Plant Protection and Quarantine, Animal Plant Health Inspection Service, U.S. Department of Agriculture, Hyattsville, MD, personal communication to D.W. Johnston, May 27, 1992.

119. Tiner, R.W., Jr., *Wetlands of the United States: Current Status and Recent Trends* (Washington, DC: U.S. Government Printing Office, 1984).

120. Tobin, R.J., *The Expendable Future: U.S. Politics and the Protection of Biological Diversity* (Durham, NC: Duke University Press, 1990), pp. 216-219.

121. Toland, B., "Use of Forested Spoil Islands by Nesting American Oystercatchers in Southeast Florida," *Journal of Field Ornithology*, vol. 63, No. 2, spring 1992, pp. 155-58.

122. Toops, C. and Dilley, W.E., *Birds of South Florida: An Interpretive Guide* (Conway, AR: River Road Press, 1986).

123. U.S. Congress, House Committee on Science, Space, and Technology, Subcommittee on Natural Resources, Agriculture Research and Environment, *Hearings on H.R. 1268—The National Biological Diversity Conservation and Environmental Research Act, No. 30*, May 17, 1989, pp. 259-334.

124. U.S. Congress, Senate Committee on Governmental Affairs, Subcommittee on Federal Services, Post Office, and Civil Service, *Hearings on Postal Implementation of the Agricultural Quarantine Enforcement Act*, June 5, 1991.

125. U.S. Department of Agriculture, Agricultural Research Service, Tropical Fruit and Vegetable Research Laboratory, "I. ARS Perspective for Fruit Fly Eradication in Hawaii and Pilot Test Requirements for Demonstration of Technology," and "II. Pilot Test to Eliminate Mediterranean Fruit Fly from the Islands of Kauai and Niihau: Detailed Work Plan," drafts (Honolulu, HI: December 1989).

126. U.S. Department of Agriculture, Animal and Plant Health Inspection Service, Animal Damage Control, Honolulu, HI, "Annual Report," FY 1989.

127. U.S. Department of Commerce, Bureau of the Census, *Statistical Abstract of the United States: 1990* (110th ed.) (Washington, DC: U.S. Government Printing Office, 1990).

128. Vitousek, P.M., Loope, L.L., and Stone, C.P., "Introduced Species in Hawaii: Biological Effects and Opportunities for Ecological Research," *Trends in Ecology and Evolution*, vol. 2, No. 7, July 1987, pp. 224-227.

129. Wagner, W.L., Herbst, D.R., and Sohmer, S.H., *Manual of the Flowering Plants of Hawaii* (Honolulu, HI: University of Hawaii Press, Bishop Museum Press, 1990).

130. Ward, D.B., "How Many Plant Species Are Native to Florida?" *Palmetto*, winter, 1989-90, pp. 3-5.

131. Warner, R.E., "The Role of Introduced Diseases in the Extinction of the Endemic Hawaiian Avifauna," *The Condor*, vol. 70, No. 2, spring, 1968, pp. 101-120.

132. Webb, J., "Managing Nature in the Everglades," *EPA Journal*, vol. 16, No. 6 November/December 1990, pp. 48-50.

133. Wester, L., "Origin and Distribution of Adventive Alien Flowering Plants in Hawaii," *Alien Plant Invasions in Native Ecosystems of Hawaii: Management and Research*, C.P. Stone et al. (eds.) (Honolulu, HI: University of Hawaii Press, in press), pp. 99-114.

134. Wiley, M.J., Tazek, P.P., and Sobaski, S.T., "Controlling Aquatic Vegetation With Triploid Grass Carp," *Illinois Natural History Survey*, Circular No. 57, 1987, pp. 1-16.

135. Williams, R.N., "Bulbul Introductions on Oahu," *Elepaio*, vol. 43, No. 11, May 1983.

136. Wilson, L.D. and Porras, L., *The Ecological Impact of Man on the South Florida Herpetofauna*, Special Publ. No. 9 (Lawrence, KS: University of Kansas, Museum of Natural History, 1983).

137. Winsberg, M.D., *Florida Weather* (Orlando, FL: University of Central Florida Press, 1990).

138. Yee, R.S.N. and Gagné, W.C., "Activities and Needs of the Horticulture Industry in Relation to Alien Plant Problems in Hawaii," *Alien Plant Invasions in Native Ecosystems of Hawaii: Management and Research*, C.P. Stone et al. (eds.) (Honolulu, HI: University of Hawaii Press, in press), pp. 712-725.

CHAPTER 9: GENETICALLY ENGINEERED ORGANISMS AS A SPECIAL CASE

1. Arntzen, C.J., "Regulation of Transgenic Plants," *Science*, vol. 257, No. 5075, Sept. 4, 1992, p. 1327.

2. Berry, D.F. and Hagedorn, C., "Soil and Groundwater Transport of Microorganisms," *Assessing Ecological Risks of Biotechnology*, L.R. Ginzburg (ed.) (Boston, MA: Butterworth-Heinemann, 1991), pp. 57-74.

3. Eisner, R., "State Legislators Seek to Broaden Regulation of Biotech Products," *The Scientist*, vol. 5, No. 4, Feb. 18, 1991, pp. 1, 6, 19.

4. Ellstrand, N.C. and Hoffman, C.A., "Hybridization as an Avenue of Escape for Engineered Genes: Strategies for Risk Reduction," *BioScience*, vol. 40, No. 6, June 1990, pp. 438-442.

5. Fuchs, R.L. and Serdy, F.S., "Genetically Modified Plants: Evaluation of Field Test Biosafety Data," *International Symposium on the Biosafety Results of Field Tests of Genetically*

Modified Plants and Microorganisms, D.R. MacKenzie and S.C. Henry (eds.) (Bethesda, MD: Agricultural Research Institute, 1990), pp. 25-30.

6. Giamporcaro, D.E., Section Chief, TSCA Biotechnology Program, U.S. Environmental Protection Agency, Washington, DC, personal communication to P.N. Windle, Office of Technology Assessment, Apr. 29, 1993.

7. Goldburg, R.J., Senior Scientist, and Hopkins, D.D., Senior Attorney, Environmental Defense Fund, New York, NY, memorandum to "Persons Interested in State Oversight of Intentional Releases of Genetically Engineered Organisms," Nov. 30, 1992.

8. Goldburg, R.J., Senior Scientist, and Hopkins, D.D., Senior Attorney, Environmental Defense Fund, New York, NY, letter to Chief, Regulatory Analysis and Development, Animal and Plant Health Inspection Service, U.S. Department of Agriculture, Jan. 4, 1993.

9. Grossman, B., Special Assistant to the Director, Hawaii Department of Health, Honolulu, HI, personal communication to E.A. Chornesky, Office of Technology Assessment, Apr. 8, 1993.

10. Hails, R.S. et al., "The Invasive Potential of Transgenic and Conventional Oilseed Rape in Natural Habitats," *Bulletin of the Ecological Society of America*, vol. 73, No. 2, June 1992, p. 197.

11. Hallerman, E.M., Assistant Professor, Virginia Polytechnic Institute and State University, Blacksburg, VA, letter to P.N. Windle, Office of Technology Assessment, Nov. 29, 1992.

12. Hallerman, E.M. and Kapuscincki, A.R., "Ecological and Regulatory Uncertainties Associated With Transgenic Fish," *Transgenic Fish*, C.L. Hew and G. Fletcher (eds.) (Singapore: World Science Publishing, 1992), pp. 209-228.

13. Herrett, R.A., Statement on behalf of the Association for Biotechnology Companies, Hearings before the U.S. House Committee on Agriculture, Subcommittee on Department Operations, Research and Foreign Agriculture, "Review of Statutes of the Animal and Plant Health Inspection Service," July 10, 1990.

14. Hess, C.E., "Opening Remarks," *International Symposium on the Biosafety Results of Field Tests of Genetically Modified Plants and Microorganisms*, D.R. MacKenzie and S.C. Henry (eds.) (Bethesda, MD: Agricultural Research Institute, 1990), pp. 25-30.

15. Industrial Biotechnology Association, "Year-End Survey of State Government Legislation on Biotechnology: 1991," Washington, DC, 1991.

16. Istock, C.A., "Genetic Exchange and Genetic Stability in Bacterial Populations," *Assessing Ecological Risks of Biotechnology*, L.R. Ginzburg (ed.) (Boston, MA: Butterworth-Heinemann, 1991), pp. 123-150.

17. Kludy, D., Consultant, Richmond, VA, personal communication to E.A. Chornesky, Office of Technology Assessment, May 20, 1993.

18. Krimsky, S. and Wetzler, R., "Bioengineered Organisms and Non-Indigenous Species," contractor report prepared for the Office of Technology Assessment, November 1991.

19. Maclean, N. and Penman, D., "The Application of Gene Manipulation to Aquaculture," *Aquaculture*, vol. 85, 1990, pp. 1-20.

20. Medley, T.L., Director, Biotechnology, Biologics, and Environmental Protection, Animal and Plant Health Inspection Service, U.S. Department of Agriculture, Hyattsville, MD, letter to W.S. Wallace, Director of Policy and Program Development, Animal and Plant Health Inspection Service, U.S. Department of Agriculture, Hyattsville, MD, Feb. 11, 1992.

21. Mol, J.N.M. et al., "Manipulation of Floral Pigmentation Genes in Petuna: "Sense and Antisense Make Sense," *Horticultural Biotechnology*, A.B. Bennett and S.D. O'Neill (eds.) (New York, NY: John Wiley & Sons, Inc., 1990), p. 192.

22. Muench, S.R., "Requirements and Considerations in Successful Field Releases of Genetically Engineered Plants," *International Symposium on the Biosafety Results of Field Tests of Genetically Modified Plants and Microorganisms*, D.R. MacKenzie and S.C. Henry (eds.) (Bethesda, MD: Agricultural Research Institute, 1990), pp. 3-8.

23. National Research Council, *Field Testing Genetically Modified Organisms: Framework for Decisions* (Washington, DC: National Academy Press, 1989).

24. National Wildlife Federation, Biotechnology Policy Center, ''Reports to Animal and Plant Health Inpsection Service, U.S. Department of Agriculture, on Completed Field Tests of Transgenic Plants,'' information sheet, Washington, DC, August 1992.

25. Office of Science and Technology Policy, Executive Office of the President, Washington, DC, ''Coordinated Framework for Regulation of Biotechnology,'' 51 *Federal Register* 23302-23308 (June 26, 1986).

26. Payne, J.H., Associate Director, Biotechnology, Biologics, and Environmental Protection, Animal and Plant Health Inspection Service, U.S. Department of Agriculture, Hyattsville, MD, letter to P.N. Windle, Office of Technology Assessment, Nov. 10, 1992.

27. Postgate, J., ''The Malleable Microbe,'' *New Scientist*, vol. 129, No. 1756, Feb. 16, 1991, pp. 38-41.

28. President's Council on Competitiveness, ''Report on National Biotechnology Policy,'' Washington, DC, February 1991.

29. Regal, P.J., presentation on environmental issues, Transgenic Plant Conference, Annapolis, MD, Sept. 7-9, 1988, co-sponsored by U.S. Food and Drug Administration, U.S. Environmental Protection Agency, and the U.S. Department of Agriculture, 1988.

30. Rogul, M. and Levin, M., ''Regulation of Biotechnology by the Environmental Protection Agency,'' *Assessing Ecological Risks of Biotechnology*, L.R. Ginzburg (ed.) (Boston, MA: Butterworth-Heinemann, 1991), pp. 233-265.

31. Sagoff, M., ''On Making Nature Safe for Biotechnology,'' *Assessing Ecological Risks of Biotechnology*, L.R. Ginzburg (ed.) (Boston, MA: Butterworth-Heinemann, 1991), pp. 341-365.

32. Sharples, F.E., ''Applications of Introduced Species Models to Biotechnology Assessment,'' *Application of Biotechnology to Environmental and Policy Issues*, J.R. Fowle III (ed.) (Boulder, CO: Westview Press, Inc., 1987), pp. 93-98.

33. Shore, S., Biotechnologist, Division of Plant Industry, North Carolina Department of Agriculture, Raleigh, NC, personal communication to E.A. Chornesky, Office of Technology Assessment, May 25, 1993.

34. Slutsky, B., ''Pesticidal Transgenic Plants: Risk Issues,'' *Pesticidal Transgenic Plants: Product Development, Risk Assessment, and Data Needs*, U.S. EPA Conference Proceedings (EPA/21T-1024), Annapolis, MD, Nov. 6 and 7, 1990, pp. 127-132.

35. Springer, W.D., testimony before the U.S. House Committee on Agriculture, Subcommittee on Department Operations, Research, and Foreign Agriculture, *Adequacy of Current Biotechnology Regulation and the Omnibus Biotechnology Act of 1990 (HR 5312)*, Serial No. 101-75, Oct. 2, 1990.

36. Suter, G.W., ''Exotic Organisms,'' *Ecological Risk Assessment*, G.W. Suter (ed.) (Boca Raton, FL: Lewis Publishers, 1993), pp. 391-401.

37. Tiedje, J.M. et al., ''The Planned Introduction of Genetically Engineered Organisms: Ecological Considerations and Recommendations,'' *Ecology*, vol. 70, No. 2, 1989, pp. 298-315.

38. U.S. Congress, General Accounting Office, *Biotechnology: Role of Institutional Biosafety Committees*, GAO/RCED-88-64BR (Washington, DC: U.S. Government Printing Office, December 1987).

39. U.S. Congress, General Accounting Office, *Biotechnology: Delays in and Status of EPA's Efforts to issue a TSCA Regulation*, GAO/RCED-92-167, Washington, DC, June 1992.

40. U.S. Congress, House Committee on Agriculture, Subcommittee on Department Operations, Research, and Foreign Agriculture, *Review of Current and Proposed Agriculture Biotechnology Regulatory Authority and the Omnibus Biotechnology Act of 1990*, Serial No. 101-75 (Washington, DC: U.S. Government Printing Office, 1991).

41. U.S. Congress, House Committee on Merchant Marines and Fisheries, request letter to J.H. Gibbons Director, Office of Technology Assessment, July 24, 1990.

42. U.S. Congress, House Committee on Science and Technology, Subcommittee on Investigations and Oversight, *Issues in the Federal Regulation of Biotechnology: From Research to Release* (Washington, DC: U.S. Government Printing Office, December 1986).

43. U.S. Congress, Office of Technology Assessment, *A New Technological Era for American Agriculture*, OTA-F-474 (Washington, DC: U.S. Government Printing Office, August 1992).

44. U.S. Department of Agriculture, Animal and Plant Health Inspection Service, Biotechnology, Biologics, and Environmental Protection, "User's Guide for Introducing Genetically Engineered Plants and Microorganisms," technical bulletin No. 1783, Hyattsville, MD, June 1991.

45. U.S. Department of Agriculture, Animal and Plant Health Inspection Service, Biotechnology, Biologics, and Environmental Protection, Conference Report of the Workshop on Safeguards for Planned Introduction of Transgenic Potatoes, St. Andrews, Scotland, Aug. 16-17, 1991, J.P. Hegelson and H.V. Davis (eds.), Hyattsville, MD, 1991.

46. U.S. Department of Agriculture, Animal and Plant Health Inspection Service, Biotechnology, Biologics, and Environmental Protection, Conference Report on Workshop on Safeguards for Planned Introduction of Transgenic Corn and Wheat, Keystone, CO, Dec. 6-8, 1990, L.V. Giddings et al. (eds.), April 1992.

47. U.S. Department of Agriculture, Office of Agricultural Biotechnology, "Proposed USDA Guidelines for Research Involving the Planned Introduction Into the Environment of Organisms With Deliberately Modified Hereditary Traits," 56 *Federal Register* 4134-4151 (Feb. 1, 1991).

48. Vaituzis, Z., "Product Characterization and Toxicological Evaluation of Pesticidal Transgenic Plants," *Pesticidal Transgenic Plants: Product Development, Risk Assessment, and Data Needs*, Proceedings of the U.S. EPA Conference (EPA/21T-1024), Annapolis, MD, Nov. 6 and 7, 1990, pp. 135-140.

49. Wrubel, R.P., Krimsky, S., and Wetzler, R.E., "Field Testing Transgenic Plants: An Analysis of the U.S. Department of Agriculture's Environmental Assessments," *BioScience*, vol. 42, No. 4, April 1992, pp. 280-289.

CHAPTER 10: THE CONTEXT OF THE FUTURE: INTERNATIONAL LAW AND GLOBAL CHANGE

1. Anonymous, "Extremely Rare Fruit Worm Blocks Apple Export," *Seattle Times*, Nov. 30, 1991, p. A11.

2. Anonymous, "Floral Arrangement," *Journal of Commerce*, June 27, 1991, p. A3.

3. Anonymous, "Pest Infestation Spurs Doubt Over Standards in Mexico," *Christian Science Monitor*, May 4, 1992, p. 19.

4. Anonymous, "Potatoes at the Root of Another U.S.-Canada Trade Embroilment," *Journal of Commerce*, Feb. 19, 1991, p. A7.

5. Asselin, P., Assistant Deputy Minister, Fisheries and Oceans Department, Government of Canada, internal memorandum to Associate Deputy Minister, Policy and Program Planning, July 31, 1987.

6. Bangsund, D.A., and Leistritz, F.L., *Economic Impact of Leafy Spurge in Montana, South Dakota, and Wyoming*, Agricultural Economics Report No. 275 (Fargo, ND: Department of Agricultural Economics, North Dakota State University, October 1991).

7. Birmingham, M.J., "The State of Forest Health in New York," New York State Department of Environmental Conservation, Division of Lands and Forests, Albany, NY, Dec. 31, 1990.

8. Breckenridge, R.P. and Otis, M.D., "Monitoring Concepts Useful in the Assessment of Climate Change Effects on U.S. Fish and Wildlife Resources," *Coping With Climate Change*, J.C. Topping (ed.) (Washington, DC: The Climate Inst., 1989), pp. 268-272.

9. Britton, J.C., "Pathways and Consequences of the Introduction of Non-indigenous Freshwater, Terrestrial, and Estuarine Mollusks in the United States," contractor report prepared for the Office of Technology Assessment, Washington, DC, October 1991.

10. Brooks, D.J., "International Dimensions of U.S. Forestry," *Agriculture Outlook '93, Prospects for Fruit and Vegetables, Floriculture and Forest Products*, booklet 2, U.S. Department of Agriculture, Washington, DC, Dec. 12, 1992, pp. 19-33.

11. Campbell, D.S., Director, Operational Support, International Services, Animal and Plant Health Inspection Service, U.S. Department of Agricul-

ture, letter to Phyllis N. Windle, OTA, Aug. 21, 1992.

12. Campbell, F.T., Natural Resources Defense Council, "Legal Avenues for Controlling Exotics," presentation at Indiana Academy of Science Conference on Biological Pollution, Indianapolis, IN, Oct. 26, 1991.

13. Carlton, J.T., "Man's Role in Changing the Face of the Ocean," *Conservation Biology*, vol. 3, No. 3, September 1989, pp. 270-72.

14. Chi, K.S. "Foresight Activities in State Government," *Futures Research Quarterly*, vol. 7, No. 4, 1991, pp. 47-60.

15. Chock, A.K., "International Cooperation on Controlling Exotic Pests," *Exotic Plant Pests and North American Agriculture*, C.L. Wilson and C.L. Graham (eds.) (London, England: Academic Press, 1983), pp. 479-498.

16. Coates, J.F., "Factors Shaping and Shaped by the Environment, 1990-2010," *Futures Research Quarterly*, vol. 7, No. 3, fall 1991, pp. 5-55.

17. Coates, J.F., President, Coates & Jarratt, Inc., Washington, DC, personal communication to Office of Technology Assessment, Dec. 21, 1992.

18. Coates, J.F. and Jarratt, J., "What Futurists Believe," *Futures*, vol. 24, No. 6, November-December 1990, pp. 22-28.

19. Courtenay, W.R., Jr., "Pathways and Consequences of the Introduction of Non-Indigenous Fishes in the United States," contractor report prepared for the Office of Technology Assessment, September 1991.

20. Cramer, J. and Zegveld, W.C.L., "The Future Role of Technology in Environmental Management," *Futures*, vol. 23, No. 5, June 1991, pp. 451-468.

21. Crawley, M.J., "The Responses of Terrestrial Ecosystems to Global Climate Change," *Global Climate and Ecosystem Change*, G. MacDonald and L. Sertorio (eds.) (New York, NY: Plenum Press, 1990).

22. Crosson, P., "United States Agriculture and the Environment: Perspectives on the Next Twenty Years," Resources for the Future, Washington, DC, April 1992.

23. Dahlsten, D.L., Garcia, R. and Lorraine, H., "Eradication as a Pest Management Tool: Concepts and Contexts," *Eradication of Exotic Pests: Analysis with Case Histories*, D.L. Dahlsten and R. Garcia (eds.) (Dexter, MI: Thompson-Shore, Inc., 1989), pp. 3-15.

24. Davis, M.B. and Zabinski, C., "Changes in Geographical Range Resulting from Greenhouse Warming: Effects on Biodiversity in Forests," *Global Warming and Biological Diversity*, P.L. Peters and T.E. Lovejoy (eds.) (New Haven, CN: Yale University Press, 1992), pp. 297-308.

25. Delfosse, E.S., Director, National Biological Control Institute, Animal and Plant Health Inspection Service, U.S. Department of Agriculture, Hyattsville, MD, personal communication to Office of Technology Assessment, Nov. 19, 1991.

26. Di Castri, F., "On Invading Species and Invaded Ecosystems: The Interplay of Historical Chance and Biological Necessity," *Biological Invasions in Europe and the Mediterranean Basin*, F. di Castri, A.J. Hansen, and M. Debussche (eds.) (Boston, MA: Kluwer Academic Publishers, 1990), pp. 3-16.

27. Dobb, E., "Cultivating Nature," *The Sciences*, vol. 32, No. 1, January/February 1992, pp. 44-50.

28. Dobson, A., "Climate Change and Parasitic Diseases of Man and Domestic Livestock in the United States," *Coping With Climate Change*, J.C. Topping (ed.) (Washington, DC: The Climate Inst., 1989), pp. 147-152.

29. Duvick, D.N., "Plant Breeding in the 21st Century," *Choices*, vol. 7, fourth quarter 1992, pp. 26-29.

30. Ehler, L., Professor, Department of Entomology, University of California, Berkeley, CA, personal communication to Office of Technology Assessment, Dec. 23, 1992.

31. Elston, R.A., "Effective Applications of Aquaculture Disease-Control Regulations: Recommendations From an Industry Viewpoint," *Dispersal of Living Organisms into Aquatic Ecosystems*, A. Rosenfield and R. Mann (eds.) (College Park, MD: Maryland Sea Grant, 1992), pp. 353-359.

32. Elton, C.S., *The Ecology of Invasions by Animals and Plants* (London, England: Chapman and Hall, Ltd., 1958).

33. Fetterolf, C.M., "Why a Great Lakes Fisheries Commission and Why a Sea Lamprey International Symposium," *Canadian Journal of Fisheries and Aquatic Sciences*, vol. 37, 1980, pp. 1588-1593.

34. Fletcher, S.R. and Tiemann, M.E., "Trade and the Environment," *CRS Review*, Congressional Research Service, Washington, DC, February-March, 1992, pp. 29-31.

35. Forcella, F., "Final Distribution Is Related to Rate of Spread in Alien Weeds," *Weed Research*, vol. 25, 1985, pp. 181-191.

36. Fox, M.D., "Mediterranean Weeds: Exchanges of Invasive Plants Between the Five Mediterranean Regions of the World," *Biological Invasions in Europe and the Mediterranean Basin*, F. di Castri, A.J. Hansen, and M. Debussche (eds.) (Dordrecht: Kluwer Academic Publishers, 1990), pp. 179-200.

37. Gasser, C.S. and Fraley, R.T., "Genetically Engineering Plants for Crop Improvement," *Science*, vol. 244, No. 4910, June 16, 1989, pp. 1293-1299.

38. Great Lakes Fisheries Commission, Ann Arbor, Michigan, "A Joint Strategic Plan for Management of Great Lake Fisheries," December 1980.

39. Great Lakes Water Quality Agreement of 1978 (revised), as amended by Protocol signed Nov. 18, 1987, International Joint Commission (Washington, DC and Ottawa, Canada: January 1988).

40. Grooms, L., "Marketers Address Today's Flower Trends," *Seed World*, vol. 130, No. 1, January 1992, pp. 19-22.

41. Hanrahan, C.E., "Agriculture in the North American Free Trade Agreement: A Preliminary Assessment," *CRS Report for Congress*, 92-716 S, Congressional Research Service, Library of Congress, Washington, DC, Sept. 21, 1992.

42. Harville, J.P. (ed.), *Proceedings of the North American Fisheries Leadership Workshop*, Snowbird, UT, May 21-23 (Bethesda, MD: American Fisheries Society, Fisheries Administrators Section, 1991).

43. Houghton, J.T., Jenkins, G.J., Ephraums, J.J. (eds.), *Climate Change—The Intergovernmental Panel on Climate Change Scientific Assessment*, World Meteorological Organization/U.N. Environment Programme (Cambridge, England: Cambridge Univ. Press, 1990).

44. International Union for the Conservation of Nature and Natural Resources, Environmental Law Centre, Bonn, Germany, "Biological Diversity Convention Draft," June 30, 1989.

45. Jablonski, D., "Extinctions: A Paleontological Perspective," *Science*, vol. 253, No. 5021, Aug. 16, 1991, pp. 754-757.

46. Johnson, D.C., "Recession Impacts and Economic Outlook for the U.S. Nursery, Greenhouse, and Turfgrass Industries, *Agriculture Outlook '93. Prospects for Fruit and Vegetables, Floriculture and Forest Products*, booklet 2, U.S. Department of Agriculture, Washington, DC, 1992, pp. 3-12.

47. Kahn, R.P., "Exclusion as a Plant Disease Control Strategy," *Annual Review of Phytopathology*, vol. 29, 1991, pp. 219-246.

48. Kahn, R.P., Consultant, Rockville, MD, personal communication to the Office of Technology Assessment, Dec. 4, 1991.

49. Kareiva, P.M., Kingsolver, J.G., and R.B. Huey (eds.), *Biotic Interactions and Global Change* (Sunderland, MA: Sinauer Assoc., 1993).

50. Kennedy, V.S., "Potential Climate Effects of Climate Change on Chesapeake Bay Animals and Fisheries," *Coping With Climate Change*, J.C. Topping (ed.) (Washington, DC: The Climate Inst., 1989), pp. 509-513.

51. Kim, K.C. and Wheeler, A.G., Jr., "Pathways and Consequences of the Introduction of Non-Indigenous Insects and Arachnids in the United States," contractor report prepared for the Office of Technology Assessment, December 1991.

52. Kludy, D.H., "Federal Policy on Non-Indigenous Species—The Role of the United States Department of Agriculture's Animal and Plant Health Inspection Service," contractor report prepared for the Office of Technology Assessment, December 1991.

53. Lausche, B., "International Laws and Associated Programs for *In-Situ* Conservation of Wild Species," contractor report prepared for the Office of Technology Assessment, September 1985.

54. Lederberg, J., Shope, R.E., and Oaks, S.C., Jr. (eds.), *Emerging Infections: Microbial Threats*

to Health in the United States (Washington, DC: National Academy Press, 1992).

55. Longstreth, J. and Wiseman, J., "The Potential Impact of Climate Change on Patterns of Infectious Disease in the United States," *The Potential Effects of Global Climate Change on the United States, Appendix G, Health,* J.B. Smith and D. Tirpak (eds.), EPA-230-05-89-057 (Washington, DC: Office of Policy, Planning, and Evaluation, U.S. Environmental Protection Agency, May 1989), pp. 3-1—3-41.

56. Lubchenco, J. et al., "The Sustainable Biosphere Initiative: An Ecological Research Agenda," *Ecology,* vol. 72, No. 2, 1991, pp. 371-412.

57. Lyster, S., *International Wildlife Law* (Cambridge, England: Grotius Publications, 1985).

58. MacArthur, R.H., *Geographical Ecology* (New York, NY: Harper and Row, 1972).

59. Macdonald, I.A.W., Loope, L.L., Usher, M.B., and Hamann, O., "Wildlife Conservation and the Invasion of Nature Reserves by Introduced Species: A Global Perspective," *Biological Invasions,* J.A. Drake et al. (eds.), Scope 37 (Chichester, UK: John Wiley & Sons, Ltd., 1989), pp. 215-255.

60. Mack, R.N., "Additional Information on Non-Indigenous Plants in the United States," contractor report prepared for the Office of Technology Assessment, Washington, DC, October 1992.

61. Magnuson, J.J. et al., "Potential Responses of Great Lakes Fishes and Their Habitat to Global Climate Warming," *The Potential Effects of Gloabl Climate Change on the United States, Appendix E, Aquatic Resources,* J.B. Smith and D. Tirpak (eds.), EPA-230-05-89-055 (Washington, DC: U.S. Environmental Protection Agency, June 1989), pp. 2-1—2-42.

62. Marien, M. (ed.), "A Future Survey Guide to 50 Overviews and Agenda's," *Futures Research Quarterly,* vol. 6, No. 1, spring 1990, pp. 103-112.

63. M'Boob, S., U.N. Food and Agriculture Organization—Africa, presentation at the North American Plant Protection Organization Workshop on International Pest Risk Analysis, Alexandria, VA, Oct. 23, 1991.

64. Miller, W.F, Dougherty, P.M. and Switzer, G.L., "Rising CO2 and Changing Climate: Major Southern Forest Management Implications," *The Greenhouse Effect, Climate Change, and U.S. Forests* (Washington, DC: The Conservation Foundation, 1987).

65. Minnesota Interagency Exotic Species Task Force, *Report and Recommendations,* final edit (St. Paul, MN: State of Minnesota, Department of Natural Resources, July 1991).

66. Moulton, W., "Constraints to International Exchange of Animal Germplasm," contractor report prepared for the Office of Technology Assessment, October 1985.

67. Munton, P., "Problems Associated With Introduced Species," paper presented at the Workshop on Feral Animals at the Third International Theriological Conference, Helsinki, Finland, August 1982.

68. National Research Council, *Biological Contamination of Mars: Issues and Recommendations,* Space Studies Board (Washington, DC: National Academy Press, 1992).

69. North American Plant Protection Organization, Pest Risk Analysis Panel, "A Process for Analyzing the Risk to Plants and Plant Products Posed by the Introduction and/or Spread of Biotic Agents," discussion paper, Nov. 22, 1991.

70. Noss, R.F., "From Endangered Species to Biodiversity," *Balancing on the Brink of Extinction,* K.A. Kohm (ed.) (Washington, DC: Island Press, 1991), pp. 227-242.

71. O'Rourke, D., Director, Impact Center, Washington State University, Pullman, WA, personal communication to K.E. Bannon, Office of Technology Assessment, July 12, 1993.

72. Parker, N.C., "Economic Pressures Driving Genetic Changes in Fish," Introductions and Transfers of Marine and Anadromous Species," *Dispersal of Living Organisms Into Aquatic Ecosystems,* A. Rosenfield and R. Mann (eds.) (College Park, MD: Maryland Sea Grant, 1992), pp. 415-419.

73. Parry, M.L., *Climate Change and World Agriculture* (London, England: Earthscan Publishers, 1990).

74. Peters, R.L., "Consequences of Global Warming for Biological Diversity," *Global Climate Change and Life on Earth,* R.L. Wyman (ed.)

(New York, NY: Chapman and Hall, 1991), pp. 99-118.

75. Reed, C.F. and Hughes, R.O., *Economically Important Foreign Weeds*, USDA Agricultural Handbook Number 498 (Washington, DC: U.S. Department of Agriculture, 1977).

76. Rensberger, B., "Marine Biology: Blue Crabs to the Rescue?" *The Washington Post*, Feb. 24, 1992, p. A2.

77. Rinderer, T.E. et al., "The Proposed Honey Bee Regulated Zone in Mexico," *American Bee Journal*, March 1987, pp. 160-164.

78. Ritchie, M., "GATT, Agriculture and the Environment: The U.S. Double Zero Plan," *The Ecologist*, vol. 20, No. 6, November-December 1990, pp. 214-220.

79. Schneider, K., "Ranges of Animals and Plants Head North," *The New York Times*, Aug. 13, 1991, p. C1.

80. Schneider, S.H., Mearns, L. and P.H. Gleick, "Climate-Change Scenarios for Impact Assessment," *Global Warming and Biological Diversity*, R.L. Peters and T.E. Lovejoy (eds.) (New Haven, CT: Yale University Press, 1992), pp. 38-55.

81. Scott, D.C., "Mexico's Avocado Growers Grumble Over Unlikely Beachhead—Alaska," *Christian Science Monitor*, Nov. 9, 1992, p. 6.

82. Shetler, S.G., "Three Faces of Eden," *Seeds of Change* (Washington, DC: Smithsonian Press, 1991), pp. 227-247.

83. Shope, R.E., "Global Climate Change and Infectious Disease," *Environmental Health Perspectives*, vol. 96, 1991, pp. 171-174.

84. Shrybman, S., "International Trade: In Search of an Environmental Conscience," *EPA Journal*, July-August 1990, pp. 17-19.

85. Simberloff, D. and Cox, J., "Consequences and Costs of Conservation Corridors," *Conservation Biology*, vol. 1, No. 1, 1987, pp. 63-71.

86. Soulé, M.E., "Conservation Biology in the Twenty-first Century: Summary and Outlook," *Conservation for the Twenty-first Century*, D. Western and M.C. Pearl (eds.) (New York, NY: Oxford Univ. Press, 1989), pp. 297-303.

87. Soulé, M.E., "The Onslaught of Alien Species, and Other Challenges in the Coming Decades,"

Conservation Biology, vol. 4, No. 3, September 1990, pp. 233-239.

88. Soulé, M.E., "Conservation: Tactics for a Constant Crisis," *Science*, vol. 253, No. 5021, Aug. 16, 1991, pp. 744-750.

89. Steadman, D.W., "Extinction of Species: Past, Present, and Future," *Global Climate Change and Life on Earth*, R.L. Wyman (ed.) (New York, NY: Chapman and Hall, 1991), pp. 156-169.

90. Stinner, B.R. et al., "Potential Effects of Climate Change on Plant-Pest Interactions," *The Potential Effects of Global Climate Change on the United States, Appendix C, Agriculture*, J.B. Smith and D. Tirpak (eds.), EPA-230-05-89-053 (Washington, DC: Office of Policy, Planning, and Evaluation, U.S. Environmental Protection Agency, June 1989), pp. 8-1—8-35.

91. Strobel, G.A., "Biological Control of Weeds," *Scientific American*, vol. 265, No. 1, July 1991, pp. 72-78.

92. Tayama, H.K., "New Developments in World Floriculture Markets," *Agriculture Outlook '93, Late Speeches (1)*, Booklet 9, U.S. Department of Agriculture, Washington, DC, Dec. 2, 1992, pp. 44-57.

93. Uher, R.A., "Levitating Trains," *The Futurist*, vol. 24, No. 5, September/October 1990, pp. 28-32.

94. United Nations, *Convention on the Law of the Sea and Resolutions I-IV*, Third United Nations Conference on the Law of the Sea, Eleventh Session, New York, March—April, 1982.

95. United Nations Environment Programme, "Draft Convention on Biological Diversity," May 22, 1992, Art. 8.

96. U.S. Congress, General Accounting Office, *International Environment: International Agreements Are Not Well Monitored*, RCED-92-43 (Gaithersburg, MD: U.S. General Accounting Office, January 1992).

97. U.S. Congress, Office of Technology Assessment, *Changing by Degrees: Steps to Reduce Greenhouse Gases*, OTA-482 (Washington, DC: U.S. Government Printing Office, February 1991).

98. U.S. Congress, Office of Technology Assessment, *Trade and Environment: Conflicts and Opportunities*, OTA-BP-ITE-94 (Washington,

DC: U.S. Government Printing Office, May 1992).

99. U.S. Congress, Office of Technology Assessment, *Preparing for an Uncertain Climate* (Washington, DC: U.S. Government Printing Office, in press).

100. U.S. Department of Agriculture, *Agriculture Statistics 1990* (Washington, DC: U.S. Government Printing Office, 1990).

101. U.S. Department of Agriculture, Animal and Plant Health Inspection Service, Veterinary Services, Import-Export Program, "Program Monitoring Review" (undated).

102. U.S. Executive Office of the President, Office of the Press Secretary, "Statement by the President on United States Ocean Policy, Mar. 10, 1983," Washington, DC.

103. U.S. Executive Office of the President, *Science and Technology* (Washington, DC: Office of Science and Technology Policy, 1993).

104. U.S. Interagency Task Force, "Review of U.S.-Mexico Environmental Issues," Office of the U.S. Trade Representative, Washington, DC, February 1992, as cited in U.S. Congress, Office of Technology Assessment, *Trade and Environment: Conflicts and Opportunities*, OTA-BP-ITE-94 (Washington, DC: U.S. Government Printing Office, May 1992).

105. Usher, M.B., "Biological Invasions of Nature Reserves: A Search for Generalizations," *Biological Conservation*, vol. 44, Nos. 1/2, 1988, pp. 119-135.

106. Vermeij, G.J., "When Biotas Meet: Understanding Biotic Interchange," *Science*, vol. 253, No. 5024, Sept. 6, 1991, pp. 1099-1104.

107. Vogt, D.U., Congressional Research Service, Library of Congress, Washington, DC, memorandum entitled "GATT Sanitary and Phytosanitary Proposals," Nov. 14, 1990.

108. Vogt, D.U., "Sanitary and Phytosanitary Measures Pertaining to Food in International Trade Negotiations," Congressional Research Service, Library of Congress, Washington, DC, Sept. 11, 1992.

109. Waldrop, J., "Garden Variety Customers," *American Demographics*, vol. 15, No. 4, April 1993, pp. 44-49.

110. Warrick, R.A. et al., "The Effects of Increased CO_2 and Climatic Change on Terrestrial Ecosystems," *The Greenhouse Effect, Climatic Change, and Ecosystems*, SCOPE 29, B. Bolin et al. (eds.) (Chichester, UK: John Wiley & Sons, 1986), pp. 363-392.

111. Welcomme, R.L., "International Measures for the Control of Introduction of Aquatic Organisms," *Fisheries*, vol. 11, No. 2, May-April 1986, pp. 4-9.

112. Westman, W.E., "Managing for Biodiversity— Unresolved Science and Policy Questions," *BioScience*, vol. 40, No. 1, January 1990, pp. 26-33.

113. Wyman, R.L. (ed.), *Global Climate Change and Life on Earth* (New York, NY: Chapman and Hall, 1991).

114. Ziegler, W., "Envisioning the Future," *Futures*, vol. 23, No. 5, June 1991, pp. 516-527.

Index to Common and Scientific Names of Species

Common Name[1] (Scientific Name)

Disease Pathogens

African swine fever[2], Iridoviridae, 304

annosus root disease (*Heterobasidion annosum = Fomes annosus*), 6, 301

black stem rust (*Puccinia graminis*), 207

bluetogue virus, Reoviridae (Orbivirus), 176

bT (*Bacillus thuringiensis*), 153-6, 160, 196, 198, 284

burrowing nematode (*Radopholus similis*), 262

chestnut blight (*Cryphonectria parasitica*), 58, 66, 74, 118, 179

chrysanthemum rust (*Puccinia chrysanthemi = P. tanaceti*), 37

chrysanthemum white rust (*Puccinia horiana*), 82, 105

citrus canker (*Xanthomonas campestris pv. citri*), 104, 139, 175, 207, 262, 264

citrus nematode (*Tylenchulus semipenetrans*), 262

corn cyst nematode (*Heterodtera zeae*), 105

crown gall disease (*Agrobacterium tumefaciens*), 198

dengue fever, Flaviviridae (Flavivirus), 70, 81, 262

dogwood anthracnose (*Discula destructiva*), 52

Dutch elm disease (*Ceratocystis ulmi = Ophiostoma ulmi*), 66, 88, 98, 118, 179

eastern equine encephalitis, Togaviridae (Alphavirus), 81, 70, 262

Eastern filbert blight (*Anisogramma anomala*), 184

foot and mouth disease, Picornaviridae (Aphthovirus), 6, 126, 141, 301

golden nematode (*Globodera rostochiensis*), 81, 91, 176

gypsy moth nuclear polyhedrosis virus, Baculoviridae (Nuclear Polyhedrosis Virus), 154, 198

heliothis nuclear polyhedrosis virus, Baculoviridae (Nuclear Polyhedrosis Virus), 154, 198

hog cholera (European swine fever), Flaviviridae (Pestivirus), 68

Karnal bunt fungus (*Tilletia indica*), 175

LaCrosse encephalitis, Bunyaviridae (Bunyavirus), 81

larch canker (*Lachnellula* spp.), 6, 301

malaria, caused by (*Plasmodia* spp.), 253, 304

needle cast (*Mycospaerella laricina*), 105

oak wilt disease (*Ceratocystis fagacearum*), 74

peanut stripe virus, Potyviridae (Potyvirus), 50, 84, 180

pine sawfly nuclear polyhedrosis virus, Baculoviridae (Nuclear Polyhedrosis Virus), 198

pine wood nematodes (*Bursaphelenchus* spp. *B. lignicolus*), 6, 288, 301

Port-Orford cedar root disease (*Phytophthora lateralis*), 179

potato virus y—necrotic strain (n), Potyviridae (Potyvirus), 105

Rift valley fever virus, Bunyaviridae (Phlebovirus), 304

rust fungus (on chrysanthemum)—see chrysanthemum white rust

smut fungus (on rice) (*Ustilago esculenta*), 82, 105, 227

soybean rust fungus (*Phakopsora pachyrhizi*), 6, 115, 301

swine flu, Orthomyxoviridae (Orthomyxovirus), 52

tomato spotted wilt virus, Bunyaviridae (Tospovirus), 160

tussock moth nuclear polyhedrosis virus, Baculoviridae (Nuclear Polyhedrosis Virus), 198

wheat rust fungus (*Puccinia recondita*), 65

white pine blister rust (*Cronartium ribicola*), 66, 118, 179

yellow fever, Flaviviridae (Flavivirus), 70

---

[1] Only scientific names are listed for species that do not have generally accepted common names and for most of the species for which no common names were provided in the original references.

[2] For most viruses, scientific names at the species level are not formally recognized, nor have problems in nomenclature been resolved. OTA lists viruses by their family, followed by genus when it is available.

Plants

Index

Superintendent of Documents **Publications** Order Form

Order Processing Code:
*7171

P3
Telephone orders (202) 783-3238
To fax your orders (202) 512-2250
Charge your order.
It's Easy!

☐ **YES**, please send me the following:

_____ copies of *Harmful Non-Indigenous Species in the United States (400 pages)*, S/N 052-003-01347-9 at $21.00 each.

The total cost of my order is $_____. International customers please add 25%. Prices include regular domestic postage and handling and are subject to change.

_____
(Company or Personal Name)   (Please type or print)

_____
(Additional address/attention line)

_____
(Street address)

_____
(City, State, ZIP Code)

_____
(Daytime phone including area code)

_____
(Purchase Order No.)

Please Choose Method of Payment:

☐ Check Payable to the Superintendent of Documents

☐ GPO Deposit Account [ ][ ][ ][ ][ ][ ] — [ ]

☐ VISA or MasterCard Account

[ ]

[ ][ ][ ][ ] (Credit card expiration date)   *Thank you for your order!*

_____
(Authorizing Signature)   (9/93)

                                     YES NO
May we make your name/address available to other mailers? ☐ ☐

Mail To: New Orders, Superintendent of Documents, P.O. Box 371954, Pittsburgh, PA 15250-7954

THIS FORM MAY BE PHOTOCOPIED

Superintendent of Documents **Publications** Order Form

Order Processing Code:
*7171

P3
Telephone orders (202) 783-3238
To fax your orders (202) 512-2250
Charge your order.
It's Easy!

☐ **YES**, please send me the following:

_____ copies of *Harmful Non-Indigenous Species in the United States (400 pages)*, S/N 052-003-01347-9 at $21.00 each.

The total cost of my order is $_____. International customers please add 25%. Prices include regular domestic postage and handling and are subject to change.

_____
(Company or Personal Name)   (Please type or print)

_____
(Additional address/attention line)

_____
(Street address)

_____
(City, State, ZIP Code)

_____
(Daytime phone including area code)

_____
(Purchase Order No.)

Please Choose Method of Payment:

☐ Check Payable to the Superintendent of Documents

☐ GPO Deposit Account [ ][ ][ ][ ][ ][ ] — [ ]

☐ VISA or MasterCard Account

[ ]

[ ][ ][ ][ ] (Credit card expiration date)   *Thank you for your order!*

_____
(Authorizing Signature)   (9/93)

                                     YES NO
May we make your name/address available to other mailers? ☐ ☐

Mail To: New Orders, Superintendent of Documents, P.O. Box 371954, Pittsburgh, PA 15250-7954

THIS FORM MAY BE PHOTOCOPIED